AF526361

STATISTICAL THEORY OF COMMUNICATION

Y. W. LEE

Professor of Electrical Engineering
Department of Electrical Engineering
and Research Laboratory of Electronics
Massachusetts Institute of Technology

DOVER PUBLICATIONS, INC.
MINEOLA, NEW YORK

DOVER PHOENIX EDITIONS

Bibliographical Note

This Dover edition, first published in 2005, is an unabridged republication of the edition published by John Wiley & Sons, Inc., New York, in 1960.

Library of Congress Cataloging-in-Publication Data

Lee, Yuk Wing, 1904–
Statistical theory of communication / Y. W. Lee.
p. cm. — (Dover phoenix editions)
Includes index.
ISBN 0-486-43890-2
1. Telecommunication. 2. Information theory. 3. Probabilities. 4. Harmonic analysis. I. Title. II. Series.

TK5101.L4 2004
621.382—dc22

2004058234

Manufactured in the United States of America
Dover Publications, Inc., 31 East 2nd Street, Mineola, N.Y. 11501

To My Mother and My Wife
in gratitude and affection.

Preface

In the spring of 1946 I was invited by the Massachusetts Institute of Technology to be a visiting assistant professor in the Department of Electrical Engineering. Leaving thirteen years of a chaotic life in China behind, I was indeed happy to renew my association with Professor Norbert Wiener of the Department of Mathematics. He immediately directed my attention to his new work entitled, *The Extrapolation, Interpolation and Smoothing of Stationary Time Series.* Since we had worked together on network synthesis in the 1930's, he knew that I would be interested in following up the work that he had done during the war. From other sources I heard that Professor Wiener's report was often referred to as "The Yellow Peril" because of its yellow covers and its difficult mathematics. When Professor K. L. Wildes of the Department of Electrical Engineering suggested the possibility of my offering a graduate course based upon this new work, I was somewhat concerned that I might not be able to handle this formidable document. However, realizing that I could not afford to miss the opportunity, and realizing the great value of Professor Wiener's work, I took up the venture.

In October, 1947, I offered the course "Optimum Linear Systems" to a group of graduate students in the Department of Electrical Engineering. I believe that it was the first time in this country that a course of study on the modern application of statistical theory to communication problems was offered. With the inclusion of additional material as I went along, the title of the course was soon changed to "Statistical Theory of Communication." Besides presenting the course in the Department of Electrical Engineering, I have been carrying on research in this field in the Research Laboratory of Electronics at M.I.T.

This book is the outgrowth of my teaching the graduate course for many years at M.I.T. and of my research work in the Research Laboratory of Electronics. Actually, the book contains a large amount of material which I could not take up in class without the book.

The development of the statistical theory of communication is a landmark in the history of communication theory. Since the central idea in the statistical theory is that messages and noise should be considered as random phenomena, the theory incorporates probability theory and generalized harmonic analysis in its foundation. With the aid of statistical concepts many communication and control problems are seen in better light, and effective statistical methods and techniques have been developed for their solution. Among the outstanding contributions of the theory is the formulation and solution of statistical filtering, prediction, and other associated problems. The principles and techniques of the theory have now found extensive application in engineering and research. Moreover, the application is not confined to communication engineering. The same principles and techniques have gained their way into many problems in other fields of scientific and engineering work such as mechanical engineering, geophysics, biophysics, acoustics, meteorology, feedback control, aeronautics, and oceanography.

This is an introductory book. The object of the book is to present clearly and rigorously a physically motivated and systematic account of the statistical theory of communication—an account that includes essentially all of the basic elements. In writing this book I have been guided by the idea that a teacher should not attempt to cover the subject of study but should attempt to uncover it for the student.

Since it is my plan to confine the scope of this book to linear theory and to limit its length to a one-semester course, the book does not include material on nonlinear systems. However, because of the great importance of the orthonormal representation of a linear system in the statistical theory of optimum nonlinear systems, as discussed by Professor Norbert Wiener in his book *Nonlinear Problems in Random Theory*, the last chapter of the book is devoted to the synthesis of linear systems by orthonormal functions.

The book is primarily written for the first-year graduate student in electrical engineering as a one-semester course. It is also suitable for those seniors whose work is considerably better than the average of the class. I hope that students in other fields of activity such as those I mentioned in a preceding paragraph will find the book helpful. Furthermore, engineers who have not had the time or the opportunity to go into this field in the past may find this book interesting. For the engineers and physicists who are in this field of work I hope that it has some value as a reference.

I have taught this subject to a large number of students. Through their questions and discussions in class, and through their thesis work with me, I have derived considerable benefit. As I mentioned earlier, it was at the suggestion of Professor Karl L. Wildes that I introduced the graduate course. I am thankful to him not only for his suggestion but also for his support in making the course a part of the curriculum and for his continued interest in my work. I wish to thank Professor Lan J. Chu for his help in initiating my association with the Research Laboratory of Electronics, where my students and I have been carrying on research since 1947. Professor Jerome B. Wiesner, Director of the Research Laboratory of Electronics and Acting Head of the Department of Electrical Engineering, shared with me a number of research problems during the earlier years in the development of the theory. I am indebted to him for the opportunity he gave me to carry on research in the Laboratory and for his constant support. To Professor Gordon S. Brown, formerly Head of the Department of Electrical Engineering and now Dean of Engineering, I am grateful for his interest in my work and in the publication of this book. Mr. Martin Schetzen, member of the Research Laboratory of Electronics, read the entire galley proof of the book. I wish to thank him for making many suggestions and corrections. For the typing of the manuscript I extend to Miss Jeannette Pollard and Miss Noreen McSorley my sincere appreciation.

My association with Professor Norbert Wiener began in 1929, when I wrote my doctoral thesis under his supervision. My admiration for his work is quite evident from the extent to which this book is based upon it. I cannot adequately express the satisfaction and value I have found in working with him continually and having him for a friend all of these years.

Y. W. Lee

Belmont, Massachusetts
June, 1960

Contents

chapter 1

Introduction

Our primary concern in a communication or control problem is the flow of messages. Messages appear in a great variety of forms. Some examples are: spoken or written words, music, wind gusts on an aircraft equipped with automatic control, temperature fluctuations in an industrial process, electrical impulses in the nervous system of an animal, barometric changes in weather forecast, seismic waves in seismographic oil exploration, and the irregular path of an aircraft in a radar automatic tracking system. Associated with the flow of messages is noise. In an electrical system a noise is a disturbance such as shot noise or thermal noise. However, the term need not be restricted to mean a disturbance in the ordinary sense. It may be generalized to include any unwanted message in the presence of a wanted message. Thus a message in one channel of transmission is a noise in another where it is not wanted.

Since messages and noise occupy the central position in a communication problem, it is of paramount importance that they be appropriately described and represented. Before the development of statistical communication theory, classical theory had been established on the basis that a message or a noise could be considered either as a periodic phenomenon or as a transient phenomenon. In the absence of a suitable method of representation for the actual messages and noise, classical theory replaced them by periodic or transient functions depending upon which one was the more "reasonable" substitute. Such a dilemma could not have been avoided since the only existing tools, then, for message representation were the classical Fourier series and Fourier integral. Indeed, occasionally messages are practically periodic. An example is the received signal in a pulse radar set when the object is stationary and the noise is negligible. Again, occasionally some messages are transients. Consider the path of a projectile as an example. If we assume certain essential quantities, such as the initial angle and the maximum height, to be known, the path of the projectile can be expressed by a Fourier integral. We note with emphasis that according to classical theory both the Fourier series and Fourier integral expressions for messages are exact expressions which hold for all time, extending into the infinite future. The concept of messages and noise in classical theory is therefore a deterministic one.

To the large class of continuing messages and noise (some examples of which have been given earlier) that are neither periodic nor transient, the classical Fourier series is obviously inapplicable. The classical Fourier integral is also inapplicable since the continuing characteristic—the characteristic of the messages and noise that they do not approach zero in any sense as time tends to infinity—violates the condition of validity.

The fluctuations of these messages and noise as functions of time are extremely complex. Furthermore, frequently the underlying causes of the fluctuations are not completely understood. We call these functions random functions or random processes. Because of the complexity or the incomplete understanding of the phenomena, or because of both, no simple laws are available to describe their behavior exactly. Hence an important characteristic of a random process is that in the course of its fluctuations the future cannot be precisely predicted. At any moment (or in any time interval in a sequence of time intervals) during the flow of a message, a set of possible values exists, and upon this set of values the sender makes his (or its) choice. In fact, a message is valuable only when it has this characteristic. Just as in any game of chance, it is the impossibility of exact prediction of the outcome that makes the game interesting. Even in the example of the received radar signal that is practically periodic, the location of the echo pulse is not exactly predictable when the signal is valuable, that is, when the object is at an unknown distance. Similarly, in the example of the path of a projectile the valuable information for locating the terminal point of the projectile may consist of the initial angle and the maximum height which to the observer are not exactly predictable quantities. A fundamental idea in statistical communication theory is that messages and noise should be considered statistically and should be described in accordance with the concepts in probability theory.

The difference between the probabilistic concept of messages and noise and the deterministic concept of these processes in classical theory is of major importance. Out of this difference has grown a modern communication theory. The theory has developed in two main directions. In one direction it has led to the theory of correlation and its application to a large class of problems including signal detection and to the theory of statistical filtering and prediction with a host of associated problems. This book concerns the development in this direction only, and we shall refer to it as the statistical theory of communication. The other direction of development has led to the theory of information which is based upon the quantitative definition of the fundamental concepts associated with the production, transmission, and reception of information. It includes the study of various types of communication channels, the implementation of these channels, and the theory of coding.

To a large measure the development of the statistical theory of communication owes its success to the merging of the tools and techniques of the communication engineer and those of the statistician. Although the description and analysis of messages and noise with the aid of concepts and methods in probability theory is a powerful idea in the modern theory, the theory could not have progressed to its present state had its basis been confined to probability theory. Through generalized harmonic analysis it has been possible to relate the statistical description of a random process to its spectrum. The establishment of this important link has made both probability theory and generalized harmonic analysis the foundation of the statistical theory of communication and control.

Generalized harmonic analysis is not a new development; it began in about 1925. In a series of papers on the harmonic analysis of irregular motion and the study of Brownian motion, culminating in the paper entitled "Generalized Harmonic Analysis," published in *Acta Mathematica* in 1930, N. Wiener laid a solid mathematical foundation for the harmonic analysis of random processes. It is interesting to note that in the twelve years that elapsed after the publication of this classic paper no significant application to communication problems appeared.

Fortunately, in 1942 another classic work of Wiener emerged under the title *Extrapolation, Interpolation, and Smoothing of Stationary Time Series* (1). In this work Wiener presented basic ideas and concepts that have deeply influenced a large part of all modern development in communication theory and information theory. The theory of statistical filtering and prediction and a number of associated problems constituted an important portion of this paper. Wiener formulated these problems in such a manner that their solutions depended upon the results in his earlier work on generalized harmonic analysis.

Besides the theory of probability and generalized harmonic analysis, statistical theory of communication is heavily indebted to the statistical mechanics of Willard Gibbs. The concept of the ensemble and the ergodic hypothesis in statistical mechanics are an integral part of modern communication theory.

REFERENCE

1. Wiener, N., *The Extrapolation, Interpolation, and Smoothing of Stationary Time Series with Engineering Applications*, John Wiley and Sons, New York, 1949. Original work appeared as a report in 1942. Published as a book by John Wiley and Sons in 1949.

chapter 2

Generalized Harmonic Analysis

Harmonic analysis is familiar to every communication engineer since it is the most important and fruitful tool for the solution of his problems which involve periodic and transient phenomena. Through a modern development to include the handling of random phenomena this tool has gained even greater power. The extension is extremely important, for random functions are the chief carriers of the commodity of a communication or control system, namely, information. Furthermore, accompanying the transmission of information is the ever-present disturbance or noise which can be appropriately handled only as a random phenomenon. Therefore, the primary objective of this chapter is the exposition of the extension of Fourier theories to include random functions.

Since it is neither possible nor desirable to discuss the harmonic analysis of random functions alone, the extension is preceded by a discussion of the Fourier series and Fourier integral with special emphasis on those concepts and techniques which are useful in the analysis of all three types of phenomena. In particular, the concept of correlation will receive major attention because it is of the utmost importance in handling a random phenomenon or one in which periodic, transient, and random functions are intermixed.

The generalization of Fourier theories for the inclusion of random phenomena is possible only through the aid of the theories of statistics and probability. This indispensable part of the generalization requires considerable study. For effectiveness in presentation, it is not taken up in this chapter, but it will receive full consideration in several succeeding chapters.

Harmonic analysis is capable of extension in many directions. But, for the needs of this book, only that portion of it which relates to periodic functions, transient functions, and random functions of the type most frequently encountered in engineering is presented.

A. PERIODIC FUNCTIONS

1. Exponential Form of Fourier Series

A periodic function $f(t)$ of the independent variable t, which in general represents time (seconds), when expressed as a Fourier series is

$$f(t) = \frac{a_0}{2} + \sum_{n=1}^{\infty} (a_n \cos n\omega_1 t + b_n \sin n\omega_1 t) \tag{1}$$

where ω_1 (radians per second) is the fundamental angular frequency which is related to the period T_1 (seconds) of the function by the formula $T_1 = 2\pi/\omega_1$. The fundamental angular frequency is equal to 2π times the fundamental frequency (cycles per second). The constant a_0 and the coefficients a_n and b_n for the cosine and sine series respectively are given by the integrals

$$a_n = \frac{2}{T_1} \int_{-T_1/2}^{T_1/2} f(t) \cos n\omega_1 t \, dt \qquad \text{for } n = 0, 1, 2, 3, \ldots \tag{2}$$

and

$$b_n = \frac{2}{T_1} \int_{-T_1/2}^{T_1/2} f(t) \sin n\omega_1 t \, dt \qquad \text{for } n = 1, 2, 3, \ldots \tag{3}$$

Since the present discussion is not a purely mathematical development, but one which has engineering application as its ultimate objective, we shall not state precisely and fully the conditions of validity for the Fourier representation of a periodic function; it is sufficient to state that

$$\int_{-T_1/2}^{T_1/2} |f(t)|\, dt$$

is finite.

By applying the formulas

$$\cos n\omega_1 t = \tfrac{1}{2}(e^{jn\omega_1 t} + e^{-jn\omega_1 t}) \tag{4}$$

and

$$\sin n\omega_1 t = \frac{1}{2j}(e^{jn\omega_1 t} - e^{-jn\omega_1 t}) \tag{5}$$

we may write expression (1) in the form

$$f(t) = \frac{a_0}{2} + \frac{1}{2} \sum_{n=1}^{\infty} (a_n - jb_n) e^{jn\omega_1 t} + \frac{1}{2} \sum_{n=1}^{\infty} (a_n + jb_n) e^{-jn\omega_1 t} \tag{6}$$

To simplify this equation, negative values of n are introduced. These values will make

$$a_{-n} = \frac{2}{T_1}\int_{-T_1/2}^{T_1/2} f(t)\cos n\omega_1 t\,dt$$

$$= a_n \qquad \text{for } n = 1, 2, 3, \ldots \tag{7}$$

and

$$b_{-n} = -\frac{2}{T_1}\int_{-T_1/2}^{T_1/2} f(t)\sin n\omega_1 t\,dt$$

$$= -b_n \qquad \text{for } n = 1, 2, 3, \ldots \tag{8}$$

since $\cos(-n\omega_1 t) = \cos n\omega_1 t$ and $\sin(-n\omega_1 t) = -\sin n\omega_1 t$. The last term of (6) is therefore

$$\frac{1}{2}\sum_{n=1}^{\infty}(a_n + jb_n)e^{-jn\omega_1 t} = \frac{1}{2}\sum_{n=-1}^{-\infty}(a_n - jb_n)e^{jn\omega_1 t} \tag{9}$$

so that (6) simplifies to

$$f(t) = \sum_{n=-\infty}^{\infty} F(n)e^{jn\omega_1 t} \tag{10}$$

in which

$$F(n) = \tfrac{1}{2}(a_n - jb_n) \qquad \text{for } n = 0, \pm1, \pm2, \pm3, \ldots \tag{11}$$

By combining (2) and (3) according to (11), we find that

$$F(n) = \frac{1}{T_1}\int_{-T_1/2}^{T_1/2} f(t)e^{-jn\omega_1 t}\,dt \qquad \text{for } n = 0, \pm1, \pm2, \pm3, \ldots \tag{12}$$

The expression of the Fourier series (1) in the exponential form (10) and the coefficients (2) and (3) in the form (12) is a great convenience in analysis. Clearly, equations (10) and (12) are much more compact than their original forms. Furthermore, in mathematical manipulations, exponentials are in general much simpler than trigonometric functions.

Equation (12) is known as the *Fourier transform* of the periodic function $f(t)$; it is a function of the harmonic order n and is a representation of the periodic time function in the frequency domain. The representation is complete since it uniquely determines $f(t)$ by expansion (10). We observe that multiplying $f(t)$ by $e^{-jn\omega_1 t}$ and taking the average value of the product over the period for every value of n as indicated by (12) constitute an analysis of $f(t)$ for the amplitudes and phase angles of the sinusoids into which $f(t)$ is resolved. Full information concerning harmonic amplitudes and phases is contained in the function $F(n)$ which, in general, is complex and is called the *complex spectrum* of $f(t)$. Because the harmonic order n assumes only discrete values, the spectrum is a *line spectrum*.

Inasmuch as equation (10) is a summation of sinusoids in accordance with the amplitude and phase information contained in the complex spectrum $F(n)$ for regaining the original function, it is a synthesis in opposition to (12) which is an analysis. Again, whereas (12) is a transformation of $f(t)$ into its frequency-domain representation $F(n)$, (10) is an inverse transformation of $F(n)$ into its time-domain representation $f(t)$ and is called the *Fourier transform* of $F(n)$. Therefore, it follows from the Fourier representation of a periodic function $f(t)$ that the reciprocal relations between $f(t)$ and its complex line spectrum $F(n)$ are, to repeat for the purpose of putting them together as a pair,

$$f(t) = \sum_{n=-\infty}^{\infty} F(n)e^{jn\omega_1 t} \tag{13}$$

and

$$F(n) = \frac{1}{T_1}\int_{-T_1/2}^{T_1/2} f(t)e^{-jn\omega_1 t}\,dt \tag{14}$$

each being the Fourier transform of the other.

To separate the amplitude and phase characteristics in the the complex spectrum, we write $F(n)$ given by (11) as

$$F(n) = \tfrac{1}{2}\sqrt{a_n^2 + b_n^2}\exp\left[j\tan^{-1}\left(-\frac{b_n}{a_n}\right)\right] \tag{15}$$

and denote the absolute value of $F(n)$ as a function of n by $|F(n)|$, that is,

$$|F(n)| = \tfrac{1}{2}\sqrt{a_n^2 + b_n^2} \tag{16}$$

and the phase angle of $F(n)$ as a function of n by θ_n, that is,

$$\theta_n = \tan^{-1}\left(-\frac{b_n}{a_n}\right) \tag{17}$$

so that (10) reads

$$f(t) = \sum_{n=-\infty}^{\infty} |F(n)|e^{j(n\omega_1 t+\theta_n)} \tag{18}$$

The absolute value of $F(n)$ given by (16) as a function of the harmonic order n is the *amplitude spectrum* of $f(t)$, and the phase function (17) is the *phase spectrum* of $f(t)$.

Example 1. As an illustration of the application of (10) and (12) to a particular function, consider the wave $f(t)$ of Fig. 1, which is a series of rectangular pulses of height E_m and duration b and which is periodic of period T_1.

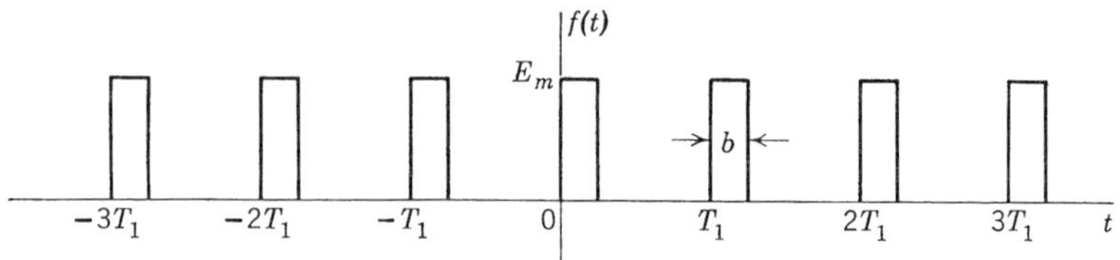

Fig. 1. Periodic wave of rectangular pulses.

This simple example may seem superfluous, but the result to be obtained is helpful in Example 2, Sec. A-4.

The complex spectrum of the wave is, by (12),

$$F(n) = \frac{1}{T_1}\int_o^b E_m e^{-jn\omega_1 t}\,dt$$

$$= \frac{E_m b}{T_1}\left(\frac{\sin n\pi \dfrac{b}{T_1}}{n\pi \dfrac{b}{T_1}}\right) e^{-jn\omega_1(b/2)} \tag{19}$$

Therefore, according to (10)

$$f(t) = \sum_{n=-\infty}^{\infty} \frac{E_m b}{T_1}\left(\frac{\sin n\pi \dfrac{b}{T_1}}{n\pi \dfrac{b}{T_1}}\right) \exp\left[jn\omega_1\left(t - \frac{b}{2}\right)\right] \tag{20}$$

If $b = T_1/2$, we shall have

$$F(n) = \frac{E_m}{2}\left(\frac{\sin n\dfrac{\pi}{2}}{n\dfrac{\pi}{2}}\right) e^{-jn\omega_1(T_1/4)} \tag{21}$$

and

$$f(t) = \sum_{n=-\infty}^{\infty} \frac{E_m}{2}\left(\frac{\sin n\dfrac{\pi}{2}}{n\dfrac{\pi}{2}}\right) \exp\left[jn\omega_1\left(t - \frac{T_1}{4}\right)\right] \tag{22}$$

$$= \frac{E_m}{2} + \frac{2E_m}{\pi}\left[\cos\omega_1\left(t - \frac{T_1}{4}\right) - \tfrac{1}{3}\cos 3\omega_1\left(t - \frac{T_1}{4}\right) + \tfrac{1}{5}\cos 5\omega_1\left(t - \frac{T_1}{4}\right) - \cdots\right] \tag{23}$$

A graph of this function is drawn in Fig. 2.

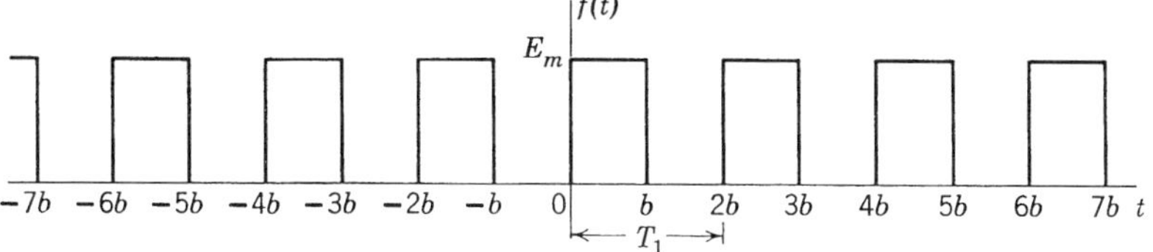

Fig. 2. Periodic wave of rectangular pulses.

2. Correlation

In the general theory of harmonic analysis, an expression of considerable importance and interest is, in the case of periodic functions,

$$\frac{1}{T_1}\int_{-T_1/2}^{T_1/2} f_1(t)f_2(t+\tau)\,dt \tag{24}$$

where $f_1(t)$ and $f_2(t)$ have the same fundamental angular frequency ω_1 and τ is a continuous time of displacement in the range $(-\infty, \infty)$, independent of t. One important property of this expression is the fact that its Fourier transform is

$$\bar{F}_1(n)F_2(n) \tag{25}$$

if $f_1(t)$ has the complex spectrum $F_1(n)$, and $f_2(t)$, $F_2(n)$. The bar indicates the complex conjugate of the quantity over which it is placed.

To establish this fact, let the Fourier expansions for $f_1(t)$ and $f_2(t)$ be

$$f_1(t) = \sum_{n=-\infty}^{\infty} F_1(n)e^{jn\omega_1 t} \tag{26}$$

and

$$f_2(t) = \sum_{n=-\infty}^{\infty} F_2(n)e^{jn\omega_1 t} \tag{27}$$

where the complex spectrums are

$$F_1(n) = \frac{1}{T_1}\int_{-T_1/2}^{T_1/2} f_1(t)e^{-jn\omega_1 t}\,dt$$

and

$$F_2(n) = \frac{1}{T_1}\int_{-T_1/2}^{T_1/2} f_2(t)e^{-jn\omega_1 t}\,dt \tag{29}$$

(28)

By writing $f_2(t+\tau)$ of (24) in the form (27), we find that

$$\frac{1}{T_1}\int_{-T_1/2}^{T_1/2} f_1(t)f_2(t+\tau)\,dt = \frac{1}{T_1}\int_{-T_1/2}^{T_1/2} f_1(t)\,dt \sum_{n=-\infty}^{\infty} F_2(n)e^{jn\omega_1(t+\tau)} \tag{30}$$

The manner in which the right-hand member of the equation is written indicates that summation with respect to n comes first and is followed by integration with respect to t. An inversion of the order of summation and integration leads to

$$\frac{1}{T_1}\int_{-T_1/2}^{T_1/2} f_1(t)f_2(t+\tau)\,dt = \sum_{n=-\infty}^{\infty} F_2(n)e^{jn\omega_1\tau}\frac{1}{T_1}\int_{-T_1/2}^{T_1/2} f_1(t)e^{jn\omega_1 t}\,dt \tag{31}$$

where the last integral is recognized, by comparison with (28), as the complex conjugate of $F_1(n)$. We have, therefore,

$$\frac{1}{T_1}\int_{-T_1/2}^{T_1/2} f_1(t)f_2(t+\tau)\,dt = \sum_{n=-\infty}^{\infty} [\overline{F}_1(n)F_2(n)]e^{jn\omega_1\tau} \tag{32}$$

This result is in the form (10) in being a periodic function of fundamental angular frequency ω_1 with the complex spectrum $\overline{F}_1(n)F_2(n)$. Application of (12) gives us

$$\overline{F}_1(n)F_2(n) = \frac{1}{T_1}\int_{-T_1/2}^{T_1/2} e^{-jn\omega_1\tau}\,d\tau\,\frac{1}{T_1}\int_{-T_1/2}^{T_1/2} f_1(t)f_2(t+\tau)\,dt \tag{33}$$

Here, as in (30), the convention is that the extreme right integration is first performed with respect to t; the resulting function of τ is then multiplied by $e^{-jn\omega_1\tau}$ and the product integrated with respect to τ to obtain a function of n. We have thus shown that

$$\frac{1}{T_1}\int_{-T_1/2}^{T_1/2} f_1(t)f_2(t+\tau)\,dt \quad \text{and} \quad \overline{F}_1(n)F_2(n) \tag{34}$$

are Fourier transforms of each other, and we shall call this relation the *correlation theorem* for periodic functions.

The integral (24) involves a combination of three important operations:

1. One of the periodic functions concerned, $f_2(t)$, is given a time displacement τ.
2. The displaced function is multiplied by the other periodic function of the same fundamental frequency.
3. The product is averaged by integration over a complete period.

These steps are repeated for every value of τ in the interval $(-\infty, \infty)$ so that a function is generated. This combination of the three operations, namely, *displacement*, *multiplication*, and *integration*, is termed *correlation*. Correlation is not limited to periodic functions, as indeed it is applicable to aperiodic and random functions.

3. Autocorrelation

In the correlation theorem for periodic functions we find that, under the special condition $f_1(t) = f_2(t)$, equation (32) is

$$\frac{1}{T_1}\int_{-T_1/2}^{T_1/2} f_1(t)f_1(t+\tau)\,dt = \sum_{n=-\infty}^{\infty} |F_1(n)|^2 e^{jn\omega_1\tau} \tag{35}$$

Since this relation holds for all values of τ in the interval $(-\infty, \infty)$, the equation for $\tau = 0$ is obviously

$$\frac{1}{T_1}\int_{-T_1/2}^{T_1/2} f_1^{\,2}(t)\,dt = \sum_{n=-\infty}^{\infty} |F_1(n)|^2 \tag{36}$$

which states that the mean square value of the function $f_1(t)$ is equal to the sum, over the entire range of harmonics of the square of the absolute value of its spectrum. If $f_1(t)$ represents a voltage or a current and a load of 1 ohm pure resistance is assumed, we have the well-known fact that the mean power consumed is the sum of all the powers contributed by the harmonics into which the voltage or current wave has been resolved. In mathematical work, relation (36) is known as a *Parseval theorem* for periodic functions.

Let the left-hand member of (35) be called the *autocorrelation function* of $f_1(t)$, and let it be represented by the symbol $\varphi_{11}(\tau)$. That is,

$$\varphi_{11}(\tau) = \frac{1}{T_1}\int_{-T_1/2}^{T_1/2} f_1(t)f_1(t+\tau)\,dt \tag{37}$$

by definition. Let the symbol $\Phi_{11}(n)$ denote the *power spectrum* in (35), so that

$$\Phi_{11}(n) = |F_1(n)|^2 \tag{38}$$

The subscript 11 indicates that the autocorrelation function is obtained by correlating a function with itself; and since the power spectrum is the representation of $\varphi_{11}(\tau)$ in the frequency domain, we keep the same subscript. In terms of the new symbols, (35) becomes

$$\varphi_{11}(\tau) = \sum_{n=-\infty}^{\infty} \Phi_{11}(n)e^{jn\omega_1\tau} \tag{39}$$

and inversely

$$\Phi_{11}(n) = \frac{1}{T_1}\int_{-T_1/2}^{T_1/2} \varphi_{11}(\tau)e^{-jn\omega_1\tau}\,d\tau \tag{40}$$

These reciprocal relations express a particular form of the correlation theorem which we shall call the *autocorrelation theorem* for periodic functions. Expressed in words, the theorem states that the autocorrelation function and the power spectrum of a periodic function are Fourier transforms of each other, so that given one the other is uniquely determined.

Since the spectrum of the autocorrelation function is the square of the absolute value of the complex spectrum of the given periodic function, its phase spectrum is always zero for all harmonics. Consequently, one of the important properties of an autocorrelation function is that, even though retaining all harmonics of the given function and containing no new ones, it discards all their phase angles. In other words, all periodic functions having the same harmonic amplitudes but differing in their initial phase angles have the same autocorrelation function, which amounts to saying that the power spectrum of a periodic function is independent of the phase angles of its harmonics.

The autocorrelation function is an even function of τ. This fact is clear from an examination of (35) and may further be demonstrated by considering that

$$\begin{aligned}\varphi_{11}(-\tau) &= \frac{1}{T_1}\int_{-T_1/2}^{T_1/2} f_1(t)f_1(t-\tau)\,dt \\ &= \frac{1}{T_1}\int_{-T_1/2-\tau}^{T_1/2-\tau} f_1(x)f_1(x+\tau)\,dx\end{aligned} \tag{41}$$

when a change of variable, $x = t - \tau$, is made. Since $\varphi_{11}(\tau)$ is a periodic function of period T_1, integration over $(-T_1/2 - \tau,\ T_1/2 - \tau)$ is the same as that over $(T_1/2,\ -T_1/2)$. Hence

$$\begin{aligned}\varphi_{11}(-\tau) &= \frac{1}{T_1}\int_{-T_1/2}^{T_1/2} f_1(x)f_1(x+\tau)\,dx \\ &= \varphi_{11}(\tau)\end{aligned} \tag{42}$$

which is an expression of the property that $\varphi_{11}(\tau)$ is even.

Inasmuch as $\varphi_{11}(\tau)$ is an even periodic function, it is expressible as a cosine series with zero initial phase angles. Therefore, alternative forms of (39) and (40) are

$$\varphi_{11}(\tau) = \sum_{n=-\infty}^{\infty} \Phi_{11}(n)\cos n\omega_1\tau \tag{43}$$

and

$$\Phi_{11}(n) = \frac{1}{T_1}\int_{-T_1/2}^{T_1/2} \varphi_{11}(\tau)\cos n\omega_1\tau\,d\tau \tag{44}$$

It seems appropriate to remark here that the purpose of the concept of correlation of periodic functions is not to facilitate the calculation of power spectrums of known periodic functions, for, obviously, if the complex spectrum of $f_1(t)$ is $F_1(n)$, its power spectrum is $|F_1(n)|^2$. An important reason for introducing the concept is that it plays a central role in the analysis of random functions which may contain periodic components. A situation of this sort occurs when a periodic signal is masked by random noise. Another reason for the consideration of correlation is its general applicability to all three types of functions, namely, periodic, aperiodic, and random, which enables us to see clearly the unity in the general theory of harmonic analysis.

A formula for the autocorrelation function in terms of the coefficients of the Fourier expansion of a given periodic function is often desirable. To find this formula, we write (43) as

$$\varphi_{11}(\tau) = \Phi_{11}(0) + 2 \sum_{n=1}^{\infty} \Phi_{11}(n) \cos n\omega_1\tau \tag{45}$$

With the knowledge that

$$f_1(t) = \frac{a_{10}}{2} + \sum_{n=1}^{\infty} (a_{1n} \cos n\omega_1 t + b_{1n} \sin n\omega_1 t) \tag{46}$$

and

$$F_1(n) = \tfrac{1}{2}(a_{1n} - jb_{1n}) \tag{47}$$

so that

$$\Phi_{11}(n) = \frac{a_{1n}^2 + b_{1n}^2}{4} \qquad \text{for } n = 0, \pm 1, \pm 2, \pm 3, \ldots \tag{48}$$

we obtain as the required expression

$$\varphi_{11}(\tau) = \frac{a_{10}^2}{4} + \frac{1}{2} \sum_{n=1}^{\infty} (a_{1n}^2 + b_{1n}^2) \cos n\omega_1\tau \tag{49}$$

It is significant that, in (49), $a_{10}^2/4$ is the mean square value of the constant in the given periodic function $f_1(t)$ and similarly $(a_{1n}^2 + b_{1n}^2)/2$ is the mean square value of the nth harmonic of $f_1(t)$. For the purpose of expressing (49) with particular emphasis on mean square values of the given periodic function, let $C_{10}^2 = a_{10}^2/4$ and $C_{1n}^2 = (a_{1n}^2 + b_{1n}^2)/2$ where n does not include zero. We then have (49) condensed to the form

$$\varphi_{11}(\tau) = \sum_{n=0}^{\infty} C_{1n}^2 \cos n\omega_1\tau \tag{50}$$

The autocorrelation function is therefore a cosine series with zero initial phase angles for all harmonics that are present in the given function and whose coefficients are equal to the mean square values of the corresponding harmonics in the given function; its constant term, likewise, is the mean square of the given constant.

Finally, let us be reminded that the mean power of $f_1(t)$ is

$$\left.\begin{aligned} \varphi_{11}(0) &= \frac{a_{10}^2}{4} + \frac{1}{2}\sum_{n=1}^{\infty}(a_{1n}^2 + b_{1n}^2) \\ &= \sum_{n=0}^{\infty} C_{1n}^2 \\ &= \frac{1}{T_1}\int_{-T_1/2}^{T_1/2} f_1{}^2(t)\,dt \end{aligned}\right\} \tag{51}$$

4. Illustrative Examples

Example 1. Autocorrelation of a Sinusoid. A simple example is

$$f_1(t) = A\cos(\omega_1 t + \theta) \tag{52}$$

The autocorrelation function is, according to (49),

$$\varphi_{11}(\tau) = \frac{A^2}{2}\cos\omega_1\tau \tag{53}$$

and the power spectrum is

$$\Phi_{11}(n) = \frac{A^2}{4} \qquad \text{for } n = \pm 1 \tag{54}$$

The graphs of $f_1(t)$ and $\varphi_{11}(\tau)$ are given in Figs. 3 and 4.

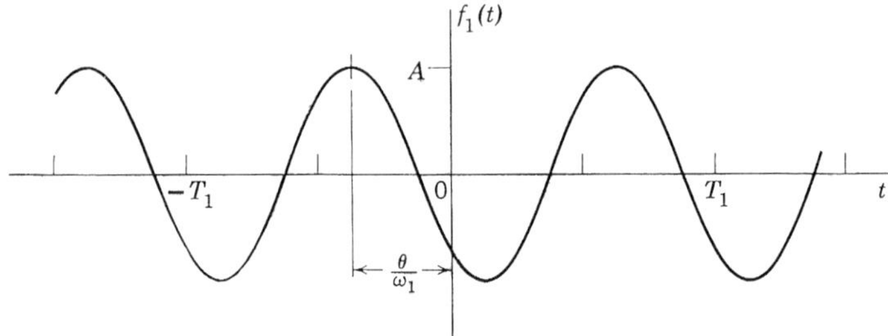

Fig. 3. Sine wave.

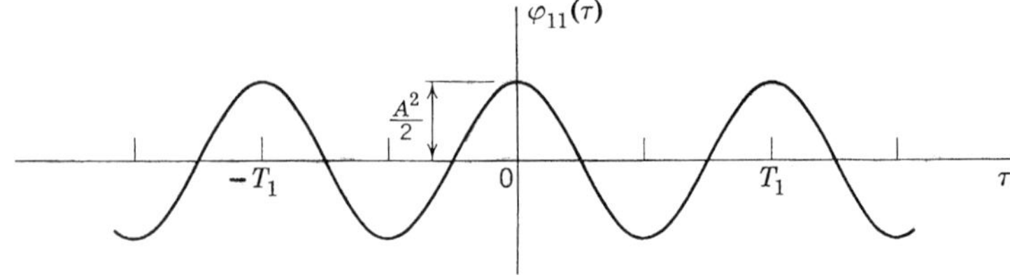

Fig. 4. Autocorrelation function of the sine wave of Fig. 3.

Example 2. Autocorrelation with Graphical Interpretation. Another example is the periodic rectangular wave of Fig. 2. Application of formula (49) or (50) to equation (23) which represents this function gives

$$\varphi_{11}(\tau) = \frac{E_m^2}{4} + \frac{2E_m^2}{\pi^2}\left[\cos \omega_1\tau + \frac{1}{3^2}\cos 3\omega_1\tau + \frac{1}{5^2}\cos 5\omega_1\tau + \cdots\right] \tag{55}$$

whose graph is shown in Fig. 5. This graph has not been determined by computation in accordance with formula (55), because such a procedure would be laborious indeed. The construction of the graph of $\varphi_{11}(\tau)$ is greatly facilitated by a graphical interpretation of the definition (37) for autocorrelation. With the aid of Fig. 6, we first consider the value of the autocorrelation function when

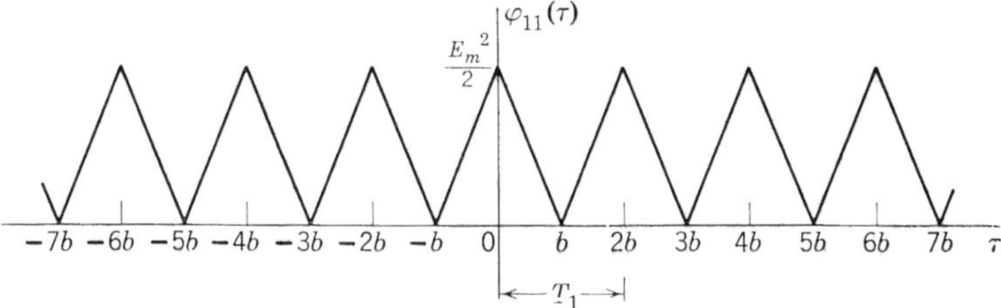

Fig. 5. Autocorrelation function of the wave of Fig. 2.

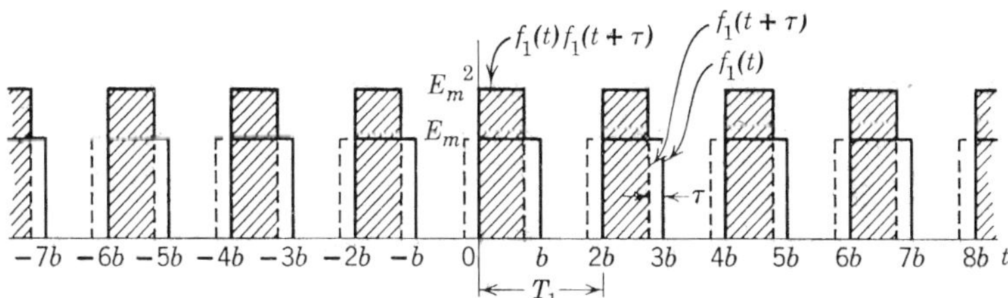

Fig. 6. Pertaining to the graphical interpretation of the autocorrelation of a rectangular wave.

$\tau = 0$. At this point, $f_1(t)f_1(t + \tau)$ is simply the square of the given function so that

$$\varphi_{11}(0) = \frac{1}{T_1}\frac{E_m^2 T_1}{2} = \frac{E_m^2}{2} \tag{56}$$

As τ assumes a value such as is shown, the area of the product $f_1(t)f_1(t + \tau)$ is less than that of $f_1^2(t)$ and the mean of the product is of course less than $E_m^2/2$. The area under $f_1(t)f_1(t + \tau)$ in a complete period gives the numerical value of the integral in (37) for a given value of τ, and the mean value in the period is the value of the autocorrelation function at time τ. Since the pulses are rectangular, the autocorrelation function decreases linearly in the range $(0, b)$, and at $\tau = b$ the function is clearly zero. Proceeding in this manner, we find a linear rise to a maximum point at $\tau = 2b$ which has the value $E_m^2/2$; and, having thus completed a cycle of the curve, we have sufficient information for the rest of it.

Example 3. Piecemeal Determination of Autocorrelation Function. When a periodic function is given graphically as a combination of straight lines and simple curves such as the function illustrated in Fig. 7, the autocorrelation

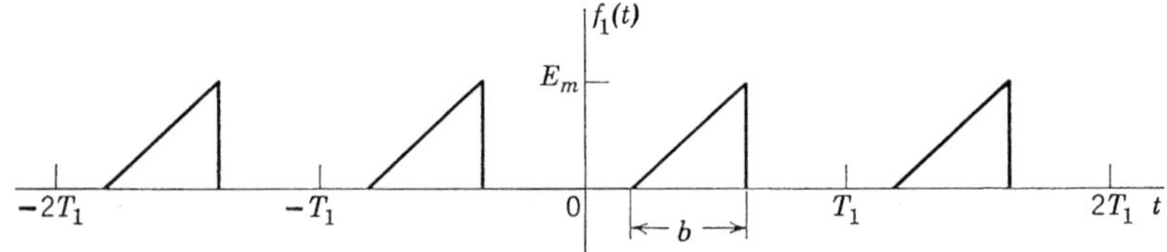

Fig. 7. Periodic wave of triangular pulses.

function can often be determined piecemeal. Taking only one period of the function as in Fig. 8 and assuming that $T_1 \geq 2b$, we represent the solid straight line by the expression $(E_m/b)t$ and the dotted line, which is the displaced solid line, by $(E_m/b)(t + \tau)$. Since autocorrelation of a periodic function is independent of the phase angles of its harmonics, the location of the reference point $t = 0$ in Fig. 7 is immaterial in our problem. Therefore, for simplicity we have arbitrarily placed the solid line through the point $t = 0$ as in Fig. 8. The portion of the autocorrelation curve $\varphi_{11}(\tau)$ from $\tau = 0$ to $\tau = b$, as shown in Fig. 9, is first determined. Application of definition (37) immediately yields

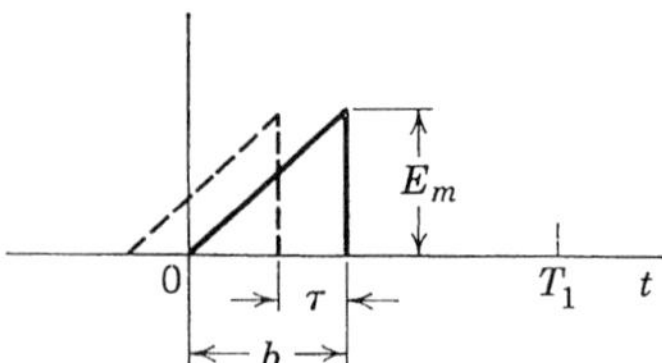

Fig. 8. Pertaining to the piecemeal determination of the autocorrelation of the wave of Fig. 7.

$$\begin{aligned}
\varphi_{11}(\tau) &= \frac{1}{T_1}\int_0^{b-\tau} \frac{E_m{}^2}{b^2}\, t(t + \tau)\, dt \\
&= \frac{E_m{}^2}{6b^2T_1}\,[2(b - \tau)^3 + 3\tau(b - \tau)^2] \\
&= \frac{E_m{}^2}{6b^2T_1}\,(\tau^3 - 3b^2\tau + 2b^3) \qquad \text{for } 0 \leq \tau \leq b
\end{aligned} \tag{57}$$

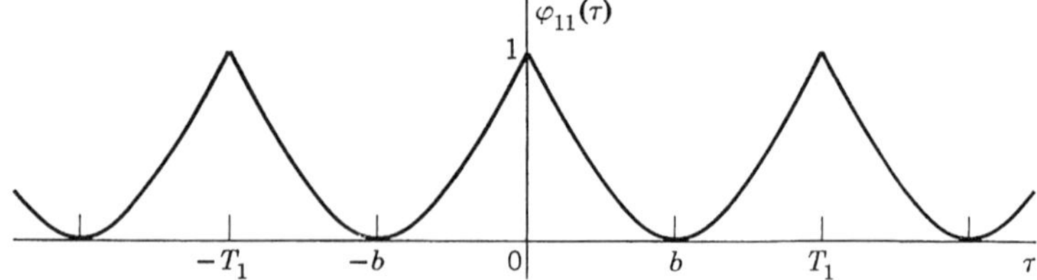

Fig. 9. Autocorrelation function of the wave of Fig. 7. $E_m{}^2 = 6$, $T_1 = 2b$.

Because $\varphi_{11}(\tau)$ is an even function, the curve in the interval $(-b, 0)$ should be symmetrical to that in the interval $(0, b)$; and, because the function is periodic of period T_1, the remainder of the function is determined without further calculation. It is observed that the wave of triangular pulses has the mean power

$$\varphi_{11}(0) = \frac{E_m{}^2 b}{3T_1} \tag{58}$$

5. Crosscorrelation

If, in the correlation theorem for periodic functions, $f_1(t)$ and $f_2(t)$ are different periodic functions (of the same fundamental frequency), we define a *crosscorrelation function* of $f_1(t)$ and $f_2(t)$ as

$$\varphi_{12}(\tau) = \frac{1}{T_1}\int_{-T_1/2}^{T_1/2} f_1(t)f_2(t+\tau)\,dt \tag{59}$$

and a *cross-power spectrum* of the functions as

$$\Phi_{12}(n) = \bar{F}_1(n)F_2(n) \tag{60}$$

The subscript 12 indicates that the crosscorrelation involves $f_1(t)$ and $f_2(t)$ with the second numeral 2 (or letter, if letters are used) referring to the fact that $f_2(t)$ has the displacement τ. It is important to keep the numerals, or letters, in the double subscript in their proper order, for when we interchange them we refer to the definitions

$$\varphi_{21}(\tau) = \frac{1}{T_1}\int_{-T_1/2}^{T_1/2} f_2(t)f_1(t+\tau)\,dt \tag{61}$$

and

$$\Phi_{21}(n) = \bar{F}_2(n)F_1(n) \tag{62}$$

which are in general different from (59) and (60) respectively although related to each other in a simple manner. In fact, by changing τ in (59) to $-\tau$ and letting $x - t - \tau$, wc have

$$\begin{aligned}\varphi_{12}(-\tau) &= \frac{1}{T_1}\int_{-T_1/2}^{T_1/2} f_1(t)f_2(t-\tau)\,dt \\ &= \frac{1}{T_1}\int_{-T_1/2-\tau}^{T_1/2-\tau} f_2(x)f_1(x+\tau)\,dx \end{aligned} \tag{63}$$

Since the crosscorrelation function of two periodic functions of a common fundamental frequency is another periodic function of the same fundamental frequency, the limits of x in (63) may be replaced by $(-T_1/2, T_1/2)$ so that

$$\varphi_{12}(-\tau) = \frac{1}{T_1}\int_{-T_1/2}^{T_1/2} f_2(x)f_1(x+\tau)\,dx \tag{64}$$

By definition the right-hand member of (64) is $\varphi_{21}(\tau)$, and the relation between $\varphi_{12}(\tau)$ and $\varphi_{21}(\tau)$ is therefore

$$\varphi_{12}(-\tau) = \varphi_{21}(\tau) \tag{65}$$

Graphically we have here the result that, with respect to the vertical axis, the function $\varphi_{12}(\tau)$ is the mirror image of $\varphi_{21}(\tau)$, and vice versa. This property of the crosscorrelation function expressed in the time domain corresponds to the property of the cross-power spectrum, which is the representation of the crosscorrelation function in the frequency domain, that

$$\Phi_{12}(n) = \overline{\Phi}_{21}(n) \tag{66}$$

That is, $\Phi_{12}(n)$ and $\Phi_{21}(n)$ are related to each other as complex conjugate quantities. This statement can be easily verified by referring to the proof of the correlation theorem and interchanging the subscripts of $f_1(t)$ and $f_2(t)$.

In terms of the symbols introduced in (59) through (62), the reciprocal relations (32) and (33) appear as

$$\varphi_{12}(\tau) = \sum_{n=-\infty}^{\infty} \Phi_{12}(n)e^{jn\omega_1\tau} \tag{67}$$

and

$$\Phi_{12}(n) = \frac{1}{T_1}\int_{-T_1/2}^{T_1/2} \varphi_{12}(\tau)e^{-jn\omega_1\tau}\,d\tau \tag{68}$$

and, similarly,

$$\varphi_{21}(\tau) = \sum_{n=-\infty}^{\infty} \Phi_{21}(n)e^{jn\omega_1\tau} \tag{69}$$

and

$$\Phi_{21}(n) = \frac{1}{T_1}\int_{-T_1/2}^{T_1/2} \varphi_{21}(\tau)e^{-jn\omega_1\tau}\,d\tau \tag{70}$$

Here we have the correlation theorem for periodic functions in a new set of symbols. To distinguish it from the special form when a single periodic function is involved, we shall call it the *crosscorrelation theorem* or simply the correlation theorem when emphasis of the difference is unnecessary.

Since $f_1(t)$ and $f_2(t)$ are, in general, different functions of the same fundamental frequency, each with a complex spectrum, their cross-power spectrum is a complex quantity as (60) shows. Although reduction of the reciprocal relations (39) and (40) to the trigonometric forms (43) and (44), because of the elimination of sine terms, is a simplification for some applications of autocorrelation, a similar reduction applied to (67) through (70) will offer no corresponding advantage since both sine and cosine terms must now be retained.

For a general formula giving the crosscorrelation function in terms of the Fourier coefficients of the given periodic functions $f_1(t)$ and $f_2(t)$, we consider the expansions

$$f_1(t) = \frac{a_{10}}{2} + \sum_{n=1}^{\infty} (a_{1n} \cos n\omega_1 t + b_{1n} \sin n\omega_1 t) \tag{71}$$

and

$$f_2(t) = \frac{a_{20}}{2} + \sum_{n=1}^{\infty} (a_{2n} \cos n\omega_1 t + b_{2n} \sin n\omega_1 t) \tag{72}$$

The crosscorrelation function is, when (60) is substituted into (67),

$$\varphi_{12}(\tau) = \sum_{n=-\infty}^{\infty} \overline{F}_1(n)F_2(n)e^{jn\omega_1\tau} \tag{73}$$

where

$$F_1(n) = \frac{a_{1n} - jb_{1n}}{2} \qquad \text{for } n = 0, \pm 1, \pm 2, \pm 3, \ldots \tag{74}$$

and

$$F_2(n) = \frac{a_{2n} - jb_{2n}}{2} \qquad \text{for } n = 0, \pm 1, \pm 2, \pm 3, \ldots \tag{75}$$

If, for either $f_1(t)$ or $f_2(t)$, we let

$$c_n = \sqrt{a_n^{\ 2} + b_n^{\ 2}} \tag{76}$$

and

$$\theta_n = \tan^{-1}\left(-\frac{b_n}{a_n}\right) \tag{77}$$

then

$$F(n) = \frac{c_n}{2}\, e^{j\theta_n} \tag{78}$$

From the relations $a_{-n} = a_n$ and $b_{-n} = -b_n$, as given by (7) and (8), we find that

$$\theta_{-n} = -\theta_n \tag{79}$$

Making use of these details in the reduction of (73) to the required form, we have

$$\varphi_{12}(\tau) = \sum_{n=-\infty}^{\infty} \frac{c_{1n}c_{2n}}{4}\, e^{j(n\omega_1\tau + \theta_{2n} - \theta_{1n})} \tag{80}$$

or

$$\varphi_{12}(\tau) = \frac{a_{10}a_{20}}{4} + \frac{1}{2}\sum_{n=1}^{\infty} c_{1n}c_{2n} \cos (n\omega_1\tau + \theta_{2n} - \theta_{1n}) \tag{81}$$

Similarly,

$$\varphi_{21}(\tau) = \frac{a_{20}a_{10}}{4} + \frac{1}{2}\sum_{n=1}^{\infty} c_{2n}c_{1n} \cos (n\omega_1\tau + \theta_{1n} - \theta_{2n}) \tag{82}$$

Two comments concerning this result should be made. First, the constant and harmonic coefficients in the crosscorrelation function appear as products of the corresponding quantities in the given functions $f_1(t)$ and $f_2(t)$. Consequently if the constant or any harmonic is absent in either $f_1(t)$ or $f_2(t)$, the constant term or the corresponding harmonic term will be absent in the crosscorrelation function. Second, a notable difference between autocorrelation and crosscorrelation is that, whereas autocorrelation discards all phase information in the given function, crosscorrelation retains the phase differences of the harmonics which are present in both periodic functions.

The crosscorrelation function (81) may be written in terms of the rms values of the harmonics involved. To do this let us note that c_{1n} is the maximum amplitude of the nth harmonic of $f_1(t)$, the rms value of which is $c_{1n}/\sqrt{2}$; similarly $c_{2n}/\sqrt{2}$ is the rms value of the nth harmonic of $f_2(t)$. Obviously, the constants $a_{10}/2$ and $a_{20}/2$ of $f_1(t)$ and $f_2(t)$ respectively remain the same when their rms values are taken. Writing $C_{10} = a_{10}/2$, $C_{20} = a_{20}/2$, $C_{1n} = c_{1n}/\sqrt{2}$ and $C_{2n} = c_{2n}/\sqrt{2}$ where n is not zero, we have, according to (81),

$$\varphi_{12}(\tau) = \sum_{n=0}^{\infty} C_{1n}C_{2n} \cos (n\omega_1\tau + \theta_{2n} - \theta_{1n}) \tag{83}$$

which expresses the fact that the coefficients of the harmonics in a crosscorrelation function of two periodic functions of the same fundamental frequency are the products of the rms values of the corresponding harmonics in the two given functions. Since the rms value of a constant is the constant itself, the constant of the crosscorrelation function is the product of the constants of the given functions.

When the order of the subscripts is reversed, (83) becomes

$$\varphi_{21}(\tau) = \sum_{n=0}^{\infty} C_{2n}C_{1n} \cos (n\omega_1\tau + \theta_{1n} - \theta_{2n}) \tag{84}$$

Needless to say, if $f_1(t) = f_2(t)$, then (83) and (84) reduce to the autocorrelation function (50) where mean square values take the place of the product of two identical rms values.

The significance of a crosscorrelation function cannot be given properly unless the physical system and the problem with which it is concerned are known. For a simple illustration, consider a two-terminal linear circuit with $f_1(t)$ as the periodic voltage wave across the terminals and $f_2(t)$ as the periodic current wave in them. Suppose that we are interested in the spectrum of the power which the circuit consumes. Let this spectrum be determined through the application of crosscorrelation.

In this situation the voltage-current crosscorrelation $\varphi_{12}(\tau)$ for $\tau = 0$ gives the mean power; that is

$$\varphi_{12}(0) = \sum_{n=0}^{\infty} C_{1n}C_{2n} \cos (\theta_{2n} - \theta_{1n}) \tag{85}$$

Putting this expression in conventional symbols for voltage and current, we have

$$\varphi_{ei}(0) = \sum_{n=0}^{\infty} E_n I_n \cos (\theta_{en} - \theta_{in}) \tag{86}$$

in which $E_0 = a_{10}/2$ and $I_0 = a_{20}/2$ are the d-c components of the voltage and current respectively, $E_n = C_{1n}$ and $I_n = C_{2n}$ are the rms values of the nth harmonics of the voltage and current respectively, $(\theta_{en} - \theta_{in})$ is the phase difference between voltage and current of harmonic order n, and $\varphi_{ei}(0)$ is the voltage-current crosscorrelation for $\tau = 0$ which is the mean power. If the frequency range is sufficiently low so that a wattmeter is suitable for power measurement, the reading of the meter is the value of $\varphi_{ei}(0)$.

If now a continuous time displacement is introduced between voltage and current, by a delay device, and the mean value of the product of voltage and current is measured by the wattmeter for every value of the time displacement τ, we shall have the crosscorrelation function

$$\varphi_{ei}(\tau) = \sum_{n=0}^{\infty} E_n I_n \cos (n\omega_1\tau + \theta_{en} - \theta_{in}) \tag{87}$$

This function, according to the crosscorrelation theorem, has the property that its Fourier transform is the cross-power spectrum of the voltage and current, which, in this particular example, is the spectrum of the power taken by the circuit. In other words, when a Fourier analysis is performed on the crosscorrelation curve which has been determined in the manner described, we shall have the power spectrum given theoretically as

$$\Phi_{ei}(n) = \begin{cases} E_0 I_0 & \text{for } n = 0 \\ E_n I_n \cos (\theta_{en} - \theta_{in}) & \text{for } n = 1, 2, 3, \ldots \end{cases} \tag{88}$$

Or, if $\varphi_{ei}(\tau)$ is expressed as

$$\varphi_{ei}(\tau) = E_0 I_0 + \frac{1}{2} \sum_{n=-\infty,\cdots,-1,1,\cdots,\infty} E_n I_n e^{j(n\omega_1\tau + \theta_{en} - \theta_{in})} \tag{89}$$

then

$$\Phi_{ei}(n) = \begin{cases} E_0 I_0 & \text{for } n = 0 \\ \frac{1}{2} E_n I_n \cos (\theta_{en} - \theta_{in}) & \text{for } n = \pm 1, \pm 2, \pm 3, \ldots \end{cases} \tag{90}$$

Although the example given here illustrates a possible method for power spectrum determination through the measurement of the voltage-current crosscorrelation as a function of their time displacement, the primary objective has been the interpretation of crosscorrelation in accordance with the actual situation and the problem concerned. In some instances, as in this example, the crosscorrelation function yields, after transformation, a power spectrum which corresponds to an actual dissipation of power, but it is not always true that a crosscorrelation function represents, in some way, a power transfer. For example, voltages and currents at different locations of a communication system are crosscorrelated in a detection problem, and the input and output of a linear system are crosscorrelated for system characteristic determination. In these and similar problems the spectrum of crosscorrelation cannot be interpreted as a transfer of power between different parts of a system although it may have the same unit of measure as that of power. Some of these problems are considered when sufficient basic material has been presented. Since it is convenient to have a name for the transform of a crosscorrelation function analogous to that for the transform of an autocorrelation function, we call it the *cross-power spectrum.*

6. Illustrative Examples

Example 1. Crosscorrelation of Triangular and Rectangular Periodic Waves. We are given a wave $f_1(t)$ of triangular pulses with time $t = 0$ at the leading point of a pulse as shown in Fig. 10, and a wave $f_2(t)$ of rectangular pulses with $t = 0$ at the leading edge of a pulse as shown in Fig. 11. Both waves are periodic of period T_1. The maximum value of $f_1(t)$ is E_{m1} and that of $f_2(t)$

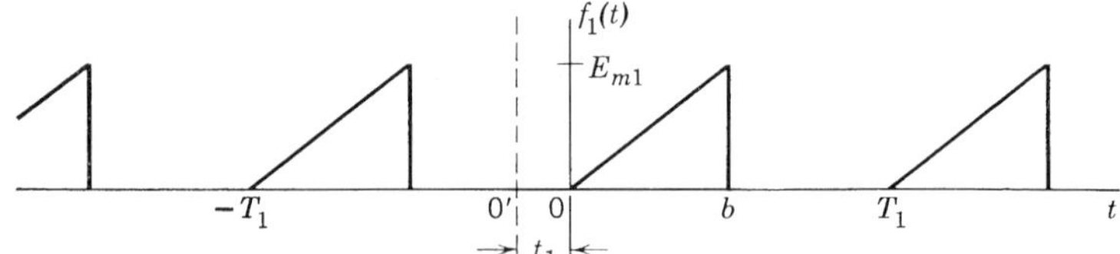

Fig. 10. Periodic wave of triangular pulses for crosscorrelation.

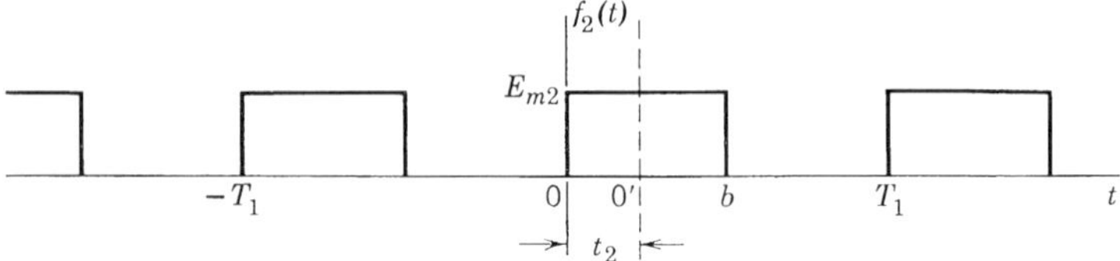

Fig. 11. Periodic wave of rectangular pulses for crosscorrelation.

is E_{m2}. For simplicity, let the duration of the pulses be $b = T_1/2$. In the determination of the crosscorrelation function $\varphi_{12}(\tau)$, the method of Example 3, Sec. 4, is applicable. Taking one period of each function, we hold the triangle in a fixed position and allow the rectangle to be displaced as shown in Fig. 12. For the portion of $\varphi_{12}(\tau)$ in the interval $(0, b)$ (see Fig. 13) we have, in accordance with definition (59),

$$\varphi_{12}(\tau) = \frac{1}{T_1}\int_0^{b-\tau} \frac{E_{m1}E_{m2}}{b}\, t\, dt$$

$$= \frac{E_{m1}E_{m2}}{4b^2}(b-\tau)^2 \qquad \text{for } 0 \le \tau \le b \tag{91}$$

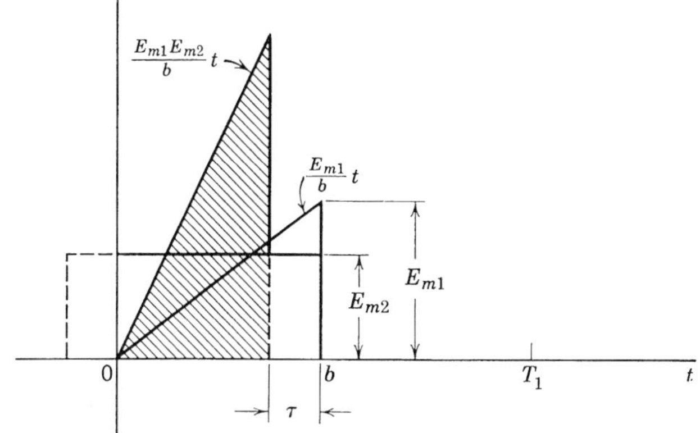

Fig. 12. Illustrating the crosscorrelation of the waves of Figs. 10 and 11.

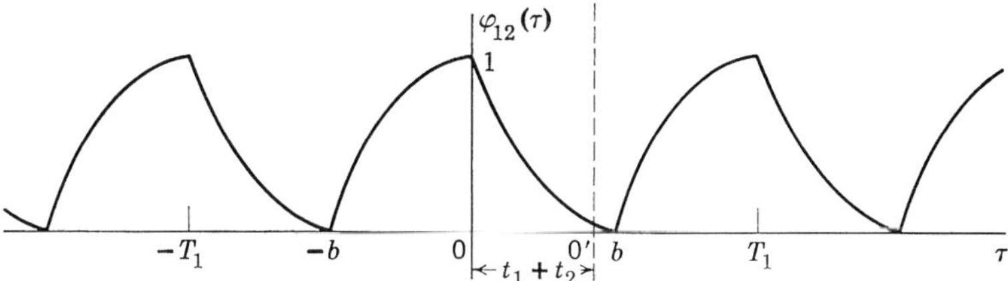

Fig. 13. Crosscorrelation function of the waves of Figs. 10 and 11. $E_{m1} = 4$, $E_{m2} = 1$.

We note here that the interval $(0, b)$ is a half period of the crosscorrelation function inasmuch as the period of $f_1(t)$ and $f_2(t)$ is $2b$. In a similar manner, the crosscorrelation curve in the interval $(-b, 0)$ may be found. With the aid of Fig. 14, which is drawn for a displacement in the opposite direction as compared with that of Fig. 12, we see that

$$\varphi_{12}(\tau) = \frac{1}{T_1}\int_\tau^{b} \frac{E_{m1}E_{m2}}{b}\, t\, dt$$

$$= \frac{E_{m1}E_{m2}}{4b^2}(b^2-\tau^2) \qquad \text{for } -b \le \tau \le 0 \tag{92}$$

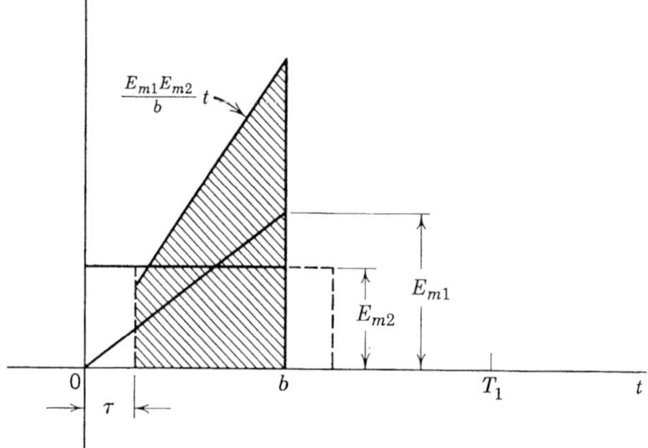

Fig. 14. Illustrating the crosscorrelation of the waves of Figs. 10 and 11.

Having thus determined one whole period of the crosscorrelation function, we have sufficient information for the complete construction of the function which is shown in Fig. 13.

To illustrate that crosscorrelation retains phase differences, let the origin of $f_1(t)$ be shifted from 0 to 0′, a duration of t_1 seconds to the left of 0 as indicated in Fig. 10, and let the origin of $f_2(t)$ be likewise displaced t_2 seconds to the right of 0 as in Fig. 11. The effect of these changes in the origins of time in $f_1(t)$ and $f_2(t)$ on the crosscorrelation function is simply the relocation of the point $\tau = 0$ in Fig. 13. In accordance with definition (59) the origin should now be at 0′ which is $t_1 + t_2$ seconds to the right of 0.

Example 2. Correlation by Parts. A function $f_1(t)$ consisting of periodic pulses in the form of an isosceles triangle with the maximum value E_m and duration $2b$ is shown in Fig. 15. For simplicity, let the period T_1 for $f_1(t)$ be equal to

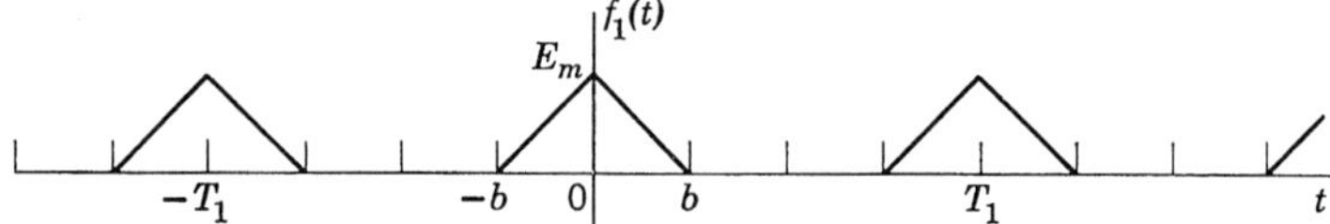

Fig. 15. Periodic wave of triangular pulses for illustrating correlation by parts.

$4b$. In the determination of the autocorrelation function, two algebraic expressions are necessary for the two straight lines which form the triangular pulse. The total algebraic work involved is not excessive; but, since human errors frequently occur whenever a single equation contains a considerable amount of algebra, it is desirable to resolve the given function into elementary components so that the autocorrelation function is expressed as the sum of simple parts. In accordance with this idea $f_1(t)$ is resolved into $f_a(t)$ and $f_b(t)$ as given in Fig. 16*a* and *b*.

Writing $f_1(t)$ as

$$f_1(t) = f_a(t) + f_b(t) \tag{93}$$

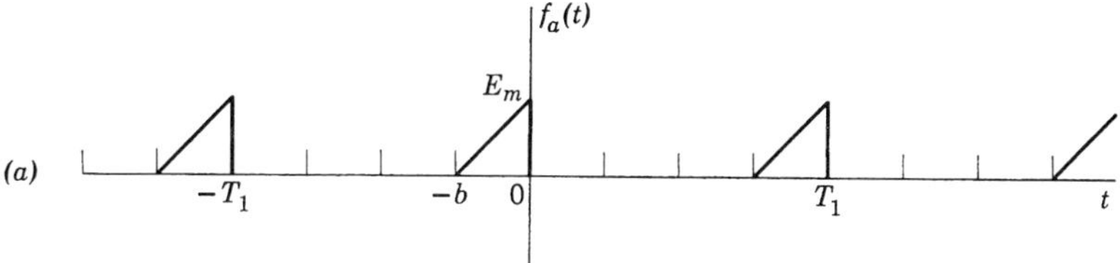

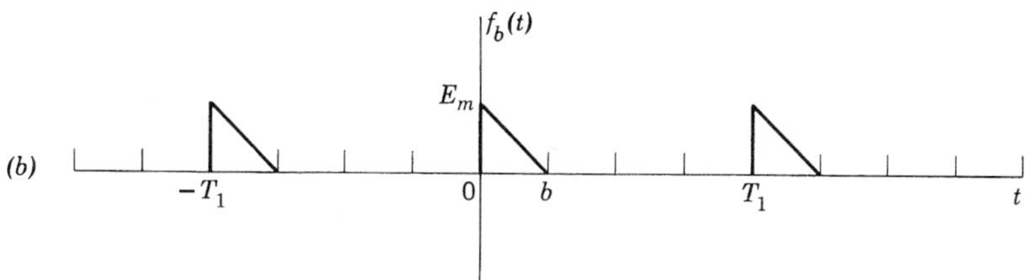

Fig. 16. Components of the wave of Fig. 15.

and inserting it in the defining equation for autocorrelation (37), and then expanding the result, we have

$$\begin{aligned}\varphi_{11}(\tau) &= \frac{1}{T_1}\int_{-T_1/2}^{T_1/2} [f_a(t) + f_b(t)][f_a(t+\tau) + f_b(t+\tau)]\,dt \\ &= \frac{1}{T_1}\int_{-T_1/2}^{T_1/2} f_a(t)f_a(t+\tau)\,dt + \frac{1}{T_1}\int_{-T_1/2}^{T_1/2} f_b(t)f_b(t+\tau)\,dt \\ &\quad + \frac{1}{T_1}\int_{-T_1/2}^{T_1/2} f_a(t)f_b(t+\tau)\,dt + \frac{1}{T_1}\int_{-T_1/2}^{T_1/2} f_b(t)f_a(t+\tau)\,dt \end{aligned} \tag{94}$$

which indicates that the autocorrelation of $f_1(t)$ is expressible as the sum of two autocorrelation functions and two crosscorrelation functions of the components $f_a(t)$ and $f_b(t)$. In short,

$$\varphi_{11}(\tau) = \varphi_{aa}(\tau) + \varphi_{bb}(\tau) + \varphi_{ab}(\tau) + \varphi_{ba}(\tau) \tag{95}$$

The first of the four component correlation functions of $\varphi_{11}(\tau)$ is $\varphi_{aa}(\tau)$, the autocorrelation function of $f_a(t)$. Since $f_a(t)$ is the same as the periodic function of Fig. 7 except for the location of the origin, we have the complete evaluation of its autocorrelation under Example 3, Sec. 4, the result of which is given graphically in Fig. 9. It is not difficult to see that the next component correlation function $\varphi_{bb}(\tau)$ and the function $\varphi_{aa}(\tau)$ are identical, for the difference between $f_a(t)$ and $f_b(t)$ is entirely due to differences in initial phase angles of their harmonics of which autocorrelation is independent. The remaining components are crosscorrelation functions of $f_a(t)$ and $f_b(t)$, but only one of them needs detailed evaluation because their simple relation is $\varphi_{ab}(\tau) = \varphi_{ba}(-\tau)$.

The determination of $\varphi_{ab}(\tau)$ follows the method of the preceding example. For the portion of $\varphi_{ab}(\tau)$ in the interval $(0, b)$ in Fig. 17, we have indicated the

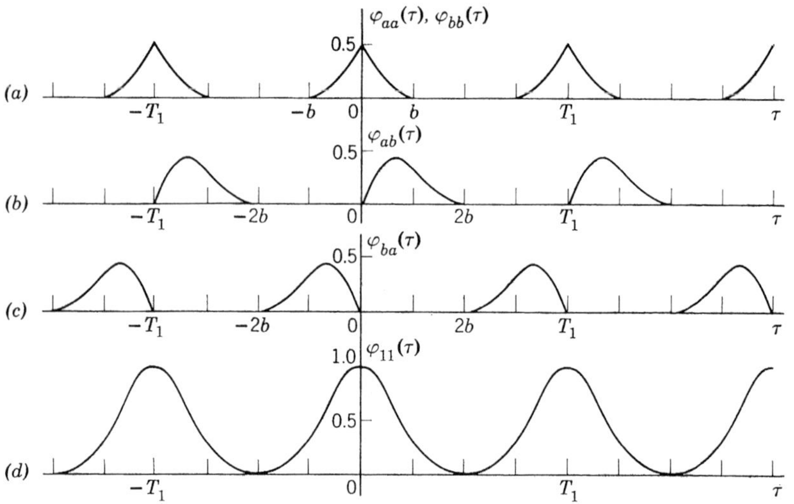

Fig. 17. Autocorrelation of the wave of Fig. 15 and its components. $E_m = \sqrt{6}$, $T_1 = 4b$.

relation between $f_a(t)$ and $f_b(t)$ in Fig. 18. The solid triangle is a pulse of $f_a(t)$, and the dotted one is a pulse of $f_b(t + \tau)$. Representing the fixed pulse by $E_m + (E_m/b)t$ in the duration of the pulse and similarly the displaced pulse by $E_m - (E_m/b)(t + \tau)$, we have, as in Example 1,

$$\varphi_{ab}(\tau) = \frac{1}{T_1} \int_{-\tau}^{0} E_m{}^2 \left(1 + \frac{t}{b}\right) \left[1 - \frac{1}{b}(t + \tau)\right] dt$$

$$= \frac{E_m{}^2}{6b^2T_1} \tau(6b^2 - 6b\tau + \tau^2) \qquad \text{for } 0 \le \tau \le b \tag{96}$$

Beyond $\tau = b$ but within the range $(b, 2b)$, the overlapping of the pulses of $f_a(t)$ and $f_b(t)$ takes the form shown in Fig. 19. Therefore, that portion of $\varphi_{ab}(\tau)$ within the limits b and $2b$ has the expression

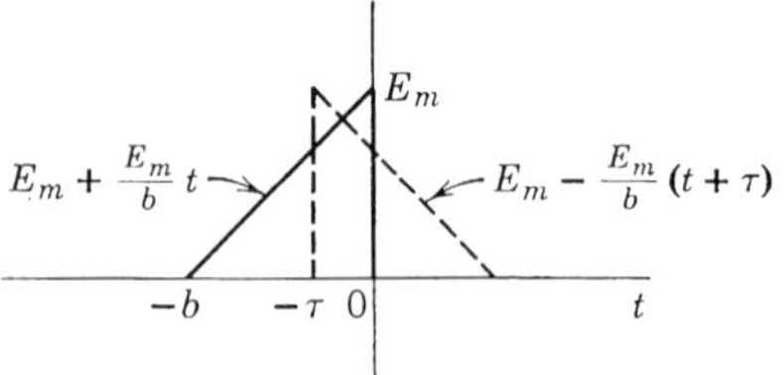

Fig. 18. Illustrating the calculation of a component of the autocorrelation function of the wave of Fig. 15.

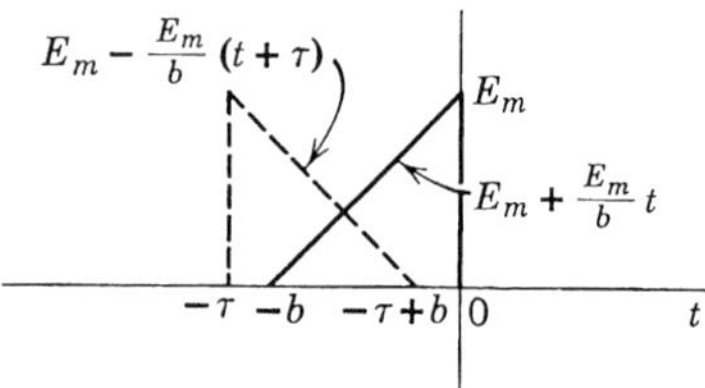

Fig. 19. Illustrating the calculation of a component of the autocorrelation function of the wave of Fig. 15.

$$\varphi_{ab}(\tau) = \frac{1}{T_1}\int_{-b}^{-\tau+b} E_m{}^2\left(1+\frac{t}{b}\right)\left[1-\frac{1}{b}(t+\tau)\right]dt$$

$$= \frac{E_m{}^2}{6b^2T_1}(2b-\tau)^3 \qquad \text{for } b \le \tau < 2b \tag{97}$$

To complete the determination of $\varphi_{ab}(\tau)$ we proceed from $\tau = 2b$ to the end of the period $\tau = T_1$. It is clear that $\varphi_{ab}(\tau)$ is zero in this interval, and with this information the construction of the periodic wave for $\varphi_{ab}(\tau)$ follows immediately. The curve of $\varphi_{ba}(\tau)$ is a mirror image of $\varphi_{ab}(\tau)$ with respect to the vertical axis. This curve is given in Fig. 17*c*.

Summing up the components for $\varphi_{11}(\tau)$ in accordance with (95), we have

$$\varphi_{11}(\tau) = 2\varphi_{aa}(\tau) + \varphi_{ab}(\tau) + \varphi_{ba}(\tau) \qquad \text{for } 0 \le \tau \le b \tag{98}$$

in which $\varphi_{ba}(\tau) = 0$ and $\varphi_{aa}(\tau)$ is given by (57) so that

$$\varphi_{11}(\tau) = \frac{E_m{}^2}{3b^2T_1}(\tau^3 - 3b^2\tau + 2b^3) + \frac{E_m{}^2}{6b^2T_1}\tau(6b^2 - 6b\tau + \tau^2)$$

$$= \frac{E_m{}^2}{6b^2T_1}(3\tau^3 - 6b\tau^2 + 4b^3) \qquad \text{for } 0 \le \tau \le b \tag{99}$$

The graph of this equation is given in Fig. 17*d*. In the interval $(b, 2b)$, $\varphi_{11}(\tau) = \varphi_{ab}(\tau)$; that is,

$$\varphi_{11}(\tau) = \varphi_{ab}(\tau)$$

$$= \frac{E_m{}^2}{6b^2T_1}(2b-\tau)^3 \qquad \text{for } b \le \tau \le 2b \tag{100}$$

The autocorrelation function has the value zero from $\tau = 2b$ to $\tau = T_1/2$. Having obtained the autocorrelation function for a half period $(0, T_1/2)$ and knowing that it is an even function, we can proceed to complete the function as indicated in Fig. 17*d*.

Although the construction of the components as shown in Fig. 17*a*–*c* is instructive and helps clarify the problem, many of the details become unnecessary when the idea of correlation by parts is well understood, for clearly all the necessary data for the determination of $\varphi_{11}(\tau)$ are: (1) the portion of $\varphi_{aa}(\tau)$ in the interval $(0, b)$ and (2) the portion of $\varphi_{ab}(\tau)$ in $(0, 2b)$. Knowledge of the properties of autocorrelation and crosscorrelation enables us to complete the evaluation without further data.

7. Convolution and Correlation

An expression having the general appearance of the crosscorrelation function (59) but a significant difference is

$$\frac{1}{T_1}\int_{-T_1/2}^{T_1/2} f_1(t)f_2(\tau - t)\,dt \tag{101}$$

The difference lies in the sign of t in the displaced function $f_2(t)$. In parallel with the derivation of (31) and (33) it is easy to show that

$$\frac{1}{T_1}\int_{-T_1/2}^{T_1/2} f_1(t)f_2(\tau - t)\,dt = \sum_{n=-\infty}^{\infty} F_1(n)F_2(n)e^{jn\omega_1\tau} \tag{102}$$

and

$$F_1(n)F_2(n) = \frac{1}{T_1}\int_{-T_1/2}^{T_1/2} e^{-jn\omega_1\tau}\,d\tau\,\frac{1}{T_1}\int_{-T_1/2}^{T_1/2} f_1(t)f_2(\tau - t)\,dt \tag{103}$$

which is the *convolution theorem* for periodic functions. A simpler statement of these relations is that

$$\frac{1}{T_1}\int_{-T_1/2}^{T_1/2} f_1(t)f_2(\tau - t)\,dt \quad \text{and} \quad F_1(n)F_2(n) \tag{104}$$

are a Fourier transform pair.

In the terminology generally applied to aperiodic functions, the expression (101) is called the *convolution* of $f_1(t)$ and $f_2(t)$. Since the negative sign of t in $f_2(\tau - t)$ of (101) represents a folding back operation on $f_2(\tau + t)$ when interpreted graphically, convolution involves this step in addition to those in a correlation. A similarity between the convolution and correlation of two periodic functions of the same fundamental frequency is that they are periodic functions having the same fundamental frequency and retaining those harmonics that are present in both given functions. On the other hand, a difference between them, as a result of the folding back operation in convolution, is that the spectrum of convolution is the product of the spectrums of the individual functions convolved whereas that of crosscorrelation is the same product with a conjugation of the spectrum belonging to the undisplaced function. Furthermore, we note that, whereas a distinction should be made between the results of crosscorrelation of two functions when the displacement τ is given to one function and then the other, such distinction need not be made in convolution because the spectrum of convolution is always $F_1(n)F_2(n)$ irrespective of which function, $f_1(t)$ or $f_2(t)$, has the displacement if the function which is displaced is also folded.

If we let

$$\rho_{12}(\tau) = \frac{1}{T_1}\int_{-T_1/2}^{T_1/2} f_1(t)f_2(\tau - t)\,dt \tag{105}$$

and

$$\rho_{21}(\tau) = \frac{1}{T_1}\int_{-T_1/2}^{T_1/2} f_2(t)f_1(\tau - t)\,dt \tag{106}$$

and form its Fourier series from (102) similar to (83) for crosscorrelation, we readily find that

$$\rho_{12}(\tau) = \rho_{21}(\tau)$$
$$= \sum_{n=0}^{\infty} C_{1n}C_{2n} \cos (n\omega_1\tau + \theta_{1n} + \theta_{2n}) \tag{107}$$

The only difference between expansions (83) and (107) is that the phase angles of (83) are $\theta_{2n} - \theta_{1n}$ whereas those of (107) are $\theta_{1n} + \theta_{2n}$.

The reader should note that the meanings of the subscripts of ρ and φ are slightly different. The second subscript of ρ refers to the function which is displaced and folded back. Since no folding operation is involved in correlation, the second subscript of φ refers to the function which is displaced.

Inasmuch as the only difference between convolution and correlation is that convolution involves the folding back of one of the functions convolved whereas correlation does not, it is clear that the correlation of two functions, one of which has been folded back, is in fact a convolution and, conversely, the convolution of two functions, one of which takes the folded form prior to convolution, is actually a correlation. Thus the convolution of $f_1(t)$ and $f_2(t)$ is

$$\rho_{12}(\tau) = \frac{1}{T_1}\int_{-T_1/2}^{T_1/2} f_1(t)f_2(\tau - t)\, dt$$
$$= \frac{1}{T_1}\int_{-T_1/2}^{T_1/2} f_1(-t)f_2(t + \tau)\, dt \tag{108}$$

The second form is the crosscorrelation of $f_1(-t)$ and $f_2(t)$. On the other hand, the crosscorrelation of $f_1(t)$ and $f_2(t)$ is

$$\varphi_{12}(\tau) = \frac{1}{T_1}\int_{-T_1/2}^{T_1/2} f_1(t)f_2(t + \tau)\, dt$$
$$= \frac{1}{T_1}\int_{-T_1/2}^{T_1/2} f_1(-t)f_2(\tau - t)\, dt \tag{109}$$

and may be regarded as the convolution of $f_1(-t)$ and $f_2(t)$.

A special situation which merits comment is the one in which even functions are concerned in convolution or correlation. For instance, if $f_2(t)$ is an even periodic function, its spectrum $F_2(n)$ is real, so that conjugation leaves it unchanged. Now, for the crosscorrelation $\varphi_{21}(\tau)$, the spectrum is $\overline{F}_2(n)F_1(n)$, which is the same as $F_2(n)F_1(n)$. Clearly, under this circumstance, crosscorrelation and convolution are equivalent; that is,

$$\begin{aligned}\varphi_{21}(\tau) &= \rho_{21}(\tau) \\ &= \rho_{12}(\tau)\end{aligned} \tag{110}$$

An interesting application of convolution is found in the interpretation of Fourier formulas (13) and (14) from a viewpoint other than that of analysis and synthesis. Written as one single formula, the Fourier expression for a periodic function is

$$f(t) = \sum_{n=-\infty}^{\infty} e^{jn\omega_1 t} \frac{1}{T_1} \int_{-T_1/2}^{T_1/2} f(\tau) e^{-jn\omega_1 \tau}\, d\tau \tag{111}$$

A change in the order of summation and integration gives us

$$f(t) = \int_{-T_1/2}^{T_1/2} f(\tau)\, d\tau \sum_{n=-\infty}^{\infty} \frac{1}{T_1} e^{jn\omega_1(t-\tau)} \tag{112}$$

For the identification of the periodic wave which the summation represents, we refer to the results of Example 1, Sec. 1. Let the function of Fig. 1 be displaced to the left by time $b/2$, and let it be $g(t)$ so as to distinguish it from $f(t)$ of (112). If $G(n)$ is the spectrum of $g(t)$, then according to (19)

$$G(n) = \frac{E_m b}{T_1} \left(\frac{\sin n\pi \dfrac{b}{T_1}}{n\pi \dfrac{b}{T_1}} \right) \tag{113}$$

Suppose that the area of each rectangular pulse remains fixed while we allow its width to become infinitely small. As a result of this limiting process we have a function in the form of periodic impulses of infinite amplitude but whose areas are finite. In particular, if the area of each impulse is unity, the function is a *periodic unit-impulse function* $u(t)$. Thus, keeping $E_m b = 1$, we have

$$\lim_{b \to 0} g(t) = u(t) \tag{114}$$

whose spectrum, being a limiting form of (113), is

$$\lim_{b \to 0} G(n) = \frac{1}{T_1} \tag{115}$$

Since

$$g(t) = \sum_{n=-\infty}^{\infty} G(n) e^{jn\omega_1 t} \tag{116}$$

its limiting form is

$$u(t) = \sum_{n=-\infty}^{\infty} \frac{1}{T_1} e^{jn\omega_1 t} \tag{117}$$

This result states that the periodic unit-impulse function when represented as a Fourier series has harmonics of all orders each of the same maximum amplitude $1/T_1$ and each with a zero initial phase angle. It may be written as a cosine series with zero phase angles since the function is even.

By comparison with formula (117) we identify the summation in (112) as the periodic unit-impulse function $u(t - \tau)$ which is $u(t)$ delayed by time τ. Note that, with respect to τ, $u(t - \tau)$ is $u(\tau)$ advanced by time t and folded back with respect to the point $\tau = 0$. By substitution, we write (112) as

$$f(t) = \int_{-T_1/2}^{T_1/2} f(\tau)u(t - \tau)\, d\tau \tag{118}$$

Except for the factor $1/T_1$, this is a convolution integral which expresses the fact that the convolution of a given periodic function and a periodic unit-impulse function of the same fundamental frequency leaves the given function unchanged. When the factor $1/T_1$ and the notations in (105) are introduced, (118) is

$$\begin{aligned} \frac{1}{T_1} f(t) &= \frac{1}{T_1} \int_{-T_1/2}^{T_1/2} f(\tau)u(t - \tau)\, d\tau \\ &= \rho_{fu}(t) \end{aligned} \tag{119}$$

Because $u(t)$ is an even function, the equation of equivalence (110) for crosscorrelation and convolution applies, so that

$$\begin{aligned} \frac{1}{T_1} f(t) &= \rho_{fu}(t) \\ &= \varphi_{uf}(t) \end{aligned} \tag{120}$$

Thus from the point of view of convolution or crosscorrelation, in contrast with that of harmonic analysis, a periodic function $f(t)$ is reproduced by the determination of its values one at a time over the whole range of the argument. The determination of a value is done by first multiplying the periodic function by a periodic unit-impulse function whose period is that of $f(t)$ and then integrating the product for a mean value over a period. The functions involved in this process are shown in Fig. 20. As the process is repeated for all values of the displacement t in $u(t - \tau)$, the original function is duplicated except for a scale factor $1/T_1$.

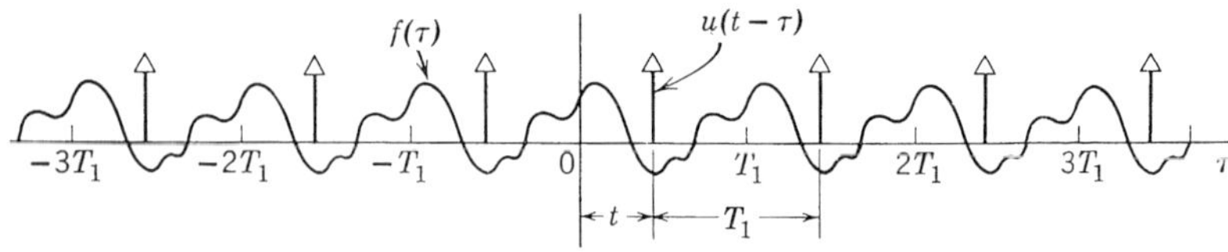

Fig. 20. Periodic function and periodic unit-impulse function of the same period in convolution.

This is a rather complex method for expressing the simple process of observing the values of a function and then recording them so that the function is reproduced; nevertheless, it is a mathematical expression of the process and is an effective tool which we shall have occasion to apply. Furthermore, the concept of convolution or correlation often provides helpful and interesting viewpoints on certain circuit operations such as the sampling of a voltage or current wave by a gating circuit. In terms of correlation this sampling procedure is a crosscorrelation of the wave and a function with unit impulses occurring at the points of sampling. Thus sampling, interpreted in this manner, implies multiplication and integration.

The function $u(t - \tau)$ is like a series of extremely narrow slits through which the function $f(\tau)$ is observed. It is called a *scanning function.* An observation corresponds to the multiplication of $f(\tau)$ by the scanning function and the integration of the product over an appropriate interval for an average value over that interval. The displacement t in $u(t - \tau)$ gives the locations where observations are made on the periodic function. Any portion of $f(\tau)$ or its entirety may be reproduced by scanning it over the corresponding range of the variable τ.

The waveform of a scanning function is not restricted to the one described. As a matter of fact, in the general expression for crosscorrelation or convolution, one of the two functions correlated or convolved may be considered as a scanning function for the other. When the two functions are identical in form as in autocorrelation, or a similar situation in convolution, a function is scanned by itself.

B. APERIODIC FUNCTIONS (OR TRANSIENT FUNCTIONS)

1. Fourier Transforms

The tool for the analysis of an aperiodic function, such as a transient voltage or current, is the Fourier integral. The basic idea in the derivation of this integral for aperiodic functions from the Fourier series for periodic functions is that a function represented as a Fourier series may be made to approach an aperiodic function if the period of the series is allowed to approach infinity. To begin with, the combination of (13) and (14) into a single expression for a periodic function is

$$f(t) = \sum_{n=-\infty}^{\infty} e^{jn\omega_1 t} \frac{1}{T_1} \int_{-T_1/2}^{T_1/2} f(\sigma) e^{-jn\omega_1 \sigma} \, d\sigma \tag{121}$$

Since $1/T_1 = \omega_1/2\pi$, expression (121) is

$$f(t) = \frac{1}{2\pi} \sum_{n=-\infty}^{\infty} e^{jn\omega_1 t} \omega_1 \int_{-T_1/2}^{T_1/2} f(\sigma) e^{-jn\omega_1 \sigma} \, d\sigma \tag{122}$$

Now, if the period T_1 grows without limit, the periodic function $f(t)$ tends to an aperiodic one, and in so doing (122) approaches a limiting form. As T_1 tends to infinity, the fundamental angular frequency ω_1 becomes a differential of angular frequency $d\omega$, and $n\omega_1$, which is the nth harmonic angular frequency, becomes the *continuous* angular frequency ω, and the summation over all harmonics becomes an integration over the entire continuous frequency range $(-\infty, \infty)$. Therefore, the limiting form of (122) for an *aperiodic* function $f(t)$ is

$$f(t) = \frac{1}{2\pi} \int_{-\infty}^{\infty} e^{j\omega t} \, d\omega \int_{-\infty}^{\infty} f(\sigma) e^{-j\omega\sigma} \, d\sigma \tag{123}$$

This is the *Fourier integral* for an aperiodic function. Rigorous considerations show that the validity of the integral requires that

$$\int_{-\infty}^{\infty} |f(t)| \, dt \tag{124}$$

be finite. This derivation of the Fourier integral is formal and heuristic. Rigorous demonstrations for the integral can be found in many standard works on Fourier analysis. Incidentally, of historical interest is the fact that Fourier's own argument in establishing his form of the integral was essentially the same as the one followed here.

Writing two reciprocal relations for $f(t)$ in (123) as it is done in the development of Fourier series, we have

$$f(t) = \int_{-\infty}^{\infty} F(\omega) e^{j\omega t} \, d\omega \tag{125}$$

in which

$$F(\omega) = \frac{1}{2\pi} \int_{-\infty}^{\infty} f(t) e^{-j\omega t} \, dt \tag{126}$$

Here $F(\omega)$ is a continuous function of the angular frequency ω and is in general complex. It is the *complex continuous spectrum* of the aperiodic function, and the reciprocal relations (125) and (126) are known as *Fourier transforms*.

Since $F(\omega)$ is in general complex, its real and imaginary parts when separated will appear as

$$F(\omega) = P(\omega) + jQ(\omega) \tag{127}$$

in which $P(\omega)$ and $Q(\omega)$ are real and are the real and imaginary parts respectively of $F(\omega)$. It follows that the *amplitude density spectrum* of $f(t)$ is

$$|F(\omega)| = \sqrt{P^2(\omega) + Q^2(\omega)} \tag{128}$$

and its *phase density spectrum* is

$$\theta(\omega) = \tan^{-1}\frac{Q(\omega)}{P(\omega)} \tag{129}$$

That $F(\omega)$ is the spectrum of $f(t)$ is apparent when transform (125) is written in the form

$$f(t) = \int_{-\infty}^{\infty} [F(\omega)\, d\omega]e^{j\omega t} \tag{130}$$

It tells us that the aperiodic function $f(t)$ is synthesized by an infinite aggregate of sinusoids $e^{j\omega t}$ of *all* angular frequencies ω in the *continuous* infinite range $(-\infty, \infty)$. It is worth remarking for emphasis that the transform is an integration of *periodic* components to give rise to an *aperiodic* function. Consequently, an aperiodic phenomenon may be studied as the sum of periodic ones by the application of Fourier series and Fourier integral theories. The sinusoids in (130), as represented by $e^{j\omega t}$, have infinitesimal amplitudes. In fact, the maximum amplitudes are given by the expression

$$|F(\omega)|\, d\omega \tag{131}$$

Moreover, the phase angles of the infinitesimal sinusoids are contained in the phase spectrum (129). It is apparent that the amplitude density spectrum $|F(\omega)|$ is not the actual amplitude characteristic of $f(t)$ because all amplitudes are of infinitesimal magnitude; it is rather a characteristic which shows *relative* magnitudes only.

Example 1. The spectrum of the rectangular pulse shown in Fig. 21 by formula (126) is

$$\begin{aligned} F(\omega) &= \frac{1}{2\pi}\int_{-b/2}^{b/2} E_m e^{-j\omega t}\, dt \\ &= \frac{E_m b}{2\pi}\left(\frac{\sin \omega \dfrac{b}{2}}{\omega \dfrac{b}{2}}\right) \end{aligned} \tag{132}$$

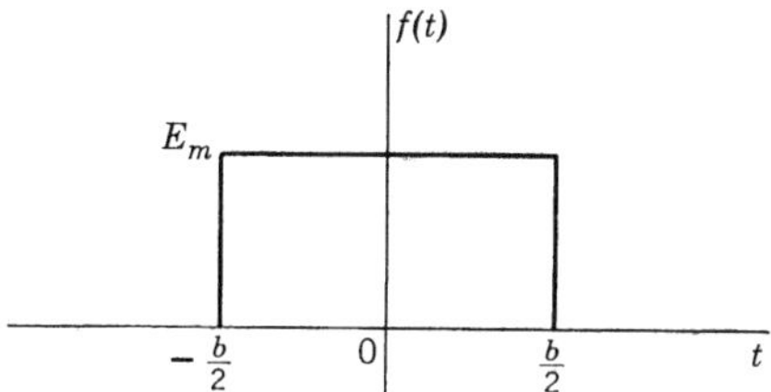

Fig. 21. Rectangular pulse.

A graph of this spectrum is given in Fig. 22. In this example it is simpler to represent the spectrum in its entirety by one graph instead of two graphs, one for the amplitude and the other for the phase. Here the negative regions shown indicate a phase shift of 180 degrees for all frequencies within those regions.

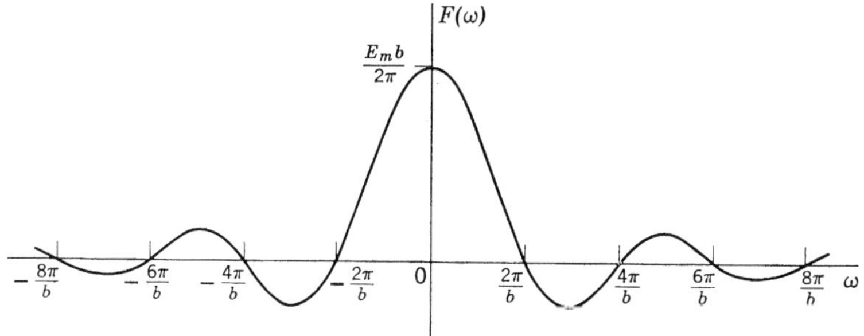

Fig. 22. Spectrum of the rectangular pulse of Fig. 21.

The interpretation of (130) just given when applied to this illustration is that the single rectangular pulse is resolvable into an infinity of infinitesimal sinusoids (cosines) whose relative amplitudes are given by

$$\frac{E_m b}{2\pi}\left|\frac{\sin \omega \dfrac{b}{2}}{\omega \dfrac{b}{2}}\right| \tag{133}$$

over the continuous frequency range $(-\infty, \infty)$. The phase angle for each frequency component is either 0 or 180 degrees out of phase depending on whether $[\sin \omega(b/2)]/[\omega b/2]$ is positive or negative respectively. With these periodic components obtained from analysis, the original function may be reproduced by synthesis. In synthesizing the function according to (130), the sinusoidal components have just the right relative magnitudes and phase angles for complete cancellation in the intervals $(-\infty, -b/2)$ and $(b/2, \infty)$ and, at the same time, for adding up to precisely the value E_m in the interval $(-b/2, b/2)$.

2. Correlation

Analogous to expression (24) for periodic functions, the integral

$$\int_{-\infty}^{\infty} f_1(t)f_2(t+\tau)\,dt \tag{134}$$

for aperiodic functions $f_1(t)$ and $f_2(t)$ is important in the harmonic analysis of aperiodic functions. Here we assume that the integrals $\int_{-\infty}^{\infty}|f_1(t)|\,dt$ and $\int_{-\infty}^{\infty}|f_2(t)|\,dt$ exist and are finite. If $F_1(\omega)$ and $F_2(\omega)$ are the spectrums of $f_1(t)$ and $f_2(t)$ respectively, the Fourier transform of (134) is

$$2\pi\overline{F}_1(\omega)F_2(\omega) \tag{135}$$

This property of the function (134) serves as the basis of much of the discussion that is to follow.

To show this relationship, we replace $f_2(t+\tau)$ in (134) by its expression as a Fourier transform in the manner of (125) so that

$$\int_{-\infty}^{\infty} f_1(t)f_2(t+\tau)\,dt = \int_{-\infty}^{\infty} f_1(t)\,dt \int_{-\infty}^{\infty} F_2(\omega)e^{j\omega(t+\tau)}\,d\omega \tag{136}$$

In the way the double integration is indicated, the first integration is the one on the extreme right which is carried out with respect to ω; then the integration with respect to t follows. By an inversion in the order of integration, (136) becomes

$$\int_{-\infty}^{\infty} f_1(t)f_2(t+\tau)\,dt = \int_{-\infty}^{\infty} F_2(\omega)e^{j\omega\tau}\,d\omega \int_{-\infty}^{\infty} f_1(t)e^{j\omega t}\,dt \tag{137}$$

Since, by application of (126), the conjugate of the spectrum of $f_1(t)$ is

$$\overline{F}_1(\omega) = \frac{1}{2\pi}\int_{-\infty}^{\infty} f_1(t)e^{j\omega t}\,dt \tag{138}$$

it is in the form of the extreme right integral of (137). Hence we have

$$\int_{-\infty}^{\infty} f_1(t)f_2(t+\tau)\,dt = \int_{-\infty}^{\infty} 2\pi\overline{F}_1(\omega)F_2(\omega)e^{j\omega\tau}\,d\omega \tag{139}$$

This is an equation of the form (125) showing that the aperiodic function $\int_{-\infty}^{\infty} f_1(t)f_2(t+\tau)\,dt$ is the transform of $2\pi\overline{F}_1(\omega)F_2(\omega)$. Inverse transformation in accordance with (126) yields

$$2\pi\overline{F}_1(\omega)F_2(\omega) = \frac{1}{2\pi}\int_{-\infty}^{\infty} e^{-j\omega\tau}\,d\tau \int_{-\infty}^{\infty} f_1(t)f_2(t+\tau)\,dt \tag{140}$$

Therefore

$$\int_{-\infty}^{\infty} f_1(t)f_2(t+\tau)\,dt \quad \text{and} \quad 2\pi\overline{F}_1(\omega)F_2(\omega) \tag{141}$$

are a Fourier transform pair. We shall call this reciprocal relationship which is expressed by (139) and (140) the *correlation theorem* for aperiodic functions.

The integral (134) is the *correlation function* of aperiodic functions. The fundamental operations displacement, multiplication and integration are involved. By definition, the correlation of periodic functions includes the factor $1/T_1$ so that when integration is completed the result is a mean value for a period, which is the same for the infinite interval. On the other hand, the definition of the correlation of aperiodic functions does not include taking the mean value over a finite interval or the infinite interval. Such a difference in definition is essential since the integral of the square of an aperiodic function of the type under consideration is assumed to be finite so that its mean square value over the infinite interval vanishes.

Let us note that in equation (139), if τ takes the value zero, we have

$$\int_{-\infty}^{\infty} f_1(t)f_2(t)\,dt = 2\pi\int_{-\infty}^{\infty} \overline{F}_1(\omega)F_2(\omega)\,d\omega \tag{142}$$

This relation concerning the integral of the product of two aperiodic functions is of considerable importance in Fourier analysis and is known as a *Parseval theorem* for aperiodic functions.

3. Autocorrelation

The condition that the two functions are identical makes (139) and (140) take the forms

$$\int_{-\infty}^{\infty} f_1(t)f_1(t+\tau)\,dt = \int_{-\infty}^{\infty} 2\pi|F_1(\omega)|^2 e^{j\omega\tau}\,d\omega \tag{143}$$

and

$$2\pi|F_1(\omega)|^2 = \frac{1}{2\pi}\int_{-\infty}^{\infty} e^{-j\omega\tau}\,d\tau\int_{-\infty}^{\infty} f_1(t)f_1(t+\tau)\,dt \tag{144}$$

We let

$$\varphi_{11}(\tau) = \int_{-\infty}^{\infty} f_1(t)f_1(t+\tau)\,dt \tag{145}$$

be the *autocorrelation function* of the aperiodic function $f_1(t)$ and let

$$\Phi_{11}(\omega) = 2\pi|F_1(\omega)|^2 \tag{146}$$

be the *energy density spectrum* of $f_1(t)$. That $2\pi|F_1(\omega)|^2$ should be expressed in units of energy density may be seen by letting τ in (143) be zero. At this value of τ,

$$\int_{-\infty}^{\infty} f_1^2(t)\,dt = \int_{-\infty}^{\infty} 2\pi\,|F_1(\omega)|^2\,d\omega \tag{147}$$

If $f_1(t)$ represents a voltage or a current and if a 1-ohm load of pure resistance is assumed, the total energy in watt-seconds consumed by the resistance is clearly expressed by the left-hand member of the equation. It is clear that, to obtain the same amount of energy by integration with respect to the angular frequency ω as indicated by the right-hand member of the equation, the integrand should be measured in watt-seconds per unit angular frequency.

It is interesting to point out that the equality (147) can be derived without the intermediate step of correlation: through the idea that the components of the aperiodic function are periodic and such components are obtainable as a limiting form of a Fourier series. Starting from a periodic function $f_1(t)$, we know that the average power is, by (36),

$$\frac{1}{T_1}\int_{-T_1/2}^{T_1/2} f_1^2(t)\,dt = \sum_{n=-\infty}^{\infty} |F_1(n)|^2 \tag{148}$$

from which we see that the energy in one fundamental period T_1 is

$$\int_{-T_1/2}^{T_1/2} f_1^2(t)\,dt = \sum_{n=-\infty}^{\infty} |F_1(n)|^2 T_1 \tag{149}$$

In accordance with the limiting process for establishing the Fourier integral for an aperiodic function, $f_1(t)$ becomes the aperiodic function as T_1 increases without limit; and $|F_1(n)|$, which is the amplitude spectrum of the periodic function, becomes $|F_1(\omega)|\,d\omega$ for the aperiodic function as stated in (131). The spectrum $F_1(\omega)$ giving the relative amplitudes of the periodic components of the aperiodic function is

$$F_1(\omega) = \lim_{T_1\to\infty} \frac{1}{2\pi}\int_{-T_1/2}^{T_1/2} f_1(t)e^{-jn\omega_1 t}\,dt \tag{150}$$

This fact may be verified by reference to (121) through (123) and (126).

The total energy which the aperiodic function represents may now be found as the limit of the energy expression (149) for one period of the periodic function as the period tends to infinity. Hence the total energy is

$$\begin{aligned}\lim_{T_1\to\infty}\sum_{n=-\infty}^{\infty} |F_1(n)|^2 T_1 &= \lim_{T_1\to\infty}\sum_{n=-\infty}^{\infty}\left|\frac{1}{T_1}\int_{-T_1/2}^{T_1/2} f_1(t)e^{-jn\omega_1 t}\,dt\right|^2 T_1 \\ &= \lim_{T_1\to\infty}\sum_{n=-\infty}^{\infty}\left|\int_{-T_1/2}^{T_1/2} f_1(t)e^{-jn\omega_1 t}\,dt\right|^2 \frac{1}{T_1}\end{aligned} \tag{151}$$

Applying (150) and knowing that $1/T_1$ tends to $(1/2\pi)\, d\omega$ as T_1 tends to infinity, we find the total energy to be

$$\int_{-\infty}^{\infty} 2\pi \,|\, F_1(\omega)\,|^2 \, d\omega \tag{152}$$

which agrees with expression (147).

Since the total energy which the aperiodic function represents is a finite quantity, the average power on the basis of the infinite interval should be infinitesimal; that is to say, since the integral square of the function over the infinite interval is finite, the mean square of the function over the same interval is an infinitesimal quantity. We might think of the periodic components of the aperiodic function as being so small in amplitude that not only the individual component but the aggregate as well can maintain only an infinitesimal rate of flow of energy, so that even for an infinitely long time the total accumulation of energy is finite.

Autocorrelation is not restricted to functions that represent voltages or currents. For instance, in certain mathematical operations an autocorrelation function may be obtained from another autocorrelation function. This is a repeated autocorrelation function to which the association of energy is meaningless. Under this and similar circumstances where the term "energy density" may be a source of misunderstanding, the function $\Phi_{11}(\omega)$ should be called the transform or spectrum of $\varphi_{11}(\tau)$.

With the symbols $\varphi_{11}(\tau)$ and $\Phi_{11}(\omega)$ as defined by (145) and (146), we abbreviate the reciprocal relations (143) and (144) to

$$\varphi_{11}(\tau) = \int_{-\infty}^{\infty} \Phi_{11}(\omega) e^{j\omega\tau} \, d\omega \tag{153}$$

and

$$\Phi_{11}(\omega) = \frac{1}{2\pi} \int_{-\infty}^{\infty} \varphi_{11}(\tau) e^{-j\omega\tau} \, d\tau \tag{154}$$

These equations relate the autocorrelation function of an aperiodic function to its energy density spectrum as Fourier transforms of each other. It is assumed that the physical terminology is appropriate; otherwise, the reciprocal relationship is said to be existing between the autocorrelation function of an aperiodic function and the spectrum of the autocorrelation function. We shall name this theorem the *autocorrelation theorem* for aperiodic functions.

Sometimes it is desirable to express the theorem in a slightly different form in which $\cos \omega\tau$ replaces $e^{-j\omega\tau}$. This is made possible by the fact that $\varphi_{11}(\tau)$ and $\Phi_{11}(\omega)$ are real functions, as we can readily verify by inspection of their definitions (145) and (146). When the transform (153) is written in the trigonometric form

$$\varphi_{11}(\tau) = \int_{-\infty}^{\infty} \Phi_{11}(\omega) \cos \omega\tau \, d\omega + j \int_{-\infty}^{\infty} \Phi_{11}(\omega) \sin \omega\tau \, d\omega \tag{155}$$

and real quantities are equated, it is necessary that

$$\varphi_{11}(\tau) = \int_{-\infty}^{\infty} \Phi_{11}(\omega) \cos \omega\tau \, d\omega \tag{156}$$

By inverse transformation, or a similar argument, we find

$$\Phi_{11}(\omega) = \frac{1}{2\pi} \int_{-\infty}^{\infty} \varphi_{11}(\tau) \cos \omega\tau \, d\tau \tag{157}$$

In addition to the autocorrelation theorem, a few properties of autocorrelation pertaining to aperiodic functions should be brought out. One of these properties is that the value of an autocorrelation function at the origin is the integral square of the aperiodic function; that is, according to (145),

$$\varphi_{11}(0) = \int_{-\infty}^{\infty} f_1^{\,2}(t) \, dt \tag{158}$$

This is the total energy of $f_1(t)$. We recall that $\varphi_{11}(0)$ is the mean power for periodic functions.

Similar to the autocorrelation of a periodic function, $\varphi_{11}(\tau)$ for an aperiodic function is even. This property may be shown simply by changing τ to $-\tau$ in the definition (145) and noting that as a consequence $\varphi_{11}(\tau) = \varphi_{11}(-\tau)$, which is a statement that the function is even. Thus we have

$$\begin{aligned} \varphi_{11}(-\tau) &= \int_{-\infty}^{\infty} f_1(t) f_1(t - \tau) \, dt \\ &= \int_{-\infty}^{\infty} f_1(x) f_1(x + \tau) \, dx \\ &= \varphi_{11}(\tau) \end{aligned} \tag{159}$$

The spectrum of the autocorrelation function of an aperiodic function is $2\pi|F_1(\omega)|^2$ which is 2π times the product of the spectrum $F_1(\omega)$ of the aperiodic function and its conjugate $\overline{F}_1(\omega)$ as indicated in (143) and (144). Therefore the amplitude spectrum $|F_1(\omega)|$ of the aperiodic function is obtainable from the spectrum of its autocorrelation function by simply taking its square root and dividing by $\sqrt{2\pi}$. On the other hand, the phase spectrum of the aperiodic function is completely lost in autocorrelation since multiplication of $F_1(\omega)$ by $\overline{F}_1(\omega)$ incurs exact cancellation of their phase angles. Because of this fact, functions with the same amplitude spectrum but different phase spectrums have the same autocorrelation function. These remarks are similar to those made with respect to periodic functions although here we are concerned with continuous spectrums instead of discrete ones.

Still another property of the autocorrelation function concerns its value at the origin. It is that this value is the maximum which no other value of the function may exceed or even equal in absolute magnitude. To show this property, we consider the expression

$$\int_{-\infty}^{\infty} [f_1(t) \pm f_1(t+\tau)]^2 \, dt \qquad \text{for } \tau \neq 0 \tag{160}$$

The integral (160) is always a positive nonzero quantity. The fact that it is positive is obvious. It is nonzero because $f_1(t)$ is aperiodic, a restriction which precludes the possibility of having $f_1(t) = f_1(t+\tau)$ for certain values of τ other than zero. This equality can be satisfied only by periodic functions. Hence

$$\int_{-\infty}^{\infty} [f_1(t) \pm f_1(t+\tau)]^2 \, dt > 0 \qquad \text{for } \tau \neq 0 \tag{161}$$

Expansion of this expression is

$$\int_{-\infty}^{\infty} f_1{}^2(t) \, dt + \int_{-\infty}^{\infty} f_1{}^2(t+\tau) \, dt \pm 2\int_{-\infty}^{\infty} f_1(t) f_1(t+\tau) \, dt > 0 \qquad \text{for } \tau \neq 0 \tag{162}$$

in which the first two terms have the same value, it's being $\varphi_{11}(0)$, and the third term is $2\varphi_{11}(\tau)$. Therefore, we have

$$\varphi_{11}(0) > \pm\varphi_{11}(\tau) \qquad \text{for } \tau \neq 0 \tag{163}$$

which clearly means that

$$\varphi_{11}(0) > |\varphi_{11}(\tau)| \qquad \text{for } \tau \neq 0 \tag{164}$$

Example 1. To find the autocorrelation function of the rectangular pulse of Fig. 23, in accordance with equation (145), the pulse is displaced by time τ as illustrated in Fig. 24. The product of the given pulse and the displaced pulse is represented by the shaded rectangle whose height is $E_m{}^2$. In the interval $(0 \leq \tau \leq b)$ the autocorrelation function is

$$\int_{-t_0}^{b-t_0-\tau} E_m{}^2 \, dt = E_m{}^2(b - |\tau|) \qquad \text{for } 0 \leq \tau \leq b \tag{165}$$

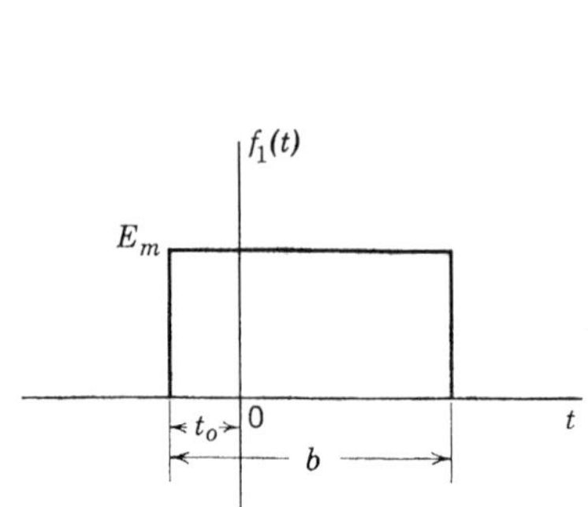

Fig. 23. Rectangular pulse.

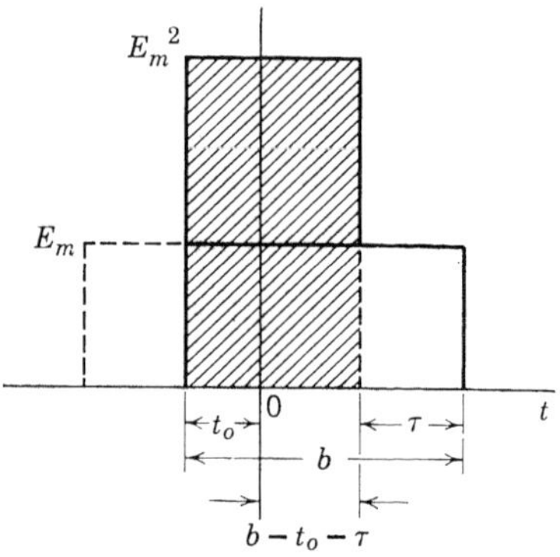

Fig. 24. Illustrating the autocorrelation of the rectangular pulse of Fig. 23.

Since the function is even and vanishes for $\tau > b$, the complete expression is

$$\varphi_{11}(\tau) = \begin{cases} E_m^{\ 2}(b - |\tau|) & \text{for } -b \leq \tau \leq b \\ 0 & \text{elsewhere} \end{cases} \tag{166}$$

which is graphically represented in Fig. 25.

Graphically we see that in Fig. 24 the area of the shaded rectangle, as expressed by (165), remains the same no matter where the pulse may have its initial value. Also, we see that an autocorrelation function is even, for displacing a pulse of general configuration in one direction and multiplying it by the original pulse will produce a curve differing from that obtained through the same displacement in the opposite direction only by a displacement. Since both curves give the same area, the value of the autocorrelation function for a positive displacement is equal to that for a negative displacement of the same magnitude.

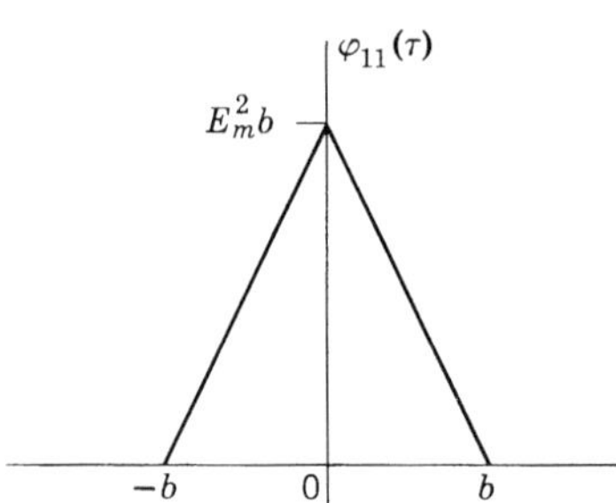

Fig. 25. Autocorrelation function of the rectangular pulse of Fig. 23.

If $f_1(t)$ of Fig. 23 represents a voltage or a current, its energy density spectrum should be

$$\Phi_{11}(\omega) = \frac{1}{2\pi} \int_{-b}^{b} E_m^{\ 2}(b - |\tau|) \cos \omega\tau \, d\tau$$

$$= \frac{E_m^{\ 2} b^2}{2\pi} \left(\frac{\sin \omega \dfrac{b}{2}}{\omega \dfrac{b}{2}} \right)^2 \tag{167}$$

This is a well-known curve and is given in Fig. 26.

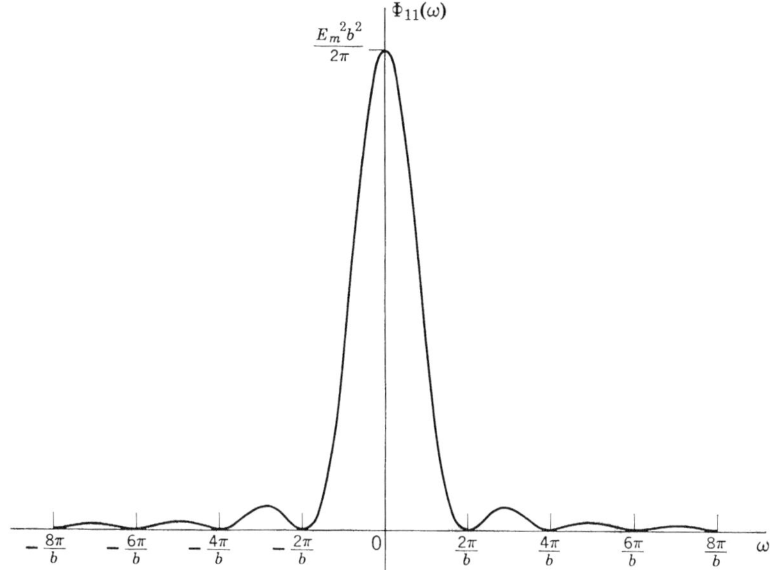

Fig. 26. Energy density spectrum of the rectangular pulse of Fig. 23.

We have pointed out that autocorrelation is introduced in harmonic analysis not entirely for the purpose of power spectrum calculation in dealing with periodic functions. Remarks similar to those made in connection with periodic functions are appropriate here while the matter of energy density spectrum is under consideration. Clearly, if calculation of energy density spectrum is the sole objective, autocorrelation is not indispensable since this spectrum is easily obtained through formula (146).

4. Crosscorrelation

When $f_1(t)$ and $f_2(t)$ in (139) and (140) are different aperiodic functions, we define a *crosscorrelation function* $\varphi_{12}(\tau)$ of $f_1(t)$ and $f_2(t)$ as

$$\varphi_{12}(\tau) = \int_{-\infty}^{\infty} f_1(t) f_2(t + \tau)\, dt \tag{168}$$

Analogous to the energy density spectrum, the *cross-energy density spectrum* $\Phi_{12}(\omega)$ is

$$\Phi_{12}(\omega) = 2\pi \overline{F}_1(\omega) F_2(\omega) \tag{169}$$

If the term *energy* is confusing in a particular situation, we shall call it the *spectrum* of $\varphi_{12}(\tau)$. Expressed in the symbols $\varphi_{12}(\tau)$ and $\Phi_{12}(\omega)$, equations (139) and (140) now read

$$\varphi_{12}(\tau) = \int_{-\infty}^{\infty} \Phi_{12}(\omega)e^{j\omega\tau}\, d\omega \tag{170}$$

and

$$\Phi_{12}(\omega) = \frac{1}{2\pi}\int_{-\infty}^{\infty} \varphi_{12}(\tau)e^{-j\omega\tau}\, d\tau \tag{171}$$

We shall call this reciprocal relation the *crosscorrelation theorem* for aperiodic functions.

If $f_1(t)$ instead of $f_2(t)$ is given the displacement τ, we have

$$\varphi_{21}(\tau) = \int_{-\infty}^{\infty} f_2(t)f_1(t + \tau)\, dt \tag{172}$$

by definition. As we have seen in connection with periodic functions, an interchange with respect to displacement has the effect of folding on the crosscorrelation function. This effect is shown by noting that

$$\varphi_{12}(-\tau) = \int_{-\infty}^{\infty} f_1(t)f_2(t - \tau)\, dt$$

$$= \int_{-\infty}^{\infty} f_2(x)f_1(x + \tau)\, dx \tag{173}$$

where a change of variable $x = t - \tau$ has been made. Thus from (172) and (173) we have the relation

$$\varphi_{12}(-\tau) = \varphi_{21}(\tau) \tag{174}$$

which graphically means that $\varphi_{21}(\tau)$ is the same as $\varphi_{12}(\tau)$ when it is folded back so that the positive half along the τ-axis becomes the negative half, and vice versa. In the frequency domain it is clear from (169) that

$$\Phi_{12}(\omega) = \overline{\Phi}_{21}(\omega) \tag{175}$$

Unlike autocorrelation, a crosscorrelation function is in general not an even function, except under special circumstances. For example, if both $f_1(t)$ and $f_2(t)$ are even or both are odd, the crosscorrelation function is even. The reason is that the function is even when its spectrum (169) is real; this is true either when both $F_1(\omega)$ and $F_2(\omega)$ are real or when both are pure imaginary. These two conditions correspond to the situation that both $f_1(t)$ and $f_2(t)$ are even and the situation that both are odd respectively.

With respect to the upper and lower bounds of a crosscorrelation function consider the integral

$$\int_{-\infty}^{\infty} [f_1(t) \pm f_2(t + \tau)]^2 \, dt \tag{176}$$

If $f_1(t) \neq \pm f_2(t + \tau)$ for all values of τ, this integral is nonzero and positive, so that, upon expansion, we have

$$\int_{-\infty}^{\infty} f_1{}^2(t) \, dt + \int_{-\infty}^{\infty} f_2{}^2(t + \tau) \, dt \pm 2 \int_{-\infty}^{\infty} f_1(t) f_2(t + \tau) \, dt > 0 \qquad \text{for } -\infty < \tau < \infty \tag{177}$$

Since

$$\varphi_{11}(0) = \int_{-\infty}^{\infty} f_1{}^2(t) \, dt \tag{178}$$

and

$$\varphi_{22}(0) = \int_{-\infty}^{\infty} f_2{}^2(t + \tau) \, dt \tag{179}$$

expression (177) is

$$\varphi_{11}(0) + \varphi_{22}(0) > \pm 2\varphi_{12}(\tau) \qquad \text{for } -\infty < \tau < \infty \tag{180}$$

which is equivalent to the inequality

$$\varphi_{11}(0) + \varphi_{22}(0) > 2|\varphi_{12}(\tau)| \qquad \text{for } -\infty < \tau < \infty \tag{181}$$

This is a statement to the effect that the sum of the integral squares of $f_1(t)$ and $f_2(t)$, under the restriction stated earlier, exceeds twice the magnitude of their crosscorrelation function for any value of τ including zero.

5. Correlation and Convolution

An integral of great importance both in theoretical work and physical application is the *convolution integral*

$$\int_{-\infty}^{\infty} f_1(t) f_2(\tau - t) \, dt \tag{182}$$

In essence this integral resembles crosscorrelation but differs from it in one important respect. It resembles the crosscorrelation function in that displacement, multiplication, and integration are operations in both expressions. It differs from crosscorrelation because convolution includes folding, or reflection, of either $f_1(t)$ or $f_2(t)$, whichever has been displaced, whereas crosscorrelation does not. Thus the function $f_2(\tau - t)$

in (182) is $f_2(t)$ first displaced to the left by τ and then folded back with respect to the point $t = 0$.

Because of this difference the Fourier transform of (182) is

$$2\pi F_1(\omega)F_2(\omega) \tag{183}$$

without conjugation of either $F_1(\omega)$ or $F_2(\omega)$ as in crosscorrelation. To show this, $f_2(\tau - t)$ in (182) is replaced by its Fourier transform so that

$$\int_{-\infty}^{\infty} f_1(t)f_2(\tau - t)\,dt = \int_{-\infty}^{\infty} f_1(t)\,dt \int_{-\infty}^{\infty} F_2(\omega)e^{j\omega(\tau-t)}\,d\omega \tag{184}$$

By an inversion of the order of integration we have

$$\int_{-\infty}^{\infty} f_1(t)f_2(\tau - t)\,dt = \int_{-\infty}^{\infty} F_2(\omega)e^{j\omega\tau}\,d\omega \int_{-\infty}^{\infty} f_1(t)e^{-j\omega t}\,dt \tag{185}$$

where the integral on the extreme right is $2\pi F_1(\omega)$ with the result that

$$\int_{-\infty}^{\infty} f_1(t)f_2(\tau - t)\,dt = \int_{-\infty}^{\infty} 2\pi F_1(\omega)F_2(\omega)e^{j\omega\tau}\,d\omega \tag{186}$$

This equation is in the form of a Fourier transform indicating that

$$\int_{-\infty}^{\infty} f_1(t)f_2(\tau - t)\,dt \quad \text{and} \quad 2\pi F_1(\omega)F_2(\omega) \tag{187}$$

are Fourier transforms of each other. In other words the inverse of (186) is

$$2\pi F_1(\omega)F_2(\omega) = \frac{1}{2\pi}\int_{-\infty}^{\infty} e^{-j\omega\tau}\,d\tau \int_{-\infty}^{\infty} f_1(t)f_2(\tau - t)\,dt \tag{188}$$

These reciprocal relations will be referred to as the *convolution theorem* for aperiodic functions.

In a manner similar to correlation, let us represent the convolution integral (182) by $\rho_{12}(\tau)$; that is,

$$\rho_{12}(\tau) = \int_{-\infty}^{\infty} f_1(t)f_2(\tau - t)\,dt \tag{189}$$

In this definition the second subscript of ρ refers to the function which has been displaced and folded. When the order of the subscripts are reversed,

$$\rho_{21}(\tau) = \int_{-\infty}^{\infty} f_2(t)f_1(\tau - t)\,dt \tag{190}$$

It is simple to show that

$$\rho_{12}(\tau) = \rho_{21}(\tau) \tag{191}$$

That is to say, in the convolution of $f_1(t)$ and $f_2(t)$, the result remains the same no matter which function is folded if the folded function is also the one displaced. To bring out the similarity and difference between convolution and correlation, $\rho_{12}(\tau)$ and $\varphi_{12}(\tau)$ are graphically represented in Fig. 27.

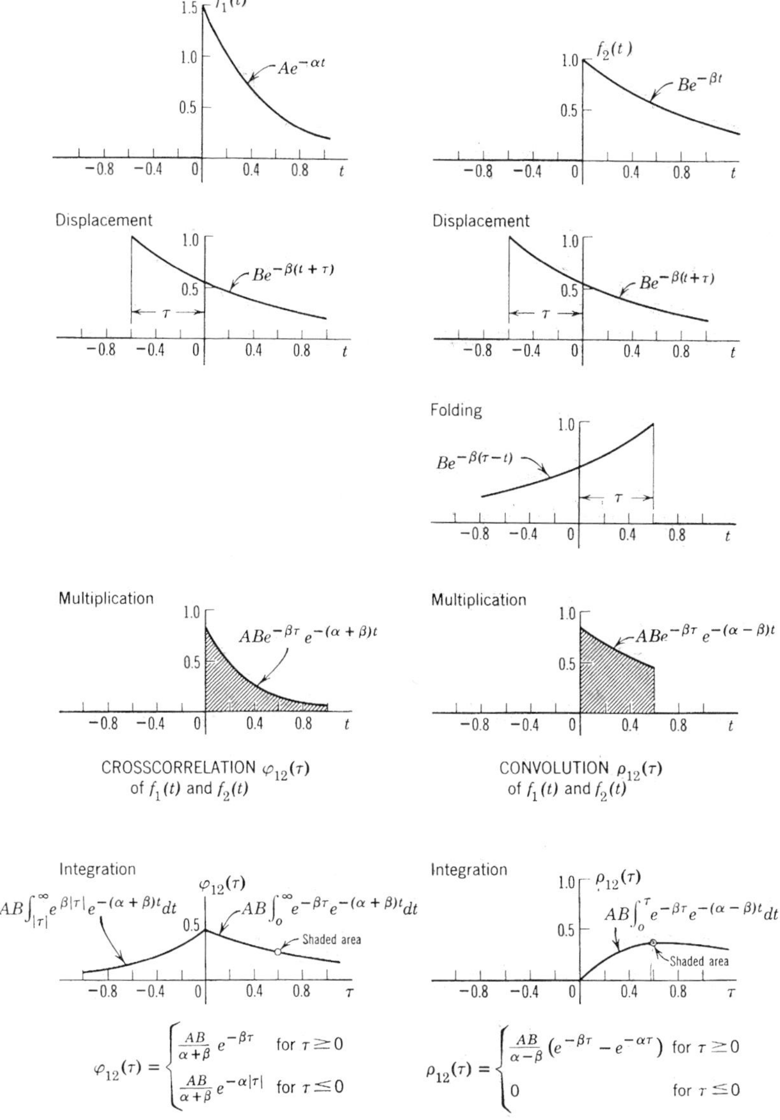

Fig. 27. Illustrating the operations involved in crosscorrelation and convolution of two aperiodic functions. $A = 1.5$, $B = 1.0$, $\alpha = 2$, $\beta = 1$.

Many remarks made with reference to the crosscorrelation and convolution of periodic functions are appropriate in the present discussion. In particular we refer to those concerning the expression of one in terms of the other. In a manner similar to (108), we write for aperiodic functions

$$\rho_{12}(\tau) = \int_{-\infty}^{\infty} f_1(t) f_2(\tau - t)\, dt$$

$$= \int_{-\infty}^{\infty} f_1(-t) f_2(t + \tau)\, dt \tag{192}$$

showing that the convolution of $f_1(t)$ and $f_2(t)$ may be regarded as the crosscorrelation of $f_1(-t)$ and $f_2(t)$. When (109) is written for aperiodic functions,

$$\varphi_{12}(\tau) = \int_{-\infty}^{\infty} f_1(t) f_2(t + \tau)\, dt$$

$$= \int_{-\infty}^{\infty} f_1(-t) f_2(\tau - t)\, dt \tag{193}$$

which indicates that the crosscorrelation of $f_1(t)$ and $f_2(t)$ is equivalent to the convolution of $f_1(-t)$ and $f_2(t)$.

If $f_1(t)$ is an even function, it is clear that

$$\varphi_{12}(\tau) = \rho_{12}(\tau) \tag{194}$$

and, similarly, if $f_2(t)$ is even

$$\varphi_{21}(\tau) = \rho_{21}(\tau) \tag{195}$$

An interesting application of convolution is in the interpretation of the Fourier integral (123). With a change of the order of integration it reads

$$f(t) = \int_{-\infty}^{\infty} f(\sigma)\, d\sigma \int_{-\infty}^{\infty} \frac{1}{2\pi} e^{j\omega(t-\sigma)}\, d\omega \tag{196}$$

Analogous to (112) for periodic functions, in which the summation is a periodic unit-impulse function, the extreme right integral in (196) represents a function which has a single impulse whose integral is unity. To show this we return to Example 1 of Sec. B-1, which gives the spectrum of the rectangular pulse of Fig. 21 as

$$F(\omega) = \frac{E_m b}{2\pi} \left(\frac{\sin \omega \frac{b}{2}}{\omega \frac{b}{2}} \right) \tag{197}$$

By keeping the area $E_m b$ of the pulse equal to unity and allowing the width of the pulse to become infinitesimal, thus causing the height of the pulse to approach infinity, we obtain a single impulse at the origin whose integral is unity and whose spectrum is uniformly $1/2\pi$. This conclusion follows directly from the limiting form of (197). Denoting the unit-impulse function by $u(t)$ and its spectrum by $U(\omega)$, we have, for $E_m b = 1$,

$$\begin{aligned} U(\omega) &= \lim_{b \to 0} \frac{E_m b}{2\pi} \left(\frac{\sin \omega \frac{b}{2}}{\omega \frac{b}{2}} \right) \\ &= \frac{1}{2\pi} \end{aligned} \tag{198}$$

and

$$\begin{aligned} u(t) &= \int_{-\infty}^{\infty} U(\omega) e^{j\omega t} \, d\omega \\ &= \int_{-\infty}^{\infty} \frac{1}{2\pi} e^{j\omega t} \, d\omega \end{aligned} \tag{199}$$

If the impulse is delayed by time σ, then

$$u(t - \sigma) = \int_{-\infty}^{\infty} \frac{1}{2\pi} e^{j\omega(t-\sigma)} \, d\omega \tag{200}$$

which is precisely the first (extreme right) integral of (196). In terms of the unit-impulse function, (196) is

$$f(t) = \int_{-\infty}^{\infty} f(\sigma) u(t - \sigma) \, d\sigma \tag{201}$$

This is the Fourier integral in the form of a convolution. We have here an expression of the fact that the convolution of an aperiodic function and the unit-impulse function leaves the function unchanged.

From the point of view of correlation, (201) is equivalent to the cross-correlation of $u(t)$ and $f(t)$ because $u(t)$ is an even function, so that (195) applies, making it possible to state that

$$\begin{aligned} f(t) &= \rho_{uf}(t) \\ &= \varphi_{uf}(t) \end{aligned} \tag{202}$$

C. RANDOM FUNCTIONS

1. Introduction

Periodic and aperiodic functions are associated with physical experiments which produce the same instantaneous variations in the output upon repetition of the experiments. For example, a square-wave generator maintains the same periodic waveform whenever it is called upon to operate if, of course, the conditions of its associated circuits are not disturbed. A transient phenomenon, such as the output response as a function of time to a sudden excitation at the input of a system, is clearly reproducible with great precision whenever the same excitation is applied. A delay in the time of occurrence of the cause of a transient phenomenon merely translates the response by the same delay.

When experiments carried out under essentially the same conditions do not produce identical results even though time delay has been taken into consideration we have a random phenomenon. Examples of random phenomena are given in Chapter 1.

At the moment we are concerned with the development of a mathematical theory capable of handling random phenomena under the general concept of harmonic analysis. When we speak of a random phenomenon, we refer to one in which the fluctuations of the quantity under observation as a function of time cannot be precisely predicted. No mathematical expression that relates uniquely the effect to the cause is available. The complexity of the cause is beyond our full understanding and our ability to formulate it.

In order to establish a simple basic theory suitable for application to physical problems, unnecessary complication and refinement are avoided. Certain restrictions and idealizations are introduced. First, we shall assume that the random functions have an infinite duration extending from the infinite past to the infinite future. Next, we realize that our interest in a communication problem concerns not just a single random time function but rather an aggregate of random functions which are produced by similarly prepared sources. The theory is of little consequence if it is tied to a particular waveform in an application, since a random wave does not repeat in repeated experiments. Clearly, any phenomenon in which waveforms recur is not a random one and should not be treated as such. A communication or control system is designed for effective operation on a class of messages and a class of noise, not a particular message and a particular noise. The class of all possible message waves (which may contain a disturbance) or noise waves, as the case may be, generated by sources of similar character under essentially the same conditions form an infinite aggregate of random functions.

In a practical situation, the aggregate may be the message or noise waves obtainable from a single source upon repeated experiments under essentially the same conditions even though these experiments are necessarily limited in duration. The infinite aggregate of messages or noise, or of their combination, is called the *ensemble.* Frequently, when we speak of a message or noise, it is understood that we refer to an ensemble of messages or noise. When only one function in the ensemble is meant, the function is called a *member function* of the ensemble.

There are many physical problems in which random functions are intermixed with periodic or aperiodic functions. For example, in a pulse radar set, thermal noise is combined with a periodic signal if the reflecting object is stationary. On the other hand, seismic waves in seismic methods of geophysical exploration are, under certain circumstances, a mixture of random and aperiodic functions. The problem of the separation of signal and noise is extremely important, and some of its theoretical aspects will be considered. But, for the present, we are concerned with the analysis of random functions by themselves, so that periodic and aperiodic components as well as constants are assumed absent.

2. Autocorrelation

In the analysis of an ensemble where the variety of waveforms is infinite so that characterization by waveform is contrary to the inherent feature of the ensemble, characteristics that are common to all member functions on the basis of average behavior are introduced. An extremely important characteristic of this type is the autocorrelation function defined in the following manner: If $f_1(t)$ is a member function of an ensemble, its *autocorrelation function* is

$$\varphi_{11}(\tau) = \lim_{T \to \infty} \frac{1}{2T} \int_{-T}^{T} f_1(t) f_1(t + \tau)\, dt \tag{203}$$

This function is assumed to exist for every value of τ. We make a further important assumption that the autocorrelation function is the same for every member function of the ensemble, and it is therefore a characteristic of the ensemble.

With the treatment of periodic and aperiodic functions as background, we can readily see the reason for giving the autocorrelation function (203) a central role in the harmonic analysis of random functions. The autocorrelation of periodic and aperiodic functions is independent of the phase spectrums of the functions correlated so that an infinite variety of waveforms may have the same autocorrelation. We have discussed this property in some detail. It is therefore clear that autocorrelation is of the utmost importance in the harmonic analysis of an ensemble of random functions where variety in waveform is an inherent attribute.

The analytical determination of (203) is not as simple as that of the autocorrelation of periodic or aperiodic functions. Periodic or aperiodic functions are functions known for all time, and their expression by simple formulas and graphs is frequently possible so that the autocorrelation function is not difficult to determine. Random functions, on the contrary, are functions whose instantaneous values do not permit precise specification. These functions should be described quantitatively in accordance with the theory of statistics and probability. Consequently, the matter of finding autocorrelation functions for specific ensembles cannot be taken up at the present moment but must await the introduction of the necessary statistical tools and techniques. Meanwhile some of the important properties of the autocorrelation function (203) should be examined.

Knowing that an autocorrelation function for a periodic or an aperiodic function is even, we expect it to have the same property for a random function. This may be shown to be true by considering that

$$\begin{aligned}\varphi_{11}(-\tau) &= \lim_{T\to\infty} \frac{1}{2T} \int_{-T}^{T} f_1(t) f_1(t-\tau)\, dt \\ &= \lim_{T\to\infty} \frac{1}{2T} \int_{-T-\tau}^{T-\tau} f_1(x) f_1(x+\tau)\, dx \qquad (204)\end{aligned}$$

when we put $x = t - \tau$. Since both limits of integration in the last integral are displaced by τ in the same direction, the interval $2T$ between limits has not been changed except for its location. The range of the integral $(-T-\tau, T-\tau)$ in the limit is $(-\infty, \infty)$ as $T \to \infty$. We may therefore write (204) as equivalent to

$$\varphi_{11}(-\tau) = \lim_{T\to\infty} \frac{1}{2T} \int_{-T}^{T} f_1(x) f_1(x+\tau)\, dx \qquad (205)$$

By definition, the right-hand member of this equation is $\varphi_{11}(\tau)$ so that we have the result

$$\varphi_{11}(-\tau) = \varphi_{11}(\tau) \qquad (206)$$

which tells us that the autocorrelation function is even.

Concerning the values of $\varphi_{11}(\tau)$ for $\tau = 0$ and $\tau \to \infty$, we find that, inasmuch as (203) holds for all values of τ in the range $(-\infty, \infty)$, its expression for $\tau = 0$ is

$$\varphi_{11}(0) = \lim_{T\to\infty} \frac{1}{2T} \int_{-T}^{T} f_1^2(t)\, dt \qquad (207)$$

This is the mean square of $f_1(t)$, or the mean power if $f_1(t)$ represents a voltage or a current for a 1-ohm load. In a later chapter it is shown that the other extreme value of $\varphi_{11}(\tau)$, when $\tau \rightarrow \infty$, is

$$\varphi_{11}(\infty) = 0 \tag{208}$$

if $f_1(t)$ contains no d-c or periodic components, which is the assumption already made.

Along with the discussion on properties of the autocorrelation function, it seems desirable to have an example of this function where those properties find concrete expression. A random wave which will receive considerable attention in Chapter 8 is shown in Fig. 28. It has two

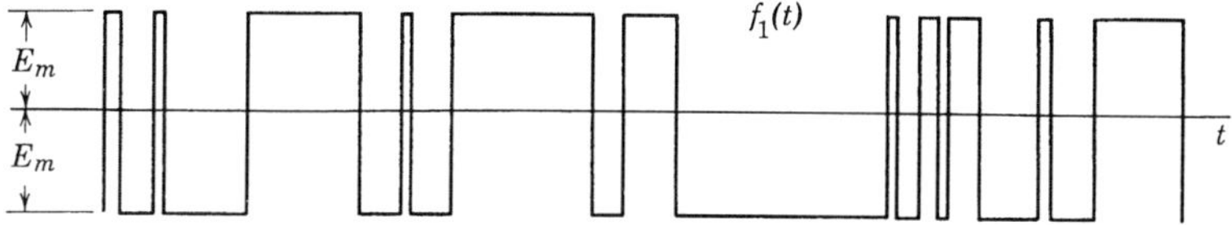

Fig. 28. Random wave with two possible values E_m and $-E_m$.

possible values, E_m and $-E_m$, and alternates between them at random intervals of time. Under the assumption that the zero crossings of the wave, where it shifts from one value to the other, follow a probability distribution known as the Poisson distribution, it is possible to show that the autocorrelation function is

$$\varphi_{11}(\tau) = {E_m}^2 e^{-2k|\tau|} \tag{209}$$

in which k is the average number of zero crossings per second. The graph of this function is given in Fig. 29. It is an even function with the

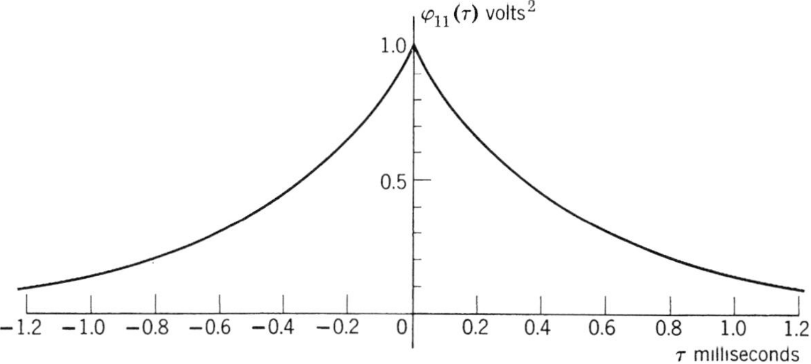

Fig. 29. Autocorrelation function of the wave of Fig. 28. $k = 1000$ per second, $E_m = 1$ volt.

maximum value $E_m{}^2$ at $\tau = 0$ and the value zero as τ tends to infinity. It is clear from the graph that the mean square value is

$$\lim_{T\to\infty} \frac{1}{2T} \int_{-T}^{T} f_1{}^2(t)\,dt = E_m{}^2$$
$$= \varphi_{11}(0) \tag{210}$$

Another important property of the autocorrelation function is that the function is continuous everywhere if it is continuous at the origin. Wiener has shown this property by an application of the *Schwarz inequality*. This theorem, when applied to functions $f(t)$ and $g(t)$, for which

$$\lim_{T\to\infty} \frac{1}{2T} \int_{-T}^{T} f^2(t)\,dt \tag{211}$$

and

$$\lim_{T\to\infty} \frac{1}{2T} \int_{-T}^{T} g^2(t)\,dt \tag{212}$$

exist and are finite, states that

$$\left| \lim_{T\to\infty} \frac{1}{2T} \int_{-T}^{T} f(t)g(t)\,dt \right|^2 \le \lim_{T\to\infty} \left(\frac{1}{2T}\right)^2 \left\{ \left[\int_{-T}^{T} f^2(t)\,dt\right] \left[\int_{-T}^{T} g^2(t)\,dt\right] \right\} \tag{213}$$

Now, to show that the autocorrelation function is continuous, consider the expression

$$|\varphi_{11}(\tau) - \varphi_{11}(\tau \pm \epsilon)| \tag{214}$$

where ϵ is a positive increment. By definition, we have

$$|\varphi_{11}(\tau) - \varphi_{11}(\tau \pm \epsilon)| = \left| \lim_{T\to\infty} \frac{1}{2T} \int_{-T}^{T} f_1(t)[f_1(t+\tau) - f_1(t+\tau\pm\epsilon)]\,dt \right| \tag{215}$$

and, by comparing it with the Schwarz inequality, we find that $f(t)$ in (213) corresponds to $f_1(t)$ in (215) and $g(t)$ corresponds to $f_1(t+\tau) - f_1(t+\tau\pm\epsilon)$. For expression (214), the inequality is therefore

$$|\varphi_{11}(\tau) - \varphi_{11}(\tau \pm \epsilon)|^2 \le \lim_{T\to\infty} \left(\frac{1}{2T}\right)^2$$
$$\left\{ \left[\int_{-T}^{T} f_1{}^2(t)\,dt\right] \left[\int_{-T}^{T} [f_1(t+\tau) - f_1(t+\tau\pm\epsilon)]^2\,dt\right] \right\} \tag{216}$$

The last factor of the inequality upon expansion is

$$\lim_{T\to\infty} \frac{1}{2T} \int_{-T}^{T} [f_1(t+\tau) - f_1(t+\tau\pm\epsilon)]^2\,dt$$
$$= \lim_{T\to\infty} \frac{1}{2T} \int_{-T}^{T} f_1{}^2(t+\tau)\,dt + \lim_{T\to\infty} \frac{1}{2T} \int_{-T}^{T} f_1{}^2(t+\tau\pm\epsilon)\,dt$$
$$- 2\lim_{T\to\infty} \frac{1}{2T} \int_{-T}^{T} f_1(t+\tau) f_1(t+\tau\pm\epsilon)\,dt \tag{217}$$

where the first term of the expansion is

$$\lim_{T\to\infty} \frac{1}{2T} \int_{-T}^{T} f_1{}^2(t+\tau)\,dt = \lim_{T\to\infty} \frac{1}{2T} \int_{-T+\tau}^{T+\tau} f_1{}^2(x)\,dx$$
$$= \lim_{T\to\infty} \frac{1}{2T} \int_{-T}^{T} f_1{}^2(x)\,dx$$
$$= \varphi_{11}(0) \tag{218}$$

by an argument similar to that given in connection with (204) and (205). Similarly, the second term is

$$\lim_{T\to\infty} \frac{1}{2T} \int_{-T}^{T} f_1{}^2(t+\tau\pm\epsilon)\,dt = \varphi_{11}(0) \tag{219}$$

and in the last term,

$$\lim_{T\to\infty} \frac{1}{2T} \int_{-T}^{T} f_1(t+\tau) f_1(t+\tau\pm\epsilon)\,dt = \varphi_{11}(\pm\epsilon) \tag{220}$$

Substitution of (218), (219), and (220) in (216) gives

$$|\varphi_{11}(\tau) - \varphi_{11}(\tau\pm\epsilon)|^2 \le \varphi_{11}(0)[2\varphi_{11}(0) - 2\varphi_{11}(\pm\epsilon)] \tag{221}$$

Since the autocorrelation function of a message or noise in a physical problem is generally continuous at the origin, that is

$$\lim_{\epsilon\to 0} \varphi_{11}(\pm\epsilon) = \varphi_{11}(0) \tag{222}$$

we conclude that, except for unusual circumstances where (222) does not hold, (221) in the limit as $\epsilon \to 0$ is

$$\lim_{\epsilon\to 0} |\varphi_{11}(\tau) - \varphi_{11}(\tau\pm\epsilon)| = 0 \tag{223}$$

for all values of τ. This conclusion is equivalent to the statement that $\varphi_{11}(\tau)$ is continuous everywhere if it is continuous at the origin.

Like the autocorrelation function for an aperiodic function, the peak value of $\varphi_{11}(\tau)$ for a random function occurs at $\tau = 0$. We show this by considering the positive function

$$\lim_{T\to\infty} \frac{1}{2T} \int_{-T}^{T} [f_1(t) \pm f_1(t+\tau)]^2 \, dt \tag{224}$$

Because periodic functions are excluded,

$$\lim_{T\to\infty} \frac{1}{2T} \int_{-T}^{T} [f_1(t) \pm f_1(t+\tau)]^2 \, dt > 0 \qquad \text{for } \tau \neq 0 \tag{225}$$

Upon expansion, we have

$$\lim_{T\to\infty} \frac{1}{2T} \int_{-T}^{T} f_1{}^2(t) \, dt + \lim_{T\to\infty} \frac{1}{2T} \int_{-T}^{T} f_1{}^2(t+\tau) \, dt \\ \pm 2 \lim_{T\to\infty} \frac{1}{2T} \int_{-T}^{T} f_1(t) f_1(t+\tau) \, dt > 0 \qquad \text{for } \tau \neq 0 \tag{226}$$

Since the first term is $\varphi_{11}(0)$, the second term is also $\varphi_{11}(0)$ as in (218), and the third term is $\pm 2\varphi_{11}(\tau)$, this expression is simply

$$\varphi_{11}(0) > \pm\varphi_{11}(\tau) \qquad \text{for } \tau \neq 0 \tag{227}$$

or, otherwise stated,

$$\varphi_{11}(0) > |\varphi_{11}(\tau)| \qquad \text{for } \tau \neq 0 \tag{228}$$

This is an expression of the fact that the maximum value of the autocorrelation function is its value at $\tau = 0$ which no other value of the function can equal in magnitude. The maximum value referred to is the mean square value of the random function.

3. Wiener Theorem for Autocorrelation

We have already indicated the great importance of the autocorrelation function of a random function because, aside from other reasons, it is a key to the spectrum. Let us derive the relation. To do this, we follow a limiting process and start with the autocorrelation theorem for periodic functions. The reciprocal relations (39) and (40) of the theorem when written as a single expression is

$$\varphi_{11}(\tau) = \sum_{n=-\infty}^{\infty} e^{jn\omega_1\tau} \frac{1}{T_1} \int_{-T_1/2}^{T_1/2} \varphi_{11}(\mu) e^{-jn\omega_1\mu} \, d\mu \tag{229}$$

Associated with this Fourier expression for $\varphi_{11}(\tau)$ of a periodic function

$f_1(t)$ is the definition of $\varphi_{11}(\tau)$. To repeat for convenience, it is

$$\varphi_{11}(\tau) = \frac{1}{T_1} \int_{-T_1/2}^{T_1/2} f_1(t) f_1(t + \tau)\, dt \tag{230}$$

As we allow the period T_1 to approach infinity, $f_1(t)$ is made to approach the random function. In a manner similar to the derivation of (123), $1/T_1$ in (229) tends to the differential of angular frequency $(1/2\pi)\, d\omega$, $n\omega_1$ tends to the continuous angular frequency ω, and the summation becomes an integration over the range $(-\infty, \infty)$. The result of this limiting process is that $\varphi_{11}(\tau)$ loses its periodic form and becomes an aperiodic function. We have thus obtained formally and heuristically the autocorrelation function of a random function in the form of a Fourier integral

$$\varphi_{11}(\tau) = \frac{1}{2\pi} \int_{-\infty}^{\infty} e^{j\omega\tau}\, d\omega \int_{-\infty}^{\infty} \varphi_{11}(\mu) e^{-j\omega\mu}\, d\mu \tag{231}$$

Here we assume that

$$\int_{-\infty}^{\infty} |\varphi_{11}(\tau)|\, d\tau \tag{232}$$

is finite. Furthermore, with a minor change $(T = T_1/2)$, the limiting form of (230) is

$$\varphi_{11}(\tau) = \lim_{T \to \infty} \frac{1}{2T} \int_{-T}^{T} f_1(t) f_1(t + \tau)\, dt \tag{233}$$

which is the autocorrelation function of the random function $f_1(t)$. This equation agrees with the definition (203), as it should.

Proceeding in parallel with our treatment of the Fourier integral, we define a function $\Phi_{11}(\omega)$ in such a manner that, when written as a pair of reciprocal relations, (231) is

$$\varphi_{11}(\tau) = \int_{-\infty}^{\infty} \Phi_{11}(\omega) e^{j\omega\tau}\, d\omega \tag{234}$$

and

$$\Phi_{11}(\omega) = \frac{1}{2\pi} \int_{-\infty}^{\infty} \varphi_{11}(\tau) e^{-j\omega\tau}\, d\tau \tag{235}$$

Since, by (235), $\Phi_{11}(\omega)$ is defined as the Fourier transform of $\varphi_{11}(\tau)$, which is an aperiodic function, it is the continuous spectrum of $\varphi_{11}(\tau)$.

Sometimes it is preferable to write (234) and (235) as cosine transforms. This is possible because $\varphi_{11}(\tau)$ is a real and even function so that its transform $\Phi_{11}(\omega)$ should also be a real and even function. We have shown this in connection with the autocorrelation theorem for aperiodic functions. Rewriting (234) and (235) as cosine transforms, we have

$$\varphi_{11}(\tau) = \int_{-\infty}^{\infty} \Phi_{11}(\omega) \cos \omega\tau \, d\omega \tag{236}$$

and

$$\Phi_{11}(\omega) = \frac{1}{2\pi} \int_{-\infty}^{\infty} \varphi_{11}(\tau) \cos \omega\tau \, d\tau \tag{237}$$

It is important that we investigate the physical significance of the function $\Phi_{11}(\omega)$. By letting $\tau = 0$ in (236), we have the simple relation

$$\varphi_{11}(0) = \int_{-\infty}^{\infty} \Phi_{11}(\omega) \, d\omega \tag{238}$$

which is an expression of $\Phi_{11}(\omega)$ in terms of the mean square value of the random function, for, by (207),

$$\lim_{T \to \infty} \frac{1}{2T} \int_{-T}^{T} f_1{}^2(t) \, dt = \int_{-\infty}^{\infty} \Phi_{11}(\omega) \, d\omega \tag{239}$$

If $f_1(t)$ is a voltage or current for a 1-ohm load, the mean square of $f_1(t)$ is the mean power taken by the load. Now, (239) states that the integral of $\Phi_{11}(\omega)$ with respect to the angular frequency over the entire range of angular frequencies $(-\infty, \infty)$ gives the mean power of $f_1(t)$. Consequently, it appears that $\Phi_{11}(\omega)$ is the *power density spectrum* of $f_1(t)$. A proof of this result is given in Sec. D-10. The spectrum should be expressed in units of watts per unit of angular frequency, or volts squared (or amperes squared) per unit of angular frequency, when the load is not considered. The relation (239) should be compared with (36) for periodic functions.

Having established the physical meaning of $\Phi_{11}(\omega)$, we may state that *the autocorrelation function of a random function and the power density spectrum of the random function are related to each other by a Fourier cosine transformation as given by* (236) *and* (237). This theorem has been rigorously established by Norbert Wiener and is generally called the *Wiener theorem* for autocorrelation. A rigorous proof of the theorem is given in Sec. D-10.

The fundamental assumption underlying the Wiener theorem is that the autocorrelation function exists and that it is common to all member functions of the ensemble. Because of the unique relationship between a function and its Fourier transform, it is evident that the Wiener theorem establishes the fact that the assumption is tantamount to the requirement that every member of the ensemble has the same power density spectrum irrespective of waveform.

Looking back briefly, we find in the analysis of periodic functions that the autocorrelation function and power spectrum are related by a pair of Fourier transforms (39) and (40). The autocorrelation function is periodic, and the power spectrum discrete. Through these transforms we obtain analogous expressions for random functions with a significant difference in the fact that the autocorrelation function is now aperiodic and the spectrum continuous, giving not power at various harmonics but rather density of power with respect to angular frequency so that power contribution by a continuous band of frequencies is represented by the area of the spectrum function within the limits of the band. The power due to any infinitesimal band, $d\omega$, of sinusoidal components into which the random function has been resolved is an infinitesimal quantity, since it is

$$\Phi_{11}(\omega)\,d\omega \tag{240}$$

A finite amount of power exists only within a finite band of frequencies; and the total power of $f_1(t)$ is represented by the total area under the power density spectrum curve.

As a simple illustration of the application of the Wiener theorem, we refer to the random wave of Fig. 28 for which the autocorrelation function is given by (209). If the determination of the power density spectrum of the random wave is our problem, it follows immediately from the Wiener theorem that the spectrum is

$$\begin{aligned}\Phi_{11}(\omega) &= \frac{1}{2\pi}\int_{-\infty}^{\infty} E_m{}^2 e^{-2k|\tau|}\cos\omega\tau\,d\tau \\ &= \frac{E_m{}^2}{\pi}\,\frac{2k}{(2k)^2+\omega^2}\end{aligned} \tag{241}$$

The graph of this function is shown in Fig. 30.

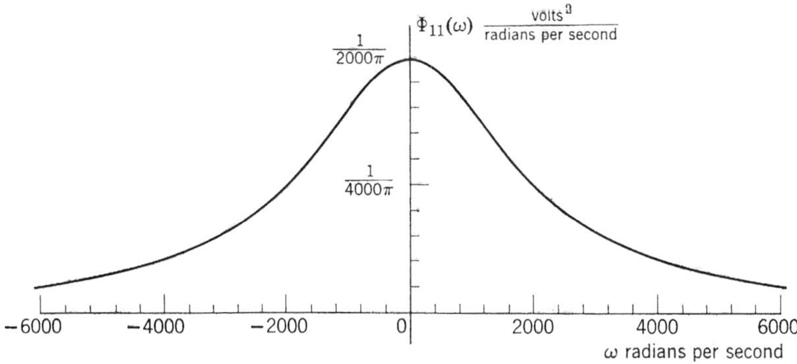

Fig. 30. Power density spectrum of the random wave of Fig. 28. $k = 1000$ per second, $E_m = 1$ volt.

4. Power Density Spectrum, Integrated Power Spectrum, and Their Interpretation

In order to make clear the physical meaning of the power density spectrum, we consider one simple method by which theoretically it may be measured. Suppose that the power density spectrum of the random voltage wave $f_1(t)$ shown in Fig. 31 is required. For simplicity, assume

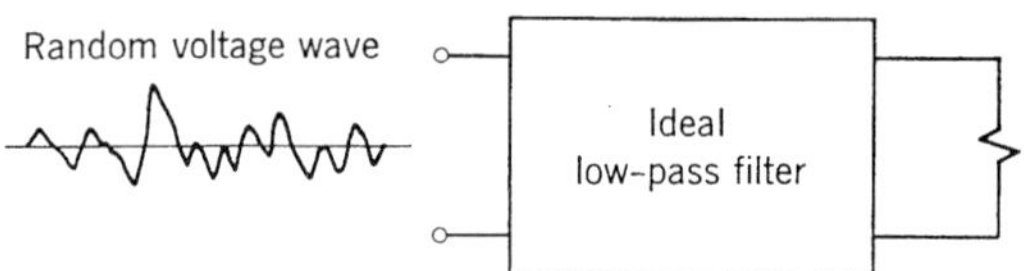

Fig. 31. Ideal low-pass filter for the determination of the power density spectrum of a random wave.

that the random wave contains no periodic or d-c components. To determine the spectrum the random wave is applied to an ideal low-pass filter of the conventional type whose output is a 1-ohm resistor as shown. Let the cutoff frequency of the filter be continuously adjustable over the important range of frequencies of the random wave. Since the filter is ideal, all components of the random wave below the cutoff frequency pass the network unattenuated, and all others are almost completely eliminated by the network. By means of an appropriate power-measuring device the power consumed by the resistor is determined at various cutoff angular frequencies $\omega_1, \omega_2, \omega_3, \ldots$ and plotted as in Fig. 32. Clearly,

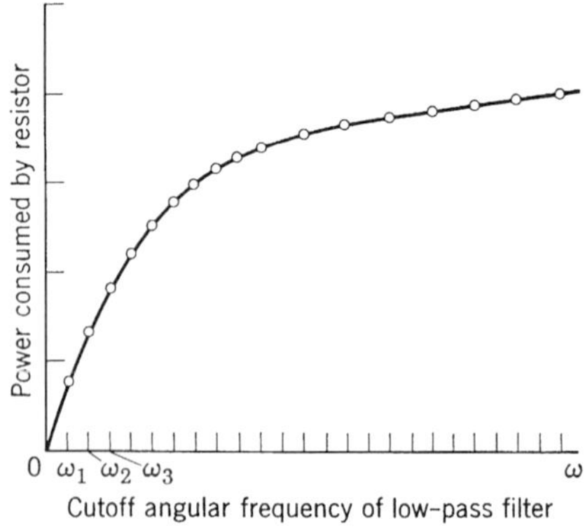

Fig. 32. Curve of power consumed by resistor in circuit of Fig. 31 versus cutoff angular frequency of ideal low-pass filter.

a point on this curve at a given angular frequency ω indicates power consumption in the frequency range between zero and the given frequency. If the slope of this curve is determined graphically as a function of ω, the result would appear as the dotted curve of Fig. 33. This curve of slope versus angular frequency is a spectrum of power density because its integral from zero to a given value of ω is the power due to the aggregate of components within the range of frequencies under consideration. In other words, the measured curve of Fig. 32 is an integrated power spectrum so that its derivative, as given in Fig. 33, is a power

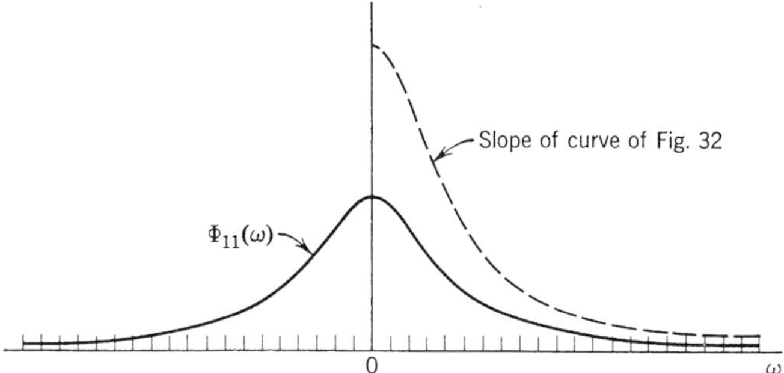

Fig. 33. Slope of curve of Fig. 32 and power density spectrum of random wave of Fig. 31.

density spectrum. Finally, in order that the power density spectrum thus obtained be in agreement with its theoretical definition (235), the amplitude of the curve for positive frequencies is halved and the upper half is drawn for negative frequencies of corresponding values. The result is the curve $\Phi_{11}(\omega)$ of Fig. 33 which is an even function, as it should be. Negative frequencies, although physically nonexistent, have been introduced in harmonic analysis for the reason that they make possible the expression of Fourier series and integrals in exponential forms which are simple and compact. It is, of course, necessary that expressions involving negative frequencies in a physical problem be correctly interpreted.

The power contributed by components of the voltage wave of all frequencies between zero and a given angular frequency such as ω_a, which the ordinate of Fig. 32 at $\omega = \omega_a$ represents, is

$$\int_{-\omega_a}^{\omega_a} \Phi_{11}(\omega)\, d\omega \tag{242}$$

Numerically this power is equal to the area under $\Phi_{11}(\omega)$ from $-\omega_a$ to ω_a. Similarly, the power due to all components in a band (ω_a, ω_b) is given by

$$2\int_{\omega_a}^{\omega_b} \Phi_{11}(\omega)\,d\omega \tag{243}$$

which is the same as

$$\left[\int_{-\omega_b}^{-\omega_a} + \int_{\omega_a}^{\omega_b}\right] \Phi_{11}(\omega)\,d\omega \tag{244}$$

Although at $\omega = 0$ the curve of $\Phi_{11}(\omega)$ in Fig. 32 has a finite value, it does not mean that the random voltage has a finite d-c component. This is explained by the fact that $\Phi_{11}(\omega)$ is a spectrum of power density so that the d-c power, according to (239), should be

$$\Phi_{11}(0)\,d\omega \tag{245}$$

which is an infinitesimal quantity. As a matter of fact, this amount includes the contributions of all components in the band $d\omega$ in the neighborhood of $\omega = 0$. Since the area under the curve of $\Phi_{11}(\omega)$, not its amplitude, represents power, the amount of power in any band of frequencies, finite or infinite, is finite, but that in an infinitesimal band $d\omega$ is infinitesimal. Furthermore, the power at a particular frequency is zero when the spectrum is a continuous density spectrum.

If the random wave has a periodic component, the power density spectrum is the sum of two parts—a continuous curve for the random component, and a series of impulses for the periodic component as illustrated in Fig. 34. At the fundamental and harmonic frequencies of

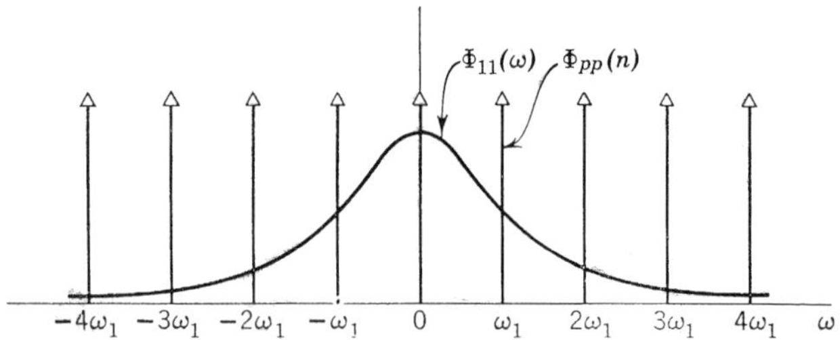

Fig. 34. Power density spectrum of a random wave with a periodic component.

the periodic component, the impulses are infinite in amplitude but the integrals (areas) of the impulses correspond to the power at these frequencies. In particular, if the power spectrum of the periodic component $f_p(t)$ is

$$\Phi_{pp}(n) = \frac{a_n^{\,2} + b_n^{\,2}}{4} \qquad \text{for } n = 0, \pm 1, \pm 2, \pm 3, \ldots \tag{246}$$

which is similar to (48), then the integrals of the impulses at $\omega = 0$, $\pm\omega_1$, $\pm 2\omega_1$, $\pm 3\omega_1$, ... should have the values given by (246).

When a random wave has a periodic component, it is clear from this discussion that either a power spectrum or a power density spectrum might be inadequate in representing the composite wave. In a power spectrum the periodic component consists of a series of lines of finite heights, but the random component cannot be appropriately represented. On the other hand, the power density spectrum, although well suited for characterizing a random wave, fails to give a periodic wave a convenient representation because of the infinite discontinuities involved. To overcome this difficulty, the use of an integrated spectrum is often desirable. This is simply the integral of the power density spectrum. For the random component the *integrated power spectrum* is

$$S_{11}(\omega) = \int_{-\infty}^{\omega} \Phi_{11}(\omega)\, d\omega \tag{247}$$

which is a positive increasing monotonic function, since $\Phi_{11}(\omega)$ is a positive function; and for the periodic component, it is

$$\begin{aligned} S_{pp}(\omega) &= \sum_{m=-\infty}^{n} \Phi_{pp}(m\omega_1) \\ &= \frac{1}{4} \sum_{m=-\infty}^{n} (a_{m\omega_1}^2 + b_{m\omega_1}^2) \qquad \text{for } n\omega_1 < \omega < (n+1)\omega_1 \\ &\qquad\qquad \text{and} \quad n = 0, \pm 1, \pm 2, \pm 3, \ldots \end{aligned} \tag{248}$$

Here we have written $a_{m\omega_1}$, $b_{m\omega_1}$, and $\Phi_{pp}(m\omega_1)$ for a_n, b_n, and $\Phi_{pp}(n)$ respectively, which are numerically the same when $m = n$, to indicate that the integrated power spectrum of the periodic function is a function of ω ascending in steps at intervals of length ω_1. An example of an integrated power spectrum for a random function is given in Fig. 35, and

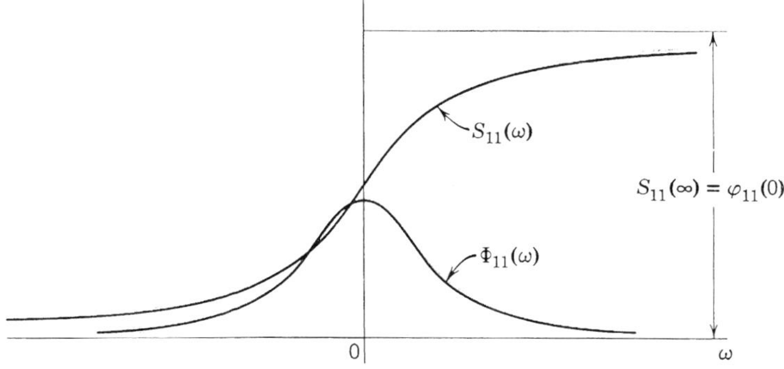

Fig. 35. Power density spectrum and integrated power spectrum of a random wave.

one for a periodic function in Fig. 36. The integrated power spectrum $S(\omega)$ for the composite wave is shown in Fig. 37. In the integrated spec-

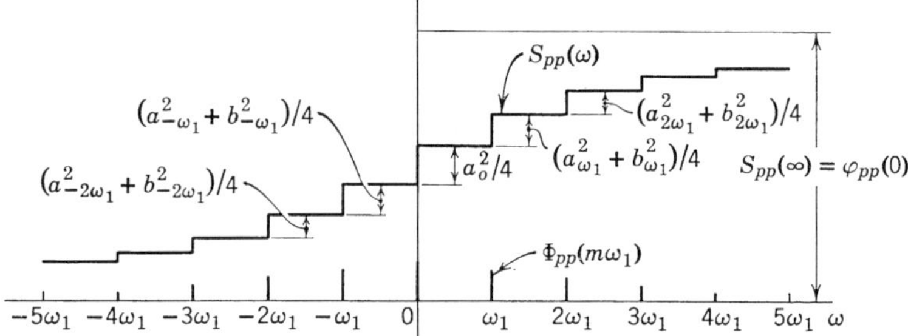

Fig. 36. Power spectrum and integrated power spectrum of a periodic wave.

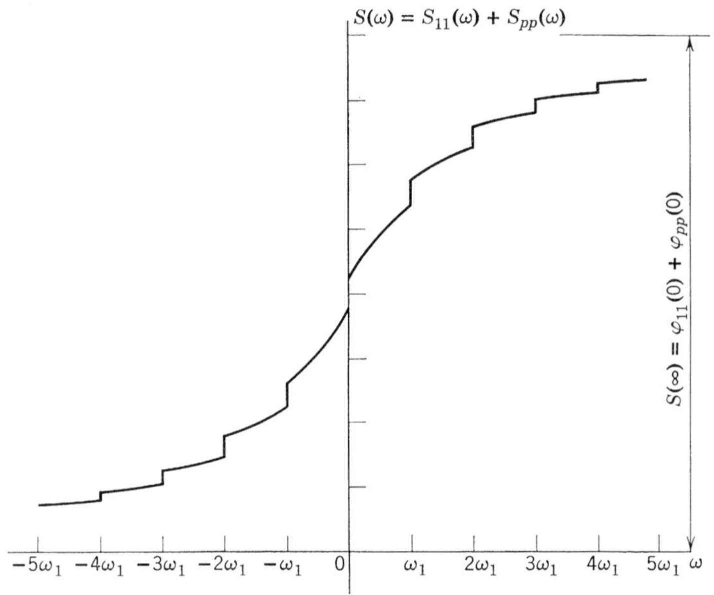

Fig. 37. Integrated power spectrum of a random wave with a periodic component.

trum for a random wave, the power due to a band of frequencies such as (ω_a, ω_b) is twice the difference between the ordinates at ω_a and ω_b, that is,

$$2[S_{11}(\omega_b) - S_{11}(\omega_a)] \tag{249}$$

which is the same as

$$[S_{11}(\omega_b) - S_{11}(\omega_a)] + [S_{11}(-\omega_a) - S_{11}(-\omega_b)] \tag{250}$$

and the total power is clearly

$$\varphi_{11}(0) = S_{11}(\infty) \tag{251}$$

Similarly, twice the sum of the jumps of the integrated spectrum for a periodic function within a given band of frequencies (ω_a, ω_b) is the power within the band. This power is

$$2[S_{pp}(\omega_b) - S_{pp}(\omega_a)] \tag{252}$$

or

$$[S_{pp}(\omega_b) - S_{pp}(\omega_a)] + [S_{pp}(-\omega_a) - S_{pp}(-\omega_b)] \tag{253}$$

The total power of the periodic wave is

$$\varphi_{pp}(0) = S_{pp}(\infty) \tag{254}$$

When it is desirable to express the spectrum for positive values of ω only, the integrated power spectrum for a random wave becomes

$$S_{11}^*(\omega) = 2\int_0^\omega \Phi_{11}(\omega)\,d\omega \tag{255}$$

and that for a periodic wave becomes

$$\begin{aligned} S_{pp}^*(\omega) &= \Phi_{pp}(0) + \sum_{m=1}^{n} \Phi_{pp}(m\omega_1) \\ &= \frac{a_0{}^2}{4} + \frac{1}{2}\sum_{m=1}^{n}(a_{m\omega_1}^2 + b_{m\omega_1}^2) \qquad \text{for } n\omega_1 < \omega < (n+1)\omega_1 \text{ and } n = 1, 2, 3, \ldots \end{aligned} \tag{256}$$

According to this representation, $S_{11}(\omega)$ of Fig. 35 becomes $S_{11}^*(\omega)$ of Fig. 38. This curve corresponds to that of Fig. 32 which is theoretically

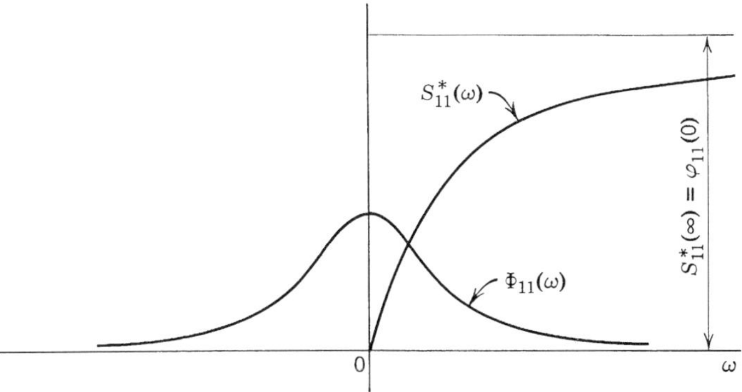

Fig. 38. Power density spectrum of a random wave and the integrated power spectrum for only positive values of ω.

possible to obtain by direct measurement. The power in the band (ω_a, ω_b) is

$$S_{11}^*(\omega_b) - S_{11}^*(\omega_a) \tag{257}$$

which is numerically the same as (249). The total power is

$$\varphi_{11}(0) = S_{11}^*(\infty) \tag{258}$$

For the periodic wave, $S_{pp}(\omega)$ of Fig. 36 becomes $S_{pp}^*(\omega)$ of Fig. 39. In

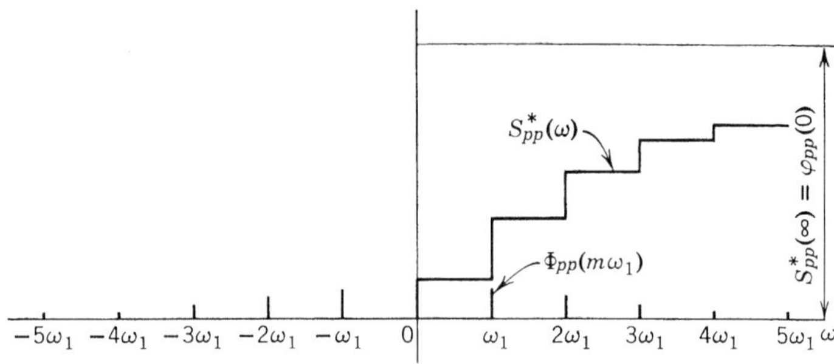

Fig. 39. Power spectrum of a periodic wave and the integrated power spectrum for only positive values of ω.

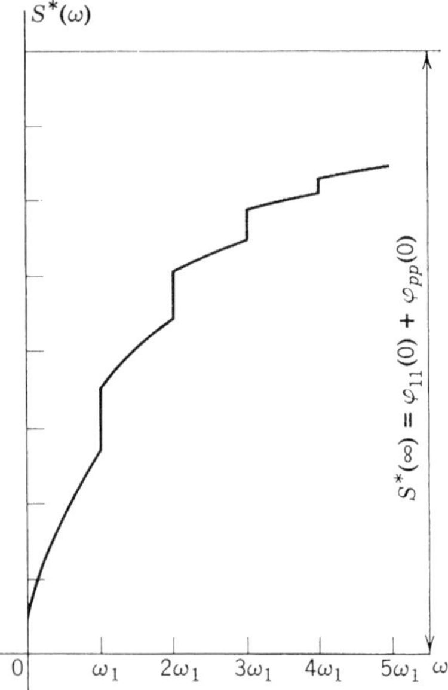

Fig. 40. Integrated power spectrum of a random wave with a periodic component for only positive values of ω.

the band (ω_a, ω_b) the power is

$$S_{pp}^*(\omega_b) - S_{pp}^*(\omega_a) \tag{259}$$

and the total power is

$$\varphi_{pp}(0) = S_{pp}^*(\infty) \tag{260}$$

The integrated power spectrum for the sum of the two components as a function of only positive ω is $S^*(\omega)$ of Fig. 40.

5. Graphical Determination of Autocorrelation Function

One exceedingly important application of the Wiener theorem is the determination of the power density spectrum through autocorrelation. As mentioned earlier, the analytical method for finding the autocorrelation function for specific ensembles of random functions cannot be presented until appropriate tools and techniques of probability and statistics are introduced, but the method of calculation based upon experimental data should be discussed at this point.

In accordance with the Wiener theorem, the power density spectrum is obtainable as a cosine transform of the autocorrelation function. We shall not discuss at this time the experimental and analytical techniques for carrying out a cosine transformation but shall only consider the evaluation of the autocorrelation function. Since observation time, of necessity, is finite, let $f_T(t)$ in Fig. 41 with an arbitrary reference point

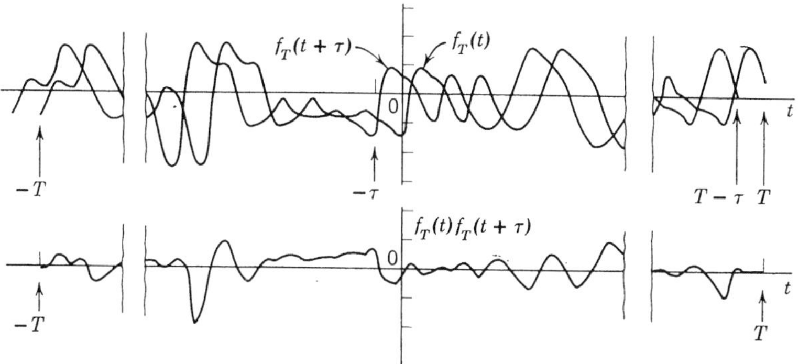

Fig. 41. Graphical determination of autocorrelation curve.

$t = 0$ as indicated, represent the recorded random voltage wave $f_1(t)$ of Fig. 31 for the duration $2T$; that is,

$$f_T(t) = \begin{cases} f_1(t) & \text{for } -T < t < T \\ 0 & \text{elsewhere} \end{cases} \tag{261}$$

According to definition (203) the approximate expression for the autocorrelation function of $f_1(t)$ is

$$\varphi_{11}(\tau) \cong \varphi_{TT}(\tau) \tag{262}$$

and

$$\varphi_{TT}(\tau) = \frac{1}{2T}\int_{-T}^{T} f_T(t)f_T(t+\tau)\,dt \tag{263}$$

To evaluate this function graphically, the first step is to form the function $f_T(t + \tau_1)$ by displacing $f_T(t)$ by the amount $\tau = \tau_1$ as indicated in Fig. 41. Then the product $f_T(t)f_T(t + \tau_1)$, represented by the lower curve, is obtained. Finally, the average of the product curve is taken by integration and division by the interval $2T$. This average value is

$\varphi_{TT}(\tau_1)$ and is plotted in Fig. 42. The procedure is repeated for a series of values of τ so that a curve is determined.

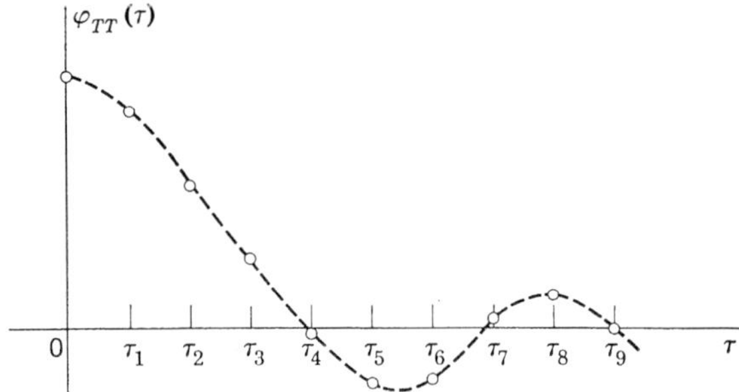

Fig. 42. Autocorrelation curve obtained by graphical method.

Clearly, the accuracy of this curve depends upon the length of the record. Just how long the record should be for a given tolerance can be determined if the autocorrelation that we seek is known. Since in general this condition cannot be satisfied, except when experiments and calculations are made for the purpose of verification, the necessary length of the record can only be estimated. The method of estimation is discussed in a later chapter.

The calculation of the approximate autocorrelation curve from given data according to formula (263) involves continuous multiplication and integration since $f_T(t)$ is continuous. Although these operations may be carried out in various ways with the aid of machines, it is frequently desirable to replace continuous operations by discrete ones. To reduce formula (263) to one requiring only discrete operations, let $f_T(t)$ be divided into N equally spaced small intervals as shown in Fig. 43, so that

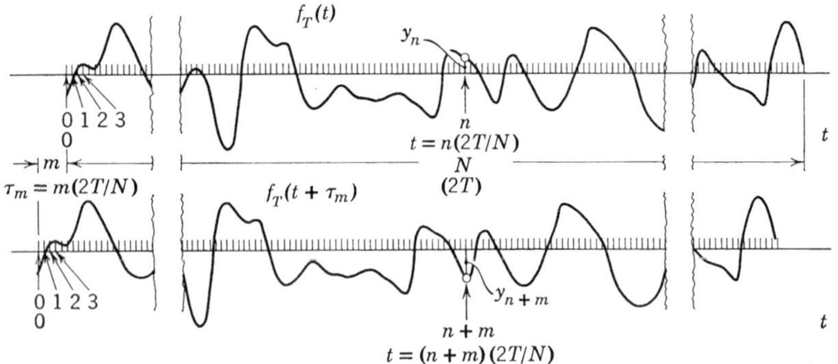

Fig. 43. Graphical determination of autocorrelation curve. Note that origin of $f_T(t)$ differs from that of Fig. 41. Also, subscripts of τ do not correspond to those of Figs. 41 and 42.

each division corresponds to $2T/N$ seconds, if t is in seconds. The curve $f_T(t + \tau_m)$ is $f_T(t)$ displaced m divisions to the left, corresponding to a shift of $\tau_m = m(2T/N)$ seconds. At a point n divisions $[n(2T/N)$ seconds] from the origin of $f_T(t)$, the value of the curve is y_n and the value of $f_m(t + \tau_m)$ is y_{n+m}. The approximate expression for (263) at discrete values τ_m of τ is therefore

$$\varphi_{TT}(\tau_m) \cong \frac{1}{N+1} \sum_{n=0}^{N-m} y_n y_{n+m} \tag{264}$$

6. Further Discussion on Properties of Autocorrelation and Power Density Spectrum

The power density spectrum is always a real, even, and non-negative function. To show this heuristically let us first note that from equations (229), (231), and (235) we find that the limiting form of

$$\frac{1}{2\pi} \int_{-T_1/2}^{T_1/2} \varphi_{11}(\mu) e^{-jn\omega_1\mu} \, d\mu \tag{265}$$

(in which $\varphi_{11}(\mu)$ is periodic) for a periodic function is

$$\Phi_{11}(\omega) = \frac{1}{2\pi} \int_{-\infty}^{\infty} \varphi_{11}(\tau) e^{-j\omega\tau} \, d\tau \tag{266}$$

(in which $\varphi_{11}(\tau)$ is aperiodic) for a random function. For the periodic function $f_1(t)$, whose spectrum is $F_1(n)$, equation (265) is, according to (38) and (40)

$$\frac{1}{2\pi} \int_{-T_1/2}^{T_1/2} \varphi_{11}(\mu) e^{-jn\omega_1\mu} \, d\mu = \frac{T_1}{2\pi} |F_1(n)|^2 \tag{267}$$

To emphasize the fact that (267) is a function of ω_1, we replace n by $n\omega_1$ and T_1 by $2\pi/\omega_1$. The power density spectrum (266) for a random function is now expressible as

$$\Phi_{11}(\omega) = \lim_{\omega_1 \to 0} \frac{1}{\omega_1} |F_1(n\omega_1)|^2 \tag{268}$$

Since $|F_1(n\omega_1)|^2$ is real, even, and non-negative, the power density spectrum $\Phi_{11}(\omega)$ for a random function should have these properties. Proof that the power density spectrum has these properties can be obtained easily from the material in Sec. D-10, particularly the rigorous expression for the power density spectrum (380). We have seen that the power spectrum (38) of a periodic function and the energy density spectrum (146) of an aperiodic function are also nonnegative functions.

We find, therefore, that an important property of an autocorrelation function is that its Fourier transform is a non-negative function. This statement applies to the autocorrelation function of a periodic, an aperiodic, or a random function. A function having the general appearance of an autocorrelation function for an aperiodic or a random function with the properties that its maximum value is at $\tau = 0$, that it is continuous, and that it is even may still not be a realizable autocorrelation function; for the Fourier transform of such a function may have certain ranges in which it becomes negative. When we say that a function is not a realizable autocorrelation function, we mean that a real time function whose autocorrelation is as given does not exist.

A question now arises as to what condition or conditions a function must satisfy if it is a realizable autocorrelation function. In answering this question, we shall at this time discuss the simpler situation where the autocorrelation of an aperiodic function is concerned and consider first the condition under which a real, even, and non-negative function is a realizable energy density spectrum. Let $[G(\omega)]^2$ be the real, even, and non-negative function, and assume that

$$\int_{-\infty}^{\infty} [G(\omega)]^2 \, d\omega \tag{269}$$

is finite. This assumption permits the existence of the Fourier transform of $[G(\omega)]^2$ as well as that of $G(\omega)$. Let the Fourier transform of $G(\omega)$ be $g(t)$; that is,

$$g(t) = \int_{-\infty}^{\infty} G(\omega) e^{j\omega t} \, d\omega \tag{270}$$

Since $G(\omega)$ is real and even, $g(t)$ is real and even. The function $g(t)$ is real if $G(\omega)$ is real and even because

$$\begin{aligned} g(t) &= \int_{-\infty}^{\infty} G(\omega) \cos \omega t \, d\omega + j \int_{-\infty}^{\infty} G(\omega) \sin \omega t \, d\omega \\ &= \int_{-\infty}^{\infty} G(\omega) \cos \omega t \, d\omega \end{aligned} \tag{271}$$

which is real, and it is even because

$$g(-t) = \int_{-\infty}^{\infty} G(\omega) \cos \omega t \, d\omega \tag{272}$$

showing that

$$g(-t) = g(t) \tag{273}$$

which is an expression of the stated property. Now, according to equation (143), the Fourier transform of $[G(\omega)]^2$ is

$$\int_{-\infty}^{\infty} [G(\omega)]^2 e^{j\omega\tau}\, d\omega = \frac{1}{2\pi} \int_{-\infty}^{\infty} g(t)g(t+\tau)\, dt \tag{274}$$

This is the autocorrelation of $(1/\sqrt{2\pi}\,)g(t)$. By application of the autocorrelation theorem in (153) and (154), the function $[G(\omega)]^2$ is found to be the energy density spectrum of $(1/\sqrt{2\pi}\,)g(t)$. We have therefore shown that *a real, even, and non-negative function* $\Phi_{11}(\omega)$ *is a realizable energy density spectrum if*

$$\int_{-\infty}^{\infty} \Phi_{11}(\omega)\, d\omega \tag{275}$$

is finite, and a function which has such an energy density spectrum is

$$g(t) = \frac{1}{\sqrt{2\pi}} \int_{-\infty}^{\infty} \sqrt{\Phi_{11}(\omega)}\, \cos \omega t\, d\omega \tag{276}$$

It is noted that we have determined only one of an infinite number of functions for which the energy density spectrum is given, and we have selected, for simplicity, the special function which has a zero phase spectrum so that it is an even function. Since the present discussion concerns primarily the conditions for a realizable autocorrelation function, we shall not go into the method for determining other possible functions in the problem.

Because the Fourier transform of the autocorrelation function of an aperiodic function is the energy density spectrum, as given by (153), the condition that

$$\int_{-\infty}^{\infty} \Phi_{11}(\omega)\, d\omega \tag{277}$$

is finite with respect to $\Phi_{11}(\omega)$ corresponds to the condition that $\varphi_{11}(0)$ is finite with respect to $\varphi_{11}(\tau)$, since, for $\tau = 0$, (153) reduces to

$$\varphi_{11}(0) = \int_{-\infty}^{\infty} \Phi_{11}(\omega)\, d\omega \tag{278}$$

Furthermore, the autocorrelation $\varphi_{11}(\tau)$, as we recall, is a real and even function.

Thus we conclude that, *for a real and even function, whose value at the origin is finite and positive, to be a realizable autocorrelation function of an aperiodic function, it is necessary and sufficient that it have a non-negative Fourier transform.* When this condition is satisfied by a given function $\varphi_{11}(\tau)$, its transform $\Phi_{11}(\omega)$ will yield the function $g(t)$, according to (276), whose autocorrelation is the given function. Since an autocorrelation function has a positive value at the origin, which is the one and only maximum value in amplitude, as expressed by (164), it is evident that the condition for the realizability of an autocorrelation function cannot be satisfied by any real and even function without such a peak at the origin. This fact is not explicitly stated in the condition; nevertheless, it is implied.

A complete derivation of the realizability condition for the autocorrelation function of a random function cannot be undertaken at this moment. We simply state here that, for a real and even function, whose value at the origin is finite and positive, to be a realizable autocorrelation function of a random function, it is necessary and sufficient that it have a non-negative Fourier transform that can be factorized in the manner described in Chapter 14, Sec. 8. When the Fourier transform (which is the power density spectrum of the random function) can be factorized in this manner, a random function called a Poisson wave with the given power density spectrum can be determined by the method presented in Chapter 13, Sec. 7(*a*).

7. Derivatives of the Autocorrelation Function

Starting with the definition (203)

$$\varphi_{11}(\tau) = \lim_{T\to\infty} \frac{1}{2T} \int_{-T}^{T} f_1(t) f_1(t+\tau)\, dt \tag{279}$$

and differentiating formally under the integral sign, we have

$$\varphi_{11}'(\tau) = \lim_{T\to\infty} \frac{1}{2T} \int_{-T}^{T} f_1(t) f_1'(t+\tau)\, dt \tag{280}$$

where $\varphi_{11}'(\tau) = (d/d\tau)\varphi_{11}(\tau)$ and $(d/dx)f_1(x) = f_1'(x)$. We find, therefore, that the derivative of the autocorrelation function of $f_1(t)$ is the correlation of $f_1(t)$ and its derivative.

With a change of variable $x = t + \tau$, equation (280) is

$$\varphi_{11}'(\tau) = \lim_{T\to\infty} \frac{1}{2T} \int_{-T+\tau}^{T+\tau} f_1'(x) f_1(x-\tau)\, dx \tag{281}$$

and, by the argument given in connection with the limits of integral (204), we write

$$\varphi_{11}'(\tau) = \lim_{T\to\infty} \frac{1}{2T} \int_{-T}^{T} f_1'(x) f_1(x-\tau)\, dx \tag{282}$$

A second formal differentiation under the integral sign yields the result

$$\varphi_{11}''(\tau) = -\lim_{T\to\infty} \frac{1}{2T} \int_{-T}^{T} f_1'(x)f_1'(x-\tau)\,dx \tag{283}$$

or, with a change of variable $t = x - \tau$ as before,

$$\varphi_{11}''(\tau) = -\lim_{T\to\infty} \frac{1}{2T} \int_{-T-\tau}^{T-\tau} f_1'(t)f_1'(t+\tau)\,dt \tag{284}$$

which reduces to

$$\varphi_{11}''(\tau) = -\lim_{T\to\infty} \frac{1}{2T} \int_{-T}^{T} f_1'(t)f_1'(t+\tau)\,dt \tag{285}$$

Thus we have shown that the second derivative of the autocorrelation function is another autocorrelation function. It is the autocorrelation of the derivative of $f_1(t)$, except for a negative sign.

To find the corresponding relation in terms of the power density spectrum of $f_1(t)$, we begin with the Wiener theorem

$$\varphi_{11}(\tau) = \int_{-\infty}^{\infty} \Phi_{11}(\omega)e^{j\omega\tau}\,d\omega \tag{286}$$

Formally, the first derivative of $\varphi_{11}(\tau)$ is

$$\varphi_{11}'(\tau) = \int_{-\infty}^{\infty} j\omega\Phi_{11}(\omega)e^{j\omega\tau}\,d\omega \tag{287}$$

and the second derivative is

$$\varphi_{11}''(\tau) = \int_{-\infty}^{\infty} (j\omega)^2\Phi_{11}(\omega)e^{j\omega\tau}\,d\omega \tag{288}$$

By inverse transformation, we have for (287)

$$j\omega\Phi_{11}(\omega) = \frac{1}{2\pi} \int_{-\infty}^{\infty} \varphi_{11}'(\tau)e^{-j\omega\tau}\,d\tau \tag{289}$$

and for (288)

$$(j\omega)^2\Phi_{11}(\omega) = \frac{1}{2\pi} \int_{-\infty}^{\infty} \varphi_{11}''(\tau)e^{-j\omega\tau}\,d\tau \tag{290}$$

8. Summary of Properties of Autocorrelation

The following points concerning the autocorrelation function, defined as

$$\varphi_{11}(\tau) = \lim_{T\to\infty} \frac{1}{2T} \int_{-T}^{T} f_1(t)f_1(t+\tau)\,dt \tag{291}$$

for a random function $f_1(t)$, have been considered:

1. The autocorrelation function is assumed to exist for every value of the argument.
2. It is an even function. $\varphi_{11}(\tau) = \varphi_{11}(-\tau)$.
3. Its value at the origin is the mean square value of the random function.

$$\varphi_{11}(0) = \lim_{T \to \infty} \frac{1}{2T} \int_{-T}^{T} f_1^2(t)\, dt$$

4. It tends to zero as the argument tends to infinity if the random function contains no d-c or periodic components. $\varphi_{11}(\infty) = 0$.
5. It is continuous everywhere if it is continuous at the origin.
6. Its value at the origin is the one and only maximum value in magnitude. $\varphi_{11}(0) > |\varphi_{11}(\tau)| \qquad \text{for } \tau \neq 0$.
7. It is reciprocally related to the power density spectrum of the random function by the Wiener theorem.
8. It has a necessary and sufficient condition of realizability.
9. Its first derivative is expressible as the correlation between the random function and its derivative.
10. Its second derivative is expressible as the negative of the autocorrelation of the derivative of the random function.

9. Crosscorrelation

The crosscorrelation of random functions enters into communication problems in various ways. One exceedingly important crosscorrelation function is that between the input and the output of a linear system. Another equally important one is required in the design of a linear system for the optimum extraction of a message from a corrupted message. It is the crosscorrelation between the desired message and the input to the system which contains the message in the presence of a noise or an unwanted message. The theory of this problem is the subject matter of several later chapters. Other examples of crosscorrelation are the crosscorrelation among different channels in a communication system as a measure of interference and the crosscorrelation of electrical impulses in different parts of the nervous system in the study of their interrelation. In short, the crosscorrelation between two random functions that are in some way related to each other is an important expression of their relationship in many problems concerning them.

In the crosscorrelation of two random functions we consider two ensembles of which they are members, one from each ensemble. The members of one ensemble, as in the discussion on autocorrelation, are produced by mechanisms that are essentially similar in every respect. Since to each member function of this ensemble there corresponds one and only one member function in the other ensemble, it is necessary that the member functions in the ensembles be paired. For instance, when the crosscorrelation between the input and the output of a linear system is considered, the input may be a thermal noise from a certain circuit. We then assume that an infinite number of circuits of this type, essentially the same in every respect, are available. Each circuit supplies a thermal noise to a separate linear system which is identical to all the others. Therefore, to each thermal noise in the first ensemble there corresponds an output noise in the second ensemble. Similar to the situation in autocorrelation, the ensemble may be produced, in actual experiment, by only one thermal noise source and one linear system. The first ensemble is formed from repeated experiments, each of considerable duration, and the second from the corresponding outputs.

For the definition of crosscorrelation we let $f_1(t)$ be a member function of the first ensemble and $f_2(t)$ be the corresponding member function in the second ensemble. The *crosscorrelation function* of the random functions is then defined as

$$\varphi_{12}(\tau) = \lim_{T\to\infty} \frac{1}{2T} \int_{-T}^{T} f_1(t) f_2(t + \tau)\, dt \tag{292}$$

This function is assumed to exist for every value of the argument. It is further assumed that the crosscorrelation between the functions of any pair in the ensemble is the same function. The crosscorrelation function is therefore a characteristic of the ensemble.

The definition (292) should be compared with the definition of crosscorrelation for periodic functions as given by (59). Because $f_1(t)$ and $f_2(t)$ in (59) are periodic and have the same fundamental period, the mean value of $f_1(t)f_2(t + \tau)$ taken over one complete period should be the same as that taken over the infinite interval as a limit. It is therefore clear that, for periodic functions of the same period, the crosscorrelation function can be defined either in the form (59) or the form (292).

As we have already seen in the crosscorrelation of periodic functions, the order of the subscripts of the crosscorrelation function is important, for by $\varphi_{21}(\tau)$ we mean the definition

$$\varphi_{21}(\tau) = \lim_{T\to\infty} \frac{1}{2T} \int_{-T}^{T} f_2(t) f_1(t + \tau)\, dt \tag{293}$$

The functions $\varphi_{12}(\tau)$ and $\varphi_{21}(\tau)$ are related to each other by the simple formula

$$\varphi_{12}(\tau) = \varphi_{21}(-\tau) \tag{294}$$

which is an expression of the fact that $\varphi_{12}(\tau)$ when folded back is $\varphi_{21}(\tau)$, and vice versa. This relation is easy to show. Changing the sign of τ in (292), we write

$$\varphi_{12}(-\tau) = \lim_{T\to\infty} \frac{1}{2T} \int_{-T}^{T} f_1(t)f_2(t-\tau)\,dt$$

and, with a change of variable $x = t - \tau$, it becomes

$$\varphi_{12}(-\tau) = \lim_{T\to\infty} \frac{1}{2T} \int_{-T-\tau}^{T-\tau} f_2(x)f_1(x+\tau)\,dx \tag{295}$$

By the same argument given in connection with (204), this equation is equivalent to

$$\varphi_{12}(-\tau) = \lim_{T\to\infty} \frac{1}{2T} \int_{-T}^{T} f_2(x)f_1(x+\tau)\,dx \tag{296}$$

and we then see that

$$\varphi_{12}(-\tau) = \varphi_{21}(\tau) \tag{297}$$

Unlike the autocorrelation function, $\varphi_{12}(\tau)$ is not necessarily an even function. In general, it is neither even nor odd. Furthermore, in contrast with autocorrelation, the value of $\varphi_{12}(\tau)$ at $\tau = 0$ need not be the maximum. On the other hand, similar to autocorrelation, crosscorrelation tends to zero as the displacement tends to infinity in either direction; that is,

$$\varphi_{12}(\pm\infty) = 0 \tag{298}$$

This is true provided that either one of the two random functions or both of them have no d-c components, and, in addition, if the random functions contain periodic components, those in one have no periods that are commensurable with the periods of the others. We shall show this property with the aid of statistical tools when correlation is considered as a statistical average.

Following Wiener's method in the analysis of the autocorrelation function, let us examine the continuity of the crosscorrelation function by means of the Schwarz inequality. For this, we consider the expression

$$|\varphi_{12}(\tau) - \varphi_{12}(\tau \pm \epsilon)| \tag{299}$$

in which ϵ is a positive increment. Written in terms of $f_1(t)$ and $f_2(t)$, (299) is

$$|\varphi_{12}(\tau) - \varphi_{12}(\tau \pm \epsilon)| = \left| \lim_{T\to\infty} \frac{1}{2T} \int_{-T}^{T} f_1(t)[f_2(t+\tau) - f_2(t+\tau\pm\epsilon)]\,dt \right| \tag{300}$$

Application of the inequality (213) leads to the expression

$$|\varphi_{12}(\tau) - \varphi_{12}(\tau \pm \epsilon)|^2 \leq \lim_{T\to\infty} \left(\frac{1}{2T}\right)^2$$

$$\left\{\left[\int_{-T}^{T} f_1{}^2(t)\,dt\right]\left[\int_{-T}^{T} [f_2(t+\tau) - f_2(t+\tau\pm\epsilon)]^2\,dt\right]\right\} \tag{301}$$

From the results (217) through (220) we write

$$\lim_{T\to\infty} \frac{1}{2T}\int_{-T}^{T} [f_2(t+\tau) - f_2(t+\tau\pm\epsilon)]^2\,dt = 2[\varphi_{22}(0) - \varphi_{22}(\pm\epsilon)] \tag{302}$$

so that (301) becomes

$$|\varphi_{12}(\tau) - \varphi_{12}(\tau \pm \epsilon)|^2 \leq 2\varphi_{11}(0)[\varphi_{22}(0) - \varphi_{22}(\pm\epsilon)] \tag{303}$$

Now, if $\varphi_{22}(\tau)$ is continuous at the origin, then

$$\lim_{\epsilon\to 0} \varphi_{22}(\pm\epsilon) = \varphi_{22}(0) \tag{304}$$

and, since $\varphi_{11}(0)$ is finite, (303) reduces to

$$\lim_{\epsilon\to 0} |\varphi_{12}(\tau) - \varphi_{12}(\tau\pm\epsilon)| = 0 \tag{305}$$

We conclude, therefore, that the crosscorrelation $\varphi_{12}(\tau)$ is continuous for all values of the argument in the range $(-\infty, \infty)$ if $\varphi_{22}(\tau)$ is continuous at its origin. Since inversion of the order of the subscripts merely reflects $\varphi_{12}(\tau)$ with respect to the vertical axis and in no way affects the continuity of the function, the alternative condition for the continuity of $\varphi_{12}(\tau)$ is obviously that $\varphi_{11}(\tau)$ is continuous at the origin. In other words, if the autocorrelation of either $f_1(t)$ or $f_2(t)$ is continuous at the origin, their crosscorrelation is continuous everywhere.

Concerning the upper and lower bounds of the crosscorrelation function, we consider the expression

$$\lim_{T\to\infty} \frac{1}{2T}\int_{-T}^{T} [f_1(t) \pm f_2(t+\tau)]^2\,dt \tag{306}$$

Since $f_1(t) \neq f_2(t+\tau)$ for any value of τ, this is a positive quantity for all values of τ; that is,

$$\lim_{T\to\infty} \frac{1}{2T}\int_{T}^{T} [f_1(t) \pm f_2(t+\tau)]^2\,dt > 0 \qquad \text{for } -\infty < \tau < \infty \tag{307}$$

Expansion of this gives

$$\lim_{T\to\infty} \frac{1}{2T} \int_{-T}^{T} f_1{}^2(t)\,dt + \lim_{T\to\infty} \frac{1}{2T} \int_{-T}^{T} f_2{}^2(t+\tau)\,dt$$

$$\pm 2 \lim_{T\to\infty} \frac{1}{2T} \int_{-T}^{T} f_1(t) f_2(t+\tau)\,dt > 0 \qquad \text{for } -\infty < \tau < \infty \tag{308}$$

or

$$\varphi_{11}(0) + \varphi_{22}(0) \pm 2\varphi_{12}(\tau) > 0 \qquad \text{for } -\infty < \tau < \infty \tag{309}$$

which is equivalent to

$$|\varphi_{12}(\tau)| < \tfrac{1}{2}[\varphi_{11}(0) + \varphi_{22}(0)] \qquad \text{for } -\infty < \tau < \infty \tag{310}$$

This inequality expresses the fact that the crosscorrelation of two random functions $f_1(t)$ and $f_2(t)$, with the condition that $f_1(t) \neq f_2(t + \tau)$ for all values of τ, is always less in magnitude than the average of the two mean square values of the random functions.

The crosscorrelation function is expressible as a Fourier integral. Thus

$$\varphi_{12}(\tau) = \frac{1}{2\pi} \int_{-\infty}^{\infty} e^{j\omega\tau}\,d\omega \int_{-\infty}^{\infty} \varphi_{12}(\mu) e^{-j\omega\mu}\,d\mu \tag{311}$$

Defining a function $\Phi_{12}(\omega)$ as

$$\Phi_{12}(\omega) = \frac{1}{2\pi} \int_{-\infty}^{\infty} \varphi_{12}(\tau) e^{-j\omega\tau}\,d\tau \tag{312}$$

we write integral (311) in the form

$$\varphi_{12}(\tau) = \int_{-\infty}^{\infty} \Phi_{12}(\omega) e^{j\omega\tau}\,d\omega \tag{313}$$

The function $\Phi_{12}(\omega)$ is called the *cross-power density spectrum* of the random functions $f_1(t)$ and $f_2(t)$. This term is generally appropriate since it is analogous to the term power density spectrum when only one random function is concerned. Nevertheless, it has been pointed out that in contrast with autocorrelation the physical meaning of crosscorrelation depends upon the problem and the manner in which it is applied. For this reason, the general term *spectrum of crosscorrelation* for $\Phi_{12}(\omega)$ is preferred if reference to power is unnecessary. The reciprocal relations (312) and (313) shall be called the *crosscorrelation theorem* for random functions.

D. SPECTRUMS OF AMPLITUDE AND PHASE AND SPECTRUMS OF ENERGY AND POWER

1. Introduction

We have considered the complex (amplitude and phase) line spectrum and the power spectrum of a periodic function and the relation between them. Similarly, we have considered the complex (relative amplitude and phase) continuous spectrum and the energy density spectrum of an aperiodic function and their relation. Concerning the random function, we have studied its power density spectrum. In addition to the periodic, aperiodic, and random functions we must consider functions that have a mean square value, are continuing in behavior, and can be specified for all time. Theoretically the entire past of a random function is a function of this type. The possibility of analyzing these functions for amplitude and phase spectrums should be investigated. Furthermore, if a type of amplitude and phase spectrum existed, its relationship to the power density spectrum should be examined. These matters are of great importance for the understanding of the nature and meaning of the power density spectrum of a random function.

2. Scope of the Wiener Theorem

Although the primary interest of the communication engineer in the Wiener theorem is its application to random functions that arise in the transmission of information, the theorem should not be misconstrued as being restricted to functions of unpredictable behavior only. In the derivation of the theorem, it is emphasized that the basic assumption is the existence of the autocorrelation function (203) for every value of the argument. The original form of Wiener's theorem includes periodic functions, but we have reduced it to a simpler form by excluding periodic functions and treating them separately in a similar theorem. To satisfy the basic assumption, it is entirely possible that the function is of the continuing type with its values determined for all time instead of being unpredictable. When the law governing the instantaneous values of a function is not known, it is possible to describe it in terms of probability distributions, and the autocorrelation function is then determined as a statistical average. However, should the highly complex law giving the instantaneous values of a random function be known, it will no longer be random, and, theoretically, it yields to analysis without the aid of probability theory. It is therefore clear that the Wiener theorem is applicable to functions that can be described either analytically or statistically if they satisfy the basic assumption stated.

3. Generalized Fourier Transforms

Fourier series theory is limited to periodic functions, and Fourier integral theory is applicable to transient functions whose integrals are absolutely convergent; that is, $\int_{-\infty}^{\infty} |f(t)|\, dt$ is finite. To include those functions which are neither periodic nor transient but are continuing in behavior under the general concept of amplitude and phase analysis, the Fourier transform theory needs generalization.

Consider a function $f(t)$ for which

$$\lim_{T \to \infty} \frac{1}{2T} \int_{-T}^{T} f^2(t)\, dt \tag{314}$$

exists. This quantity becomes zero if $f(t)$ is aperiodic (transient) and has a Fourier transform. It is finite if $f(t)$ is either periodic or continuing. With this condition, Wiener has shown that the integrated Fourier transform of $f(t)$ exists. We write the *integrated Fourier transform* as

$$G(\omega) = \frac{1}{2\pi} \int_{-\infty}^{\infty} f(t) \left(\frac{e^{-j\omega t}}{-jt} \right) dt \tag{315}$$

which is a much simplified form of Wiener's precise expression. For the needs of the present discussion this form is sufficient.

Let us evaluate $G(\omega)$ when the function $f(t)$ is periodic. Expressing the periodic function in the form

$$f(t) = \frac{a_0}{2} + \sum_{n=1}^{\infty} (a_n \cos n\omega_1 t + b_n \sin n\omega_1 t) \tag{316}$$

we can evaluate the integrated Fourier transform readily by substitution of (316) in (315) and with the aid of the integrals

$$\int_{-\infty}^{\infty} \frac{\cos n\omega_1 t \sin \omega t}{t}\, dt = \begin{cases} \pi & \text{for } n\omega_1 < \omega < \infty \\ \dfrac{\pi}{2} & \omega = n\omega_1 \\ 0 & -n\omega_1 < \omega < n\omega_1 \\ \dfrac{-\pi}{2} & \omega = -n\omega_1 \\ -\pi & -\infty < \omega < -n\omega_1 \end{cases} \tag{317}$$

$$\int_{-\infty}^{\infty} \frac{\sin n\omega_1 t \cos \omega t}{t}\, dt = \begin{cases} 0 & \text{for } n\omega_1 < \omega < \infty \\ \dfrac{\pi}{2} & \omega = n\omega_1 \\ \pi & -n\omega_1 < \omega < n\omega_1 \\ \dfrac{\pi}{2} & \omega = -n\omega_1 \\ 0 & -\infty < \omega < -n\omega_1 \end{cases} \tag{318}$$

and

$$\int_{-\infty}^{\infty} \frac{\sin \omega t}{t}\, dt = \begin{cases} \pi & \text{for } 0 < \omega < \infty \\ 0 & \omega = 0 \\ -\pi & -\infty < \omega < 0 \end{cases} \tag{319}$$

The graphs of (317), (318), and (319) are given in Figs. 44, 45, and 46 respectively. The result is

$$G(\omega) = G_r(\omega) + jG_i(\omega) \tag{320}$$

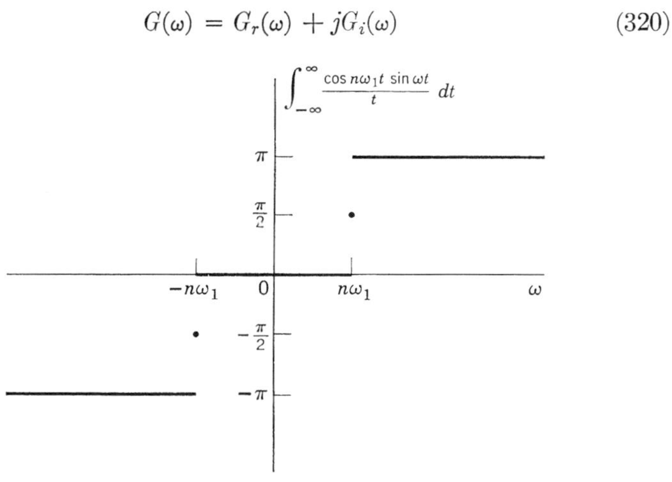

Fig. 44. Graph of (317).

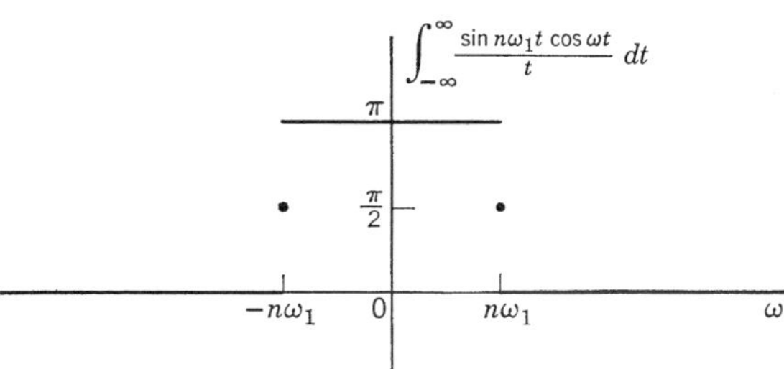

Fig. 45. Graph of (318).

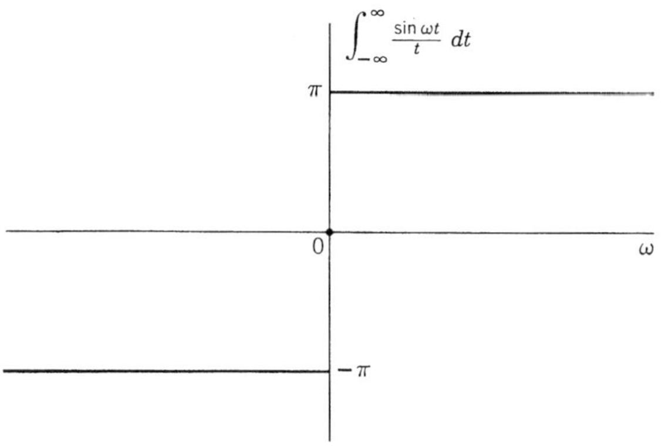

Fig. 46. Graph of (319).

where $G_r(\omega)$ denotes the real part of $G(\omega)$ and is

$$G_r(\omega) = \begin{cases} \dfrac{a_0}{4} + \dfrac{1}{2}\displaystyle\sum_{n=1}^{m} a_n & \text{for } m\omega_1 < \omega < (m+1)\omega_1 \\ -\dfrac{a_0}{4} - \dfrac{1}{2}\displaystyle\sum_{n=-1}^{-m} a_n & \text{for } -(m+1)\omega_1 < \omega < -m\omega_1 \end{cases} \tag{321}$$

and the imaginary part of $G(\omega)$ is

$$G_i(\omega) = \begin{cases} \dfrac{1}{2}\displaystyle\sum_{n=m}^{\infty} b_n & \text{for } (m-1)\omega_1 < \omega < m\omega_1 \\ \dfrac{1}{2}\displaystyle\sum_{n=-m}^{-\infty} b_n & \text{for } -m\omega_1 < \omega < -(m-1)\omega_1 \end{cases} \tag{322}$$

The real and imaginary components of $G(\omega)$ are shown in Figs. 47 and 48 respectively. These are step functions with jumps at the harmonic frequencies of the periodic function, the values of the jumps being equal to

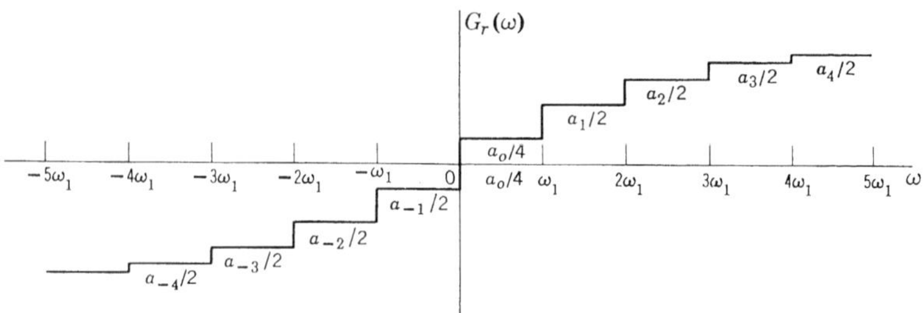

Fig. 47. Graph of real part of the integrated Fourier transform of a periodic function.

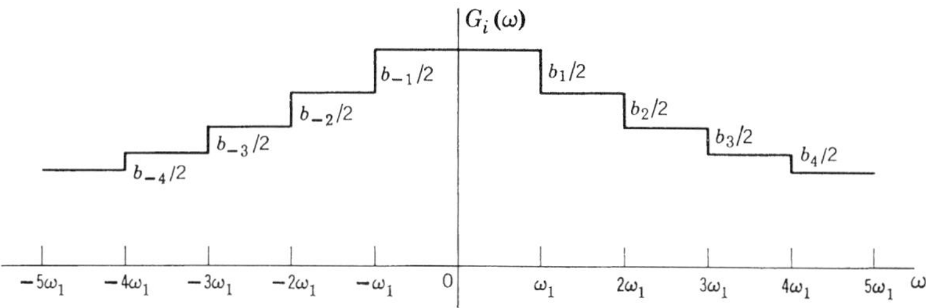

Fig. 48. Graph of the imaginary part of the integrated Fourier transform of a periodic function.

the coefficients of the harmonics. It is noted that, as a matter of convenience, the figures have been drawn for positive harmonic coefficients only.

Applying (315) to an aperiodic function, we find that, if the function has a Fourier transform $F(\omega)$, the integrated Fourier transform, except for an additive constant, is

$$G(\omega) = \int F(\omega)\, d\omega \tag{323}$$

A simple example is the function

$$f(t) = Ae^{-a|t|} \tag{324}$$

which has the Fourier transform

$$\begin{aligned} F(\omega) &= \frac{1}{2\pi} \int_{-\infty}^{\infty} Ae^{-a|t|} e^{-j\omega t}\, dt \\ &= \frac{A}{\pi} \frac{a}{a^2 + \omega^2} \end{aligned} \tag{325}$$

It follows from (323) that

$$G(\omega) = \frac{A}{\pi} \tan^{-1} \frac{\omega}{a} \tag{326}$$

To help our understanding of the integrated Fourier transform of a continuing function, we now introduce the quadratic variation.

4. Quadratic Variation

Wiener's *quadratic variation* of the integrated Fourier transform is

$$\lim_{\epsilon \to 0} \frac{1}{2\epsilon} \int_{-\infty}^{\infty} |G(\omega + \epsilon) - G(\omega - \epsilon)|^2\, d\omega \tag{327}$$

The value of this expression is equal to the mean square value of the function whose integrated Fourier transform is $G(\omega)$. To show this, we write in accordance with (315)

$$G(\omega \pm \epsilon) = \frac{1}{2\pi} \int_{-\infty}^{\infty} f(t) \left[\frac{e^{-j(\omega \pm \epsilon)t}}{-jt} \right] dt \tag{328}$$

from which we obtain

$$G(\omega + \epsilon) - G(\omega - \epsilon) = \frac{1}{\pi} \int_{-\infty}^{\infty} \left[f(t) \frac{\sin \epsilon t}{t} \right] e^{-j\omega t}\, dt \tag{329}$$

By application of the Parseval theorem (142) of which (147) is its special form, we have

$$\frac{1}{2} \int_{-\infty}^{\infty} |G(\omega + \epsilon) - G(\omega - \epsilon)|^2\, d\omega = \frac{1}{\pi} \int_{-\infty}^{\infty} f^2(t) \frac{\sin^2 \epsilon t}{t^2}\, dt \tag{330}$$

At this point we need a Tauberian theorem of Wiener, Bochner, and Hardy. This theorem states that, if

$$\frac{1}{T} \int_0^T |g(t)|\, dt \tag{331}$$

is bounded and

$$\lim_{T \to \infty} \frac{1}{T} \int_0^T g(t)\, dt = A \tag{332}$$

then

$$\lim_{\epsilon \to 0} \frac{2}{\pi \epsilon} \int_0^{\infty} g(t) \frac{\sin^2 \epsilon t}{t^2}\, dt = A \tag{333}$$

Now, if $f^2(t)$ is $g(t)$ with the range extended to $(-\infty, \infty)$, this theorem gives

$$\lim_{T \to \infty} \frac{1}{2T} \int_{-T}^{T} f^2(t)\, dt = \lim_{\epsilon \to 0} \frac{1}{\pi \epsilon} \int_{-\infty}^{\infty} f^2(t) \frac{\sin^2 \epsilon t}{t^2}\, dt \tag{334}$$

Furthermore, if, for (330), the limit

$$\lim_{\epsilon \to 0} \frac{1}{2\epsilon} \int_{-\infty}^{\infty} |G(\omega + \epsilon) - G(\omega - \epsilon)|^2\, d\omega = \lim_{\epsilon \to 0} \frac{1}{\pi \epsilon} \int_{-\infty}^{\infty} f^2(t) \frac{\sin^2 \epsilon t}{t^2}\, dt \tag{335}$$

exists, we have finally

$$\lim_{\epsilon \to 0} \frac{1}{2\epsilon} \int_{-\infty}^{\infty} |G(\omega + \epsilon) - G(\omega - \epsilon)|^2\, d\omega = \lim_{T \to \infty} \frac{1}{2T} \int_{-T}^{T} f^2(t)\, dt \tag{336}$$

This result states that *the quadratic variation of the integrated Fourier transform of a function is equal to the mean square value of the function.* We shall refer to it as *Wiener's theorem of quadratic variation.* Inasmuch as the existence of the integrated Fourier transform of a periodic, transient, or continuing function depends only upon the existence of its mean square value, the relation (336) is valid for all three types of functions.

5. Quadratic Variation for a Periodic Function

Applying this to the periodic function, we find that $G(\omega + \epsilon) - G(\omega - \epsilon)$ in (336), as determined from (321) and (322), is

$$G(\omega + \epsilon) - G(\omega - \epsilon)$$
$$= \begin{cases} \frac{1}{2}a_0 & \text{for } -\epsilon < \omega < \epsilon \\ \frac{1}{2}(a_n - jb_n) & (n\omega_1 - \epsilon) < \omega < (n\omega_1 + \epsilon) \\ \frac{1}{2}(a_{-n} + jb_{-n}) & -(n\omega_1 + \epsilon) < \omega < -(n\omega_1 - \epsilon) \\ 0 & \text{elsewhere} \end{cases} \tag{337}$$

The real part of the function, $R[G(\omega + \epsilon) - G(\omega - \epsilon)]$, is shown in Fig. 49, and the imaginary part, $I[G(\omega + \epsilon) - G(\omega - \epsilon)]$, is shown in Fig. 50.

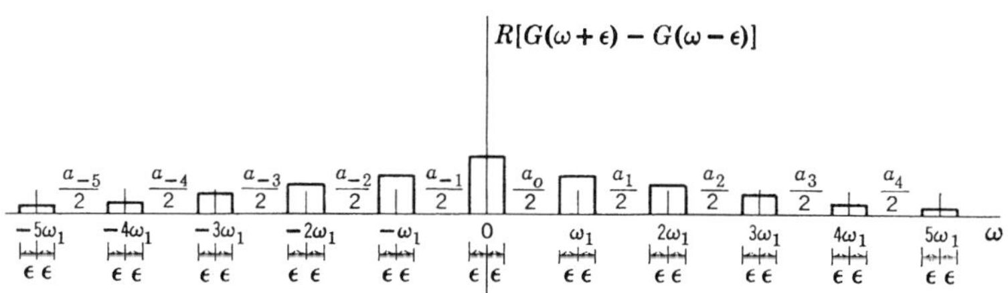

Fig. 49. Real part of $G(\omega + \epsilon) - G(\omega - \epsilon)$ in the quadratic variation for a periodic function.

Fig. 50. Imaginary part of $G(\omega + \epsilon) - G(\omega - \epsilon)$ in the quadratic variation for a periodic function.

From these figures, it is clear that the quadratic variation is

$$\lim_{\epsilon \to 0} \frac{1}{2\epsilon} \int_{-\infty}^{\infty} |G(\omega + \epsilon) - G(\omega - \epsilon)|^2 \, d\omega$$

$$= \lim_{\epsilon \to 0} \frac{1}{2\epsilon} \int_{-\infty}^{\infty} [\{R[G(\omega + \epsilon) - G(\omega - \epsilon)]\}^2$$

$$+ \{I[G(\omega + \epsilon) + G(\omega - \epsilon)]\}^2] d\omega$$

$$= \lim_{\epsilon \to 0} \frac{1}{2\epsilon} \frac{1}{4} [a_0^2 + (a_1^2 + b_1^2) + (a_2^2 + b_2^2) + (a_3^2 + b_3^2) + \cdots$$

$$+ (a_1^2 + b_1^2) + (a_2^2 + b_2^2) + (a_3^2 + b_3^2) + \cdots] 2\epsilon$$

$$= \frac{a_0^2}{4} + \frac{1}{2} \sum_{n=1}^{\infty} (a_n^2 + b_n^2) \tag{338}$$

This is the mean square value [see (51)] of the periodic function, as it should be according to (336). It is clear from this illustration that the quadratic variation is a device by which the sum of the squares of the absolute values of the jumps of the integrated Fourier transform of a function is taken. We shall pay special attention to this important property of the quadratic variation because it reveals the behavior of the Fourier transforms and integrated Fourier transforms of the various types of functions with which we are concerned.

6. Quadratic Variation for a Transient Function

When the quadratic variation is applied to a transient function $f(t)$, we note that $\int_{-\infty}^{\infty} |f(t)| \, dt$ is finite and that the Fourier transform $F(\omega)$ of the function is the derivative of the integrated Fourier transform $G(\omega)$ as pointed out in (323). With this in mind, we write the quadratic variation in the following form:

$$\lim_{\epsilon \to 0} 2\epsilon \int_{-\infty}^{\infty} \frac{|G(\omega + \epsilon) - G(\omega - \epsilon)|^2}{4\epsilon^2} \, d\omega \tag{339}$$

Since $F(\omega)$ is the derivative of $G(\omega)$ and $F(\omega)$ exists for all values of ω, we can write

$$\lim_{\epsilon \to 0} \frac{|G(\omega + \epsilon) - G(\omega - \epsilon)|^2}{4\epsilon^2} = |F(\omega)|^2 \tag{340}$$

The quadratic variation becomes

$$\lim_{\epsilon \to 0} 2\epsilon \int_{-\infty}^{\infty} |F(\omega)|^2 \, d\omega \tag{341}$$

which has the alternative form

$$\lim_{\epsilon \to 0} \frac{\epsilon}{\pi} \int_{-\infty}^{\infty} f^2(t)\, dt \tag{342}$$

as a consequence of the Parseval theorem (147). Clearly, this quantity vanishes because $\int_{-\infty}^{\infty} f^2(t)\, dt$ is finite. This result is in agreement with Wiener's theorem of quadratic variation, for, the mean square value of the transient function is zero; that is, in (336) the right-hand member is

$$\lim_{T \to \infty} \frac{1}{2T} \int_{-T}^{T} f^2(t)\, dt = 0 \tag{343}$$

7. Quadratic Variation for Some Special Functions

It is well known that there are certain functions for which $\int_{-\infty}^{\infty} |f(t)|\, dt$ diverges, and yet they have Fourier transforms. Such transforms are obtained through a limiting process and are primarily associated with transient analysis in physical systems. An important function of this class is the step function defined as

$$f(t) = \begin{cases} E & \text{for } 0 \leq t < \infty \\ 0 & \text{elsewhere} \end{cases} \tag{344}$$

The graph is in Fig. 51. The Fourier transform of this function is obtainable in the limit as

$$\begin{aligned} F(\omega) &= \lim_{\epsilon \to 0} \frac{1}{2\pi} \int_{0}^{\infty} E e^{-\epsilon t} e^{-j\omega t}\, dt \\ &= \lim_{\epsilon \to 0} \frac{E}{2\pi(\epsilon + j\omega)} \\ &= \frac{E}{2\pi j\omega} \end{aligned} \tag{345}$$

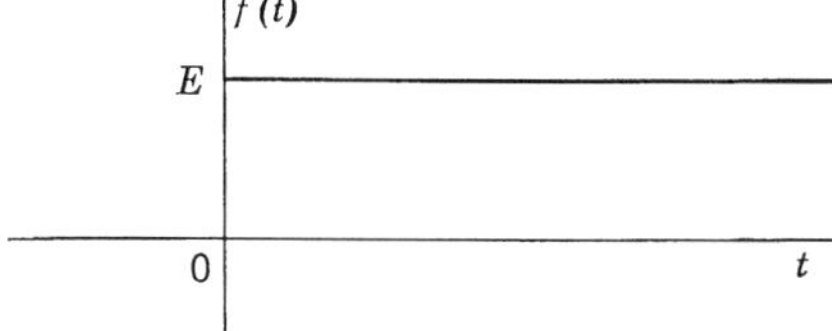

Fig. 51. Step function.

From this we find readily that the integrated Fourier transform is, with the provision given in connection with (323),

$$G(\omega) = \frac{E}{2\pi j} \int \frac{d\omega}{\omega}$$

$$= \frac{E}{2\pi j} \log \omega \tag{346}$$

The quadratic variation of $G(\omega)$ is accordingly

$$\lim_{\epsilon \to 0} \frac{1}{2\epsilon} \int_{-\infty}^{\infty} \left| \frac{E}{2\pi j} \log \frac{\omega + \epsilon}{\omega - \epsilon} \right|^2 d\omega \tag{347}$$

which reduces to

$$\frac{E^2}{8\pi^2} \int_{-\infty}^{\infty} \left| \log \frac{x+1}{x-1} \right|^2 dx \tag{348}$$

when we put $\omega = \epsilon x$.

For the evaluation of (348), let us examine the behavior of log $[(x + 1)/(x - 1)]$. We have

$$\log \frac{x+1}{x-1} = \begin{cases} \log \dfrac{x+1}{x-1} & \text{for } |x| > 1 \\ \log \dfrac{1+x}{1-x} + j\pi & \text{for } |x| < 1 \end{cases} \tag{349}$$

so that

$$\left| \log \frac{x+1}{x-1} \right|^2 = \begin{cases} \left[\log \dfrac{x+1}{x-1} \right]^2 & \text{for } |x| > 1 \\ \left[\log \dfrac{1+x}{1-x} \right]^2 + \pi^2 & \text{for } |x| < 1 \end{cases} \tag{350}$$

Now, integral (348) is

$$\frac{E^2}{8\pi^2} \int_{-\infty}^{\infty} \left| \log \frac{x+1}{x-1} \right|^2 dx$$

$$= \frac{E^2}{8\pi^2} \left\{ 2\int_1^{\infty} \left[\log \frac{x+1}{x-1} \right]^2 dx + 2\int_0^1 \left[\log \frac{1+x}{1 \quad x} \right]^2 dx + \int_{-1}^1 \pi^2 \, dx \right\} \tag{351}$$

With the change of variable $u = (x + 1)/(x - 1)$ for the first integral on the right and $u = (1 + x)/(1 - x)$ for the second integral, (351) becomes

$$\frac{E^2}{8\pi^2}\int_{-\infty}^{\infty}\left|\log\frac{x+1}{x-1}\right|^2 dx$$

$$= \frac{E^2}{4\pi^2}\left\{2\int_1^{\infty}(\log u)^2\frac{du}{(u-1)^2} + 2\int_1^{\infty}(\log u)^2\frac{du}{(u+1)^2} + \pi^2\right\}$$

$$= \frac{E^2}{4\pi^2}\left\{4\int_1^{\infty}(\log u)^2\frac{u^2+1}{(u^2-1)^2}\,du + \pi^2\right\} \tag{352}$$

If we put $v = \log u$, then

$$\frac{1}{2}\int_{-\infty}^{\infty}\left|\log\frac{x+1}{x-1}\right|^2 dx = 4\int_0^{\infty} v^2\,\frac{e^{2v}+1}{(e^{2v}-1)^2}\,e^v\,dv + \pi^2$$

$$= 2\int_0^{\infty} v^2 \coth v \operatorname{csch} v\,dv + \pi^2$$

$$= 2\left[v^2 \operatorname{csch} v\Big|_0^{\infty} + 2\int_0^{\infty} v \operatorname{csch} v\,dv\right] + \pi^2$$

$$= 8\int_0^{\infty} v(e^{-v} + e^{-3v} + e^{-5v} + \cdots)\,dv + \pi^2$$

$$= 8\left[1 + \frac{1}{3^2} + \frac{1}{5^2} + \cdots\right] + \pi^2$$

$$= 2\pi^2 \tag{353}$$

Returning to (348) and then (347), we find that (347) has the value $E^2/2$. Thus we have the result that the quadratic variation for the step function of amplitude E is $E^2/2$. According to the Wiener theorem (336), this should be equal to the mean square value of the function; and indeed we find that

$$\lim_{T\to\infty}\frac{1}{2T}\int_{-T}^{T} f^2(t)\,dt = \lim_{T\to\infty}\frac{1}{2T}(E^2T)$$

$$= \frac{E^2}{2} \tag{354}$$

Since the quadratic variation of an integrated Fourier transform which is differentiable everywhere is zero, as shown in Sec. D-6, the only contribution toward the quadratic variation for the step function comes from the logarithmic infinity of the integrated Fourier transform at $\omega = 0$. In other words, the mean power of the step function is entirely d-c power with no components at other frequencies. The logarithmic discontinuity of $G(\omega)$ at $\omega = 0$, as far as quadratic variation is concerned, may be regarded as a jump at $\omega = 0$ similar to the jumps of the integrated Fourier transforms of a periodic function.

Although both aperiodic functions with zero mean square values and some special functions with finite mean square values have Fourier transforms, there is an important difference between them. As we have seen, the Fourier transform of a function whose mean square value is zero exists for all values of the argument, and the quadratic variation vanishes. On the other hand, a function with a finite mean square value cannot have a Fourier transform which exists everywhere, because the quadratic variation is finite and must be equal to the mean square value of the function. It is only at those points where a Fourier transform tends to infinity that the quadratic variation can become finite. For the step function, the Fourier transform tends to infinity at $\omega = 0$. A similar example is the sinusoidal section shown in Fig. 52. The function is

$$f(t) = \begin{cases} E_m \sin \omega_1 t & \text{for } t \geq 0 \\ 0 & \text{for } t < 0 \end{cases} \tag{355}$$

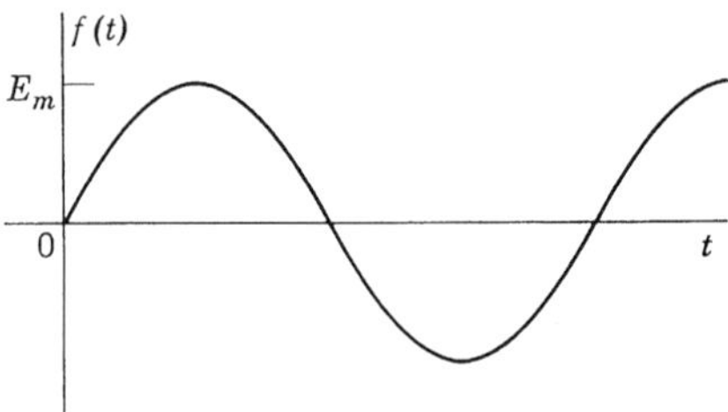

Fig. 52. Sinusoid section.

It has a finite mean square value and the Fourier transform

$$F(\omega) = \frac{E_m}{2\pi} \frac{\omega_1}{(\omega_1{}^2 - \omega^2)} \tag{356}$$

This transform, obtained by a limiting process as in the case of the step function, tends to infinity at $\omega = \pm\omega_1$.

It is interesting to study these special functions from the point of view of correlation. With the aid of Fig. 53 and in accordance with definition (203), the autocorrelation of the step function is

$$\varphi_{ff}(\tau) = \lim_{T\to\infty} \frac{1}{2T} \int_0^T E^2 \, dt = \frac{E^2}{2} \tag{357}$$

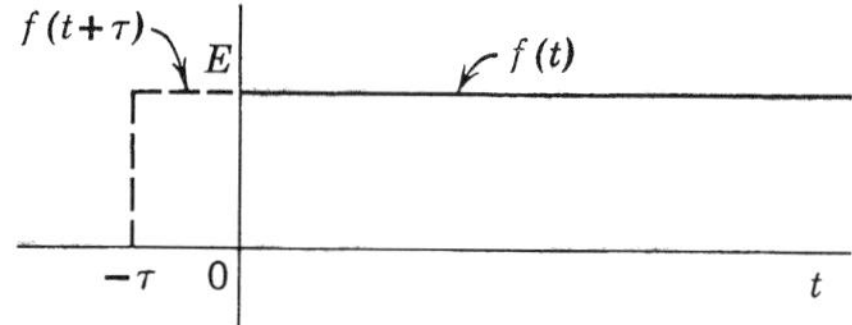

Fig. 53. Pertaining to the autocorrelation of a step function.

Obviously, this is the same as the autocorrelation of a constant of value $E/\sqrt{2}$ in the time range $(-\infty, \infty)$, and the autocorrelation theorem for periodic functions applies, under the special circumstance that the function is a constant. Accordingly, the power spectrum is simply a line at $\omega = 0$, or the power density spectrum is an impulse at $\omega = 0$ with its area equal to $E^2/2$. The autocorrelation and power density spectrum for the step function are given in Figs. 54 and 55 respectively. It is im-

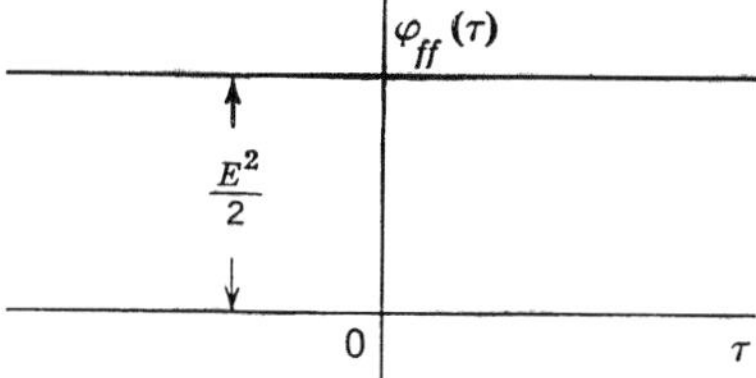

Fig. 54 Autocorrelation function of the step function of Fig. 51.

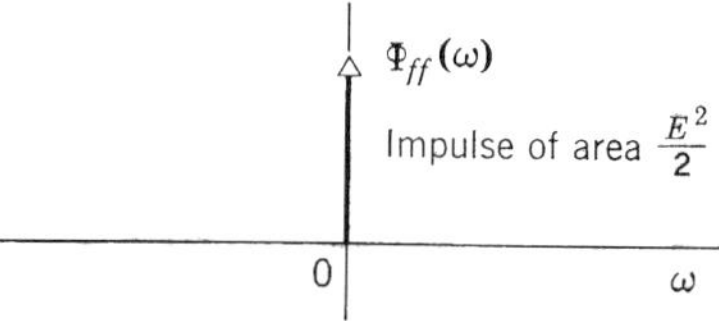

Fig. 55. Power density spectrum of the step function of Fig. 51.

portant to point out that the power density spectrum (Fig. 55) of the step function is not related to its complex (relative amplitude and phase) spectrum (345) in a simple manner as the energy density spectrum of an aperiodic function is to its complex spectrum given by (146). The step function has only d-c power so that the complex spectrum excluding the discontinuity at $\omega = 0$ has nothing to contribute to the power density spectrum. This portion of the complex spectrum corresponds to a finite amount of energy which cannot maintain a finite amount of power on the basis of the infinite interval.

Since the example given has brought out the main points of interest, we shall not continue the analysis of those functions whose mean square values are finite and whose Fourier transforms exist almost everywhere.

8. Quadratic Variation for Continuing Functions

One of the important facts concerning a continuing function brought out by the quadratic variation is that the Fourier transform of the function does not exist anywhere. In showing this, it is assumed that the continuing function contains no periodic components inasmuch as periodic functions have been treated separately. It is further assumed that functions with a discrete power density spectrum, such as the special functions discussed in the preceding section, are absent. The continuing function under consideration is, therefore, assumed to have a continuous power density spectrum. Let us write the Wiener theorem of quadratic variation (336) in the form

$$\lim_{\epsilon\to 0} 2\epsilon \int_{-\infty}^{\infty} \frac{|G(\omega+\epsilon) - G(\omega-\epsilon)|^2}{4\epsilon^2}\, d\omega = \lim_{T\to\infty} \frac{1}{2T}\int_{-T}^{T} f^2(t)\, dt \qquad (358)$$

If the derivative of $G(\omega)$ exists for all values of ω, then $f(t)$ must be a transient function with a zero mean square value, as we have shown, and cannot be the continuing function whose mean square value is finite. On the other hand, if the derivative of $G(\omega)$ becomes infinite for some discrete values of ω, then $f(t)$ is a function with a discrete power density spectrum. But the power density spectrum of $f(t)$ is continuous by assumption. It is therefore necessary that $G(\omega)$ *does not have a derivative at any value of* ω. Thus we have the important and interesting conclusion that, *if a continuing function has a continuous power density spectrum, it has an integrated Fourier transform but not a Fourier transform.*

Inasmuch as we have shown that the Fourier transform of a continuing function with a continuous power density spectrum does not exist although the integrated Fourier transform does, it is impossible to express the function as

$$f(t) = \frac{1}{2\pi}\int_{-\infty}^{\infty} [F(\omega)\, d\omega] e^{j\omega t} \qquad (359)$$

in a manner similar to the synthesis of an aperiodic function. Nevertheless, we may write the inverse of the integrated Fourier transform (315) formally as

$$f(t) = \frac{1}{2\pi}\int_{-\infty}^{\infty} [dG(\omega)] e^{j\omega t} \qquad (360)$$

In other words, the continuing function does not have a relative complex spectrum similar to that of the aperiodic function, but it has an *integrated complex spectrum* giving the relative integrated amplitudes and phases of the infinitesimal sinusoidal components from $\omega = -\infty$ to a given value of ω.

9. Relation between Integrated Fourier Transform and Autocorrelation

The discussions and illustrations of the quadratic variation have made clear that it is capable of extracting from the integrated Fourier transform of a function the power components it contains. The function under consideration need not be confined to a particular type; it may have components of the various types we have discussed. In view of the fact that the autocorrelation function, like the quadratic variation, yields similar information regarding power, a certain relationship should exist between them. To find this relationship, let us note that, when autocorrelation is interpreted as the synthesis of power components, we write

$$\varphi_{ff}(\tau) = \int_{-\infty}^{\infty} [\Phi_{ff}(\omega)\, d\omega] \cos \omega\tau \tag{361}$$

for a function $f(t)$ with the continuous power density spectrum $\Phi_{ff}(\omega)$, and a similar equation for a function with a discrete power spectrum. It is clear that, inasmuch as

$$\lim_{T\to\infty} \frac{1}{2T} \int_{-T}^{T} f^2(t)\, dt = \int_{-\infty}^{\infty} \Phi_{ff}(\omega)\, d\omega \tag{362}$$

when $f(t)$ has a continuous power density spectrum, the theorem of quadratic variation may be expressed in terms of $\Phi_{ff}(\omega)$. Thus

$$\lim_{\epsilon\to 0} \frac{1}{2\epsilon} \int_{-\infty}^{\infty} |G(\omega + \epsilon) - G(\omega - \epsilon)|^2\, d\omega = \int_{-\infty}^{\infty} \Phi_{ff}(\omega)\, d\omega \tag{363}$$

Again, there is a similar equation for the discrete power spectrum. Now, inspection of (361) and (363) shows that the relation between autocorrelation and the integrated Fourier transform is

$$\varphi_{ff}(\tau) = \lim_{\epsilon\to 0} \frac{1}{2\epsilon} \int_{-\infty}^{\infty} |G(\omega + \epsilon) - G(\omega - \epsilon)|^2 \cos \omega\tau\, d\omega \tag{364}$$

Although the derivation of this result is heuristic, it can be made rigorous. This is *Wiener's quadratic average for autocorrelation.* It reduces to the quadratic variation if we put $\tau = 0$. In the generalized harmonic analysis of Wiener, this theorem is of considerable importance.

10. Proof of the Wiener Theorem for Autocorrelation

To show (364) rigorously, we start with the definition of $\varphi_{ff}(\tau)$,

$$\varphi_{ff}(\tau) = \lim_{T\to\infty} \frac{1}{2T} \int_{-T}^{T} f(t)f(t + \tau)\, dt \tag{365}$$

In order that we may apply the Wiener theorem of quadratic variation (336), the definition is written in the form

$$\varphi_{ff}(\tau) = \frac{1}{4} \lim_{T\to\infty} \frac{1}{2T} \int_{-T}^{T} \{[f(t) + f(t+\tau)]^2 - [f(t) - f(t+\tau)]^2\} \, dt \quad (366)$$

By this theorem we have

$$\lim_{T\to\infty} \frac{1}{2T} \int_{-T}^{T} [f(t) + f(t+\tau)]^2 \, dt = \lim_{\epsilon\to 0} \frac{1}{2\epsilon} \int_{-\infty}^{\infty} |G(\omega+\epsilon) + G_\tau(\omega+\epsilon) - G(\omega-\epsilon) - G_\tau(\omega-\epsilon)|^2 \, d\omega \quad (367)$$

In this equation $G_\tau(\omega)$ is the integrated Fourier transform of $f(t+\tau)$, that is,

$$G_\tau(\omega) = \frac{1}{2\pi} \int_{-\infty}^{\infty} f(t+\tau) \frac{e^{-j\omega t}}{-jt} \, dt \quad (368)$$

Writing (367) in a form that facilitates simplification, we have

$$\begin{aligned} \lim_{T\to\infty} \frac{1}{2T} \int_{-T}^{T} & [f(t) + f(t+\tau)]^2 \, dt \\ &= \lim_{\epsilon\to 0} \frac{1}{2\epsilon} \int_{-\infty}^{\infty} |G(\omega+\epsilon) - G(\omega-\epsilon) + G_\tau(\omega+\epsilon) - G_\tau(\omega-\epsilon) \\ &\qquad - e^{j\omega\tau}[G(\omega+\epsilon) - G(\omega-\epsilon)] + e^{j\omega\tau}[G(\omega+\epsilon) - G(\omega-\epsilon)]|^2 \, d\omega \\ &= \lim_{\epsilon\to 0} \frac{1}{2\epsilon} \int_{-\infty}^{\infty} |G_\tau(\omega+\epsilon) - G_\tau(\omega-\epsilon) - e^{j\omega\tau}[G(\omega+\epsilon) - G(\omega-\epsilon)] \\ &\qquad + (1 + e^{j\omega\tau})[G(\omega+\epsilon) - G(\omega-\epsilon)]|^2 \, d\omega \end{aligned} \quad (369)$$

The simplification depends upon the fact that the value of the integral

$$\int_{-\infty}^{\infty} |G_\tau(\omega+\epsilon) - G_\tau(\omega-\epsilon) - e^{j\omega\tau}[G(\omega+\epsilon) - G(\omega-\epsilon)]|^2 \, d\omega \quad (370)$$

is of the order of ϵ^2. Concerning this point let us note that by (329)

$$\begin{aligned} G_\tau(\omega+\epsilon) - G_\tau(\omega-\epsilon) &= \frac{1}{\pi} \int_{-\infty}^{\infty} f(t+\tau) \frac{\sin \epsilon t}{t} e^{-j\omega t} \, dt \\ &= \frac{1}{\pi} \int_{-\infty}^{\infty} f(t) \frac{\sin \epsilon(t-\tau)}{t-\tau} e^{-j\omega(t-\tau)} \, dt \end{aligned} \quad (371)$$

and

$$e^{j\omega\tau}[G(\omega + \epsilon) - G(\omega - \epsilon)] = \frac{1}{\pi}\int_{-\infty}^{\infty} f(t)\,\frac{\sin \epsilon t}{t}\,e^{-j\omega(t-\tau)}\,dt \tag{372}$$

so that

$$G_\tau(\omega + \epsilon) - G_\tau(\omega - \epsilon) - e^{j\omega\tau}[G(\omega + \epsilon) - G(\omega - \epsilon)]$$
$$= \frac{1}{\pi}\int_{-\infty}^{\infty} f(t)\left[\frac{\sin \epsilon(t-\tau)}{t-\tau} - \frac{\sin \epsilon t}{t}\right] e^{-j\omega(t-\tau)}\,dt \tag{373}$$

By an application of the Parseval theorem (147), we have

$$\frac{1}{2}\int_{-\infty}^{\infty} |G_\tau(\omega + \epsilon) - G_\tau(\omega - \epsilon) - e^{j\omega\tau}[G(\omega + \epsilon) - G(\omega - \epsilon)]|^2\,d\omega$$
$$= \frac{1}{\pi}\int_{-\infty}^{\infty} f^2(t)\left[\frac{\sin \epsilon(t-\tau)}{t-\tau} - \frac{\sin \epsilon t}{t}\right]^2 dt \tag{374}$$

Wiener has shown * that the value of the integrals of (374) is of the order of ϵ^2.

At this point we need a form of the Minkowski inequality † which states that if $\int_a^b \alpha^2(x)\,dx$ and $\int_a^b \beta^2(x)\,dx$ are finite, then

$$\left[\int_a^b |\alpha(x) + \beta(x)|^2\,dx\right]^{1/2} \leq \left[\int_a^b \alpha^2(x)\,dx\right]^{1/2} + \left[\int_a^b \beta^2(x)\,dx\right]^{1/2} \tag{375}$$

Expressing (369) in accordance with (375), we obtain

$$\left\{\lim_{T\to\infty}\frac{1}{2T}\int_{-T}^{T}[f(t) + f(t+\tau)]^2\,dt\right\}^{1/2}$$
$$\leq \left\{\lim_{\epsilon\to 0}\frac{1}{2\epsilon}\int_{-\infty}^{\infty} |G_\tau(\omega + \epsilon) - G_\tau(\omega - \epsilon) - e^{j\omega\tau}[G(\omega + \epsilon) - G(\omega - \epsilon)]|^2\,d\omega\right\}^{1/2}$$
$$+ \left\{\lim_{\epsilon\to 0}\frac{1}{2\epsilon}\int_{-\infty}^{\infty} |(1 + e^{j\omega\tau})[G(\omega + \epsilon) - G(\omega - \epsilon)]|^2\,d\omega\right\}^{1/2} \tag{376}$$

* N. Wiener, *The Fourier Integral and Certain of Its Applications*, Dover Publications, New York, 1933, pp. 156–158.

† *Ibid.* p. 21

Since the integral in the first right-hand term is (370), which has a value of the order of ϵ^2, this term is of the order of $\sqrt{\epsilon}$ with ϵ tending to zero. Consequently (376) reduces to

$$\lim_{T\to\infty} \frac{1}{2T} \int_{-T}^{T} [f(t) + f(t+\tau)]^2 \, dt$$
$$= \lim_{\epsilon\to 0} \frac{1}{2\epsilon} \int_{-\infty}^{\infty} (2 + e^{j\omega\tau} + e^{-j\omega\tau}) \,|\, G(\omega+\epsilon) - G(\omega-\epsilon)\,|^2 \, d\omega \quad (377)$$

In a similar manner we find

$$\lim_{T\to\infty} \frac{1}{2T} \int_{-T}^{T} [f(t) - f(t+\tau)]^2 \, dt$$
$$= \lim_{\epsilon\to 0} \frac{1}{2\epsilon} \int_{-\infty}^{\infty} (2 - e^{j\omega\tau} - e^{-j\omega\tau}) \,|\, G(\omega+\epsilon) - G(\omega-\epsilon)\,|^2 \, d\omega \quad (378)$$

Finally, combining (377) and (378) in accordance with (366), we obtain the result

$$\varphi_{ff}(\tau) = \lim_{\epsilon\to 0} \frac{1}{2\epsilon} \int_{-\infty}^{\infty} |\, G(\omega+\epsilon) - G(\omega-\epsilon)\,|^2 \cos \omega\tau \, d\omega \quad (379)$$

This is the relation (364) that we wished to prove.

Having shown the physical significance of the quadratic variation (336) in the extraction of power components from the various types of functions, and having proved the relation (379), we can say that (379) states that the autocorrelation function of a continuing function $f(t)$ is the cosine transform of the power density spectrum of the continuing function. The power density spectrum $\Phi_{ff}(\omega)$ in terms of the integrated Fourier transform $G(\omega)$ of $f(t)$ is, in accordance with (336) or (379),

$$\Phi_{ff}(\omega) = \lim_{\epsilon\to 0} \frac{1}{2\epsilon} \,|\, G(\omega+\epsilon) - G(\omega-\epsilon)\,|^2 \quad (380)$$

so that (379) reads

$$\varphi_{ff}(\tau) = \int_{-\infty}^{\infty} \Phi_{ff}(\omega) \cos \omega\tau \, d\omega \quad (381)$$

By inverse transformation we obtain

$$\Phi_{ff}(\omega) = \frac{1}{2\pi} \int_{-\infty}^{\infty} \varphi_{ff}(\tau) \cos \omega\tau \, d\tau \quad (382)$$

This is a proof of the Wiener theorem for autocorrelation, which we heuristically derived in Sec. C-3.

PROBLEMS

1. Figure P1 shows a rectified sinusoid $f_1(t)$ of period π and maximum amplitude 10. Determine and plot on graph paper the autocorrelation function.

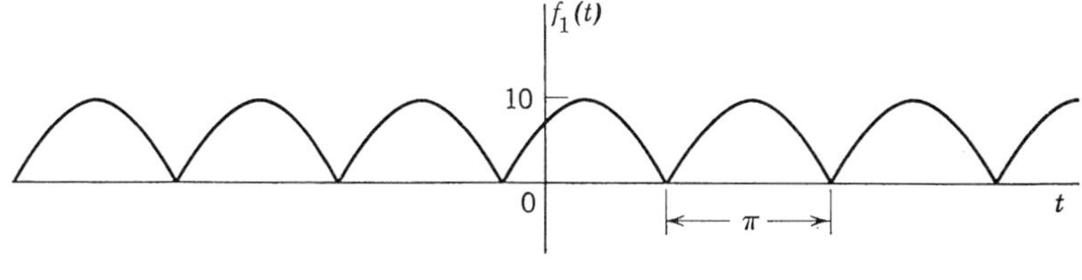

Fig. P1.

2. The periodic wave shown in Fig. P2 is formed by inverting every other cycle of a sinusoid whose maximum amplitude is 1 and angular frequency is 1 radian per second. Determine and plot on graph paper the autocorrelation function.

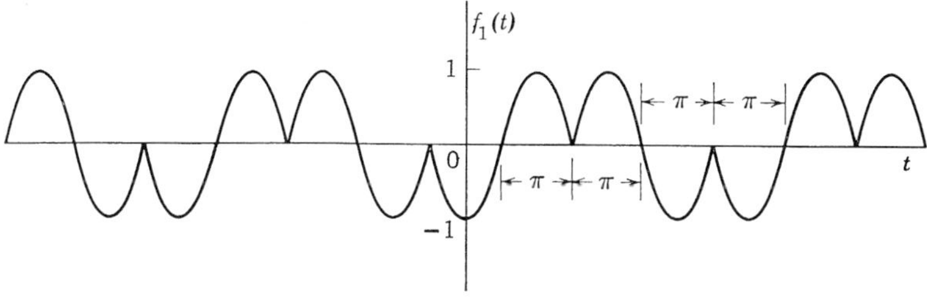

Fig. P2.

3. A periodic function is formed by eliminating every other cycle of a sinusoidal wave with the maximum amplitude 10 and angular frequency 1 radian per second as shown in Fig. P3.

(*a*) Find the equation for the waveform of the autocorrelation function.

(*b*) Plot the autocorrelation function on graph paper.

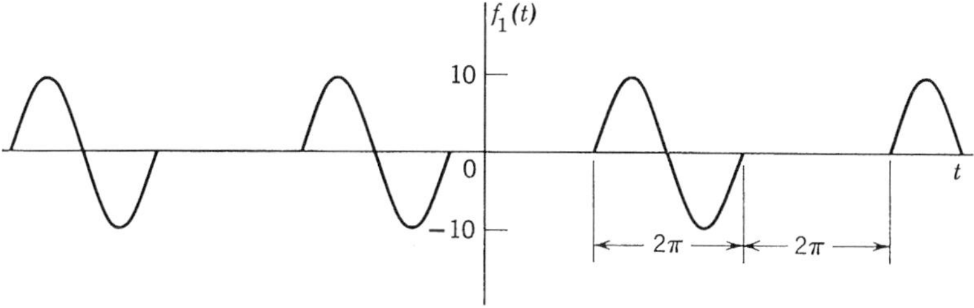

Fig. P3.

4. Determine and plot on graph paper the autocorrelation function of the periodic function shown in Fig. P4.

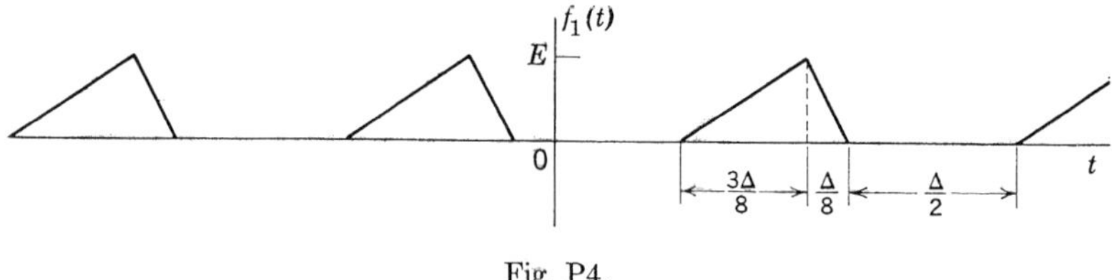

Fig. P4.

5. Determine and plot on graph paper the autocorrelation function of the periodic function shown in Fig. P5 which has the period T_1, maximum value $2E$, and minimum value $-E$. Other data are given in the figure.

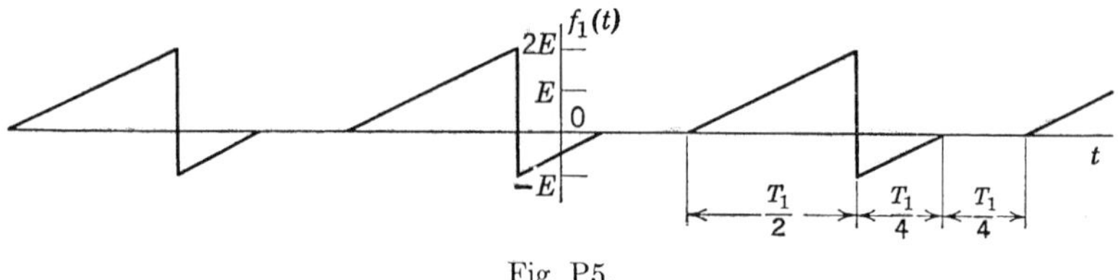

Fig. P5.

6. Determine and plot on graph paper the crosscorrelation function $\varphi_{12}(\tau)$ of the periodic functions $f_1(t)$ and $f_2(t)$ as shown in Fig. P6 for $\delta = \Delta/4$.

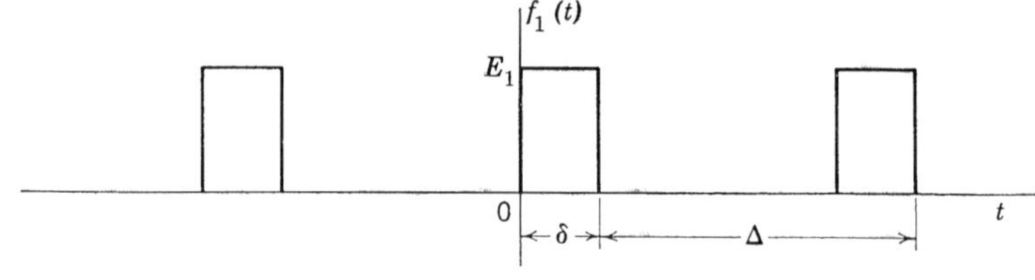

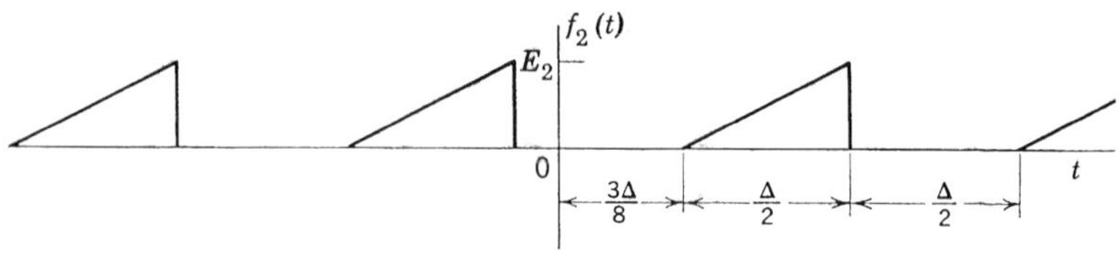

Fig. P6.

7. Determine and plot on graph paper the autocorrelation function of the periodic function shown in Fig. P7.

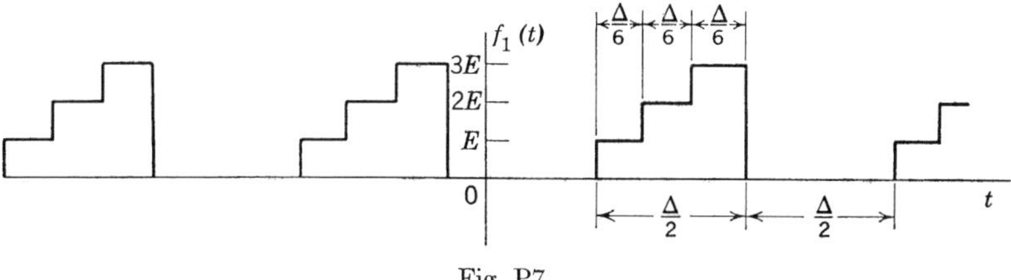

Fig. P7.

8. Determine and plot on graph paper the crosscorrelation function $\varphi_{12}(\tau)$ of the periodic functions $f_1(t)$ and $f_2(t)$ as shown in Fig. P8.

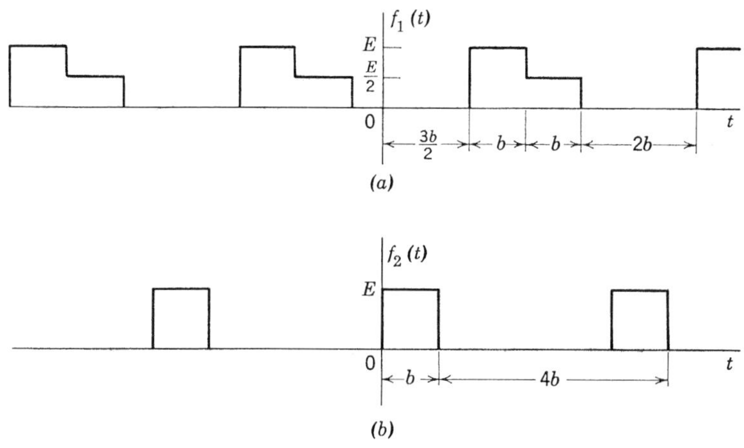

Fig. P8.

9.* (a) Determine the autocorrelation function of the periodic function $f_1(t)$ shown in Fig. P9 which has the period T_1, maximum amplitude E, and minimum amplitude $-E$. The expressions should be in terms of T_1 and E and reduced to the simplest form. State the ranges in which the expressions hold.

(b) Plot the autocorrelation function on graph paper for $T_1 = 1$, and $E = 1$.

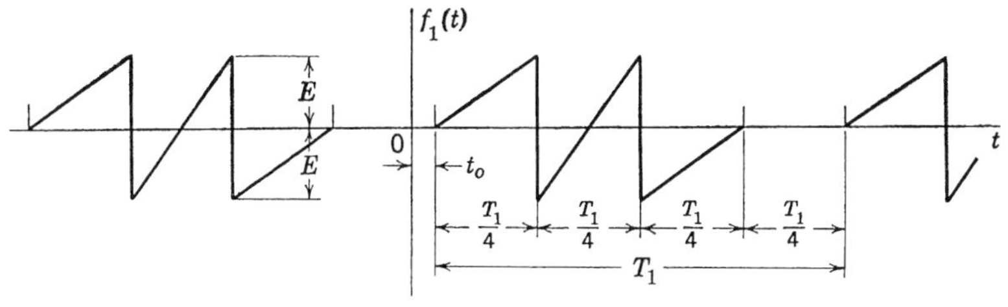

Fig. P9.

* The periodic function in this problem should be resolved into such components that would simplify the solution. Application of the autocorrelation formula by direct substitution that leads to complicated algebra should be avoided.

10.* (*a*) Determine the autocorrelation function of the periodic function $f_1(t)$ shown in Fig. P10 which has the period T_1, maximum amplitude E, and minimum amplitude $-E$. The expressions should be in terms of T_1 and E and reduced to the simplest form. State the ranges in which the expressions hold.

(*b*) Plot the autocorrelation function on graph paper for $T_1 = 1$, and $E = 1$.

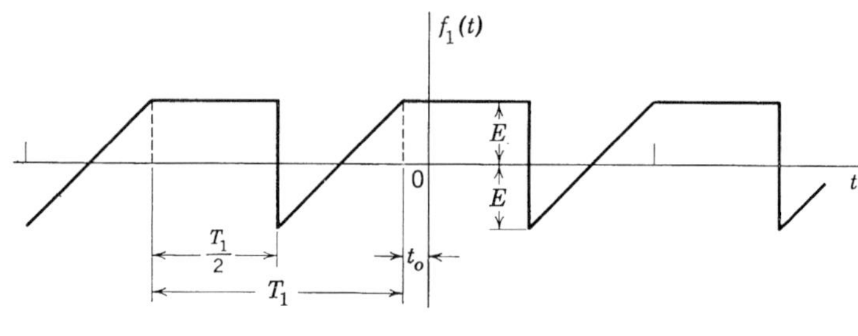

Fig. P10.

11.* (*a*) Determine the autocorrelation function of the periodic function $f_1(t)$ shown in Fig. P11. The expressions should be in terms of T_1 and E and reduced to the simplest form. State the ranges in which the expressions hold.

(*b*) Plot on graph paper the autocorrelation function for $T_1 = 1$, $E = 1$.

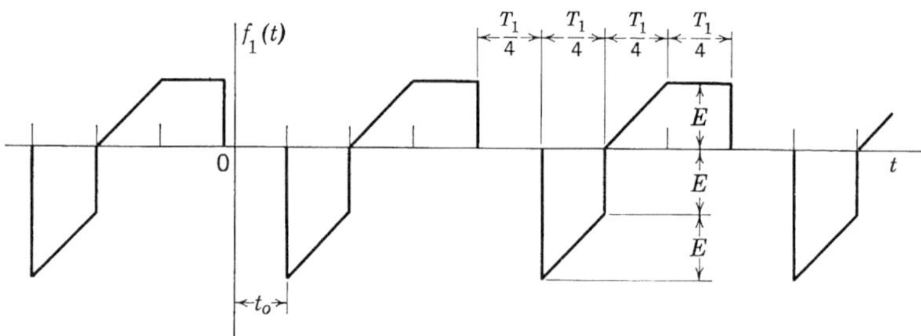

Fig. P11.

12.* (*a*) Determine the autocorrelation function of the periodic function $f_1(t)$ shown in Fig. P12 which has the period T_1, maximum amplitude E, and minimum amplitude $-E$. The expressions should be in terms of T_1 and E and reduced to the simplest form. State the ranges in which the expressions hold.

(*b*) Plot the autocorrelation function on graph paper for $T_1 = 1$, and $E = 1$.

* The periodic function in this problem should be resolved into such components that would simplify the solution. Application of the autocorrelation formula by direct substitution that leads to complicated algebra should be avoided.

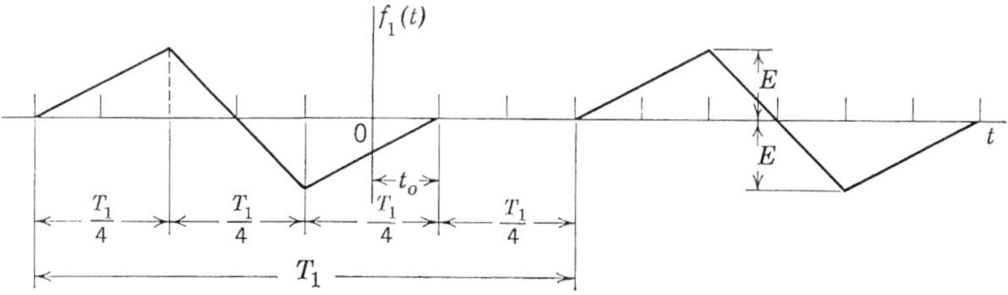

Fig. P12.

13.* (*a*) Determine the autocorrelation function of the periodic function $f_1(t)$ shown in Fig. P13. The expressions should be in terms of T_1 and E and reduced to the simplest form. State the ranges in which the expressions hold.

(*b*) Plot on graph paper the autocorrelation function for $T_1 = 1$, $E = 1$.

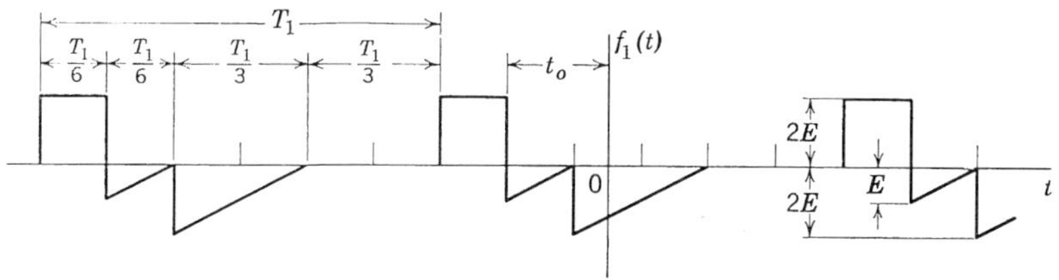

Fig. P13.

14.* (*a*) Determine the autocorrelation function of the periodic function $f_1(t)$ shown in Fig. P14. The expressions should be in terms of T_1 and E and reduced to the simplest form. State the ranges in which the expressions hold.

(*b*) Plot the autocorrelation function on graph paper for $T_1 = 1$, and $E = 1$.

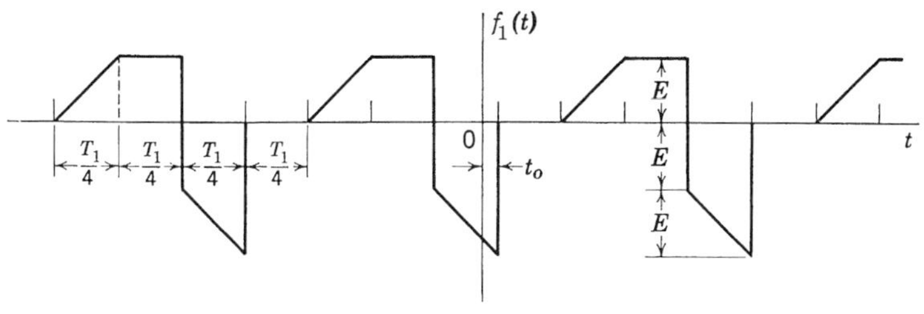

Fig. P14.

* The periodic function in this problem should be resolved into such components that would simplify the solution. Application of the autocorrelation formula by direct substitution that leads to complicated algebra should be avoided.

15. If $h(t)$ is the time response of a linear system to a unit-impulse excitation applied at $t = 0$, the output $f_o(t)$ of the system for an aperiodic input $f_i(t)$ is given by the convolution integral

$$f_o(t) = \int_{-\infty}^{\infty} h(\tau)f_i(t - \tau)\, d\tau$$

As shown in Fig. P15 $f_a(t)$ and $f_b(t)$ are aperiodic inputs of two linear systems whose unit-impulse responses are $h_a(t)$ and $h_b(t)$, and the corresponding outputs are $f_1(t)$ and $f_2(t)$. Express the crosscorrelation function $\varphi_{12}(\tau)$ of $f_1(t)$ and $f_2(t)$ in terms of $h_a(t)$, $h_b(t)$, and crosscorrelation function $\varphi_{ab}(\tau)$ of the inputs.

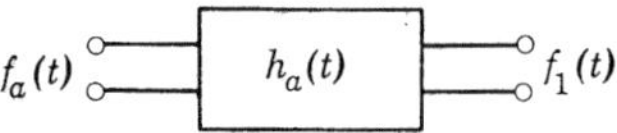

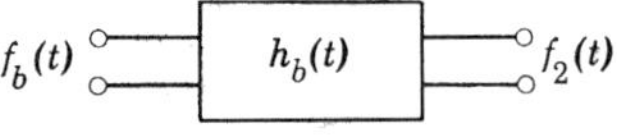

Fig. P15.

16. If $h(t)$ is the time response of a linear system to a unit-impulse excitation applied at $t = 0$, the output $f_o(t)$ of the system for an aperiodic input $f_i(t)$ is given by the convolution integral

$$f_o(t) = \int_{-\infty}^{\infty} h(x)f_i(t - x)\, dx$$

As shown in Fig. P16 $f_i(t)$ and $f_o(t)$ are the aperiodic input and output of a linear system consisting of two linear systems A and B in cascade. The unit-impulse responses of A and B are $h_a(t)$ and $h_b(t)$, and the output of A is $f_j(t)$. Express the crosscorrelation function $\varphi_{jo}(\tau)$ as a multiple integral involving $h_a(t)$, $h_b(t)$, and the crosscorrelation function $\varphi_{ij}(\tau)$.

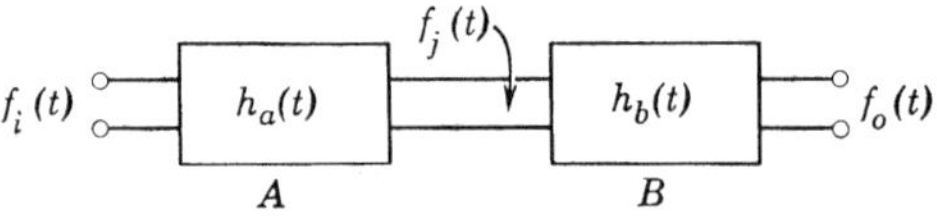

Fig. P16.

17. The function $f_1(t)$ shown in Fig. P17 is linear from $t = 2$ to $t = 3$ and zero elsewhere; and $f_2(t)$ has the expression $3e^t$ from $t = -\infty$ to $t = 0$ and is zero elsewhere. Determine and sketch the crosscorrelation function

$$\varphi_{12}(\tau) = \int_{-\infty}^{\infty} f_1(t)f_2(t + \tau)\, dt$$

and the convolution integral

$$\rho_{12}(\tau) = \int_{-\infty}^{\infty} f_1(t)f_2(\tau - t)\, dt$$

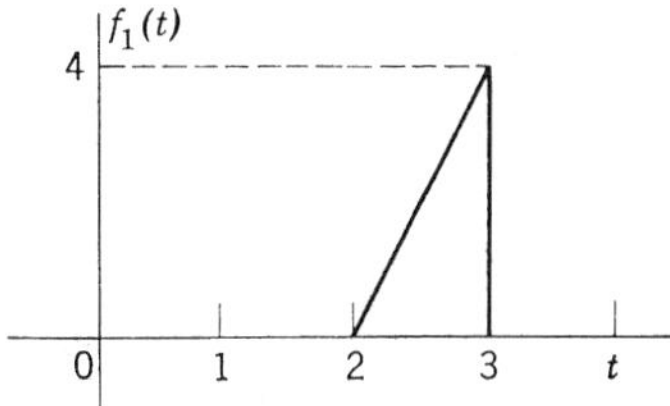

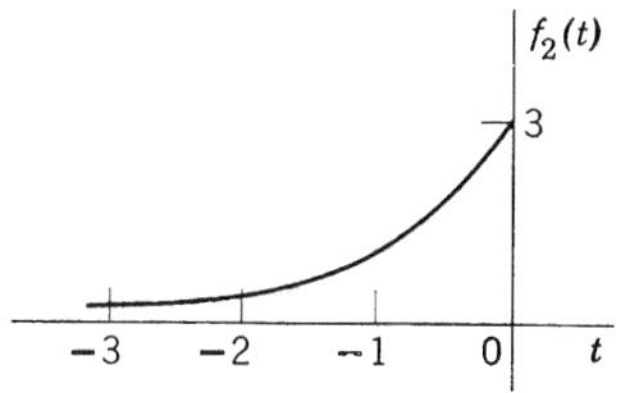

Fig. P17.

18. We have given the functions

$$f_1(t) = \begin{cases} -te^t & \text{for } t \leq 0 \\ 0 & \text{for } t > 0 \end{cases}$$

$$f_2(t) = \begin{cases} e^{-t} & \text{for } t \geq 1 \\ 0 & \text{for } t < 1 \end{cases}$$

Determine and sketch the crosscorrelation function

$$\varphi_{12}(\tau) = \int_{-\infty}^{\infty} f_1(t)f_2(t + \tau)\, dt$$

and the convolution integral

$$\rho_{12}(\tau) = \int_{-\infty}^{\infty} f_1(t)f_2(\tau - t)\, dt$$

19. We have given the functions

$$f_1(t) = \begin{cases} 3e^{2t} & \text{for } t \leq -1 \\ 0 & \text{for } t > -1 \end{cases}$$

$$f_2(t) = \begin{cases} 2e^{-t} & \text{for } t \geq 0 \\ 0 & \text{for } t < 0 \end{cases}$$

Determine and sketch the crosscorrelation function

$$\varphi_{12}(\tau) = \int_{-\infty}^{\infty} f_1(t)f_2(t + \tau)\, dt$$

and the convolution integral

$$\rho_{12}(\tau) = \int_{-\infty}^{\infty} f_1(t)f_2(\tau - t)\, dt$$

20. If $h(t)$ is the time response of a linear system to an excitation in the form of a unit impulse applied at time $t = 0$, then for an input $f_i(t)$ the output $f_o(t)$ is given by the convolution integral

$$f_o(t) = \int_{-\infty}^{\infty} h(\sigma)f_i(t - \sigma)\, d\sigma$$

(*a*) Express the crosscorrelation function $\varphi_{io}(\tau)$ between the input and output of the system in terms of $h(t)$ and the autocorrelation function of the input. Assume that $f_i(t)$ is aperiodic.

(b) We have given:

$$h(t) = \begin{cases} 10 & \text{for } 0 \leq t \leq 1 \\ 0 & \text{elsewhere} \end{cases}$$

$$f_i(t) = \begin{cases} 4e^{-t} & \text{for } 0 \leq t < \infty \\ 0 & \text{elsewhere} \end{cases}$$

Determine the crosscorrelation function $\varphi_{io}(\tau)$ making use of the expression found in part a.

21. We have given the autocorrelation function

$$\varphi_{11}(\tau) = \begin{cases} \dfrac{E^2}{12b^3}(b - |\tau|)^3 & \text{for } -b \leq \tau \leq b \\ 0 & \text{elsewhere} \end{cases}$$

for a random function, where E and b are constants. Show that the power density spectrum of the random function is

$$\Phi_{11}(\omega) = \frac{E^2 b}{4\pi(b\omega)^2}\left[1 - \left(\frac{\sin\frac{b\omega}{2}}{\frac{b\omega}{2}}\right)^2\right]$$

Find the value of $\Phi_{11}(\omega)$ at $\omega = 0$.

22. We have given the autocorrelation function

$$\varphi_{11}(\tau) = \frac{1}{4a^5} e^{-a|\tau|}(3 + 3a|\tau| + a^2\tau^2)$$

for a random function. Show that the power density spectrum of the random function is

$$\Phi_{11}(\omega) = \frac{2}{\pi}\frac{1}{(a^2 + \omega^2)^3}$$

23. By considering the properties of an energy density spectrum and of a power density spectrum show that the function

$$\varphi_{11}(\tau) = \begin{cases} 1 - \tau^2 & \text{for } -1 < \tau < 1 \\ 0 & \text{elsewhere} \end{cases}$$

is not a possible autocorrelation function of an aperiodic function or of a random function.

24. Show that

$$\varphi_{11}(\tau) = \frac{1}{a^4 + \tau^4}$$

where a is a positive constant, is not a realizable autocorrelation function of a random function or of an aperiodic function.

25. Determine whether

$$\varphi_{11}(\tau) = \frac{1}{(a^2 + \tau^2)^2}$$

where a is a positive constant, is a realizable autocorrelation function of an aperiodic function. If it is realizable, find an aperiodic function which has the given autocorrelation.

chapter 3

Probability

1. Introduction

There are many physical phenomena in which the observed quantities are uniquely related to a number of initial conditions and the forces causing the phenomena by simple laws. Examples of such phenomena are the output voltage or current in a linear system under the action of a periodic or an aperiodic voltage at the input and, in mechanics, the motion of a particle under the action of a force.

On the other hand, there is an important class of phenomena in which the observed quantities cannot be related to the initial conditions and the actuating forces by simple laws. Consider, for example, thermal noise in a circuit. If we attempt to relate the instantaneous values of the fluctuating noise voltage to the motions of the individual electrons, it would immediately become clear that we are faced with an impossible task. Even if we could achieve a solution of the problem by this method, its complexity would indeed be an obstacle in the analysis of systems where thermal noise is present. For handling noise effectively, new concepts are necessary.

In a dice game the prediction of the outcome of a throw, if it were possible, would require much detailed information which we do not possess. The physical properties of the dice and of the surface to which they are thrown and the initial conditions of motion of the dice are critical factors in the prediction. These factors are never fully known. Furthermore, if we assume that we have this information and are able to apply classical mechanics to the problem, it is easy to imagine how utterly complicated the solution would be. We must remember that since the phenomenon of chance is the essence of a dice game, we really do not seek the law for the exact outcome of the game. What we desire is a mathematical description of the game in which chance is a fundamental characteristic.

In mechanics, the analysis of the behavior of a gas in a container is a problem relevant to this discussion. If the gas is assumed to be a large number of minute particles moving under their mutual interaction, then theoretically, in accordance with the principles of dynamics, the course of any particle of the gas for the infinite future is determined once the initial position and velocity of every particle are known. But, such initial conditions are impossible to obtain. Even if they were available and the solution were possible, it is beyond imagination that this infinitely complex formulation of the problem serves a useful purpose. Among the concepts introduced, other than those of classical mechanics, for the analysis of a gas, leading to meaningful results, are those belonging to the theory of probability.

The formation of speech waves is another example of a highly complex process which is not governed by simple laws. In fact, since the purpose of speech is to transmit information, it must have the characteristic that its variation as time passes cannot be predicted exactly by the listener. The information which the listener is to receive is unknown to him before its transmission. It is made known to him through a sequence of words selected by the sender from a collection of all possible words which are known to both sender and receiver. The message, in the form of speech, is clearly under the command of the speaker, and the listener can only make observations and cannot predict exactly, at any moment, its future course. It is true that, at times, a word or two, or even a short sequence of words, are predictable because of the rules of grammar or the occurrence of common phrases and idioms. However, precise prediction is impossible. Clearly, if speech does not possess this characteristic and is determinable exactly and completely, then indeed it conveys no information. This situation is well illustrated by certain occasions when the speaker tells an old story or a stale joke.

All information-carrying functions (which we shall refer to as messages) in the form of fluctuating quantities, such as voltages, velocities, temperatures, positions, or pressures, which are to be processed, transmitted, and utilized for the attainment of an objective, necessarily possess the characteristic that they are not subject to precise prediction. In view of the foregoing discussion, thermal noise, shot noise, errors in measurement, and other similar forms of disturbances in the transmission of a message should be considered to have the same characteristic. These quantities vary with time, either continuously or at discrete points, and are referred to as *random time functions*.

Because messages and noise possess the characteristics of random phenomena for which adequate theories of statistics and probability are available, the analysis of communication and control problems should be developed as an application of these theories. Notwithstanding this basic fact, classical communication and control theories take the deterministic point of view and consider messages and noise to be either prescribed periodic functions or prescribed transients. Thus we have the familiar terminology that a problem is attacked by the "steady-state approach" or the "transient approach."

Augmented by the concepts and methods of statistics and probability, communication and control theories have gained, substantially, in power and scope. Inasmuch as in presenting generalized harmonic analysis in the last chapter the discussion on the concepts of the theory relating to statistics and probability have been brief and qualitative, it is now necessary that we consider these aspects in detail and quantitatively.

The theory of probability has had a long period of development, beginning with about the middle of the seventeenth century. Although it was first developed in association with games of chance, this mathematical theory has found its way into the heart of science and engineering. Modern statistical theory with probability theory as a basic tool has an expanding field of application. Notable developments have appeared in physics and in genetics. It is well known that the theory has been applied with success to congestion problems in telephone systems, quality control of manufactured products, and a great variety of social and biological problems. Its application to communication and control problems is a comparatively recent development.

2. Statistical Regularity

If, in a *random experiment*, some examples of which have just been given, there is no evidence of regularity of any sort, then indeed it is not subject to quantitative description and there cannot be any scientific analysis. But, it is a human experience that certain average values of a series of observations on repeated random experiments, which are carried out under essentially the same conditions, exhibit the regularity that they tend, in an irregular manner, to center about certain constant values as the number of repetitions becomes large. It is well known, for example, that in the random experiment of tossing a coin, the ratio of the number of heads (or, alternatively, the number of tails) to the total number of tosses tends to center around the value $\frac{1}{2}$ as the total number of tosses becomes greater and greater. The same sort of regularity is found in card games, dice games, and other random phenomena. In fact, it was because of such regularity in games of chance that the development of the theory of probability was initiated some 250 years ago. This fundamental physical characteristic of random experiments, which are performed under essentially uniform conditions, is known as *statistical regularity*. The repeated performances of a random experiment are called *trials*, and the trials which yield a desired result (such as heads) are called *successes*. If in a total number of M trials there are N successes, the ratio N/M is called the *frequency ratio*. It is clear that the practical application of the theory of probability to random phenomena requires the basic assumption of regularity of frequency ratios even if repeated experiments are not actually carried out.

The collection of observations made in a series of trials of a random experiment constitute the set of *statistical data* for the random experiment. Thus in the tossing of a coin, a series of trials may result in the following statistical data if H represents heads and T tails:

$$H, T, T, T, H, T, H, H, T, T, H, \ldots$$

Or, if we represent a head by 1 and a tail by 0, the sequence would be

$$1, 0, 0, 0, 1, 0, 1, 1, 0, 0, 1, \ldots$$

It is important to note that in this experiment and in many others, as, for example, the measurement of the dimensions of a manufactured machine part, the measurement of the weight and height of a group of people, or the observation of the price of a commodity at regular intervals of time, each observation is expressible as a number, or a group of numbers. In contrast with this class of random experiments, we find that experiments associated with a communication or control system generally require the variable time in the expression of an observation. In other words, the observation on a trial in an experiment of communication or control is a message or a noise which is either a continuous or a discrete function of time. It is quite possible that, for certain problems, only a finite number of values from an observation are sufficient; but, in general, the result of a trial is a continuous or discrete function of time. Consequently, in an experiment of communication or control the statistical data consist of a series of time functions such as those shown in Fig. 1. Such a collection of statistical data may be formed by repeating an experiment under essentially the same conditions with a single given system. For instance, a noise voltage from a given circuit recorded over successive intervals of time, each of considerable duration and under uniform conditions, would yield a set of data as shown in Fig. 1 if the record for each interval of time is regarded as the result of a separate trial. Under idealized conditions, this practical procedure for obtaining statistical data is equivalent to the simultaneous recording of the outputs of a series of noise circuits which are essentially the same and are operating under identical conditions. As in the game of coin tossing, a sequence of heads and tails obtained by one person tossing a given true coin repeatedly should exhibit the same sort of statistical

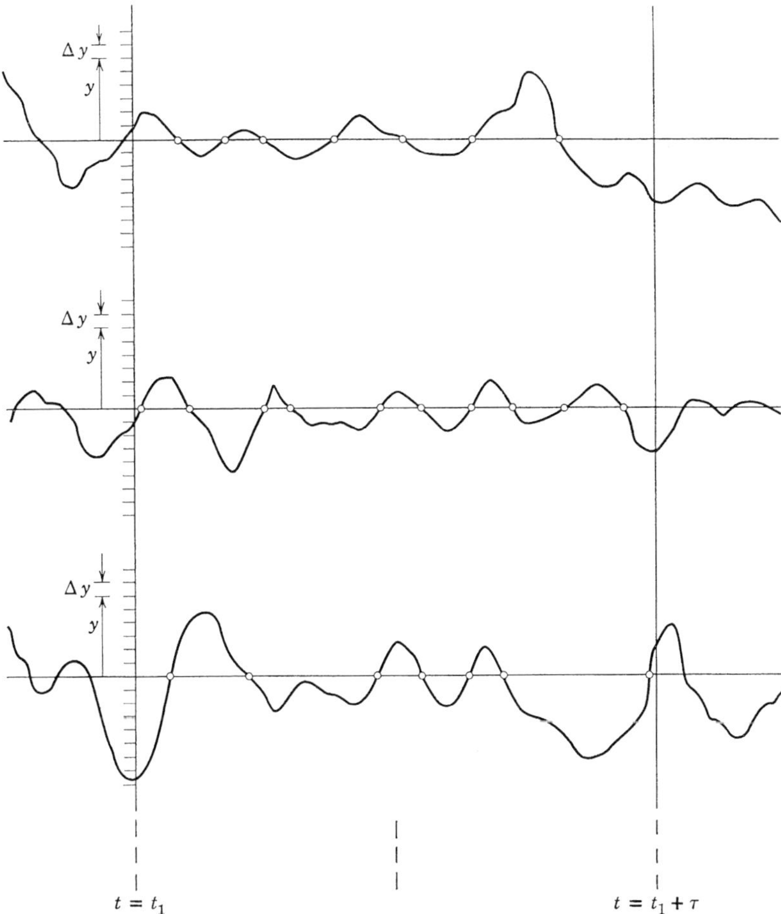

Fig. 1. A series of experimental data in a communication or control problem.

regularity which is found in a sequence obtained by a large group of people tossing, each person once, identical coins under the same conditions. In all random experiments it is assumed that the trials may be repeated indefinitely, at least in theory. We have already emphasized the fact that the trials are carried out under essentially the same conditions so that statistical regularity is maintained. The question of what conditions should be kept essentially the same in a random experiment can only be answered with the aid of experience. When the relative magnitude of the effect of certain factors are in doubt, it is quite clear that it can be experimentally determined on the basis of statistical regularity and the results deduced on this basis.

The statistical regularity of the series of random functions shown in Fig. 1 may be exhibited in many ways. For instance, if we are interested in the value of the random functions falling within a small interval Δy, from y to $y + \Delta y$, at the time $t = t_1$, as indicated in the figure, we would find that the frequency ratio of the number of times the functions are found to have this value to the total number of functions observed tends to center about some fixed value, in an irregular manner, as the number of trials increases. We are assuming, of course, that the relevant conditions for statistical regularity have been maintained. Another quantity which should show statistical regularity is the number of zero crossings in the interval τ as indicated in the figure. The number of functions which have a given number of zero crossings in the interval τ, from t_1 to $t_1 + \tau$, to the total number of trials examined for the purpose under discussion, is a frequency ratio which should tend to settle down to some constant as the number of trials becomes large.

3. Outcomes of a Random Experiment

In a random experiment, the outcomes, or observations, are called *events*. Depending upon the experiment, events are expressed in terms of the characteristics of the objects involved or as numerical values of the quantities under observation. Thus in the coin-tossing game they are heads and tails, and in measuring an instantaneous noise voltage they are the numerical values of the voltage. An event is not necessarily expressed as a single characteristic or number; it may take the form of a combination of characteristics or numbers. For example, groups of people may be observed with respect to occupation and level of formal education so that two characteristics are required for each observation; and, in the experiment with an instantaneous noise voltage, each measurement may consist of two values with a fixed time of separation. Obviously observations of greater complexity are possible.

When the conditions of a random experiment are fixed, we should be able to name all of its possible results. For instance, if the random experiment of tossing two true coins, one marked a and the other b, is considered, the possible results are

Coin a	Coin b
H	T
T	H
H	H
T	T

These are all the possible results, because, according to the conditions of the game, the two coins are distinguishable and there are two possible results for each coin. This set of possible outcomes is *exhaustive* (or *complete*), meaning that in a trial of the experiment one of the outcomes must take place. They are called the *simple events* of the experiment. Furthermore, since the occurrence of one event precludes the occurrence of any one of the others, these events are *mutually exclusive.*

On the basis of the specified conditions and the simple events, other events relating to the experiment may be considered. For example, the event that at least one head turns up refers to any one of the first three simple events; the event that both coins fall alike refers to either of the last two simple events; and the event that coin b falls heads refers to the second or third simple event. Each of these events is satisfied by two or more of the simple events. They are known as *compound events.* A simple event satisfying the specifications of a compound event is said to be *favorable* to it.

Now, suppose that in this experiment the coins are not marked so that in recording the results we can no longer distinguish the two coins; the possible results become

$$\begin{array}{cc} H & T \\ H & H \\ T & T \end{array}$$

With reference to the conditions of the modified experiment these constitute the complete set of simple events. Compound events of the experiment may be formed in a manner similar to the previous example.

In the two examples just given, the number of possible outcomes is finite, but it is possible that in some random experiments this number is infinitely large. An example of this situation is found in the counting of the number of zero crossings in the noise voltage during the interval τ as indicated in Fig. 1. In the conceptual experiment, we may allow the possibility of having an infinitely large number of crossings in a given finite duration. Under this assumption, the complete set of possible outcomes is $0, 1, 2, 3, \ldots, \infty$.

It is important to remember that, in the development of a theory of probability, we refer to a conceptual random experiment. We imagine that such an experiment may be repeated indefinitely under identical conditions; we assume that such conditions are given at the outset that a complete set of the possible outcomes are known; we assume statistical regularity in these outcomes; and we are concerned with the set of all possible outcomes of the random experiment.

4. Measure of Probability

Before we define the *measure* of probability, it is necessary that we agree upon the meaning of the expression that two events are equally likely to occur. It immediately becomes clear to us that it is an intuitive idea and cannot be defined. According to Bernoulli, one of the best-known founders of the theory of probability, the state of equal probability of occurrence of two events is considered to exist if the conditions of the conceptual random experiment are such that it is inconceivable that they influence the occurrence of one more than the other.

Thus, in an idealized experiment of throwing a true die we are intuitively willing to say that the events represented by the numbers 1, 2, 3, 4, 5, 6 are equally likely. Again, in the example of tossing two identical coins, marked a and b, the 4 simple events H_aT_b, T_aH_b, H_aH_b, T_aT_b are equally likely events, for the coins are true in an idealized experiment and it is inconceivable that any of these events is favored more than the others by the conditions of the experiment. On the basis of this intuitive reasoning, it is clear that, if the coins are indistinguishable, the events HH and TT are still equally likely since the formation of these events has not been affected by the new condition. But, because of the lack of identity of the coins, the event HT can be formed in two ways whereas each of the other two events can be formed in only one way. Therefore, in the modified experiment the events HT and HH (or TT) are not equally likely.

When actual physical problems are involved, sufficient evidence must be available before a reasonable conclusion is reached regarding the equality or inequality of the probabilities of events.

A definition of the numerical measure of *mathematical probability*, which is also known as a *classical definition of probability*, can now be given as follows: *In a conceptual random experiment under a fixed set of conditions, if there are n exhaustive, mutually exclusive, equally likely events and if m of these events are favorable to an event A, then the probability of occurrence of A is defined as the ratio m/n.*

Concerning this definition, it is important to note that:

(*a*) Probabilities are associated only with the possible outcomes of a given conceptual random experiment.

(*b*) The conditions of the conceptual random experiment are given at the outset.

(*c*) The *exhaustive* (*complete*) set of all possible outcomes is obtainable from the given conditions. (In a single trial, one of the events must occur.)

(*d*) The outcomes are *mutually exclusive.* (The occurrence of one event precludes the occurrence of any of the others.)

(*e*) The outcomes are *equally likely.*

(*f*) The event A is a compound event, which is reduced to a simple event if $m = 1$.

As a simple illustration, consider the random experiment of drawing one card from a full deck of 52 cards and noting the identity of the card drawn with respect to suit and denomination. From the given conditions of the experiment we find that there are 52 exhaustive simple events. These events are mutually exclusive. Intuitively, we say that these events are equally likely. Now, for the event A, let us consider the event of drawing an ace. Clearly, of the 52 simple events there are 4 events favorable to the event A, so that the definition of the measure of probability gives $4/52$ or $1/13$ as the probability of the event A.

Let us note that the probability of drawing an ace in a single draw of a card from a deck of 52 cards can be found in another way. We shall modify the conditions of the random experiment so that the observations are made not with respect to suit and denomination but with respect to denomination only, considering cards of the same denomination as indistinguishable. Under these conditions, we have only 13 exhaustive, mutually exclusive, and equally likely events. The event that an ace is drawn is one of the 13 simple events under the given new conditions, and its probability is therefore $1/13$.

If the probability of obtaining at least one head in tossing a coin twice is asked, the random experiment consists in tossing the coin and noting whether it falls heads or tails and repeating the operation and observation as one trial. Following an example given before, we can mark the first toss by a and the second by b. Under the conditions of the random experiment we can write, as we did before, H_aT_b, T_aH_b, H_aH_b, T_aT_b as the exhaustive simple events. Obviously they are mutually exclusive. By intuitive reasoning, we consider them to be equally likely events. The event A in this instance is any one of the first three simple events. Thus we have $n = 4$, $m = 3$, and the probability of obtaining at least one head in tossing a coin twice is $3/4$.

If, in the definition, the event A is that one of the n events occurs, then $m = n$, and we have the probability that one of the n events occurs in a single trial of the conceptual random experiment is 1. In other words, if an event is certain to occur, the probability of the event is 1. On the other hand, if an event is not a possible outcome of the random experiment, it is necessarily not included as one of the n events. Nevertheless, in the course of the analysis of a problem, it may happen that a quantity representing the probability of an event which is not a possible outcome of the random experiment under consideration appears. When such an occasion arises, we shall agree on the convention that the probability of such an event is 0.

Although the definition of the measure of probability is in terms of discrete outcomes, it is possible to extend it for application to situations where events are expressible as a continuous variable. In the measurement of a noise voltage, for example, the events are the values of the noise wave at discrete points of time, and these values are in a continuous range within the lower and the upper limits. The extension of the definition for application to continuous events is not a major difficulty, and we shall see how this is done when the computation of averages in random experiments is discussed.

A major difficult point in the definition of the measure of probability is the requirement of equal probability of the simple events. Whereas it is generally possible to find in games of chance equally likely events, the situation is entirely different in many problems of science and engineering where the theory of probability may be applied. In these problems, we are able to specify at the outset the exhaustive and mutually exclusive outcomes, but we are, more often than not, unable to subdivide these outcomes into equally likely events. Because of this difficulty, it is impossible to make use of the definition to find the probabilities of specified events. When this is possible, as in games of chance, the probabilities so determined are called *a priori probabilities*, because they are known before any trials are made. Although the a priori probability of an event in an actual physical problem is generally not available, it may be estimated when necessary. An estimate may be made by carrying out a large number of trials and relying on the regularity of frequency ratios. An estimated probability is called an *empirical probability*.

In the face of the difficulty, we shall assume the existence of the probabilities in problems where a priori probabilities are not available. We shall further assume that the theorems which are established on the basis of the definition of mathematical probability are applicable to the probabilities in these problems. The success which these assumptions have enjoyed in a great variety of physical problems is sufficient justification for these assumptions.

5. Theorem of Total Probability

The theorem of total probability concerns the probability of a compound event in a random experiment. The compound event has characteristics which are shared by a group of mutually exclusive events (compound or simple). These mutually exclusive events are called the forms of the compound event. The theorem gives the relationship between the probability of the compound event and the probabilities of its forms.

The *theorem of total probability* states that, *in a conceptual random experiment, if an event A can occur in the mutually exclusive forms* A_1, A_2, A_3, ..., A_K, *and in no other forms, then the probability of event A is the sum of the probabilities of events* A_1, A_2, A_3, ..., A_K. *Symbolically,*

$$\mathcal{P}(A) = \mathcal{P}(A_1) + \mathcal{P}(A_2) + \mathcal{P}(A_3) + \cdots + \mathcal{P}(A_K) \tag{1}$$

Here $\mathcal{P}(A)$ stands for "the probability of the event A," $\mathcal{P}(A_1)$ for "the probability of the event A_1," and so on.

To illustrate this theorem, let us consider the probability of getting heads at least once in tossing a coin twice. The experiment has been described and the outcomes are tabulated in Sec. 3. The event A in this game is "heads at least once in two throws." One possible way to state the forms of A is the following: form A_1, "heads once in two throws"; and form A_2, "heads both times." Since event A refers to any one of the first three simple events tabulated, of which A_1 refers to the first two and A_2 refers to the third, it is clear that A_1 and A_2 together include the same component simple events as A does and there are no other forms of A. Furthermore, the component simple events of A_1 and A_2 do not overlap so that A_1 and A_2 are mutually exclusive. By the definition of probability, in view of the discussion under Sec. 3, $\mathcal{P}(A_1) = 2/4$ and $\mathcal{P}(A_2) = 1/4$. Hence the theorem of total probability leads to the result $\mathcal{P}(A) = 2/4 + 1/4 = 3/4$, which is in agreement with that obtained in Sec. 4 by direct application of the definition of probability.

To prove this theorem, let n be the number of the exhaustive, mutually exclusive, equally likely events of the experiment. These are the simple events of the given experiment. Let A_1, A_2, A_3, ..., A_K be the complete set of mutually exclusive forms of A; and let m_1, m_2, m_3, ..., m_K be the respective numbers of simple events favorable to A_1, A_2, A_3, ..., A_K. The number of simple events favorable to A is therefore the sum $m_1 + m_2 + m_3 + \cdots + m_K$. In accordance with the definition of probability, the probability of event A is

$$\mathcal{P}(A) = \frac{m_1 + m_2 + m_3 + \cdots + m_K}{n} \tag{2}$$

We may write this expression in the form

$$\mathcal{P}(A) = \frac{m_1}{n} + \frac{m_2}{n} + \frac{m_3}{n} + \cdots + \frac{m_K}{n} \tag{3}$$

Clearly, by definition, the ratios in this expression are the probabilities of the events $A_1, A_2, A_3, \ldots, A_K$; that is, $\mathcal{P}(A_1) = m_1/n$, $\mathcal{P}(A_2) = m_2/n$, $\mathcal{P}(A_3) = m_3/n, \ldots, \mathcal{P}(A_K) = m_K/n$. Hence we have the relation

$$\mathcal{P}(A) = \mathcal{P}(A_1) + \mathcal{P}(A_2) + \mathcal{P}(A_3) + \cdots + \mathcal{P}(A_K) \tag{4}$$

which is the theorem of total probability.

We note that, since $A_1, A_2, A_3, \ldots, A_K$ are the forms of A, so that an occurrence of any one of these forms is an occurrence of A, we may refer to the forms of A instead of A itself in the statement of the theorem of total probability. Accordingly, the theorem, in an alternative form, is stated as follows: *In a conceptual random experiment, if $A_1, A_2, A_3, \ldots, A_K$ are mutually exclusive events, the probability of any one of these events occurring is the sum of their probabilities. Symbolically,*

$$\mathcal{P}(A_1 \text{ or } A_2 \text{ or } A_3 \ldots \text{ or } A_K) = \mathcal{P}(A_1) + \mathcal{P}(A_2) + \mathcal{P}(A_3) + \cdots + \mathcal{P}(A_K) \tag{5}$$

6. Theorem of Compound Probability

This theorem concerns the probability of the joint occurrence of two events in a random experiment. The essence of the theorem lies in the concept of *conditional probability*. Let us first make clear the meaning of this term by an illustration. A random experiment consists of drawing a card from a deck of 52 cards and, without returning this card to the deck, drawing a second card from the deck of 51 cards. The question is now asked: If it is known that the first card is a spade, what is the probability that the second card is also a spade? The required probability is a conditional probability since it is dependent upon a known event.

To express conditional probability in general terms, let n be the number of the exhaustive, mutually exclusive, equally likely events in a random experiment, of which m_A events are favorable to event A and m_{AB} events are favorable to both events A and B. Let $\mathcal{P}(A, B)$ stand for the probability of the joint occurrence of A and B, $\mathcal{P}(A)$ for the probability of A, $\mathcal{P}(B)$ for the probability of B, $\mathcal{P}(B|A)$ for the conditional probability of B given that A has occurred, and $\mathcal{P}(A|B)$ for the conditional probability of A given that B has occurred.

By definition of conditional probability,

$$\mathcal{P}(B|A) = \frac{m_{AB}}{m_A} \tag{6}$$

since we have given that A has occurred, making it necessary to consider all those events favorable to A as the exhaustive, mutually exclusive, and equally likely events in a new random experiment. Of these events there are m_{AB} events which are favorable to B. If we divide both numerator and denominator of the ratio by n, then

$$\mathcal{P}(B|A) = \frac{\dfrac{m_{AB}}{n}}{\dfrac{m_A}{n}} \tag{7}$$

Again, by definition

$$\mathcal{P}(A) = \frac{m_A}{n} \tag{8}$$

and

$$\mathcal{P}(A, B) = \frac{m_{AB}}{n} \tag{9}$$

so that (7) becomes

$$\mathcal{P}(B|A) = \frac{\mathcal{P}(A, B)}{\mathcal{P}(A)} \tag{10}$$

In the illustration just given, we find $n = 52 \times 51$, $m_A = 13 \times 51$, and $m_{AB} = 13 \times 12$. Hence $\mathcal{P}(A) = (13 \times 51)/(52 \times 51) = 1/4$, $\mathcal{P}(A, B) = (13 \times 12)/(52 \times 51) = 1/17$, and $\mathcal{P}(B|A) = (1/17)/(1/4) = 4/17$.

Similarly,

$$\mathcal{P}(A|B) = \frac{m_{AB}}{m_B} = \frac{\dfrac{m_{AB}}{n}}{\dfrac{m_B}{n}} = \frac{\mathcal{P}(A, B)}{\mathcal{P}(B)} \tag{11}$$

The results (10) and (11) are usually written as expressions for the probability of the joint occurrence of A and B and constitute the *theorem of compound probability* which states that, *in a conceptual random experiment, the probability of the joint occurrence of events A and B is equal to the product of the unconditional probability of event A and the conditional probability of event B given that event A has occurred, or the product of the unconditional probability of event B and the conditional* *probability of event A given that event B has occurred. Symbolically,*

$$\mathcal{P}(A, B) = \mathcal{P}(A)\mathcal{P}(B \mid A) \tag{12}$$

or

$$\mathcal{P}(A, B) = \mathcal{P}(B)\mathcal{P}(A \mid B) \tag{13}$$

It is easy to see that, for the probability of the joint occurrence of three events A, B, and C, repeated application of (12) yields

$$\begin{aligned} \mathcal{P}(A, B, C) &= \mathcal{P}(A, B)\mathcal{P}(C \mid A, B) \\ &= \mathcal{P}(A)\mathcal{P}(B \mid A)\mathcal{P}(C \mid A, B) \end{aligned} \tag{14}$$

In the first application the joint occurrence of A and B is considered as a single event, and, in the second, the term $\mathcal{P}(A, B)$ is expressed as a product.

Associated with the concept of conditional probability is the important concept of *statistical independence.* If the probability of occurrence of the event B is not dependent upon whether or not event A has occurred, B is statistically independent (or independent, for brevity) of A by definition. In other words, if

$$\mathcal{P}(B \mid A) = \mathcal{P}(B) \tag{15}$$

then B is said to be statistically independent of A.

In the experiment of tossing a coin twice, for instance, the probability of heads in the second toss is ½ whether or not the outcome of the first toss is heads or tails. Again, in the experiment of drawing two cards from a deck of 52 cards, if the first card drawn is returned to the deck before the second draw, the probability that the second card is a spade is ¼ and is independent of the result of the first draw.

It follows immediately that, in a conceptual random experiment, if event B is statistically independent of event A, the theorem of compound probability becomes

$$\mathcal{P}(A, B) = \mathcal{P}(A)\mathcal{P}(B) \tag{16}$$

Since the second form of the theorem of compound probability is $\mathcal{P}(A, B) = \mathcal{P}(B)\mathcal{P}(A \mid B)$, we must have, under the condition stated

$$\mathcal{P}(A)\mathcal{P}(B) = \mathcal{P}(B)\mathcal{P}(A \mid B) \tag{17}$$

From this result we find

$$\mathcal{P}(A \mid B) = \mathcal{P}(A) \tag{18}$$

which is an expression of the fact that event A is statistically independent of event B. Our conclusion is that, if B is independent of A, it follows that A is independent of B.

In view of this fact the theorem of compound probability in the symbolic form (16) is stated as follows: *If events A and B in a conceptual random experiment are statistically independent, the probability of their joint occurrence is the product of their unconditional probabilities.*

For a simple illustration of this theorem we refer to the experiment of drawing a card from a deck of 52 cards twice. We ask for the probability of getting two spades. The event A is "the first card drawn is a spade," and the event B is "the second card drawn is a spade." Since the second draw is from a full deck of 52 cards, the first card having been returned to the deck, event B is independent of A. Therefore we have $\mathcal{P}(A) = 1/4$, $\mathcal{P}(B) = 1/4$, and $\mathcal{P}(A, B) = (1/4)(1/4) = 1/16$. We recall that, if the first card drawn is not returned to the deck so that B is dependent upon A, the compound probability is $1/17$.

Extending the theorem of compound probability to three events A, B, and C, we have that, if

$$\left.\begin{aligned}\mathcal{P}(A \mid B) &= \mathcal{P}(A)\\ \mathcal{P}(A \mid C) &= \mathcal{P}(A)\\ \mathcal{P}(B \mid C) &= \mathcal{P}(B)\end{aligned}\right\} \tag{19}$$

and

$$\mathcal{P}(C \mid A, B) = \mathcal{P}(C) \tag{20}$$

then A, B, and C are statistically independent. From conditions (19) we see that

$$\left.\begin{aligned}\mathcal{P}(B \mid A) &= \mathcal{P}(B)\\ \mathcal{P}(C \mid A) &= \mathcal{P}(C)\\ \mathcal{P}(C \mid B) &= \mathcal{P}(C)\end{aligned}\right\} \tag{21}$$

since we have shown that, if A is independent of B, then B is independent of A. It can be shown that on the basis of conditions (19) and (20) we have the additional conditions

$$\left.\begin{aligned}\mathcal{P}(B \mid A, C) &= \mathcal{P}(B)\\ \mathcal{P}(A \mid B, C) &= \mathcal{P}(A)\end{aligned}\right\} \tag{22}$$

Therefore, when (19) and (20) are satisfied, it is implied that events A, B, and C are mutually independent. When A, B, and C are mutually independent, it follows immediately from (14) that their compound probability is

$$\mathcal{P}(A, B, C) = \mathcal{P}(A)\mathcal{P}(B)\mathcal{P}(C) \tag{23}$$

since, in (14), $\mathcal{P}(B \mid A) = \mathcal{P}(B)$ and $\mathcal{P}(C \mid A, B) = \mathcal{P}(C)$ which is condition (20).

7. Bernoulli's Theorem

In the introduction of this chapter it is pointed out that the application of the theory of probability to physical problems needs the basic assumption that the physical phenomenon concerned exhibits statistical regularity even if experiments are not repeated or never performed. We imagine that experiments are repeated many times under identical conditions and that the frequency ratio of certain observations would tend to come closer and closer to some fixed value as the number of repeated trials increases. When we speak of the probability of an event, we intuitively associate it with the frequency ratio of the event in the random experiment and think of the probability as the limit of the frequency ratio as the number of trials becomes infinitely large. These physical ideas should not be an integral part of a mathematical theory of probability but should have a mathematical counterpart in the form of a theorem relating the frequency ratio of an event in a conceptual random experiment to the probability of the event. Such a theorem is of great importance in the theory of probability. Jacob Bernoulli in his book *Ars Conjectandi*, published in 1713, presented this theorem.

Bernoulli's theorem states that, *if in a series of M independent trials of a conceptual random experiment the number of successes of an event is N_M and the probability of the event is p, then the probability that the frequency ratio N_M/M differs from p by less than a preassigned quantity ϵ, however small, tends to unity as M tends to infinity. In symbolic form,*

$$\mathcal{P}\left(\left|\frac{N_M}{M} - p\right| < \epsilon\right) \to 1 \tag{24}$$

as $M \to \infty$.

Although the proof of the theorem is beyond the scope of this book, some remarks concerning its interpretation should be made. The theorem does not say that the frequency ratio tends to a limit as the number of trials tends to infinity. If this were true, then the sequence of frequency ratios

$$\frac{N_1}{1}, \frac{N_2}{2}, \frac{N_3}{3}, \ldots, \frac{N_M}{M}, \ldots \tag{25}$$

where N_1, N_2, N_3, and N_M represent the numbers of successes in 1, 2, 3, and M trials, respectively, would have a limit. According to the usual definition of a limit, it is necessary that, if p is the limit, we should be able to find a value of M, say M', such that for $M \geq M'$, $|p - N_M/M| < \epsilon$ no matter how small the preassigned value of ϵ may be. But, because the trials are independent, meaning that the probability of the event concerned does not depend upon the occurrence or non-occurrence of the event in any of the previous trials, we cannot be assured of the possibility of satisfying the requirement for a sequence to approach a limit. Long sequences of successes are possible, and even success for every trial is not impossible. Although in an actual experiment, if the number of trials is sufficiently large, a limit may be reached "practically," it is impossible to prove that the sequence approaches a limit in the usual mathematical sense. What is possible to prove is that the *probability* of the frequency ratio lying within the limits $p + \epsilon$ and $p - \epsilon$ is dependent upon the number of trials. This *probability* approaches a limit, and the probability that the ratio lies within the limits $(p + \epsilon, p - \epsilon)$ tends to unity as the number of trials tends to infinity. Bernoulli's theorem establishes rigorously the probability relationship between the frequency ratio and the probability of an event and justifies the experimental determination of the probability of an event by carrying out a sufficiently large number of trials.

REFERENCES

1. Cramer, H., *Mathematical Methods of Statistics*, Princeton University Press, Princeton, 1946.
2. Feller, W., *An Introduction to Probability Theory and Its Applications*, 2nd ed., John Wiley and Sons, New York, 1957.
3. Fry, T. C., *Probability and Its Engineering Uses*, D. Van Nostrand Company, Princeton, 1928.
4. Mood, A. M., *Introduction to the Theory of Statistics*, McGraw-Hill Book Company, New York, 1950.
5. Uspensky, J. V., *Introduction to Mathematical Probability*, McGraw-Hill Book Company, New York, 1937.

chapter 4

Random Variables, Ensembles, and Distributions

Since a periodic or aperiodic function of time is completely determined for every value of the independent variable, time, the description of the function is therefore given over the whole set of values of the independent variable. This is the description of a function in the time domain.

Through the development of harmonic analysis, we are able to describe a function in another domain which we know is the domain of frequency. A function of time can be completely specified on the entire range of the frequency variable in accordance with the concepts and methods of Fourier's theory.

If the function is random, its description involves new problems, and additional concepts are necessary. However, not all of the function requires these concepts for its description, for the past of a random function is completely determined for every value of time. This portion of the function is describable in the time domain and, in a certain way, in the frequency domain as well. The future of the function, however, is not subject to precise prediction and must be described in the domain of the whole set of its possible values in conjunction with the concept of probability distributions. The complete set of possible values on which random functions are described is, in the terminology of statistical mechanics, the phase space. This chapter is devoted to the development of a method of description which we may call the description of random functions in the domain of phase space.

Inasmuch as the concept of an ensemble is of great importance in the development, much of the discussion is related to the ensemble. As we shall see, the term "phase" in this work is given to a concept which bears no relation to the phase spectrum in harmonic analysis. The similarity in terminology should not be allowed to cause confusion.

1. Discrete Random Variables

In giving a full expression of the outcome of a random experiment, we must have the exhaustive set of mutually exclusive events and the set of probabilities of occurrence. We assume that the events are expressed numerically and that, if they are generally identified by characteristics, numerical codes are given. Since the outcome taking any one of the possible values is a matter of chance, the outcome is called a *random variable*. There are several other names for this quantity, such as *chance variable*, *stochastic variable*, and *variate*. Thus if the outcome of a random experiment is represented by the random variable ξ, we express its exhaustive, mutually exclusive values as

$$x_1, x_2, \ldots, x_n \tag{1}$$

and the respective probabilities as

$$P_\xi(x_1), P_\xi(x_2), \ldots, P_\xi(x_n) \tag{2}$$

The reason for expressing the probabilities in this form instead of a sequence of numerical values, such as P_1, P_2, ..., P_n, is that presently these probabilities are expressed as a function defined for every possible value of ξ.

We express the fact that the probability of ξ taking the value x_1 is $P_\xi(x_1)$ in the symbolic form

$$\mathcal{P}(\xi = x_1) = P_\xi(x_1) \tag{3}$$

and in a similar manner we write

$$\left.\begin{array}{ccc} \mathcal{P}(\xi = x_2) & = & P_\xi(x_2) \\ \vdots & & \vdots \\ \mathcal{P}(\xi = x_n) & = & P_\xi(x_n) \end{array}\right\} \tag{4}$$

The general expression is

$$\mathcal{P}(\xi = x_i) = P_\xi(x_i) \qquad \text{for } i = 1, 2, \ldots, n \tag{5}$$

An example will clarify the symbolic expression. If a coin is marked 1 on one side and 2 on the other and if the experiment consists of tossing the coin twice and taking the sum of the numbers, this sum is a random variable ξ, and the exhaustive, mutually exclusive events are $\xi = 2$, $\xi = 3$, $\xi = 4$ with the respective probabilities $1/4$, $1/2$, $1/4$. For this experiment, (5) becomes

$$\mathcal{P}(\xi = x_i) = P_\xi(x_i) \qquad \text{for } i = 1, 2, 3 \tag{6}$$

with

and

$$\left.\begin{array}{c} x_1 = 2, x_2 = 3, x_3 = 4 \\ P_\xi(x_1) = \frac{1}{4}, P_\xi(x_2) = \frac{1}{2}, P_\xi(x_3) = \frac{1}{4} \end{array}\right\} \tag{7}$$

or, specifically

$$\left.\begin{aligned}\mathcal{P}(\xi = 2) &= P_\xi(2) = \tfrac{1}{4}\\ \mathcal{P}(\xi = 3) &= P_\xi(3) = \tfrac{1}{2}\\ \mathcal{P}(\xi = 4) &= P_\xi(4) = \tfrac{1}{4}\end{aligned}\right\} \tag{8}$$

The function $P_\xi(x_i)$ in (5) which gives the probability of ξ taking a possible value as a function of its complete set of possible values is called the *probability distribution function* of the random variable ξ. We shall frequently shorten the name to *probability distribution, probability function, distribution function,* or simply *distribution.* Since the range of fluctuation of the random variable is a discrete set of values, the variable is a *discrete random variable,* and the distribution associated with it is a *discrete distribution function.* Note that the distribution of a random variable is expressed as a function of the entire *range* of possible values of the random variable; it is a function defined for every possible value of the *range variable* x_i. Although in (1) we have indicated that the set of exhaustive, mutually exclusive events is finite, it is possible that the set is infinite. In either case the distribution is a non-negative function in accordance with the definition of probability, and

$$\sum_i P_\xi(x_i) = 1 \tag{9}$$

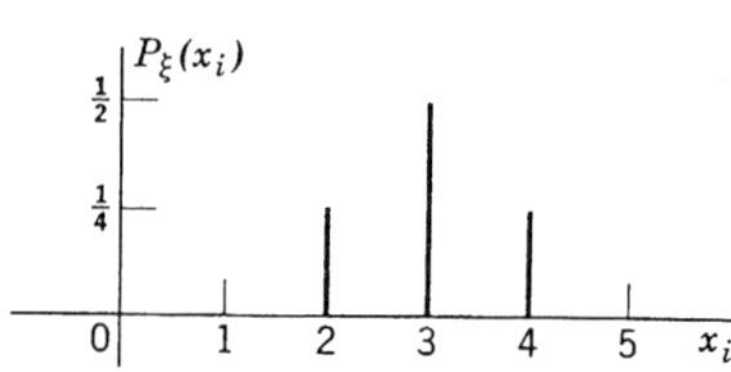

Fig. 1. Probability distribution for numerical example.

because the set of events is exhaustive, meaning that in a single trial of the experiment one of the events must be realized. The distribution function for the example is graphically represented in Fig. 1.

A more general illustration of the distribution function is that of a random voltage or some other physical quantity from a random experiment. If the voltage is ξ and can take only the discrete values

$$x_{-m}, \ldots, x_{-2}, x_{-1}, x_0, x_1, x_2, \ldots, x_n \tag{10}$$

with the respective probabilities

$$P_\xi(x_{-m}), \ldots, P_\xi(x_{-2}), P_\xi(x_{-1}), P_\xi(x_0), P_\xi(x_1), P_\xi(x_2), \ldots, P_\xi(x_n) \tag{11}$$

which we assume to be known, the graphical representation of the distribution function $P_\xi(x_i)$ is as shown in Fig. 2. Although the values of x_i in the figure are integral multiples of x_1, it is clear that they need not be so. Except for the fact that a distribution function is not necessarily even, it is similar to the normalized (to the total power) power spectrum

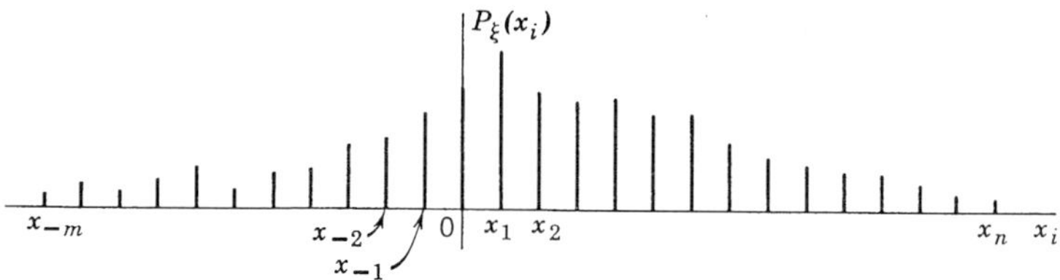

Fig. 2. A probability distribution.

of a function whose components are periodic, in the sense that both characteristics are non-negative functions of discrete variables and that the summation of either characteristic over all values of the argument is 1.

In connection with the distribution function $P_\xi(x_i)$, the probabilities of certain compound events should be pointed out. If the event A is that the voltage ξ falls within the interval (x_j, x_k), that is $x_j \leq \xi \leq x_k$, then by the theorem of total probability

$$\mathcal{P}(A) = \sum_{i=j}^{k} P_\xi(x_i) \tag{12}$$

which may be written in the form

$$\mathcal{P}(x_j \leq \xi \leq x_k) = \sum_{i=j}^{k} P_\xi(x_i) \tag{13}$$

Similarly, if event B is that the voltage is positive, then, assuming that $x_0 = 0$,

$$\mathcal{P}(B) = \sum_{i=1}^{n} P_\xi(x_i)$$

or

$$\mathcal{P}(\xi > 0) = \sum_{i=1}^{n} P_\xi(x_i) \tag{14}$$

It is easy to see how the probabilities of other compound events are expressed. According to (9), we have

$$\sum_{i=-m}^{n} P_\xi(x_i) = 1 \tag{15}$$

2. Continuous Random Variables

Random variables in physical experiments, in such forms as voltages, temperatures, velocities, and so on, are frequently not restricted to a range of discrete values but may have a range of variation which is continuous. When the range of variation of a random variable is continuous, the variable is called a *continuous random variable.* We should note that the term "continuous" refers to the range of variation of the random variable, not the values of the variable in a sequence of trials. In such a sequence the random variable takes values in this range according to chance, and it is clear that the sequence of values cannot be continuous although the range of possible values is continuous.

The concept of probability distribution should be extended for application to continuous random variables. For this extension, consider a random variable ξ which assumes values in a continuous range in the interval $(-a, b)$. Let the range variable be x, and let us divide the interval $(-a, b)$ into small elements of length Δx. Suppose that in so doing we find n elements in the interval $(0, b)$ and m elements in the interval $(-a, 0)$. We now imagine that a sufficiently large number of trials, M, of the experiment associated with ξ are made and that of this number of trials there are

N_1 trials in which ξ falls in the interval $(0, \Delta x)$

N_2 trials in which ξ falls in the interval $(\Delta x, 2\,\Delta x)$

$\vdots$

N_n trials in which ξ falls in the interval $([n - 1]\,\Delta x, n\,\Delta x)$

N_{-1} trials in which ξ falls in the interval $(0, -\Delta x)$

N_{-2} trials in which ξ falls in the interval $(-\Delta x, -2\,\Delta x)$

$\vdots$

N_{-m} trials in which ξ falls in the interval $(-[m - 1]\,\Delta x, -m\,\Delta x)$

We then construct a frequency ratio graph as shown in Fig. 3 in which the area of the rectangle above an element represents the frequency ratio for the event that ξ falls within that element. For example, in the

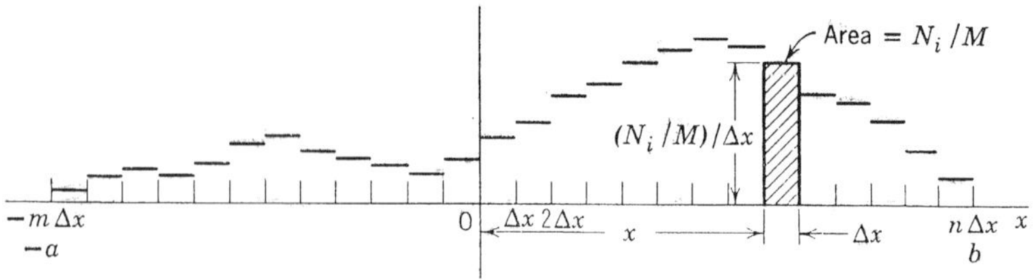

Fig. 3. Illustrating an empirical probability density.

interval $(x, x + \Delta x)$ the area of the shaded rectangle is N_i/M, or $[(N_i/M)/\Delta x]\,\Delta x$. Hence the height of the rectangle is $(N_i/M)/\Delta x$.

To characterize a random variable whose range of variation is continuous, a probability distribution function with a continuous argument representing the range becomes necessary. In this function, an element of the range is the differential dx, and with it a probability is associated. A convenient representation of the probability is the area of a rectangle with dx as its base. These considerations account for the steps we have taken to show the interpretation and the estimation of a function which satisfies our needs.

Let $P_\xi(x)$ be the function in question. We assume that it exists although for most practical situations it can only be estimated by experiment. The product $P_\xi(x)\,dx$ is then an element of area which we take to represent an element of probability. This probability is for the event that, in a single trial, ξ falls in the infinitesimal interval $(x, x + dx)$.

Now, as we make Δx small and allow M to increase, the ratios $(N_i/M)/\Delta x$ should approximately remain at certain fixed values after a sufficiently large value of M has been reached. This statement is based upon the assumption of statistical regularity. Under these conditions we can say that the ratios are an estimate of $P_\xi(x)$. As we have pointed out in connection with the Bernoulli theorem, the concept of a limit cannot be applied to a frequency ratio, so that it is impossible to say, in a mathematical sense, that $(N_i/M)/\Delta x$ approaches $P_\xi(x)$ as $\Delta x \to 0$ and $M \to \infty$. What we can say is that $P_\xi(x)$ exists by assumption and that it can be estimated on the basis of statistical regularity.

The function $P_\xi(x)$ is called the *probability density function* for ξ. When desirable, we shorten this name to *probability density* or *density*. Since the range is divided into elements each of length dx and the element of probability $P_\xi(x)\,dx$ is the probability of ξ falling in one of these elements, the density $P_\xi(x)$ is actually a function showing the *relative* magnitudes of probability for ξ to fall in an element dx at x. The probability, itself, of the event that ξ falls in an element dx at a given value of x is, of course, an infinitesimal. Figure 4 shows the curve of a

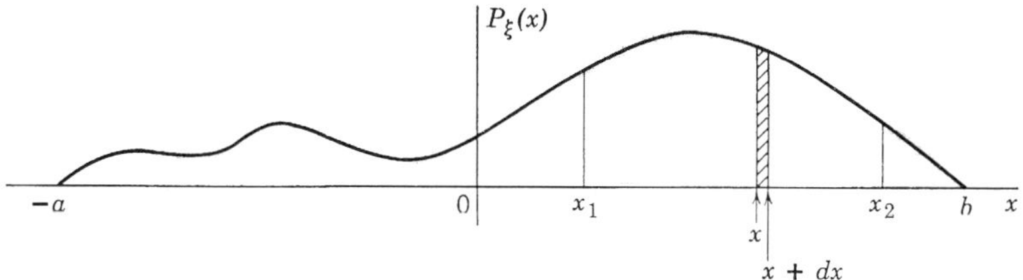

Fig. 4. A probability density.

probability density which may be considered to correspond to the frequency ratio graph of Fig. 3. These figures illustrate the general appearance of the functions and are not associated with a particular situation.

For the probability of the event that ξ falls in the infinitesimal interval $(x, x + dx)$ we write symbolically

$$\mathcal{P}(x < \xi < x + dx) = P_\xi(x)\,dx \tag{16}$$

This element of probability is the probability of a simple event of the experiment. The infinitesimal quantity $P_\xi(x)\,dx$ is represented by the area of the rectangle of width dx under the curve of $P_\xi(x)$ at x, as shown in Fig. 4.

Let us consider some examples of compound events. The probability of the event that ξ falls in the interval (x_1, x_2) is expressed as

$$\mathcal{P}(x_1 < \xi < x_2) = \int_{x_1}^{x_2} P_\xi(x)\,dx \tag{17}$$

which is represented by the area under the density curve between the limits x_1 and x_2. This expression is in agreement with the theorem of total probability, for the event that ξ falls in (x_1, x_2) is the same as the event that ξ falls in an element of all the non-overlapping elements (mutually exclusive), each of length dx, into which the interval (x_1, x_2) is divided, so that the probability of the event is the sum of the probabilities associated with the complete set of elements in (x_1, x_2). Since $P_\xi(x)$ is continuous, the summation is an integration. Similarly, if the event is that ξ is positive, then

$$\mathcal{P}(\xi > 0) = \int_{0}^{b} P_\xi(x)\,dx \tag{18}$$

We scarcely need be reminded that $P_\xi(x)$ is a non-negative function and that

$$\int_{-a}^{b} P_\xi(x)\,dx = 1 \tag{19}$$

Since $P_\xi(x)$ is defined over the interval $(-\infty, \infty)$ and, in the present instance, is zero outside the interval $(-a, b)$, it is convenient to write

$$\int_{-\infty}^{\infty} P_\xi(x)\,dx = 1 \tag{20}$$

as a fundamental property of the probability density function.

3. Variables with Continuous and Discrete Ranges

The range of fluctuation of a random variable need not be either entirely discrete or entirely continuous. It is possible that the range is partly continuous and partly discrete. An example of such a situation is a thermal noise voltage with amplitude limiting. The noise voltage has only a continuous range of possible values when there is no amplitude limiting. With limiting, we know that in addition to the continuous range a particular value has replaced a part of the range. Since at this value of the range the probability is finite, its representation in the density function for the continuous portion has the difficulty that it becomes infinite. For a better description of the function at this point we introduce an impulse whose integral (area) is numerically equal to the probability that ξ takes this value of the range. If $P_\xi(x)$ of Fig. 4 is the probability density function for the noise without limiting and if the limiting is applied at amplitude $x = x_2$, the new probability density $P_{\xi'}(x)$ for the new random variable ξ' is as shown in Fig. 5. The continuous portion of the density remains as before, but the impulse at $x = x_2$ has an area equal to the area under the curve of Fig. 4 in the interval (x_2, b), since the probability that ξ' takes the value x_2 with limiting is equal to the probability that ξ is greater than x_2 without limiting. Therefore we have

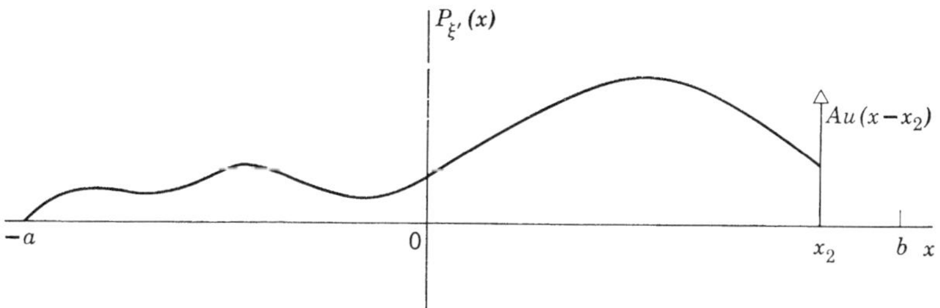

Fig. 5. A probability density with an infinite discontinuity.

$$P_{\xi'}(x) = P_\xi{}^*(x) + Au(x - x_2) \tag{21}$$

where

$$P_\xi{}^*(x) = \begin{cases} P_\xi(x) & \text{for } -\infty < x < x_2 \\ 0 & \text{elsewhere} \end{cases} \tag{22}$$

and

$$A = \int_{x_2}^{b} P_\xi(x)\, dx \tag{23}$$

and for the unit impulse $u(x - x_2)$

$$\int_{-\infty}^{\infty} u(x - x_2)\, dx = 1 \tag{24}$$

In accordance with property (20) it is necessarily true that

$$\int_{-\infty}^{\infty} P_{\xi'}(x)\, dx = 1 \tag{25}$$

Clearly, if ξ' has a set of possible values with finite probabilities additional to a continuous range of possible values, the probability density function will have a continuous component and a set of impulses at the set of given values, each impulse having an integral (area) equal to the probability that ξ takes the value concerned.

4. Representation of a Discrete Distribution as a Density

With the aid of the unit impulse function it is possible to represent a discrete distribution function as a probability density. For instance, if ξ has the possible values (10) with the respective probabilities (11) the probability density is

$$P_\xi(x) = \sum_{i=-m}^{n} P_\xi(x_i) u(x - x_i) \tag{26}$$

Obviously, whereas the discrete distribution is represented by a set of lines each having a length equal to the probability concerned, as Fig. 2 shows, the probability density (26) is represented by a set of impulses each having an area equal to the probability. The property of a discrete distribution as given by (9) corresponds to the property that the integral over $(-\infty, \infty)$ of the probability density (26) is

$$\left.\begin{aligned} \int_{-\infty}^{\infty} P_\xi(x)\, dx &= \int_{-\infty}^{\infty} \sum_{i=-m}^{n} P_\xi(x_i) u(x - x_i)\, dx \\ &= \sum_{i=-m}^{n} P_\xi(x_i) \\ &= 1 \end{aligned}\right\} \tag{27}$$

5. Integrated Probability Density

The representation of a discrete distribution by a probability density reminds us of the representation of a discrete power spectrum by a power density spectrum discussed in Chapter 2, Sec. C-4. The line characteristic does not adequately include the continuous component, and the density characteristic necessarily involves impulses. To avoid the impulses for convenience in certain analytic manipulations, the integrated power spectrum is introduced although this type of spectrum is frequently unnecessary. It is easy to see that for a similar reason a similar function is introduced in the theory of probability. Thus the *integrated probability density* is defined as

$$Q_\xi(x) = \int_{-\infty}^{x} P_\xi(x)\,dx \tag{28}$$

and is sometimes called the *cumulative distribution function* of the random variable ξ. It is a positive increasing monotonic function since $P_\xi(x)$ is non-negative. The function is continuous except at those points where ξ has finite probabilities. At these points the function jumps by amounts equal to the respective probabilities. Clearly, we have

$$Q_\xi(-\infty) = 0 \tag{29}$$

and

$$Q_\xi(\infty) = 1 \tag{30}$$

For expression (17) we now write

$$\mathcal{P}(x_1 < \xi < x_2) = Q_\xi(x_2) - Q_\xi(x_1) \tag{31}$$

assuming that $Q_\xi(x)$ is continuous at $x = x_1$ and $x = x_2$, and for (18)

$$\begin{aligned} \mathcal{P}(\xi > 0) &= Q_\xi(\infty) - Q_\xi(0) \\ &= 1 - Q_\xi(0) \end{aligned} \tag{32}$$

assuming that $Q_\xi(x)$ is continuous at $x = 0$.

6. Remarks concerning the Basic Element of Probability for a Continuous Random Variable

Concerning a continuous random variable, a question frequently asked is: What is the probability that the random variable takes a particular value in the continuous range? The answer is zero. We have, in this discussion, necessarily excluded the possibility of the random variable having a discrete component range. The probability density is therefore devoid of impulses. Since the fact that the range of ξ is continuous makes it necessary to represent probability by an integral with respect to the range, the measure of probability is therefore dependent upon the measure of the range. The measure of any interval of the range is the length of the interval. In computing the probability of an event, we see that, if the interval of the range concerned is of zero measure, the probability is consequently zero. Now, a particular value of the range is of zero measure and so our answer follows. In fact, according to the theory of measure, any *enumerable* set of points, even if it is an infinite set, always has zero measure. Hence, if we ask what the probability is that ξ takes any one of the enumerable set of values $x_1, x_2, x_3, \ldots$, our answer would still be zero.

But, we may then ask, since in a single trial ξ must take a value within its range so that the probability of the event is one, is not this fact in contradiction with the fact that the probability is zero that ξ takes a specified value of its possible values? How is it possible to obtain probability 1 for an event from probability 0 for its specific forms? To answer these questions, we must remember that, when ξ is considered to have a continuous range, it is understood that the possible values of ξ cannot be given as an enumerable set even if we allow it to be an infinite set; for it is known in measure theory that the set of all points belonging to a nonzero interval is *non-enumerable.* By a non-enumerable set we mean that the values of the set cannot be so arranged as to correspond, in a one-to-one manner, to the set of all positive integers 1, 2, 3, ... Now, if ξ can take only an enumerable set of values, which may or may not be infinite, we know that a finite probability can be associated with each value. However, if the possible values of ξ are all points in an interval, the set of points is non-enumerable, and it is known that such a set is of a higher order of infinity than the infinite enumerable set.

To handle this situation, we shift our attention from the individual points to the infinitesimal intervals. We divide the interval into elements each of length dx. With this element of the range of ξ we associate an element of probability. The event that ξ falls in the elementary interval $(x, x + dx)$ is a simple event, and the probability is $P_\xi(x)\,dx$. The totality of the elements, which are non-overlapping, each of length dx, constitute the complete set of mutually exclusive simple events associated with ξ. Generally, they are not equally likely events since $P_\xi(x)$ may assume various forms, but, if $P_\xi(x)$ is a uniform density, the events are then equally likely. Under this method, the event that ξ takes a value of its possible values is the event that ξ falls in an element dx. Applying the theorem of total probability, we have

$$\begin{aligned}\mathcal{P}(-\infty < \xi < \infty) &= \int_{-\infty}^{\infty} P_\xi(x)\,dx \\ &= 1 \end{aligned} \tag{33}$$

Furthermore, in this method the probability element $P_\xi(x)\,dx$ is associated with the aggregate of all points in dx; we do not consider how the probability is distributed with respect to the non-enumerable and enumerable sets of points which are contained in dx.

When the question of the probability of ξ taking a particular value of x is raised, we can answer according to the concept of the probability element for a continuous random variable. As we have stated earlier, this concept gives the answer zero since, in this instance, dx becomes zero. Such a result should not be disturbing because our method for evaluating the probability of compound events does not involve the summation of probabilities for individual points so that the situation of adding zeros does not arise. One can easily understand this by referring to other problems of a similar sort. In the calculation of area by calculus, for example, the fact that the area under a continuous curve at a point on the x-axis is zero does not mean that the area under a curve in a finite interval is a sum of zeros.

7. Ensembles

For mechanical problems involving a large number of particles, such as those concerning the properties and behavior of gases, J. Willard Gibbs developed a theory, known as statistical mechanics, in which he introduced the concept of a probability distribution of initial conditions while retaining classical mechanical concepts in the general formulation. As we know, the complete solution of such problems by classical mechanics alone is impossible. Associated with the probability distribution is the Gibbsian ensemble which is a conceptual aggregate of dynamical systems of identical structure. The concept of an ensemble and its modifications have been of great value not only in physical statistics but in communication theory as well.

Before going into a detailed discussion of the ensemble in communication theory, it seems desirable that we state briefly some of the ideas in statistical mechanics relating to the Gibbsian ensemble. In mechanics, the state of a system with N degrees of freedom may be specified by N generalized position coordinates and N generalized momenta. The state of the system expressed by these quantities is known as the *phase* of the system. Since the $2N$ coordinates determine a $2N$-dimensional space, it is called the *phase space* of the system. A point in the phase space describes the system at a given time, and the point travels over a path in the space as time passes. Now, for an ensemble whose member systems have random initial conditions, the points in the phase space representing the whole ensemble, at any given time, have a certain probability distribution. For the description of the initial conditions the concept of *phase probability* is introduced.

The Gibbsian ensemble leads to the concept of an ensemble of messages or of noise in communication theory. As described in Chapter 2, Sec. C-1, an ensemble of messages is the set of outcomes of an aggregate of identical message sources operating under the same conditions. The ensemble and the sources are, of course, conceptual. An ensemble of noise is formed in a similar manner. These messages and noise are conceived to have started in the infinite past and to continue into the infinite future. However, since they are not completely predictable functions, only their past is available for examination.

We should point out that, whereas the members of a Gibbsian ensemble are mechanical systems whose description involves mechanical concepts, the members of a message or noise ensemble are random time functions whose description is free from mechanical concepts.

8. Description of an Ensemble at One Point in Time; Stationary Ensembles

Let us now consider the description of message or noise ensembles. In Fig. 6 we have a graphical representation of an ensemble. The series

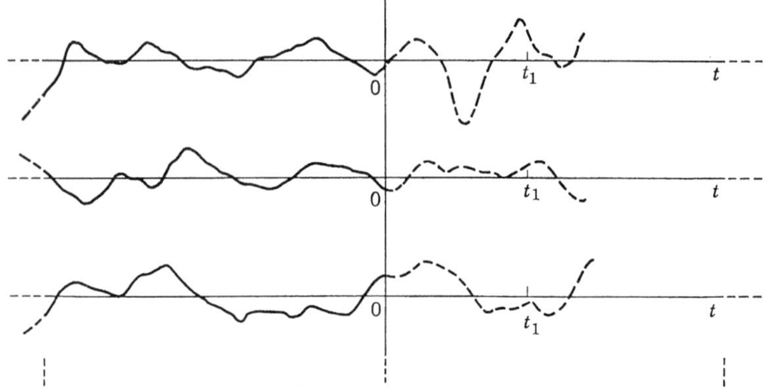

Fig. 6. An ensemble.

of continuous time functions represent the outcomes of a series of identical mechanisms producing messages or noise, as the case may be, under the same conditions. If we imagine that, with one source, a series of trials of a fixed duration T are made under the same conditions and then the records are arranged as in Fig. 7, we would have a set of values $x^{(1)}, x^{(2)}, x^{(3)}, \ldots,$ as the observations at time $t = 0$, with the beginning of each trial at $t = t'$ in the past. Thus we have a situation where a conceptual random experiment may be performed repeatedly under the same conditions.

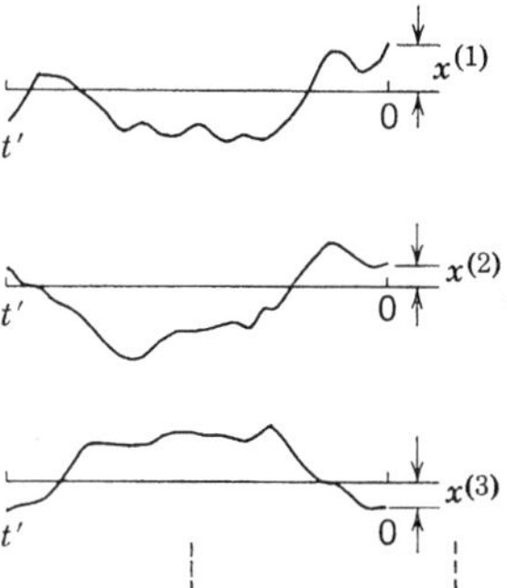

Fig. 7. Statistical data from experiments of the same duration.

If, instead of repeating an experiment with a single source, we perform simultaneously experiments with additional identical sources under the same conditions, we shall have experiments that may be considered as the repeated trials of the original experiment. The graphical representation of the results of the simultaneous experiments is the same as that shown in Fig. 7 except that the actual records would be different. Although in physical experiments the duration of the trials are necessarily finite, we may assume that theoretically the starting point of the trials is in the infinite past. Since the set of identical message or noise sources, working in the manner described, will produce random time functions which may be considered as the results of repeated trials of one experiment and this experiment may be associated with any one of the identical sources, a description of the ensemble in terms of probability can be given. Thus we can speak of the probability of finding the amplitude of any particular member function of the ensemble, at a given time, falling in a specified infinitesimal interval of its continuous range of possible values. In speaking of the amplitude of any particular member function at a given time, we actually refer to the outcome of an experiment at that time with the understanding that previous outcomes, that is, previous values of the function, are not considered. This point is important because the probability under discussion is unconditional. Furthermore, the time of occurrence of the event in question is in the future. Since the values of the time function in the past are determined as events come to pass according to chance, the description of this portion of the function is given by its actual values. If the random amplitude of any one of the member functions at time t_1 (future) as indicated in Fig. 6 is ξ_1 with the probability density $P_{\xi 1}(x_1, t_1)$, we express this symbolically as

$$\mathcal{P}(x_1 < \xi_1 < x_1 + dx_1; t = t_1) = P_{\xi 1}(x_1, t_1)\, dx_1 \tag{34}$$

where x_1 is a continuous range variable representing the finite or infinite range of ξ_1.

Although in (34) we have indicated that the probability density is a function of the time at which the random variable is located, it is possible that the distribution density is independent of this time. In a practical situation where it is necessary to consider random functions with a finite past, the probability distribution density is, under certain circumstances, a function of the time with reference to the beginning of the experiment. We have such a situation, for instance, when the random function is a message or noise from an electric circuit, and the time at which the random amplitude is considered is so close to the starting point that initial transients are not negligible and steady temperatures have not been reached. Nevertheless, after a sufficiently long delay, when the initial transients have subsided and steady temperatures have been reached, it is possible that the probability density is invariant thereafter. If the probability density of amplitude of an ensemble is the same, no matter at what time observations are to be made, the ensemble is said to be *stationary.** Accordingly, in describing a stationary ensemble, we modify our statement (34) to

$$\mathcal{P}(x_1 < \xi_1 < x_1 + dx_1; t = t_1) = P_{\xi 1}(x_1)\, dx_1 \tag{35}$$

in which $P_{\xi 1}(x_1)$ is the probability density of ξ_1, and it is independent of t_1. The density is sometimes referred to as the *first probability density* of the ensemble. Stated in words, the statement (35) is, in a stationary ensemble, the probability of finding the amplitude of any particular member function in the infinitesimal range $(x_1, x_1 + dx_1)$, at any given time $t = t_1$ in the interval $(0, \infty)$, is $P_{\xi 1}(x_1)\, dx_1$. We take $t = 0$ as the present. The quantity $P_{\xi 1}(x_1)\, dx_1$ is the *probability element* for one random variable.

9. Description of a Stationary Ensemble at Two Points in Time

Expression (35) concerns only the unconditional probability density of amplitude of the member functions of an ensemble and is by no means a complete description of the ensemble. To describe a stationary ensemble in greater detail, let us consider a set of observations made on the ensemble shown in Fig. 8 at two points of time instead of one. These points of time are necessarily in the past, for future values are as yet unattainable. Suppose that the continuous range of possible values is divided into elements each of length Δx. Let the divisions at $t = t_1'$ be marked by subscript 1 $(\ldots, -i\,\Delta x_1, \ldots, -2\,\Delta x_1, -\Delta x_1, 0, \Delta x_1, 2\,\Delta x_1, \ldots, i\,\Delta x_1, \ldots)$ and those at $t = t_2'$ by subscript 2 $(\ldots, -j\,\Delta x_2, \ldots, -2\,\Delta x_2, -\Delta x_2, 0, \Delta x_2, 2\,\Delta x_2, \ldots, j\,\Delta x_2, \ldots)$, although all elements

* Since at this point we are concerned with only the first probability density and since there are higher order probability densities, strictly speaking the ensemble in question is a first-order stationary ensemble.

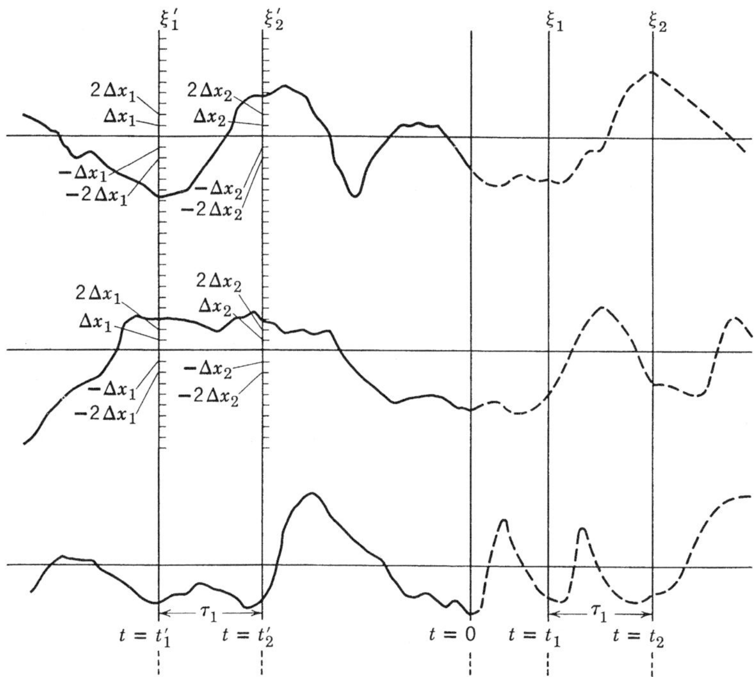

Fig. 8. Illustrating the analysis of statistical data for estimating a joint probability density.

are of the same length. The difference in time between the two points is τ_1; that is, $t_2' - t_1' = \tau_1$. Each observation consists of the value of the random variable ξ_1' at $t = t_1'$ and the value of the random variable ξ_2' at $t = t_2'$; thus, an observation consists of a pair of values from a member function. For a sufficiently large total of M observations which are made on M member functions, let us denote the number of observations in which ξ_1' is found to fall in the interval $(0, \Delta x_1)$ and ξ_2' in $(0, \Delta x_2)$ by $N_{1,1}$, the number of observations in which ξ_1' is found to fall in $(0, \Delta x_1)$ and ξ_2' in $(\Delta x_2, 2\,\Delta x_2)$ by $N_{1,2}$, and, in general, the number of observations in which ξ_1' is found in $([i-1]\,\Delta x_1, i\,\Delta x_1)$ and ξ_2' in $([j-1]\,\Delta x_2, j\,\Delta x_2)$ by $N_{i,j}$. We tabulate the observations in the accompanying table. The middle row gives the intervals in which ξ_1' is found, and the middle column gives the intervals in which ξ_2' is found. The rest of the table is self-evident. Since the total number of observations is M, we have

$$\sum_i \sum_j N_{i,j} = M \tag{36}$$

	$\vdots$		$\vdots$	$\vdots$	$\vdots$	$\vdots$	$\vdots$		$\vdots$	
$\cdots$	$N_{-i,-j}$	$\cdots$	$N_{-2,-j}$	$N_{-1,-j}$	$-j\,\Delta x_2 < \xi_2' < -(j-1)\,\Delta x_2$	$N_{1,-j}$	$N_{2,-j}$	$\cdots$	$N_{i,-j}$	$\cdots$
	$\vdots$		$\vdots$	$\vdots$	$\vdots$	$\vdots$	$\vdots$		$\vdots$	
$\cdots$	$N_{-i,-2}$	$\cdots$	$N_{-2,-2}$	$N_{-1,-2}$	$-2\,\Delta x_2 < \xi_2' < -\Delta x_2$	$N_{1,-2}$	$N_{2,-2}$	$\cdots$	$N_{i,-2}$	$\cdots$
$\cdots$	$N_{-i,-1}$	$\cdots$	$N_{-2,-1}$	$N_{-1,-1}$	$-\Delta x_2 < \xi_2' < 0$	$N_{1,-1}$	$N_{2,-1}$	$\cdots$	$N_{i,-1}$	$\cdots$
$\cdots$	$-i\,\Delta x_1 < \xi_1' < -(i-1)\,\Delta x_1$	$\cdots$	$-2\,\Delta x_1 < \xi_1' < -\Delta x_1$	$-\Delta x_1 < \xi_1' < 0$		$0 < \xi_1' < \Delta x_1$	$\Delta x_1 < \xi_1' < 2\,\Delta x$	$\cdots$	$(i-1)\,\Delta x_1 < \xi_1' < i\,\Delta x_1$	$\cdots$
$\cdots$	$N_{-i,1}$	$\cdots$	$N_{-2,1}$	$N_{-1,1}$	$0 < \xi_2' < \Delta x_2$	$N_{1,1}$	$N_{2,1}$	$\cdots$	$N_{i,1}$	$\cdots$
$\cdots$	$N_{-i,2}$	$\cdots$	$N_{-2,2}$	$N_{-1,2}$	$\Delta x_2 < \xi_2' < 2\,\Delta x_2$	$N_{1,2}$	$N_{2,2}$	$\cdots$	$N_{i,2}$	$\cdots$
	$\vdots$		$\vdots$	$\vdots$	$\vdots$	$\vdots$	$\vdots$		$\vdots$	
$\cdots$	$N_{-i,j}$	$\cdots$	$N_{-2,j}$	$N_{-1,j}$	$(j-1)\,\Delta x_2 < \xi_2' < j\,\Delta x_2$	$N_{1,j}$	$N_{2,j}$	$\cdots$	$N_{i,j}$	$\cdots$
	$\vdots$		$\vdots$	$\vdots$	$\vdots$	$\vdots$	$\vdots$		$\vdots$	

In a manner analogous to the treatment of a single random variable, we construct a three-dimensional representation of the frequency ratio of the joint occurrence of two events associated with the random variables ξ_1 and ξ_2. Since these variables have continuous ranges, it is necessary that the measure of a frequency ratio be a volume. The representation is shown in Fig. 9. The frequency ratio is $N_{i,j}/M$, where $N_{i,j}$ is

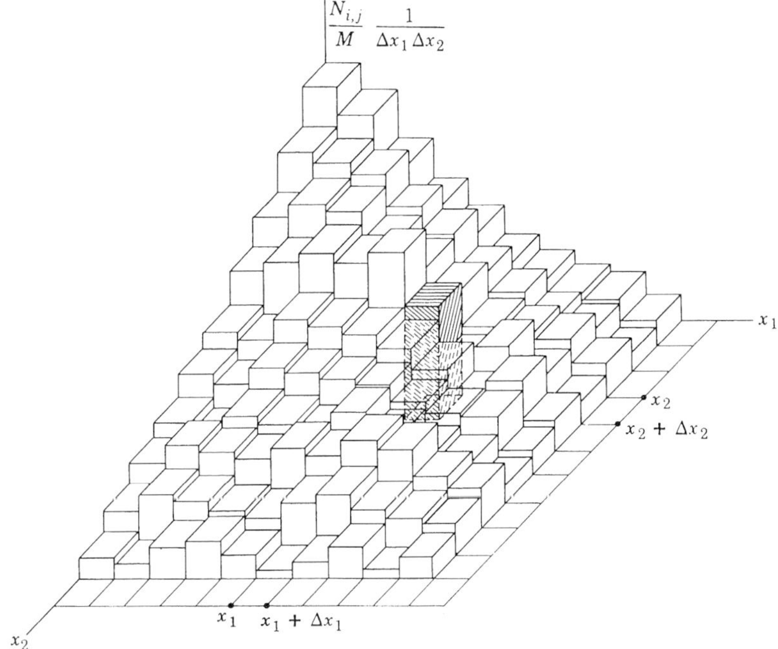

Fig. 9. Illustrating an empirical second probability density.

the number of successes in finding jointly ξ_1' in the interval $([i - 1] \Delta x_1, i \Delta x_1)$ and, at time τ_1 later, ξ_2' in the interval $([j - 1] \Delta x_2, j \Delta x_2)$ in a total of M trials. To represent this quantity, we have at the intersection of x_1 and x_2 a column whose base is a square of area $\Delta x_1 \Delta x_2$ and whose volume is $N_{i,j}/M$, thus making it necessary that the height of the column be equal to $(N_{i,j}/M)/(\Delta x_1 \Delta x_2)$. The remainder of the representation is constructed according to the given table of observed values. The representation is therefore an aggregate of columns of the same base with volumes numerically equal to the frequency ratios of the joint occurrence of two events. It is to be noted that only one value of τ_1 has been considered. A separate representation has to be constructed for each value of τ_1.

We now assume that a function $P_{\xi_1\xi_2}(x_1, x_2; \tau_1)$ exists such that, for any particular member function of a stationary ensemble, as shown in Fig. 8, $P_{\xi_1\xi_2}(x_1, x_2; \tau_1)\, dx_1\, dx_2$ is the probability that, at a given future time $t = t_1$, ξ_1 is found in the interval $(x_1, x_1 + dx_1)$ and later, at time $t = t_2 = t_1 + \tau_1$, ξ_2 is found in the interval $(x_2, x_2 + dx_2)$. On the basis of statistical regularity, we can say that, as we make Δx_1 and Δx_2 sufficiently small and allow M to become sufficiently large, the frequency ratios $(N_{i,j}/M)(\Delta x_1\, \Delta x_2)$ for the continuous random variable are an estimate of the function $P_{\xi_1\xi_2}(x_1, x_2; \tau_1)$. If the function is continuous, the surfaces which represent it have the general appearance as illustrated in Fig. 10. Each surface is drawn for a particular value of τ_1,

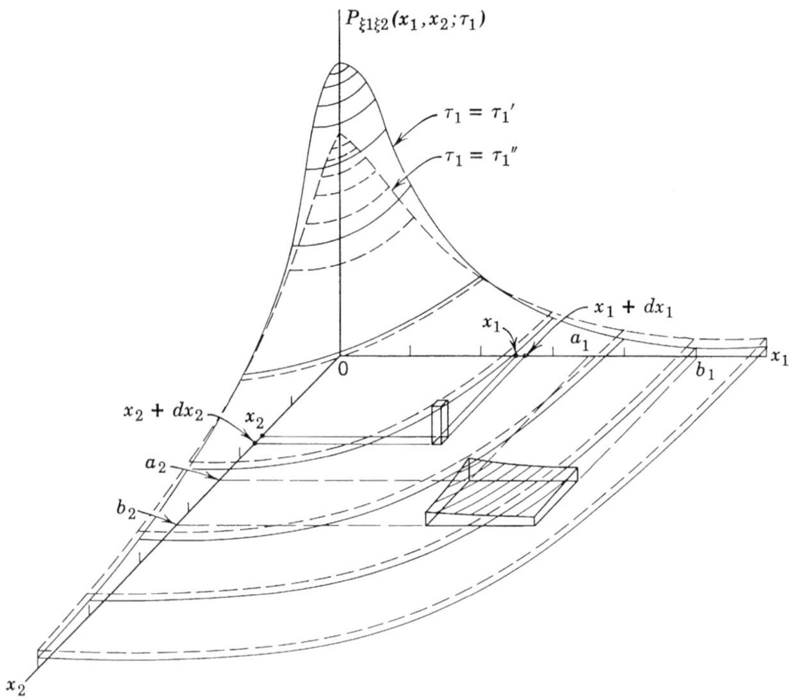

Fig. 10. A second probability density.

and one of these may be regarded as corresponding to the frequency ratio representation of Fig. 9. The function $P_{\xi_1\xi_2}(x_1, x_2; \tau_1)$ is a *joint probability density*. In conjunction with the first probability density it is called the *second probability density* of the ensemble. Since we are concerned with stationary ensembles only, the density is independent of the time t_1.* The estimate of the density is an *empirical joint probability density* or an *empirical second probability density*.

In order to avoid ambiguity, it seems necessary that we indicate by the subscripts $\xi 1 \xi 2$ that the density is for a joint occurrence of events relating to the random variables ξ_1 and ξ_2 which are located at $t = t_1$ and $t = t_2$ respectively. The density is a function of the continuous range variables x_1 and x_2 and of the continuous time τ_1 which is equal to $t_2 - t_1$.

To express in symbolic form the probability element for two random variables, we write

$$\mathcal{P}(x_1 < \xi_1 < x_1 + dx_1, x_2 < \xi_2 < x_2 + dx_2; t_2 - t_1 = \tau_1) = P_{\xi_1 \xi_2}(x_1, x_2; \tau_1)\, dx_1\, dx_2 \quad (37)$$

This probability element is graphically represented by the infinitesimal volume of the column at point (x_1, x_2) as shown in Fig. 10.

10. Some Properties of the Second Probability Density

For any particular member function of the ensemble, the probability of the joint event that, at time $t = t_1$, ξ_1 is found in the interval (a_1, b_1) and, at time $t = t_2 = t_1 + \tau_1$ later, ξ_2 is found in the interval (a_2, b_2) is

$$\mathcal{P}(a_1 < \xi_1 < b_1, a_2 < \xi_2 < b_2; t_2 - t_1 = \tau_1) = \int_{a_2}^{b_2} \int_{a_1}^{b_1} P_{\xi_1 \xi_2}(x_1, x_2; \tau_1)\, dx_1\, dx_2 \quad (38)$$

This probability is represented by the volume bounded by the x_1x_2-plane, the surface $P_{\xi_1 \xi_2}(x_1, x_2; \tau_1)$, and the planes $x_1 = a_1$, $x_1 = b_1$, $x_2 = a_2$, $x_2 = b_2$, as shown in Fig. 10.

Since probability is non-negative, the second probability density is a non-negative function. In addition, the fact that in a single trial one of the possible events must happen is expressed, as before, by the equation

$$\int_{-\infty}^{\infty} \int_{-\infty}^{\infty} P_{\xi_1 \xi_2}(x_1, x_2; \tau_1)\, dx_1\, dx_2 = 1 \quad (39)$$

It is frequently helpful in analysis to express the joint probability density in terms of a conditional probability density. This is accomplished by an application of the theorem of compound probability given

* Strictly, if only the second probability density is known to be independent of the time origin, the ensemble should be called a second-order stationary ensemble.

in Chapter 3. The *conditional probability density* is written as

$$P_{\xi 2|\xi 1}(x_2|x_1;\tau_1) \tag{40}$$

so that $P_{\xi 2|\xi 1}(x_2|x_1;\tau_1)\,dx_2$ is the probability that, given the event that ξ_1 falls in the interval $(x_1, x_1 + dx_1)$ at time $t = t_1$, ξ_2 falls in $(x_2, x_2 + dx_2)$ at $t = t_2 = t_1 + \tau_1$. The probability applies to any particular member of the stationary ensemble. To put the statement in symbolic form, we write

$$\begin{aligned}\mathcal{P}(x_2 < \xi_2 < x_2 + dx_2|x_1 < \xi_1 < x_1 + dx_1; t_2 - t_1 = \tau_1)\\ = P_{\xi 2|\xi 1}(x_2|x_1;\tau_1)\,dx_2\end{aligned} \tag{41}$$

Now, the events A and B in the theorem of compound probability are: (A) ξ_1 falls in the interval $(x_1, x_1 + dx_1)$ at time $t = t_1$, and (B) ξ_2 falls in the interval $(x_2, x_2 + dx_2)$ at time $t = t_2 = t_1 + \tau_1$. We have

$$\left.\begin{aligned}\mathcal{P}(A) &= \mathcal{P}(x_1 < \xi_1 < x_1 + dx_1; t = t_1)\\ &= P_{\xi 1}(x_1)\,dx_1\\ \mathcal{P}(B) &= \mathcal{P}(x_2 < \xi_2 < x_2 + dx_2; t = t_2 = t_1 + \tau_1)\\ &= P_{\xi 2}(x_2)\,dx_2\\ \mathcal{P}(A,B) &= \mathcal{P}(x_1 < \xi_1 < x_1 + dx_1, x_2 < \xi_2 < x_2 + dx_2;\\ &\qquad t_2 - t_1 = \tau_1)\\ &= P_{\xi 1\xi 2}(x_1, x_2; \tau_1)\,dx_1\,dx_2\\ \mathcal{P}(B|A) &= \mathcal{P}(x_2 < \xi_2 < x_2 + dx_2|x_1 < \xi_1 < x_1 + dx_1;\\ &\qquad t_2 - t_1 = \tau_1)\\ &= P_{\xi 2|\xi 1}(x_2|x_1;\tau_1)\,dx_2\end{aligned}\right\} \tag{42}$$

The first expression of (42) is (35), and the second is written in a similar manner for ξ_2. The third expression is (37) and the last is (41). According to the theorem we have

$$P_{\xi 1\xi 2}(x_1, x_2; \tau_1)\,dx_1\,dx_2 = P_{\xi 1}(x_1)\,dx_1\,P_{\xi 2|\xi 1}(x_2|x_1;\tau_1)\,dx_2 \tag{43}$$

Dividing both sides of the equation by $dx_1\,dx_2$, we have the important relation

$$P_{\xi 1\xi 2}(x_1, x_2; \tau_1) = P_{\xi 1}(x_1)P_{\xi 2|\xi 1}(x_2|x_1;\tau_1) \tag{44}$$

The conditional probability density (40) is a function of x_2 and τ_1 for every given value of x_1. Hence x_1 is a parameter of the function. As the subscripts of the symbol show, the function is for the random variable ξ_2 when ξ_1 is known to have occurred in a specified infinitesimal interval of its range. Under the condition that ξ_1 occurs in $(x_1, x_1 + dx_1)$, the random variable ξ_2 must occur in one of the range elements (dx_2) at time τ_1 later. As we know, this fact is expressed by the equation

$$\int_{-\infty}^{\infty} P_{\xi_2|\xi_1}(x_2 \,|\, x_1;\tau_1)\, dx_2 = 1 \tag{45}$$

based upon the element of conditional probability (41) and the theorem of total probability. It seems desirable to remark here that the function $P_{\xi_2|\xi_1}(x_2 \,|\, x_1;\tau_1)$ cannot be arbitrarily formed, for we see in (45) that the way the variable τ_1 enters into the expression must be such that the integration results in 1 and is entirely independent of τ_1.

Integrating (44) with respect to x_2 over the interval $(-\infty, \infty)$, we have

$$\begin{aligned}\int_{-\infty}^{\infty} P_{\xi_1\xi_2}(x_1, x_2;\tau_1)\, dx_2 &= \int_{-\infty}^{\infty} P_{\xi_1}(x_1) P_{\xi_2|\xi_1}(x_2 \,|\, x_1;\tau_1)\, dx_2 \\ &= P_{\xi_1}(x_1) \int_{-\infty}^{\infty} P_{\xi_2|\xi_1}(x_2 \,|\, x_1;\tau_1)\, dx_2 \end{aligned} \tag{46}$$

Applying (45), we obtain the result

$$P_{\xi_1}(x_1) = \int_{-\infty}^{\infty} P_{\xi_1\xi_2}(x_1, x_2;\tau_1)\, dx_2 \tag{47}$$

which states that the unconditional probability density of ξ_1 is determinable from the joint probability density of ξ_1 and ξ_2 by integration with respect to the range of ξ_2 over the entire range of possible values. Again, the variable τ_1 does not appear in the result of integration.

Instead of considering conditional probability element (41), we may consider

$$\begin{aligned}\mathcal{P}(x_1 < \xi_1 < x_1 + dx_1 \,|\, x_2 < \xi_2 < x_2 + dx_2;\, t_2 - t_1 = \tau_1) \\ = P_{\xi_1|\xi_2}(x_1 \,|\, x_2;\tau_1)\, dx_1 \end{aligned} \tag{48}$$

Then corresponding to (44) we have

$$P_{\xi_1\xi_2}(x_1, x_2;\tau_1) = P_{\xi_2}(x_2) P_{\xi_1|\xi_2}(x_1 \,|\, x_2;\tau_1) \tag{49}$$

Corresponding to (45) we have

$$\int_{-\infty}^{\infty} P_{\xi_1|\xi_2}(x_1 \,|\, x_2;\tau_1)\, dx_1 = 1 \tag{50}$$

and corresponding to (47) we have

$$P_{\xi_2}(x_2) = \int_{-\infty}^{\infty} P_{\xi_1\xi_2}(x_1, x_2;\tau_1)\, dx_1 \tag{51}$$

Inasmuch as the probability densities under consideration are independent of the origin in the time of observation, they should be symmetrical with respect to x_1 and x_2; that is, numerically,

$$P_{\xi 1}(x_1) = P_{\xi 2}(x_2) \tag{52}$$

$$P_{\xi 1 | \xi 2}(x_1 | x_2; \tau_1) = P_{\xi 2 | \xi 1}(x_2 | x_1; \tau_1) \tag{53}$$

and

$$P_{\xi 1 \xi 2}(x_1, x_2; \tau_1) = P_{\xi 2 \xi 1}(x_2, x_1; \tau_1) \tag{54}$$

11. An Example of the Second Probability Density

It is known that a thermal noise with zero mean value has the second probability density *

$$P_{\xi 1 \xi 2}(x_1, x_2; \tau_1) = \frac{1}{2\pi\sqrt{\varphi_{11}^2(0) - \varphi_{11}^2(\tau_1)}} \times \exp\left[- \frac{\varphi_{11}(0)(x_1^2 + x_2^2) - 2\varphi_{11}(\tau_1)x_1x_2}{2[\varphi_{11}^2(0) - \varphi_{11}^2(\tau_1)]}\right] \tag{55}$$

in which $\varphi_{11}(\tau_1)$ is the autocorrelation function of the noise. This second density is called a *normal density* or a *Gaussian density.*

To determine the unconditional probability density (the first probability density) $P_{\xi 1}(x_1)$, we apply (47). Letting

$$a = \sqrt{\varphi_{11}^2(0) - \varphi_1^2\,(\tau)} \tag{56}$$

$$b = \varphi_{11}(0) \tag{57}$$

$$c = \varphi_{11}(\tau) \tag{58}$$

we have

$$P_{\xi 1}(x_1) = \frac{1}{2\pi a} \exp\left[- \frac{b x_1^2}{2a^2}\right] \int_{-\infty}^{\infty} \exp\left[- \frac{b x_2^2 - 2c x_1 x_2}{2a^2}\right] dx_2 \tag{59}$$

With the aid of the definite integral

$$\int_{-\infty}^{\infty} e^{-px^2+qx}\, dx = \sqrt{\frac{\pi}{p}}\, e^{q^2/4p} \tag{60}$$

(59) becomes

$$P_{\xi 1}(x_1) = \frac{1}{2\pi a} \exp\left[- \frac{b x_1^2}{2a^2}\right] \left\{ a \sqrt{\frac{2\pi}{b}} \exp\left[\frac{c^2 x_1^2}{2ba^2}\right]\right\}$$

$$= \frac{1}{\sqrt{2\pi b}}\, e^{-x_1^2/2b} \tag{61}$$

* S. O. Rice, "Mathematical Analysis of Random Noise," *Bell System Tech. Jour.*, **24,** No. 1, 50 (1945). Rice states that the first person to obtain this result was H. Thiede, *Elec. Nachr. Tek.*, **13,** 84–95 (1936).

Returning to (57) for the value of b, we have finally

$$P_{\xi 1}(x_1) = \frac{1}{\sqrt{2\pi\varphi_{11}(0)}} \exp\left[-\frac{{x_1}^2}{2\varphi_{11}(0)}\right] \tag{62}$$

This result is independent of the function $\varphi_{11}(\tau_1)$; and, since τ_1 appears only in $\varphi_{11}(\tau_1)$, the result is independent of τ_1 as it should be. Using (52), we write

$$P_{\xi 2}(x_2) = \frac{1}{\sqrt{2\pi\varphi_{11}(0)}} \exp\left[-\frac{{x_2}^2}{2\varphi_{11}(0)}\right] \tag{63}$$

Obviously this expression is obtainable from (55) by integration with respect to x_1.

The conditional probability density can be determined readily by formula (44). Thus

$$P_{\xi 2|\xi 1}(x_2 \,|\, x_1; \tau_1) = \frac{P_{\xi 1 \xi 2}(x_1, x_2; \tau_1)}{P_{\xi 1}(x_1)} \tag{64}$$

so that for the normal density

$$P_{\xi 2|\xi 1}(x_2 \,|\, x_1; \tau_1) = \sqrt{\frac{\varphi_{11}(0)}{2\pi[\varphi_{11}^2(0) - \varphi_{11}^2(\tau_1)]}} \times \exp\left[-\frac{[\varphi_{11}(\tau_1)x_1 - \varphi_{11}(0)x_2]^2}{2\varphi_{11}(0)[\varphi_{11}^2(0) \quad \varphi_{11}^2(\tau_1)]}\right] \tag{65}$$

Similarly,

$$P_{\xi 1|\xi 2}(x_1 \,|\, x_2; \tau_1) = \frac{P_{\xi 1 \xi 2}(x_1, x_2; \tau_1)}{P_{\xi 2}(x_2)} \tag{66}$$

so that for the normal density

$$P_{\xi 1|\xi 2}(x_1 \,|\, x_2; \tau_1) = \sqrt{\frac{\varphi_{11}(0)}{2\pi[\varphi_{11}^2(0) - \varphi_{11}^2(\tau_1)]}} \times \exp\left[-\frac{[\varphi_{11}(\tau_1)x_2 - \varphi_{11}(0)x_1]^2}{2\varphi_{11}(0)[\varphi_{11}^2(0) - \varphi_{11}^2(\tau_1)]}\right] \tag{67}$$

12. Description of a Stationary Ensemble at Three Points in Time

In view of the properties of the first and the second probability densities, it is clear that a density which concerns three points of time would describe the ensemble in still greater detail. Thus we assume that the joint probability density

$$P_{\xi 1 \xi 2 \xi 3}(x_1, x_2, x_3; \tau_1, \tau_2) \tag{68}$$

exists such that $P_{\xi_1\xi_2\xi_3}(x_1, x_2, x_3; \tau_1, \tau_2)\, dx_1\, dx_2\, dx_3$ is the probability that ξ_1 is found in $(x_1, x_1 + dx_1)$ at $t = t_1$, ξ_2 in $(x_2, x_2 + dx_2)$ at $t = t_2 = t_1 + \tau_1$, and ξ_3 in $(x_3, x_3 + dx_3)$ at $t = t_3 = t_2 + \tau_2$. This is the probability element for the joint occurrence of three events, and symbolically it is

$$\begin{aligned}\mathcal{P}(x_1 < \xi_1 < x_1 + dx_1, x_2 < \xi_2 < x_2 + dx_2, x_3 < \xi_3 < x_3 + dx_3;\\ t_2 - t_1 = \tau_1, t_3 - t_2 = \tau_2) = P_{\xi_1\xi_2\xi_3}(x_1, x_2, x_3; \tau_1, \tau_2)\, dx_1\, dx_2\, dx_3 \quad (69)\end{aligned}$$

The joint probability density (68) is the *third probability density* of the ensemble.

Based upon the element (69), the probability of the joint event that, for any particular member function, ξ_1 at $t = t_1$ falls in the interval (a_1, b_1), ξ_2 at $t = t_2 = t_1 + \tau_1$ falls in (a_2, b_2), and ξ_3 at $t = t_3 = t_2 + \tau_2$ falls in (a_3, b_3) is

$$\begin{aligned}\mathcal{P}(a_1 < \xi_1 < b_1, a_2 < \xi_2 < b_2, a_3 < \xi_3 < b_3; t_2 - t_1 = \tau_1,\\ t_3 - t_2 = \tau_2) = \int_{a_3}^{b_3}\int_{a_2}^{b_2}\int_{a_1}^{b_1} P_{\xi_1\xi_2\xi_3}(x_1, x_2, x_3; \tau_1, \tau_2)\, dx_1\, dx_2\, dx_3 \quad (70)\end{aligned}$$

Since (68) is a probability density, it must be non-negative, and, similar to (39) for two random variables, expression (70) must represent probability 1 when the three intervals are extended to include the entire ranges of ξ_1, ξ_2, and ξ_3. Hence

$$\int_{-\infty}^{\infty}\int_{-\infty}^{\infty}\int_{-\infty}^{\infty} P_{\xi_1\xi_2\xi_3}(x_1, x_2, x_3; \tau_1, \tau_2)\, dx_1\, dx_2\, dx_3 = 1 \quad (71)$$

To express the third probability density in terms of a conditional probability density, we apply (14) of Chapter 3 in a manner similar to (43) and (44) and obtain

$$\begin{aligned}P_{\xi_1\xi_2\xi_3}(x_1, x_2, x_3; \tau_1, \tau_2)\, dx_1\, dx_2\, dx_3\\ = P_{\xi_1\xi_2}(x_1, x_2; \tau_1)\, dx_1\, dx_2\, P_{\xi_3|\xi_1\xi_2}(x_3 | x_1, x_2; \tau_1, \tau_2)\, dx_3 \quad (72)\end{aligned}$$

In this expression $P_{\xi_3|\xi_1\xi_2}(x_3 | x_1, x_2; \tau_1, \tau_2)\, dx_3$ is a conditional probability and is the probability of the event that ξ_3 is found in $(x_3, x_3 + dx_3)$ at $t = t_3 = t_2 + \tau_2 = t_1 + \tau_1 + \tau_2$, given that ξ_1 is found in $(x_1, x_1 + dx_1)$ at $t = t_1$ and ξ_2 is found in $(x_2, x_2 + dx_2)$ at $t = t_2 = t_1 + \tau_1$. Symbolically

$$\begin{aligned}\mathcal{P}(x_3 < \xi_3 < x_3 + dx_3 | x_1 < \xi_1 < x_1 + dx_1,\\ x_2 < \xi_2 < x_2 + dx_2; t_2 - t_1 = \tau_1, t_3 - t_2 = \tau_2)\\ = P_{\xi_3|\xi_1\xi_2}(x_3 | x_1, x_2; \tau_1, \tau_2)\, dx_3 \quad (73)\end{aligned}$$

Dividing both sides of (72) by $dx_1\,dx_2\,dx_3$ we have the desired relation

$$P_{\xi_1\xi_2\xi_3}(x_1, x_2, x_3; \tau_1, \tau_2) = P_{\xi_1\xi_2}(x_1, x_2; \tau_1)P_{\xi_3|\xi_1\xi_2}(x_3 \,|\, x_1, x_2; \tau_1, \tau_2) \tag{74}$$

In this equation the second probability density may be expressed as in (44).

By an argument similar to that given for (45) we have

$$\int_{-\infty}^{\infty} P_{\xi_3|\xi_1\xi_2}(x_3 \,|\, x_1, x_2; \tau_1, \tau_2)\, dx_3 = 1 \tag{75}$$

If both sides of (74) are integrated with respect to x_3 over $(-\infty, \infty)$, then

$$\int_{-\infty}^{\infty} P_{\xi_1\xi_2\xi_3}(x_1, x_2, x_3; \tau_1, \tau_2)\, dx_3 = P_{\xi_1\xi_2}(x_1, x_2; \tau_1)\int_{-\infty}^{\infty} P_{\xi_3|\xi_1\xi_2}(x_3 \,|\, x_1, x_2; \tau_1, \tau_2)\, dx_3 \tag{76}$$

and application of (75) shows the relation between the second and third probability densities to be

$$P_{\xi_1\xi_2}(x_1, x_2; \tau_1) = \int_{-\infty}^{\infty} P_{\xi_1\xi_2\xi_3}(x_1, x_2, x_3; \tau_1, \tau_2)\, dx_3 \tag{77}$$

Furthermore, the relation (47) when applied to (77) shows that the unconditional probability density $P_{\xi_1}(x_1)$ can be determined from the third probability density by the double integral

$$P_{\xi_1}(x_1) = \int_{-\infty}^{\infty}\int_{-\infty}^{\infty} P_{\xi_1\xi_2\xi_3}(x_1, x_2, x_3; \tau_1, \tau_2)\, dx_2\, dx_3 \tag{78}$$

13. Complete Description of a Stationary Ensemble

We shall not continue the extension of the probability density to include a large number of variables in detail, for we see that a stationary ensemble * can be described by the set of probability densities

$$\left.\begin{array}{l} P_{\xi_1}(x_1) \\ P_{\xi_1\xi_2}(x_1, x_2; \tau_1) \\ \vdots \\ P_{\xi_1\ldots\xi_n}(x_1, \ldots, x_n; \tau_1, \ldots, \tau_{n-1}) \\ \vdots \end{array}\right\} \tag{79}$$

* When the complete set of probability densities (79) are independent of the origin from which time is measured, the ensemble is said to be strictly stationary.

These functions are non-negative, and, in addition, they have the properties that

$$\left.\begin{aligned}
&\int_{-\infty}^{\infty} P_{\xi 1}(x_1)\,dx_1 = 1\\
&\int_{-\infty}^{\infty}\int_{-\infty}^{\infty} P_{\xi 1 \xi 2}(x_1, x_2; \tau_1)\,dx_1\,dx_2 = 1\\
&\qquad\vdots\\
&\int_{-\infty}^{\infty}\cdots\int_{-\infty}^{\infty} P_{\xi 1 \ldots \xi n}(x_1, \ldots, x_n; \tau_1, \ldots, \tau_{n-1})\,dx_1 \ldots dx_n = 1\\
&\qquad\vdots
\end{aligned}\right\} \tag{80}$$

and that

$$\left.\begin{aligned}
&P_{\xi 1}(x_1) = \int_{-\infty}^{\infty} P_{\xi 1 \xi 2}(x_1, x_2; \tau_1)\,dx_2\\
&P_{\xi 1 \xi 2}(x_1, x_2; \tau_1) = \int_{-\infty}^{\infty} P_{\xi 1 \xi 2 \xi 3}(x_1, x_2, x_3; \tau_1, \tau_2)\,dx_3\\
&\vdots\\
&P_{\xi 1 \ldots \xi n}(x_1, \ldots, x_n; \tau_1, \ldots, \tau_{n-1})\\
&\qquad = \int_{-\infty}^{\infty} P_{\xi 1 \ldots \xi (n+1)}(x_1, \ldots, x_{n+1}; \tau_1, \ldots, \tau_n)\,dx_{n+1}\\
&\vdots
\end{aligned}\right\} \tag{81}$$

The set of equations (81) express the property that the nth probability density determines all the probability densities from the first to the $(n-1)$th. These $(n-1)$ densities are obtainable from the nth density by successive integration.

We shall consider that a stationary ensemble is completely described when the set of probability densities (79) is given. In this set of functions each successive member furnishes additional information on the ensemble but involves rapidly increasing complexity in expression. However, it is fortunate that much of our work does not require a complete description of the ensemble. Many problems of communication can be adequately solved by having available a partial description of the ensemble. In fact, for the problems included in this book, not more than the first and the second probability densities are necessary.

Although the instantaneous outcome of any one of the identical message or noise sources in an ensemble has been referred to as a random function, it is also called a *random process.* If the ensemble is stationary, as it always is in this book unless specially noted, the outcome is a *stationary random process.*

14. Statistically Dependent Ensembles of Continuous Processes

In much of our work the most important probability density of an ensemble is the second probability density. Similar to it, a density of considerable value is the joint probability density of two statistically dependent ensembles. In Chapter 2, Sec. C-9, under the topic of cross-correlation we have given examples of such ensembles.

Let the two stationary ensembles be $\{f(t)\}$ and $\{g(t)\}$, and let them be graphically represented as in Fig. 11. Let the random variable ξ_1 be

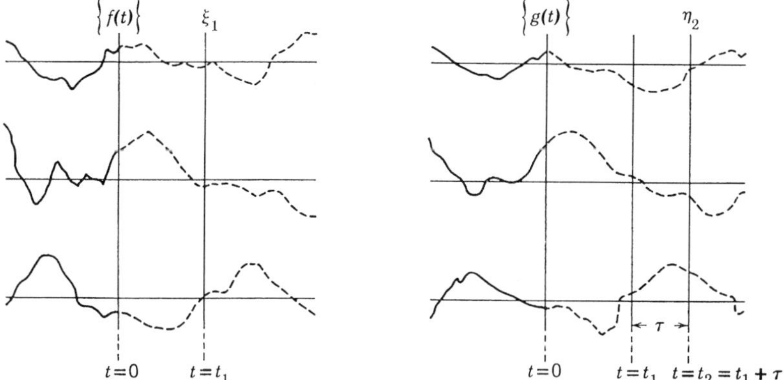

Fig. 11. Statistically dependent ensembles.

the amplitude of a member function $f(t)$ of the ensemble $\{f(t)\}$ at time $t = t_1$, and the random variable η_2 be the amplitude of the corresponding member function $g(t)$ of the ensemble $\{g(t)\}$ at time $t = t_2 = t_1 + \tau$. Reference should be made to the section cited for the pairing of $f(t)$ and $g(t)$ in the ensembles. The random variables are assumed to have continuous ranges of variation. Let x_1 be the range variable of ξ_1 and y_2 be the range variable of η_2. By analogy with the second probability density, the joint probability density for ξ_1 and η_2 is

$$P_{\xi_1\eta_2}(x_1, y_2; \tau) \tag{82}$$

so that the probability element for these variables is

$$\mathcal{P}(x_1 < \xi_1 < x_1 + dx_1, y_2 < \eta_2 < y_2 + dy_2; t_2 - t_1 = \tau) = P_{\xi 1 \eta 2}(x_1, y_2; \tau)\, dx_1\, dy_2 \quad (83)$$

Analogous to (38) we have

$$\mathcal{P}(a_1 < \xi_1 < b_1, a_2 < \eta_2 < b_2; t_2 - t_1 = \tau) = \int_{a_2}^{b_2} \int_{a_1}^{b_1} P_{\xi 1 \eta 2}(x_1, y_2; \tau)\, dx_1\, dy_2 \quad (84)$$

and to (39) we have

$$\int_{-\infty}^{\infty} \int_{-\infty}^{\infty} P_{\xi 1 \eta 2}(x_1, y_2; \tau)\, dx_1\, dy_2 = 1 \quad (85)$$

In terms of conditional probabilities, we have

$$P_{\xi 1 \eta 2}(x_1, y_2; \tau) = P_{\xi 1}(x_1) P_{\eta 2 | \xi 1}(y_2 | x_1; \tau) \quad (86)$$

or

$$P_{\xi 1 \eta 2}(x_1, y_2; \tau) = P_{\eta 2}(y_2) P_{\xi 1 | \eta 2}(x_1 | y_2; \tau) \quad (87)$$

in which the conditional probabilities have the property that

$$\int_{-\infty}^{\infty} P_{\eta 2 | \xi 1}(y_2 | x_1; \tau)\, dy_2 = 1 \quad (88)$$

and

$$\int_{-\infty}^{\infty} P_{\xi 1 | \eta 2}(x_1 | y_2; \tau)\, dx_1 = 1 \quad (89)$$

so that from (86) we obtain

$$P_{\xi 1}(x_1) = \int_{-\infty}^{\infty} P_{\xi 1 \eta 2}(x_1, y_2; \tau)\, dy_2 \quad (90)$$

and from (87)

$$P_{\eta 2}(y_2) = \int_{-\infty}^{\infty} P_{\xi 1 \eta 2}(x_1, y_2; \tau)\, dx_1 \quad (91)$$

Unlike the symmetric relations (52) through (54) for a single stationary ensemble, (82) for two different but statistically dependent stationary ensembles is not necessarily symmetrical with respect to x_1 and y_2; $P_{\eta 2 | \xi 1}(y_2 | x_1; \tau)$ is not necessarily symmetrical to $P_{\xi 1 | \eta 2}(x_1 | y_2; \tau)$; and $P_{\eta 2}(y_2)$ is not necessarily the same as $P_{\xi 1}(x_1)$ numerically.

15. Ensemble of a Discrete Random Process

If a random process (or function) has a discrete range of values, the process is a discrete random process (or a discrete random function). We should note that in both the discrete and continuous processes the time variable is continuous. Figure 12 illustrates an ensemble of a discrete process. In place of the probability densities (79), the discrete stationary random process is described by the probability distributions

$$\left.\begin{array}{l} P_{\xi 1}(x_{1i}) \\ P_{\xi 1 \xi 2}(x_{1i}, x_{2j}; \tau_1) \\ \cdot \\ \cdot \\ \cdot \\ P_{\xi 1 \ldots \xi n}(x_{1i}, \ldots, x_{nr}; \tau_1, \ldots, \tau_{n-1}) \\ \cdot \\ \cdot \\ \cdot \end{array}\right\} \qquad (92)$$

where x_{1i}, with $i = \ldots, -2, -1, 0, 1, 2, \ldots$, is the range variable of ξ_1 at $t = t_1$; x_{2j}, with $j = \ldots, -2, -1, 0, 1, 2, \ldots$, is the range variable of ξ_2 at $t = t_2 = t_1 + \tau_1$; and so on.

The distributions (92) are discrete functions of the range variables x_{1i}, x_{2j}, ... but are continuous functions of the time differences τ_1, τ_2, ... Since every value of these functions represents a probability, they are necessarily non-negative functions. Clearly the distributions must satisfy the conditions

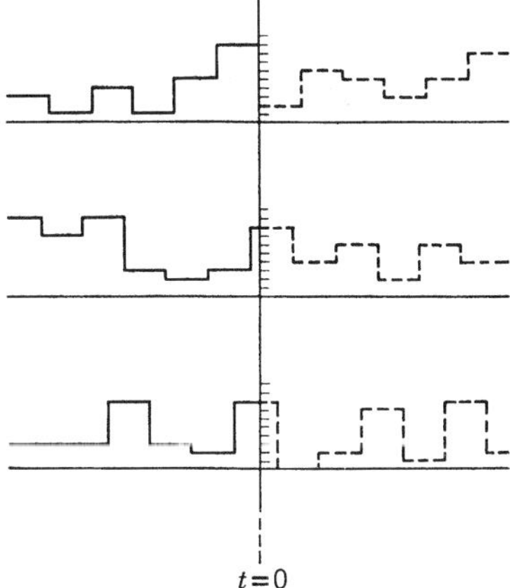

Fig. 12. An ensemble of discrete random functions.

$$\left.\begin{array}{l} \displaystyle\sum_{i=-\infty}^{\infty} P_{\xi 1}(x_{1i}) = 1 \\ \displaystyle\sum_{j=-\infty}^{\infty} \sum_{i=-\infty}^{\infty} P_{\xi 1 \xi 2}(x_{1i}, x_{2j}; \tau_1) = 1 \\ \cdot \\ \cdot \\ \cdot \\ \displaystyle\sum_{r=-\infty}^{\infty} \cdots \sum_{i=-\infty}^{\infty} P_{\xi 1 \ldots \xi n}(x_{1i}, \ldots, x_{nr}; \tau_1, \ldots, \tau_{n-1}) = 1 \\ \cdot \\ \cdot \\ \cdot \end{array}\right\} \qquad (93)$$

and the conditions

$$\left.\begin{aligned} P_{\xi 1}(x_{1i}) &= \sum_{j=-\infty}^{\infty} P_{\xi 1 \xi 2}(x_{1i}, x_{2j}; \tau_1) \\ P_{\xi 1 \xi 2}(x_{1i}, x_{2j}; \tau_1) &= \sum_{k=-\infty}^{\infty} P_{\xi 1 \xi 2 \xi 3}(x_{1i}, x_{2j}, x_{3k}; \tau_1, \tau_2) \\ &\vdots \\ P_{\xi 1 \ldots \xi n}(x_{1i}, \ldots, x_{nr}; \tau_1, \ldots, \tau_{n-1}) & \\ &= \sum_{s=-\infty}^{\infty} P_{\xi 1 \ldots \xi (n+1)}(x_{1i}, \ldots, x_{(n+1)s}; \tau_1, \ldots, \tau_n) \\ &\vdots \end{aligned}\right\} \tag{94}$$

The conditions (93) and (94) are analogous to (80) and (81) respectively. The symmetric relations (52) through (54) for the discrete stationary process are

$$P_{\xi 1}(x_{1i}) = P_{\xi 2}(x_{2j}) \tag{95}$$

$$P_{\xi 1|\xi 2}(x_{1i}|x_{2j}; \tau_1) = P_{\xi 2|\xi 1}(x_{2j}|x_{1i}; \tau_1) \tag{96}$$

and

$$P_{\xi 1 \xi 2}(x_{1i}, x_{2j}; \tau_1) = P_{\xi 2 \xi 1}(x_{2j}, x_{1i}; \tau_1) \tag{97}$$

The probability elements are

$$\mathcal{P}(\xi_1 = x_{1i}) = P_{\xi 1}(x_{1i}) \tag{98}$$

$$\mathcal{P}(\xi_1 = x_{1i}, \xi_2 = x_{2j}; t_2 - t_1 = \tau_1) = P_{\xi 1 \xi 2}(x_{1i}, x_{2j}; \tau_1) \tag{99}$$

and so on.

Analogous to (38) and (39) we find that for the discrete random process

$$\mathcal{P}(m \leq \xi_1 \leq n, m' \leq \xi_2 \leq n'; t_2 - t_1 = \tau_1) = \sum_{j=m'}^{n'} \sum_{i=m}^{n} P_{\xi 1 \xi 2}(x_{1i}, x_{2j}; \tau_1) \tag{100}$$

and

$$\begin{aligned} \mathcal{P}(-\infty < \xi_1 < \infty, -\infty < \xi_2 < \infty; t_2 - t_1 = \tau_1) & \\ &= \sum_{j=-\infty}^{\infty} \sum_{i=-\infty}^{\infty} P_{\xi 1 \xi 2}(x_{1i}, x_{2j}; \tau_1) \\ &= 1 \end{aligned} \tag{101}$$

When the second probability distribution is expressed in terms of conditional probability distributions,

$$\left.\begin{aligned} P_{\xi_1\xi_2}(x_{1i}, x_{2j}; \tau_1) &= P_{\xi_1}(x_{1i})P_{\xi_2|\xi_1}(x_{2j} \,|\, x_{1i}; \tau_1) \\ &= P_{\xi_2}(x_{2j})P_{\xi_1|\xi_2}(x_{1i} \,|\, x_{2j}; \tau_1) \end{aligned}\right\} \tag{102}$$

which are analogous to (44) and (49) for the continuous process. The conditional probability distributions satisfy the conditions

$$\sum_{j=-\infty}^{\infty} P_{\xi_2|\xi_1}(x_{2j} \,|\, x_{1i}; \tau_1) = 1 \tag{103}$$

and

$$\sum_{i=-\infty}^{\infty} P_{\xi_1|\xi_2}(x_{1i} \,|\, x_{2j}; \tau_1) = 1 \tag{104}$$

These are the analogous expressions for (45) and (50) in the continuous case.

Furthermore, like (47) and (51), the unconditional discrete probability distributions $P_{\xi_1}(x_{1i})$ and $P_{\xi_2}(x_{2j})$ are determinable from the second probability distribution by the formulas

$$P_{\xi_1}(x_{1i}) = \sum_{j=-\infty}^{\infty} P_{\xi_1\xi_2}(x_{1i}, x_{2j}; \tau_1) \tag{105}$$

and

$$P_{\xi_2}(x_{2j}) = \sum_{i=-\infty}^{\infty} P_{\xi_1\xi_2}(x_{1i}, x_{2j}; \tau_1) \tag{106}$$

16. Statistically Dependent Ensembles of Discrete Processes

If the ranges of values of the statistically dependent processes referred to in Sec. 14 are sets of discrete values, we let ξ_1 be the random amplitude of the process $f(t)$ of the ensemble $\{f(t)\}$ at $t = t_1$ and η_2 be the amplitude of $g(t)$ of $\{g(t)\}$ at $t = t_2 = t_1 + \tau$ as before, and then we follow the notations of Sec. 15 by letting x_{1i} be the range variable of ξ_1 and y_{2j} be the range variable of η_2, with $i = \ldots, -2, -1, 0, 1, 2, \ldots,$ and $j = \ldots, -2, -1, 0, 1, 2, \ldots.$ The probability element (83) becomes

$$\mathcal{P}(\xi_1 = x_{1i}, \eta_2 = y_{2j}; t_2 - t_1 = \tau) = P_{\xi_1\eta_2}(x_{1i}, y_{2j}; \tau) \tag{107}$$

The joint probability (84) becomes

$$\mathcal{P}(m \le \xi_1 \le n, m' \le \eta_2 \le n'; t_2 - t_1 = \tau) = \sum_{j=m'}^{n'} \sum_{i=m}^{n} P_{\xi_1\eta_2}(x_{1i}, y_{2j}; \tau) \tag{108}$$

and the totality of probability (85) becomes

$$\mathcal{P}(-\infty < \xi_1 < \infty, -\infty < \eta_2 < \infty; t_2 - t_1 = \tau) = \sum_{j=-\infty}^{\infty} \sum_{i=-\infty}^{\infty} P_{\xi 1 \eta 2}(x_{1i}, y_{2j}; \tau) = 1 \tag{109}$$

Expressions (86) and (87) for the relation between joint probability and conditional probability are now

$$P_{\xi 1 \eta 2}(x_{1i}, y_{2j}; \tau) = P_{\xi 1}(x_{1i}) P_{\eta 2 | \xi 1}(y_{2j} | x_{1i}; \tau) \tag{110}$$

and

$$P_{\xi 1 \eta 2}(x_{1i}, y_{2j}; \tau) = P_{\eta 2}(y_{2j}) P_{\xi 1 | \eta 2}(x_{1i} | y_{2j}; \tau) \tag{111}$$

with the conditional probability distributions satisfying the conditions

$$\sum_{j=-\infty}^{\infty} P_{\eta 2 | \xi 1}(y_{2j} | x_{1i}; \tau) = 1 \tag{112}$$

and

$$\sum_{i=-\infty}^{\infty} P_{\xi 1 | \eta 2}(x_{1i} | y_{2j}; \tau) = 1 \tag{113}$$

The relation between the first and second probability distributions is given by the formulas

$$P_{\xi 1}(x_{1i}) = \sum_{j=-\infty}^{\infty} P_{\xi 1 \eta 2}(x_{1i}, y_{2j}; \tau) \tag{114}$$

and

$$P_{\eta 2}(y_{2j}) = \sum_{i=-\infty}^{\infty} P_{\xi 1 \eta 2}(x_{1i}, y_{2j}; \tau) \tag{115}$$

REFERENCES

1. James, H. M., N. B. Nichols, and R. S. Phillips, *Theory of Servomechanisms*, McGraw-Hill Book Company, New York, 1947, Chapter 6.
2. Lawson, J. L., and G. E. Uhlenbeck, *Threshold Signals*, McGraw-Hill Book Company, New York, 1950, Chapter 3.
3. Lindsay, R. B., *Introduction to Physical Statistics*, John Wiley and Sons, New York, 1941, Chapter VI.
4. Mood, A. M., *Introduction to the Theory of Statistics*, McGraw-Hill Book Company, New York, 1950, Chapter 4.
5. Wang, M. C., and G. E. Uhlenbeck, "On the Theory of the Brownian Motion II," *Reviews of Modern Physics*, **17**, 323 (1945).

chapter 5

Averages

1. Average of a Discrete Random Variable

A useful quantity associated with a finite set of values $a_1, a_2, \ldots, a_n$ is the average or the arithmetic mean. If m denotes the average value of the set, its familiar definition is

$$m = \frac{a_1 + a_2 + \cdots + a_n}{n} \tag{1}$$

The concept of an average value is applicable to random variables, and it is of primary importance in probability theory. Consider a discrete random variable ξ with the probability distribution $P_\xi(x_i)$. Since we do not have a definite set of quantities for computing the average when the variable is random, we cannot apply definition (1). Nevertheless, a definition of the average of a random variable can be given in terms of the distribution. Before giving the definition, let us consider the experimental average which leads to the definition. Suppose that the complete set of possible values of ξ are $x_1, \ldots, x_i, \ldots, x_n$ and that of a sufficiently large total of M trials we find

$$\left.\begin{array}{l} N_1 \text{ successes for } \xi = x_1 \\ \vdots \\ N_i \text{ successes for } \xi = x_i \\ \vdots \\ N_n \text{ successes for } \xi = x_n \end{array}\right\} \tag{2}$$

where $N_1, \ldots, N_i, \ldots, N_n$ are nonzero. Then the sum of the observed values of ξ in M trials is

$$x_1 N_1 + \cdots + x_i N_i + \cdots + x_n N_n \tag{3}$$

These values being definite experimental values, definition (1) is applicable to them. Letting $\xi_{\text{em.av.}}$ be the empirical average value of ξ, per trial, in M trials, we have

$$\xi_{\text{em.av.}} = \frac{x_1 N_1 + \cdots + x_i N_i + \cdots + x_n N_n}{M} \tag{4}$$

which can be written as

$$\xi_{\text{em.av.}} = x_1 \frac{N_1}{M} + \cdots + x_i \frac{N_i}{M} + \cdots + x_n \frac{N_n}{M} \tag{5}$$

In (5) the ratios $N_1/M, \ldots, N_i/M, \ldots, N_n/M$ are frequency ratios. Now, according to Bernoulli's theorem

$$\mathcal{P}\left[\left|\frac{N_i}{M} - P_\xi(x_i)\right| < \epsilon\right] \to 1, \quad \text{as } M \to \infty \tag{6}$$

The experimental average (5) is therefore an estimate of the quantity

$$\bar{\xi} = x_1 P_\xi(x_1) + \cdots + x_i P_\xi(x_i) + \cdots + x_n P_\xi(x_n) \tag{7}$$

or

$$\bar{\xi} = \sum_{i=1}^{n} x_i P_\xi(x_i) \tag{8}$$

which is known as the *mathematical expectation* of ξ. Since the range of ξ may be negative as well as infinite, we replace (8) by the more general expression

$$\bar{\xi} = \sum_{i=-\infty}^{\infty} x_i P_\xi(x_i) \tag{9}$$

This quantity is also called the *average* or the *mean* of ξ. These terms are synonymous. The mean of ξ is the expected value of ξ per trial, and the probability that the empirical average differs from the mean by a preassigned quantity, however small, can be made as close to 1 as we please by making the number of trials sufficiently large. Various other symbols for the mean of ξ are in use, such as $E(\xi)$, $\langle\xi\rangle$, $\langle\xi\rangle_{\text{av.}}$, and μ_1.

In the example given in Sec. 1, Chapter 4, if the mean value of the outcome of the experiment is required, we find that formula (9) gives

$$\begin{aligned}\bar{\xi} &= \sum_{i=1}^{3} x_i P_\xi(x_i) \\ &= 2(\tfrac{1}{4}) + 3(\tfrac{1}{2}) + 4(\tfrac{1}{4}) \\ &= 3\end{aligned} \tag{10}$$

2. Average of a Continuous Random Variable

If a random variable ξ has a continuous range of possible values x with the probability density $P_\xi(x)$, the empirical average value of ξ per trial can be obtained as follows. Assume that the range is $(0, A)$, and let it be divided into elements each of length Δx. Let the total number of divisions be K so that the series of points marking the divisions are $0, \Delta x, 2\,\Delta x, \ldots, i\,\Delta x, \ldots, K\,\Delta x$. Suppose that a sufficiently large number of trials M are made, and of this number we find

$$\left.\begin{array}{ll} N_1 & \text{successes for } \xi \text{ falling in } (0, \Delta x) \\ N_2 & \text{successes for } \xi \text{ falling in } (\Delta x, 2\,\Delta x) \\ \cdot & \\ \cdot & \\ \cdot & \\ N_i & \text{successes for } \xi \text{ falling in } ([i-1]\,\Delta x, i\,\Delta x) \\ \cdot & \\ \cdot & \\ \cdot & \\ N_K & \text{successes for } \xi \text{ falling in } ([K-1]\,\Delta x, K\,\Delta x) \end{array}\right\} \tag{11}$$

where $N_1, N_2, \ldots, N_i, \ldots, N_K$ are nonzero. The sum of the observed values of ξ in the total of M trials will be approximately evaluated by assuming that the N_1 observations in (11) are each of the same value Δx and, similarly, the N_2 observations are each of value $2\,\Delta x$, the N_i observations are each of value $i\,\Delta x$, and the N_K observations are each of value $K\,\Delta x$. This assumption can be made if Δx is small. The sum of the observed values of ξ in a total of M trials will be

$$\Delta x\, N_1 + 2\,\Delta x\, N_2 + \cdots + i\,\Delta x\, N_i + \cdots + K\,\Delta x\, N_K \tag{12}$$

and the experimental mean value of ξ, per trial, will be

$$\xi_{\text{em.av.}} = \Delta x \frac{N_1}{M} + 2\,\Delta x \frac{N_2}{M} + \cdots + i\,\Delta x \frac{N_i}{M} + \cdots + K\,\Delta x \frac{N_K}{M} \tag{13}$$

Dividing the frequency ratios in (13) by Δx and multiplying the right-hand member of (13) by Δx, we have

$$\xi_{\text{em.av.}} = \left(\Delta x \frac{N_1}{M} \frac{1}{\Delta x} + 2\,\Delta x \frac{N_2}{M} \frac{1}{\Delta x} + \cdots + i\,\Delta x \frac{N_i}{M} \frac{1}{\Delta x} + \cdots + K\,\Delta x \frac{N_K}{M} \frac{1}{\Delta x}\right) \Delta x \tag{14}$$

or

$$\xi_{\text{em.av.}} = \sum_{i=1}^{K} i\,\Delta x \left(\frac{N_i}{M} \frac{1}{\Delta x}\right) \Delta x \tag{15}$$

In Chapter 4, Sec. 2, we have shown that $(N_i/M)/\Delta x$ is an estimate of the probability density $P_\xi(x)$. Therefore, (15) is an empirical value of the average

$$\bar{\xi} = \int_0^A x P_\xi(x)\, dx \tag{16}$$

By definition, this is the mean of ξ when its range of variation is continuous. Allowing for the possibility that ξ has an infinite range $(-\infty, \infty)$, we write (16) in the more general form

$$\bar{\xi} = \int_{-\infty}^{\infty} x P_\xi(x)\, dx \tag{17}$$

This should be compared with the analogous expression (9) for the discrete random variable.

As an example consider the probability density (62) in the illustration of Sec. 11, Chapter 4. Given this probability density for the amplitude of a random process, and assuming $\varphi_{11}(0) = 1$ for simplicity, we can evaluate the mean amplitude by (17), getting

$$\begin{aligned} \bar{\xi}_1 &= \frac{1}{\sqrt{2\pi}} \int_{-\infty}^{\infty} x_1 e^{-x_1{}^2/2}\, dx_1 \\ &= 0 \end{aligned} \tag{18}$$

This is obviously true because the probability density is an even function. However, for the purpose of illustration, let the density be shifted to the right by the amount m so that

$$P_{\xi 1}(x_1) = \frac{1}{\sqrt{2\pi}} e^{-(x_1 - m)^2/2} \tag{19}$$

With this change the mean of ξ_1 is

$$\bar{\xi}_1 = \frac{1}{\sqrt{2\pi}} \int_{-\infty}^{\infty} x_1 e^{-(x_1-m)^2/2}\, dx_1$$
$$= m \tag{20}$$

3. Moments

A quantity which appears frequently in our work is the mean square value of a random variable. The expression for this quantity can be derived easily by referring to Sec. 1. We note that, in performing the experiments for an empirical value, the random variable is now squared instead of being in its original form. But, if in a total of M trials there are N_1 successes for $\xi = x_1$, we can say that, if the function were squared, there would be the same number of successes, N_1, for $\xi^2 = x_1{}^2$. In a total of M observations made on ξ^2 and with the probability distribution of ξ as $P_\xi(x_i)$, we have

$$\left.\begin{array}{l} N_1 \text{ successes for } \xi^2 = x_1{}^2 \\ \cdot \\ \cdot \\ \cdot \\ N_i \text{ successes for } \xi^2 = x_i{}^2 \\ \cdot \\ \cdot \\ \cdot \\ N_n \text{ successes for } \xi^2 = x_n{}^2 \end{array}\right\} \tag{21}$$

Then the sum of the observed values of ξ^2 in a total of M trials is

$$x_1{}^2 N_1 + \cdots + x_i{}^2 N_i + \cdots + x_n{}^2 N_n \tag{22}$$

and the empirical average value of ξ^2, per trial, is

$$\xi^2_{\text{em. av.}} = x_1{}^2 \frac{N_1}{M} + \cdots + x_i{}^2 \frac{N_i}{M} + \cdots + x_n{}^2 \frac{N_n}{M} \tag{23}$$

It is important to note that the frequency ratios remain the same as in (5), so that the mean of ξ^2 can be written, in a manner analogous to (8), as

$$\overline{\xi^2} = \sum_{i=1}^{n} x_i{}^2 P_\xi(x_i) \tag{24}$$

or, analogous to (9), as

$$\overline{\xi^2} = \sum_{i=-\infty}^{\infty} x_i{}^2 P_\xi(x_i) \tag{25}$$

This is the definition for the mean square value of the discrete random variable ξ.

We need not further demonstrate that the mean of the mth power of ξ is

$$\overline{\xi^m} = \sum_{i=-\infty}^{\infty} x_i^m P_\xi(x_i) \tag{26}$$

Because the mechanical interpretation of this expression is a dynamical moment when $m = 1$ and is a moment of inertia when $m = 2$, the mean of ξ^m is often referred to as the mth moment of ξ. Thus $\bar{\xi}$ is the first moment, and $\overline{\xi^2}$ is the second moment.

For the continuous random variable we have, without further demonstration,

$$\overline{\xi^2} = \int_{-\infty}^{\infty} x^2 P_\xi(x)\, dx \tag{27}$$

and

$$\overline{\xi^m} = \int_{-\infty}^{\infty} x^m P_\xi(x)\, dx \tag{28}$$

4. Some General Forms of Averages

To obtain expressions of mean values of a general form involving one random variable, the procedure followed in Sec. 3 can be extended. Thus, instead of squaring or raising to some power the observed values of the random variable, we may consider an arbitrary function of these values. For example, if x_1 is observed, we may form $\sin x_1$, or $\log x_1$, or e^{-x_1}, and find the averages of these quantities. In other words, we consider the expected value of $f(\xi)$.

As we have done before, let the possible values of ξ be $x_1, \ldots, x_i, \ldots, x_n$, with the distribution function $P_\xi(x_i)$, and let M be a sufficiently large number of trials. Of this number, suppose that there are

$$\left.\begin{array}{l} N_1 \text{ successes for } f(\xi) = f(x_1) \\ \vdots \\ N_i \text{ successes for } f(\xi) = f(x_i) \\ \vdots \\ N_n \text{ successes for } f(\xi) = f(x_n) \end{array}\right\} \tag{29}$$

The sum of the observed values of $f(\xi)$ in a total of M trials is

$$f(x_1)N_1 + \cdots + f(x_i)N_i + \cdots + f(x_n)N_n \tag{30}$$

and the empirical average value of $f(\xi)$, per trial, is

$$f(\xi)\,|_{\text{em.av.}} = f(x_1)\,\frac{N_1}{M} + \cdots + f(x_i)\,\frac{N_i}{M} + \cdots + f(x_n)\,\frac{N_n}{M} \tag{31}$$

Replacing the frequency ratios by the probability distribution, we define the mean of $f(\xi)$ as

$$\overline{f(\xi)} = \sum_{i=1}^{n} f(x_i)P_\xi(x_i) \tag{32}$$

or, in a more general form, as

$$\overline{f(\xi)} = \sum_{i=-\infty}^{\infty} f(x_i)P_\xi(x_i) \tag{33}$$

We can go further and consider the presence of a parameter in $f(\xi)$ Accordingly, for the function $f(t, \xi)$ in which t is a real and continuous parameter, we desire the mean of the function with respect to the distribution of ξ. For example, the function of interest may be $A \sin (\omega_1 t + \xi)$, or $\xi \sin (\omega_1 t + \theta)$, or $A \sin (\xi t + \theta)$. The first expression represents a sine wave of fixed amplitude and frequency but random phase angle; the second represents a sine wave of fixed frequency and phase angle but random amplitude; and the third represents a sine wave of fixed amplitude and phase angle but random frequency. Other examples are $e^{j\xi t}$, $f(t + \tau + \xi)$, and so on.

Whereas mean values considered until now are single values, the mean of $f(t, \xi)$ is a function of t; that is, for each value of t, there is a single mean value with respect to the distribution of ξ. By analogy with (31), the empirical mean of $f(t, \xi)$, per trial, is

$$f(t, \xi)\,|_{\text{em.av.}} = f(t, x_1)\,\frac{N_1}{M} + \cdots + f(t, x_i)\,\frac{N_i}{M} + \cdots + f(t, x_n)\,\frac{N_n}{M} \tag{34}$$

Accordingly, with the distribution of ξ being $P_\xi(x_i)$, we define the mean of $f(t, \xi)$ with respect to the distribution by the expression

$$\overline{f(t, \xi)}^{\,\xi} = \sum_{i=-\infty}^{\infty} f(t, x_i)P_\xi(x_i) \tag{35}$$

The bar with the letter ξ over $f(t, \xi)$ indicates that the mean value is taken with respect to the distribution of ξ, not the parameter t. Later on we shall consider the average with respect to t.

When the range of ξ is continuous and the probability density of ξ is $P_\xi(x)$, definition (33) becomes

$$\overline{f(\xi)} = \int_{-\infty}^{\infty} f(x)P_\xi(x)\,dx \tag{36}$$

and definition (35) becomes

$$\overline{f(t, \xi)}^{\xi} = \int_{-\infty}^{\infty} f(t, x) P_\xi(x)\, dx \tag{37}$$

5. Variance

An important average associated with a random variable ξ is the *variance* σ_ξ^2 which is defined as

$$\sigma_\xi^2 = \overline{(\xi - \bar{\xi})^2} \tag{38}$$

The square root of the variance, that is, σ_ξ, is called the *standard deviation* of ξ.

An interpretation of the variance is shown in Fig. 1. If ξ has the

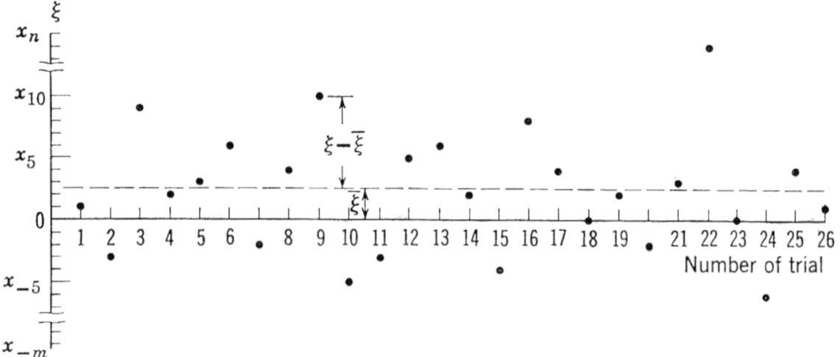

Fig. 1. Interpretation of the variance.

possible values $x_{-m}, \ldots, x_{-1}, 0, x_1, \ldots, x_n$, then in a series of experiments the outcomes are a sequence of values such as $x_1, x_{-3}, x_9, x_2, x_3, x_6, \ldots$, which are represented graphically in Fig. 1. Assuming that the mean of ξ is nonzero, we see that $\xi - \bar{\xi}$ is the random variable ξ with its reference point $\xi = 0$ shifted to $\xi = \bar{\xi}$. The mean square of this variable, that is, $\overline{(\xi - \bar{\xi})^2}$, is therefore the mean square of ξ with its mean value as the reference point. In other words, the variance of ξ is the mean square value of ξ about its mean. If the mean of ξ is zero, the variance becomes simply the mean square of ξ, and the standard deviation is then the familiar quantity, root mean square of ξ.

In statistics the variance is a measure of the spread or dispersion of distribution functions with reference to the mean. The greater the spread of statistical data, the greater is the variance. For communication problems, on the other hand, mean square values are often associated with power so that the variance of a random function is of considerable importance.

To express the variance (38) in a more convenient form, we apply (33) in which $f(\xi) = (\xi - \bar{\xi})^2$ so that

$$\sigma_\xi^2 = \sum_{i=-\infty}^{\infty} (x_i - \bar{\xi})^2 P_\xi(x_i) \tag{39}$$

In this expression, $\bar{\xi}$, as we know, is a constant. Upon expansion, (39) is

$$\begin{aligned} \sigma_\xi^2 &= \sum_{i=-\infty}^{\infty} (x_i^2 - 2\bar{\xi}x_i + \bar{\xi}^2)P_\xi(x_i) \\ &= \sum_{i=-\infty}^{\infty} x_i^2 P_\xi(x_i) - 2\bar{\xi} \sum_{i=-\infty}^{\infty} x_i P_\xi(x_i) + \bar{\xi}^2 \sum_{i=-\infty}^{\infty} P_\xi(x_i) \end{aligned} \tag{40}$$

Since by (25) the first term on the right-hand side is $\overline{\xi^2}$, by (9) the second term is $-2\bar{\xi}\bar{\xi}$, and the last term is obviously $\bar{\xi}^2$, (40) is

$$\sigma_\xi^2 = \overline{\xi^2} - \bar{\xi}^2 \tag{41}$$

If ξ is continuous, the same result can be obtained by application of (36). We have therefore shown that the variance of ξ is the mean square of ξ minus the square of the mean of ξ. The range of the random variable may be either discrete or continuous.

For an illustration consider the probability density (19) which we may regard as the first density of an ensemble of thermal noise. The mean has been found to be m by (20). To evaluate the variance, we need $\overline{\xi_1{}^2}$ which is, by (27),

$$\begin{aligned} \overline{\xi_1{}^2} &= \frac{1}{\sqrt{2\pi}} \int_{-\infty}^{\infty} x_1{}^2 e^{-(x_1-m)^2/2}\, dx \\ &= \frac{1}{\sqrt{2\pi}} \int_{-\infty}^{\infty} (u^2 + 2mu + m^2) e^{-u^2/2}\, du \\ &= \frac{1}{\sqrt{2\pi}} (\sqrt{2\pi} + m^2\sqrt{2\pi}) \\ &= 1 + m^2 \end{aligned} \tag{42}$$

The variance of the thermal noise is therefore

$$\begin{aligned} \sigma_{\xi 1}^2 &= \overline{\xi_1{}^2} - \bar{\xi}_1{}^2 = 1 + m^2 - m^2 \\ &= 1 \end{aligned} \tag{43}$$

6. Averages Involving Multiple Random Variables

Consider the stationary ensemble in Fig. 2. The random functions take only discrete values. We may regard these functions as approximations to continuous functions by steps of discrete values. Let the range of values of ξ_1' at $t = t_1'$ be $x_{1,-m}, \ldots, x_{1,-1}, x_{1,0}, x_{1,1}, \ldots, x_{1,i}, \ldots, x_{1,n}$ and the range of values of ξ_2' at $t = t_2' = t_1' + \tau_1$ be $x_{2,-m}, \ldots, x_{2,-1}, x_{2,0}, x_{2,1}, \ldots, x_{2,j}, \ldots, x_{2,n}$. The first subscript 1 of each value of the first set indicates that the time associated with the set is $t = t_1'$, and the first subscript 2 of the second set indicates that $t = t_2'$. Numerically the corresponding values of the two sets are the same; that is, $x_{1,i} = x_{2,j}$ when $i = j$.

Our problem is to find the expression for the mean value of the product of the variables ξ_1' and ξ_2'. Guided by the expression of the em-

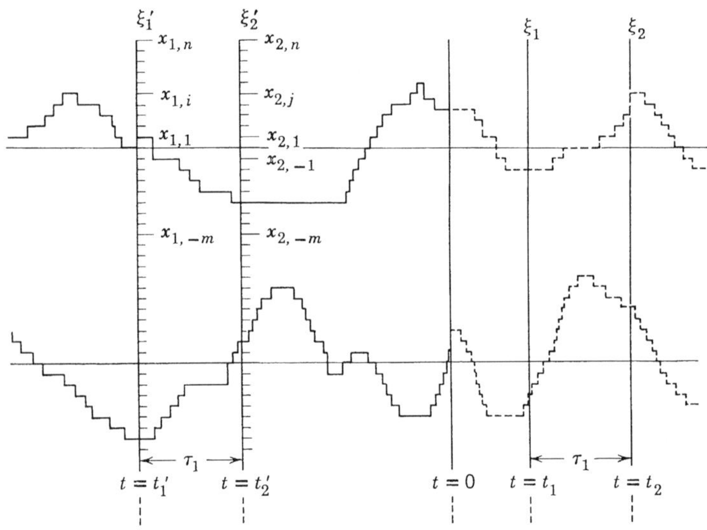

Fig. 2. An ensemble of discrete random functions.

pirical mean based upon statistical data taken from repeated experiments, we shall define the expected mean for future time in terms of a probability distribution. To do this, let a sufficiently large number of observations, M, be made on the ensemble at $t = t_1'$ and at $t = t_2' = t_1' + \tau_1$. Each observation consists of two values, one at each of the two specified times. The difference in time, τ_1, is maintained the same throughout the series of observations. Letting $N_{i,j}$ be the number of observations in which ξ_1' is found to have the value $x_{1,i}$ and ξ_2' is found to have the value $x_{2,j}$, we tabulate the observations in the accompanying table. The top row gives the range of values of ξ_1', and the extreme

	ξ_1'								
ξ_2'	$x_{1,-m}$	$\cdots$	$x_{1,-1}$	$x_{1,0}$	$x_{1,1}$	$\cdots$	$x_{1,i}$	$\cdots$	$x_{1,n}$
$x_{2,-m}$	$N_{-m,-m}$	$\cdots$	$N_{-1,-m}$	$N_{0,-m}$	$N_{1,-m}$	$\cdots$	$N_{i,-m}$	$\cdots$	$N_{n,-m}$
$\vdots$	$\vdots$		$\vdots$	$\vdots$	$\vdots$		$\vdots$		$\vdots$
$x_{2,-1}$	$N_{-m,-1}$	$\cdots$	$N_{-1,-1}$	$N_{0,-1}$	$N_{1,-1}$	$\cdots$	$N_{i,-1}$	$\cdots$	$N_{n,-1}$
$x_{2,0}$	$N_{-m,0}$	$\cdots$	$N_{-1,0}$	$N_{0,0}$	$N_{1,0}$	$\cdots$	$N_{i,0}$	$\cdots$	$N_{n,0}$
$x_{2,1}$	$N_{-m,1}$	$\cdots$	$N_{-1,1}$	$N_{0,1}$	$N_{1,1}$	$\cdots$	$N_{i,1}$	$\cdots$	$N_{n,1}$
$\vdots$	$\vdots$		$\vdots$	$\vdots$	$\vdots$		$\vdots$		$\vdots$
$x_{2,j}$	$N_{-m,j}$	$\cdots$	$N_{-1,j}$	$N_{0,j}$	$N_{1,j}$	$\cdots$	$N_{i,j}$	$\cdots$	$N_{n,j}$
$\vdots$	$\vdots$		$\vdots$	$\vdots$	$\vdots$		$\vdots$		$\vdots$
$x_{2,n}$	$N_{-m,n}$	$\cdots$	$N_{-1,n}$	$N_{0,n}$	$N_{1,n}$	$\cdots$	$N_{i,n}$	$\cdots$	$N_{n,n}$

left column gives the range of values of ξ_2'. At the intersection of a column and a row is the number of trials, in a total of M trials, in which we find the values of ξ_1' and ξ_2' to be those indicated at the heads of the column and the row respectively.

We are interested in the average value of the product of the two values in an observation. For the empirical average, first we write the expression for the sum of the products for the total of M observations, remembering that each product is that of the two values in an observation. The sum is

$$\sum_{j=-m}^{n} \sum_{i=-m}^{n} x_{1,i} x_{2,j} N_{i,j} \tag{44}$$

Dividing this quantity by M, we have the empirical average of the product of ξ_1' and ξ_2' per observation (or trial); that is,

$$\xi_1'\xi_2'\big|_{\text{em.av.}} = \sum_{j=-m}^{n} \sum_{i=-m}^{n} x_{1,i} x_{2,j} \frac{N_{i,j}}{M} \tag{45}$$

In this expression the ratio $N_{i,j}/M$ is the number of successes of finding $\xi_1' = x_{1,i}$ and $\xi_2' = x_{2,j}$ to the total number of trials. The variables ξ_1' and ξ_2' are separated by time τ_1. It is not difficult to see that the frequency ratio $N_{i,j}/M$ is an estimate of the probability distribution

$$\mathcal{P}(\xi_1 = x_{1i}, \xi_2 = x_{2j}; t_2 - t_1 = \tau_1) = P_{\xi_1\xi_2}(x_{1i}, x_{2j}; \tau_1) \tag{46}$$

which is expression (99) of Chapter 4, Sec. 15. The commas in the subscripts of (45) have been dropped for convenience. It should be noted that the joint distribution (46) concerns ξ_1 at future time $t = t_1$ and ξ_2 at future time $t = t_2 = t_1 + \tau_1$ as indicated in Fig. 2. By Bernoulli's theorem we have that

$$\mathcal{P}\left(\left|\frac{N_{ij}}{M} - P_{\xi_1\xi_2}\right| < \epsilon\right) \to 1 \quad \text{as } M \to \infty \tag{47}$$

These considerations lead us to define the mean of the product $\xi_1\xi_2$ as

$$\overline{\xi_1\xi_2} = \sum_{j=-\infty}^{\infty} \sum_{i=-\infty}^{\infty} x_{1i}x_{2j}P_{\xi_1\xi_2}(x_{1i}, x_{2j}; \tau_1) \tag{48}$$

In view of the fact that a joint distribution can be expressed in terms of a conditional distribution as given by (102), Chapter 4, Sec. 15, we write (48) in the form

$$\overline{\xi_1\xi_2} = \sum_{j=-\infty}^{\infty} \sum_{i=-\infty}^{\infty} x_{1i}x_{2j}P_{\xi_1}(x_{1i})P_{\xi_2|\xi_1}(x_{2j}|x_{1i}; \tau_1) \tag{49}$$

As we shall see, this form of (48) is easier to manage in specific applications.

If the two random variables do not belong to the same ensemble, the joint probability distribution becomes (107) of Chapter 4, Sec. 16. Referring to this section, we define the mean of the product of ξ_1 and η_2 of two ensembles, in a manner analogous to (48), as

$$\overline{\xi_1\eta_2} = \sum_{j=-\infty}^{\infty} \sum_{i=-\infty}^{\infty} x_{1i}y_{2j}P_{\xi_1\eta_2}(x_{1i}, y_{2j}; \tau_1) \tag{50}$$

which is equivalent to

$$\overline{\xi_1\eta_2} = \sum_{j=-\infty}^{\infty} \sum_{i=-\infty}^{\infty} x_{1i}y_{2j}P_{\xi_1}(x_{1i})P_{\eta_2|\xi_1}(y_{2j}|x_{1i}; \tau_1) \tag{51}$$

By analogy with (48) we define the mean of the product $\xi_1\xi_2$, when both ξ_1 and ξ_2 are continuous random variables, as

$$\overline{\xi_1\xi_2} = \int_{-\infty}^{\infty}\int_{-\infty}^{\infty} x_1x_2P_{\xi_1\xi_2}(x_1, x_2; \tau_1)\, dx_1\, dx_2 \tag{52}$$

Corresponding to (49), we have

$$\overline{\xi_1\xi_2} = \int_{-\infty}^{\infty}\int_{-\infty}^{\infty} x_1x_2P_{\xi_1}(x_1)P_{\xi_2|\xi_1}(x_2|x_1; \tau_1)\, dx_1\, dx_2 \tag{53}$$

Similarly, analogous to (50) and (51), we have

$$\overline{\xi_1\eta_2} = \int_{-\infty}^{\infty}\int_{-\infty}^{\infty} x_1 y_2 P_{\xi_1\eta_2}(x_1, y_2; \tau_1)\, dx_1\, dy_2 \tag{54}$$

and

$$\overline{\xi_1\eta_2} = \int_{-\infty}^{\infty}\int_{-\infty}^{\infty} x_1 y_2 P_{\xi_1}(x_1) P_{\eta_2|\xi_1}(y_2 | x_1; \tau_1)\, dx_1\, dy_2 \tag{55}$$

In view of the details which have been given for various analogous developments, it does not seem necessary to give further explanation for (52) through (55).

For expressions of averages which are more general than the ones we have considered in this section, we refer to Sec. 4 and see how the extensions there may be applied to multiple variables. In (29), with the numbers of successes remaining the same we can consider an arbitrary function of the observed values instead of the values themselves. Then the empirical average of a function of the observed values is expressible in terms of the frequency ratios by (31) in the same manner that (5) is written for the empirical average of the observed values themselves. It is now clear that we need not restrict the average to the product $\xi_1\xi_2$ but may consider a more general form of average, that is, that of a function of the two variables ξ_1 and ξ_2. Thus, instead of (44), we write

$$\sum_{j=-m}^{n}\sum_{i=-m}^{n} f(x_{1,i}, x_{2,j}) N_{i,j} \tag{56}$$

as the sum of the values of $f(\xi_1', \xi_2')$ for M observations and

$$f(\xi_1', \xi_2')\big|_{\text{em.av.}} = \sum_{j=-m}^{n}\sum_{i=-m}^{n} f(x_{1,i}, x_{2,j}) \frac{N_{i,j}}{M} \tag{57}$$

as the empirical average of $f(\xi_1', \xi_2')$ per observation.

This generalization leads to the definitions

$$\overline{f(\xi_1, \xi_2)} = \sum_{j=-\infty}^{\infty}\sum_{i=-\infty}^{\infty} f(x_{1i}, x_{2j}) P_{\xi_1\xi_2}(x_{1i}, x_{2j}; \tau_1) \tag{58}$$

and

$$\overline{f(\xi_1, \eta_2)} = \sum_{j=-\infty}^{\infty}\sum_{i=-\infty}^{\infty} f(x_{1i}, y_{2j}) P_{\xi_1\eta_2}(x_{1i}, y_{2j}; \tau_1) \tag{59}$$

for discrete variables and

$$\overline{f(\xi_1, \xi_2)} = \int_{-\infty}^{\infty}\int_{-\infty}^{\infty} f(x_1, x_2) P_{\xi_1\xi_2}(x_1, x_2; \tau_1)\, dx_1\, dx_2 \tag{60}$$

and

$$\overline{f(\xi_1, \eta_2)} = \int_{-\infty}^{\infty} \int_{-\infty}^{\infty} f(x_1, y_2) P_{\xi_1\eta_2}(x_1, y_2; \tau_1) \, dx_1 \, dy_2 \tag{61}$$

for continuous variables.

The extension of the formulas for mean values to three variables follows quite simply from the form of the expressions for two variables. We shall not go into details. Analogous to (52), we define the mean of the product $\xi_1\xi_2\xi_3$ as

$$\overline{\xi_1\xi_2\xi_3} = \int_{-\infty}^{\infty} \int_{-\infty}^{\infty} \int_{-\infty}^{\infty} x_1 x_2 x_3 P_{\xi_1\xi_2\xi_3}(x_1, x_2, x_3; \tau_1, \tau_2) \, dx_1 \, dx_2 \, dx_3 \tag{62}$$

and, analogous to (54), we define the product of $\xi_1\eta_2\zeta_3$, with ξ_1, η_2, and ζ_3 belonging to three stationary ensembles, as

$$\overline{\xi_1\eta_2\zeta_3} = \int_{-\infty}^{\infty} \int_{-\infty}^{\infty} \int_{-\infty}^{\infty} x_1 y_2 z_3 P_{\xi_1\eta_2\zeta_3}(x_1, y_2, z_3; \tau_1, \tau_2) \, dx_1 \, dy_2 \, dz_3 \tag{63}$$

Since expressions for three discrete variables and for $\overline{f(\xi_1, \xi_2, \xi_3)}$ and $\overline{f(\xi_1, \eta_2, \zeta_3)}$ are obvious, they will not be written.

7. Average of the Product of Independent Random Variables

In some situations the time difference τ_1 between ξ_1 and η_2 in (54) is unimportant. Such is the case when two statistically independent ensembles are under consideration. For example, we may have an ensemble of speech waves and an ensemble of thermal noise. Even if we are concerned with statistically dependent ensembles, the variables ξ_1 and η_2 become practically independent when τ_1 is sufficiently large. We assume that the ensembles do not contain periodic components. For instance, if the input to a linear system is a stationary random process, the input and output are statistically dependent; but, if the difference in time of observation with reference to a common origin is large, then in a practical situation ξ_1 and η_2 are statistically independent. By statistical independence between ξ_1 and η_2 we mean that in the relation [equation (86), Chapter 4, Sec. 14]

$$P_{\xi_1\eta_2}(x_1, y_2; \tau_1) = P_{\xi_1}(x_1) P_{\eta_2|\xi_1}(y_2 | x_1; \tau_1) \tag{64}$$

the conditional probability density becomes an unconditional density; that is,

$$P_{\eta_2|\xi_1}(y_2 | x_1; \tau_1) = P_{\eta_2}(y_2) \tag{65}$$

which is in agreement with (15) of Chapter 3, Sec. 6.

Under conditions which render ξ_1 and η_2 statistically independent,

(54) can be put into the simpler form

$$\overline{\xi_1\eta_2} = \int_{-\infty}^{\infty}\int_{-\infty}^{\infty} x_1 y_2 P_{\xi 1}(x_1)P_{\eta 2}(y_2)\, dx_1\, dy_2 \tag{66}$$

The double integral may be expressed as

$$\overline{\xi_1\eta_2} = \int_{-\infty}^{\infty} y_2 P_{\eta 2}(y_2)\, dy_2 \int_{-\infty}^{\infty} x_1 P_{\xi 1}(x_1)\, dx_1 \tag{67}$$

According to (17) the first integral is the mean of ξ_1, and the second is the mean of η_2. Hence

$$\overline{\xi_1\eta_2} = \bar{\xi}_1\bar{\eta}_2 \tag{68}$$

We have therefore shown that the mean of the product of two statistically independent random variables is the product of their mean values. Clearly this result can be extended to any finite number of mutually independent random variables. Since a similar proof can be given for discrete random variables, (68) is applicable to both continuous and discrete independent variables.

We may extend these considerations to special circumstances in the mean of two random variables in the same ensemble. Thus in an ensemble of messages or noise if the difference in time of observation, τ_1, is large, it is practically true that the random variables ξ_1 and ξ_2 at $t = t_1$ and $t = t_2 = t_1 + \tau_1$ respectively are statistically independent. In making this statement, we assume that there are no periodic components in the ensemble.

With τ_1 approaching infinity, we can state that for physical situations the relation [equation (44), Chapter 4, Sec. 10]

$$P_{\xi 1 \xi 2}(x_1, x_2; \tau_1) = P_{\xi 1}(x_1)P_{\xi 2|\xi 1}(x_2 | x_1; \tau_1) \tag{69}$$

reduces to

$$P_{\xi 1 \xi 2}(x_1, x_2) = P_{\xi 1}(x_1)P_{\xi 2}(x_2) \tag{70}$$

because by statistical independence between ξ_1 and ξ_2 we mean that

$$P_{\xi 2|\xi 1}(x_2 | x_1; \tau_1) = P_{\xi 2}(x_2) \tag{71}$$

Under the condition of statistical independence as τ_1 tends to infinity, the mean value (52) becomes

$$\begin{aligned}\overline{\xi_1\xi_2} &= \int_{-\infty}^{\infty}\int_{-\infty}^{\infty} x_1 x_2 P_{\xi 1}(x_1)P_{\xi 2}(x_2)\, dx_1\, dx_2 \\ &= \int_{-\infty}^{\infty} x_1 P_{\xi 1}(x_1)\, dx_1 \int_{-\infty}^{\infty} x_2 P_{\xi 2}(x_2)\, dx_2 \\ &= \bar{\xi}_1\bar{\xi}_2 \end{aligned} \tag{72}$$

Since the ensemble is stationary

$$\bar{\xi}_1 = \bar{\xi}_2 \tag{73}$$

so that (72) is

$$\overline{\xi_1 \xi_2} = \bar{\xi}_1{}^2 = \bar{\xi}_2{}^2 \tag{74}$$

8. Average and Average Square of the Sum of Random Variables

A random process can frequently be the sum of two or more random processes. For example, the input to a transmission system may consist of the sum of several messages and a noise. The mean of the sum of random processes is an important quantity, and its relationship to the means of the components can be found in a simple manner.

Consider the sum of two random variables $\xi + \eta$. Let the joint probability density of the variables be $P_{\xi\eta}(x, y)$. We consider τ_1 to be either zero or fixed so that it need not appear in the joint density as a variable. For the mean of $\xi + \eta$ definition (61) is applicable. Since $f(\xi, \eta) = \xi + \eta$, we have

$$\overline{\xi + \eta} = \int_{-\infty}^{\infty} \int_{-\infty}^{\infty} (x + y) P_{\xi\eta}(x, y)\, dx\, dy \tag{75}$$

which may be expressed as

$$\overline{\xi + \eta} = \int_{-\infty}^{\infty} x\, dx \int_{-\infty}^{\infty} P_{\xi\eta}(x, y)\, dy + \int_{-\infty}^{\infty} y\, dy \int_{-\infty}^{\infty} P_{\xi\eta}(x, y)\, dx \tag{76}$$

According to (47) and (51) of Chapter 4, Sec. 10, the first integrals of the double integrals are

$$\int_{-\infty}^{\infty} P_{\xi\eta}(x, y)\, dx = P_\eta(y) \tag{77}$$

and

$$\int_{-\infty}^{\infty} P_{\xi\eta}(x, y)\, dy = P_\xi(x) \tag{78}$$

so that (76) becomes

$$\begin{aligned} \overline{\xi + \eta} &= \int_{-\infty}^{\infty} x P_\xi(x)\, dx + \int_{-\infty}^{\infty} y P_\eta(y)\, dy \\ &= \bar{\xi} + \bar{\eta} \end{aligned} \tag{79}$$

Hence we have the result that the mean of the sum of two random variables, which may or may not be statistically independent, is the sum of their means. The same result can be shown to hold for discrete variables. By a similar procedure we can readily show that for a set of random variables $\xi_1, \xi_2, \ldots, \xi_n$

$$\overline{\xi_1 + \xi_2 + \cdots + \xi_n} = \bar{\xi}_1 + \bar{\xi}_2 + \cdots + \bar{\xi}_n \tag{80}$$

It follows immediately that the mean square of $\xi + \eta$ is, again by (61),

$$\begin{aligned}
\overline{(\xi + \eta)^2} &= \int_{-\infty}^{\infty}\int_{-\infty}^{\infty} (x + y)^2 P_{\xi\eta}(x, y)\, dy\, dx \\
&= \int_{-\infty}^{\infty} x^2\, dx \int_{-\infty}^{\infty} P_{\xi\eta}(x, y)\, dy + \int_{-\infty}^{\infty} y^2\, dy \int_{-\infty}^{\infty} P_{\xi\eta}(x, y)\, dx \\
&\qquad + 2\int_{-\infty}^{\infty}\int_{-\infty}^{\infty} xyP_{\xi\eta}(x, y)\, dx\, dy \\
&= \int_{-\infty}^{\infty} x^2 P_\xi(x)\, dx + \int_{-\infty}^{\infty} y^2 P_\eta(y)\, dy + 2\int_{-\infty}^{\infty}\int_{-\infty}^{\infty} xyP_{\xi\eta}(x, y)\, dx\, dy \\
&= \overline{\xi^2} + \overline{\eta^2} + 2\overline{\xi\eta}
\end{aligned} \tag{81}$$

Stated in words, the mean of the square of the sum $\xi + \eta$ is equal to the sum of the mean squares of ξ and η and twice the mean of their product. The result that we have shown can be obtained by first expanding the square of the sum of ξ and η and then taking the mean of each of the terms in the expansion. However, the validity of the procedure should be proved, as we have done. Obviously, (81) applies to discrete variables as well.

The variables ξ and η in (81) may or may not be statistically independent; but, if they are independent, then

$$\overline{(\xi + \eta)^2} = \overline{\xi^2} + \overline{\eta^2} + 2\bar{\xi}\bar{\eta} \tag{82}$$

where we have applied (68) to the last term of (81).

Extending the procedure indicated in (81), we can readily show that for the sum of n random variables

$$\begin{aligned}
\overline{\left(\sum_{i=1}^{n} \xi_i\right)^2} &= \sum_{j=1}^{n}\sum_{i=1}^{n} \overline{\xi_i\xi_j} \\
&= \sum_{i=1}^{n} \overline{\xi_i^2} + \sum_{j=1}^{n}\sum_{i=1}^{n} \overline{\xi_i\xi_j}\bigg|_{i\neq j}
\end{aligned} \tag{83}$$

and, if $\xi_1, \ldots, \xi_n$ are mutually independent random variables, then

$$\overline{\left(\sum_{i=1}^{n} \xi_i\right)^2} = \sum_{i=1}^{n} \overline{\xi_i^2} + \sum_{j=1}^{n} \sum_{i=1}^{n} \bar{\xi}_i \bar{\xi}_j \Big|_{i \neq j} \tag{84}$$

9. Characteristic Functions

A specific form of (37), which is of considerable importance and interest, is obtained by letting

$$f(t, \xi) = e^{jt\xi} \tag{85}$$

With this choice of the arbitrary function, (37) is

$$\overline{e^{jt\xi}}^{\xi} = \int_{-\infty}^{\infty} e^{jtx} P_\xi(x)\, dx \tag{86}$$

In the theory of probability this average, as a function of the parameter t (real), is called the *characteristic function* of the probability density $P_\xi(x)$. Denoting this function by $p_\xi(t)$, we can write (86) as

$$p_\xi(t) = \int_{-\infty}^{\infty} P_\xi(x) e^{jtx}\, dx \tag{87}$$

In the terminology of harmonic analysis the characteristic function, as we see it in the form (87), is the Fourier transform of the probability density function. It is a special transform because the function under transformation is real and non-negative and has an integral over $(-\infty, \infty)$ equal to 1. Clearly, if $P_\xi(x)$ is even, the characteristic function is real. But in general $p_\xi(t)$ is complex. To express $P_\xi(x)$ in terms of $p_\xi(t)$, we apply the Fourier transforms (125) and (126) of Chapter 2, Sec. B-1, and obtain

$$P_\xi(x) = \frac{1}{2\pi} \int_{-\infty}^{\infty} p_\xi(t) e^{-jtx}\, dt \tag{88}$$

One of the uses of the characteristic function is in the calculation of moments. To see this let us differentiate (87) with respect to t. The nth derivative is

$$p_\xi^{(n)}(t) = j^n \int_{-\infty}^{\infty} x^n P_\xi(x) e^{jtx}\, dx \tag{89}$$

Allowing t to take the value 0 and rearranging, we get

$$\int_{-\infty}^{\infty} x^n P_\xi(x)\, dx = \frac{1}{j^n} p_\xi^{(n)}(0) \tag{90}$$

The left-hand member of (90) is clearly the nth moment of ξ, according to (28). Hence

$$\overline{\xi^n} = \frac{1}{j^n} p_\xi^{(n)}(0) \tag{91}$$

By this result we can determine the moments of ξ through differentiation of the characteristic function.

As an example consider

$$P_\xi(x) = \frac{1}{\sqrt{2\pi}} e^{-x^2/2} \tag{92}$$

The characteristic function of this density, by (87), is

$$\begin{aligned} p_\xi(t) &= \frac{1}{\sqrt{2\pi}} \int_{-\infty}^{\infty} e^{-x^2/2} e^{jtx}\, dx \\ &= \frac{1}{\sqrt{2\pi}} \int_{-\infty}^{\infty} e^{-x^2/2} \cos tx\, dx \\ &= e^{-t^2/2} \end{aligned} \tag{93}$$

The derivatives are

$$\left.\begin{aligned} p_\xi'(t) &= -te^{-t^2/2} \\ p_\xi''(t) &= (t^2 - 1)e^{-t^2/2} \\ p_\xi'''(t) &= (-t^3 + 3t)e^{-t^2/2} \\ p_\xi''''(t) &= (t^4 - 6t + 3)e^{-t^2/2} \\ &\vdots \end{aligned}\right\} \tag{94}$$

Hence (91) gives the moments

$$\left.\begin{aligned} \bar{\xi} &= 0 \\ \overline{\xi^2} &= 1 \\ \overline{\xi^3} &= 0 \\ \overline{\xi^4} &= 3 \\ &\vdots \end{aligned}\right\} \tag{95}$$

An important application of the characteristic function is in the development of a method for determining the probability density of the sum of statistically independent variables. As a step in showing this application, let us find the relation between the characteristic functions for a set of independent random variables and that for their sum. For simplicity we consider two random variables ξ and η whose probability densities are $P_\xi(x)$ and $P_\eta(y)$ respectively and whose joint density is $P_{\xi\eta}(x, y)$. The sum of the variables is

$$\zeta = \xi + \eta \tag{96}$$

which is a random variable with probability density $P_\zeta(z)$. Let the characteristic functions of $P_\zeta(z)$, $P_\xi(x)$, and $P_\eta(y)$ be $p_\zeta(t)$, $p_\xi(t)$, and $p_\eta(t)$ respectively. According to definition

$$\begin{aligned} p_\zeta(t) &= \overline{e^{jt\zeta}}^{\zeta} \\ &= \overline{e^{jt(\xi+\eta)}}^{\xi\eta} \end{aligned} \tag{97}$$

Applying (61) with parameter t as in (37), we have

$$p_\zeta(t) = \int_{-\infty}^{\infty} \int_{-\infty}^{\infty} e^{jt(x+y)} P_{\xi\eta}(x, y)\, dx\, dy \tag{98}$$

As in Sec. 7 we consider τ_1 to be constant so that it need not appear in the joint density. Assuming that ξ and η are statistically independent, we replace the joint density by the product of the unconditional densities and obtain

$$\begin{aligned} p_\zeta(t) &= \int_{-\infty}^{\infty} \int_{-\infty}^{\infty} P_\xi(x) P_\eta(y) e^{jt(x+y)}\, dx\, dy \\ &= \int_{-\infty}^{\infty} P_\xi(x) e^{jtx}\, dx \int_{-\infty}^{\infty} P_\eta(y) e^{jty}\, dy \end{aligned} \tag{99}$$

By definition (87), the result is the product of the characteristic functions of $P_\xi(x)$ and $P_\eta(y)$; that is,

$$p_\zeta(t) = p_\xi(t) p_\eta(t) \tag{100}$$

We have therefore shown that the characteristic function for the sum of two statistically independent random variables is equal to the product of the characteristic functions for the component variables.

By an extension of the proof given, we can show that, if $\xi_1, \xi_2, \ldots, \xi_n$ are mutually independent random variables for which the characteristic functions are $p_{\xi 1}(t), p_{\xi 2}(t), \ldots, p_{\xi n}(t)$ respectively, the characteristic function $p_\xi(t)$ for their sum

$$\xi = \xi_1 + \xi_2 + \cdots + \xi_n \tag{101}$$

is

$$p_\xi(t) = p_{\xi 1}(t) p_{\xi 2}(t) \ldots p_{\xi n}(t) \tag{102}$$

To show the relation among the probability densities, we return to the simpler sum of two random variables. By taking the Fourier transform of both sides of equation (100), we have

$$\frac{1}{2\pi} \int_{-\infty}^{\infty} p_\zeta(t) e^{-jtz}\, dt = \frac{1}{2\pi} \int_{-\infty}^{\infty} p_\xi(t) p_\eta(t) e^{-jtz}\, dt \tag{103}$$

According to (88) the left-hand member is the probability density $P_\zeta(z)$. To express the right-hand member in terms of probability densities, we follow the method of Chapter 2, Sec. B-5. Accordingly, we replace $p_\xi(t)$ by (87) and then invert the order of integration. Thus equation (103) becomes

$$\begin{aligned} P_\zeta(z) &= \frac{1}{2\pi} \int_{-\infty}^{\infty} p_\eta(t) e^{-jtz}\, dt \int_{-\infty}^{\infty} P_\xi(u) e^{jtu}\, du \\ &= \frac{1}{2\pi} \int_{-\infty}^{\infty} P_\xi(u)\, du \int_{-\infty}^{\infty} p_\eta(t) e^{-jt(z-u)}\, dt \end{aligned} \tag{104}$$

Since by (88)

$$\frac{1}{2\pi} \int_{-\infty}^{\infty} p_\eta(t) e^{-jt(z-u)}\, dt = P_\eta(z - u) \tag{105}$$

equation (104) is finally

$$P_\zeta(z) = \int_{-\infty}^{\infty} P_\xi(u) P_\eta(z - u)\, du \tag{106}$$

This is the relation required for the determination of the probability density of the sum of two statistically independent random variables from the densities of the component variables. It is a convolution integral.

By repeating the convolution (106), we can express the probability density of the sum of n mutually independent random variables in terms of the probability densities of the components. For three independent variables ξ_1, ξ_2, and ξ_3 whose densities are $P_{\xi 1}(x_1)$, $P_{\xi 2}(x_2)$, and $P_{\xi 3}(x_3)$ respectively, the density $P_\xi(x)$ of the sum

$$\xi = \xi_1 + \xi_2 + \xi_3 \tag{107}$$

is

$$P_\xi(x) = \int_{-\infty}^{\infty} P_{\xi 3}(v)\, dv \int_{-\infty}^{\infty} P_{\xi 2}(u) P_{\xi 1}(x - v - u)\, du \tag{108}$$

and, for n independent variables $\xi_1, \ldots, \xi_n$ with densities $P_{\xi 1}(x_1), \ldots, P_{\xi n}(x_n)$, the density $P_\xi(x)$ of the sum

$$\xi = \xi_1 + \cdots + \xi_n \tag{109}$$

is

$$P_\xi(x) = \int_{-\infty}^{\infty} P_{\xi n}(u_n)\, du_n \int_{-\infty}^{\infty} P_{\xi n-1}(u_{n-1})\, du_{n-1} \cdots \int_{-\infty}^{\infty} P_{\xi 2}(u_2) P_{\xi 1}(x - u_n - \cdots - u_2)\, du_2 \tag{110}$$

10. Variance of a Sum of Random Variables

Since the sum of random variables is a random variable, the definition of variance given in Sec. 5 is applicable to the sum $\xi + \eta$ so that the variance of the sum is

$$\sigma_{\xi+\eta}^2 = \overline{(\xi + \eta - \overline{\xi + \eta})^2} \tag{111}$$

In this expression, $\overline{\xi + \eta} = \bar{\xi} + \bar{\eta}$ by (79). Hence

$$\begin{aligned} \sigma_{\xi+\eta}^2 &= \overline{(\xi + \eta - \bar{\xi} - \bar{\eta})^2} \\ &= \overline{[(\xi - \bar{\xi}) + (\eta - \bar{\eta})]^2} \end{aligned} \tag{112}$$

and by (81)

$$\sigma_{\xi+\eta}^2 = \overline{(\xi - \bar{\xi})^2} + \overline{(\eta - \bar{\eta})^2} + 2\overline{(\xi - \bar{\xi})(\eta - \bar{\eta})} \tag{113}$$

Letting $\sigma_\xi{}^2$ and $\sigma_\eta{}^2$ be the variances of ξ and η respectively, we have the result that the variance of the sum $\xi + \eta$ is

$$\sigma_{\xi+\eta}^2 = \sigma_\xi{}^2 + \sigma_\eta{}^2 + 2\overline{(\xi - \bar{\xi})(\eta - \bar{\eta})} \tag{114}$$

In particular, if ξ and η are independent, then, according to (68), $\overline{(\xi - \bar{\xi})(\eta - \bar{\eta})} = \overline{(\xi - \bar{\xi})}\,\overline{(\eta - \bar{\eta})}$. But $\overline{(\xi - \bar{\xi})} = 0$, and $\overline{(\eta - \bar{\eta})} = 0$. Therefore, under the condition of independence of ξ and η, (114) becomes

$$\sigma_{\xi+\eta}^2 = \sigma_\xi{}^2 + \sigma_\eta{}^2 \tag{115}$$

Our result is that, if ξ and η are independent variables, the variance of their sum is equal to the sum of their variances.

Clearly, this proof can be extended to n mutually independent variables. Thus if $\xi_1, \xi_2, \ldots, \xi_n$ are n mutually independent random variables with the variances $\sigma_1{}^2, \sigma_2{}^2, \ldots, \sigma_n{}^2$ respectively, the variance σ^2 of the sum

$$\xi = \xi_1 + \xi_2 + \cdots + \xi_n \tag{116}$$

is

$$\sigma^2 = \sigma_1{}^2 + \sigma_2{}^2 + \cdots + \sigma_n{}^2 \tag{117}$$

chapter 6

The Poisson Distribution, the Normal Density, and Other Densities

1. The Poisson Distribution

A discrete distribution function of considerable importance in probability theory is the Poisson distribution. For a better appreciation of its application to physical problems, we shall derive this function in association with a particular stationary ensemble. The ensemble is shown in Fig. 1 where each member function alternates at random inter-

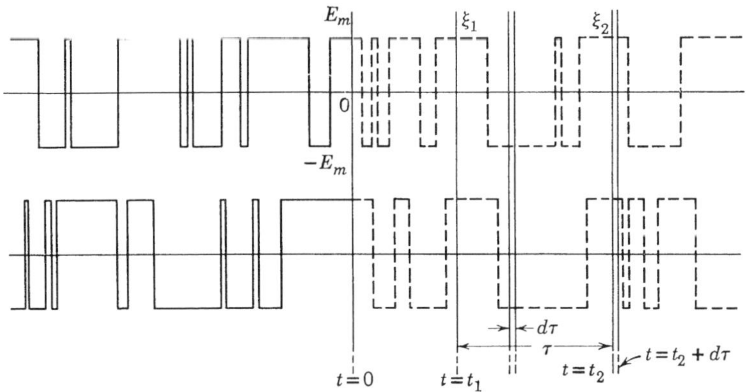

Fig. 1. An ensemble of rectangular waves, each alternating between E_m and $-E_m$ at random.

vals between the values E_m and $-E_m$, taking the value zero at the jumps. Such a process may correspond to thermal noise which has been strongly clipped and then amplified to E_m if the remaining value is positive and to $-E_m$ if it is negative. The distribution function of interest to us is the distribution of the number of zeros (or crossings) in the interval τ as indicated.

Under the conditions to be specified, it is assumed at the outset that the probability distribution exists. If we let the random variable ξ be the number of zeros in the interval τ of any particular ensemble member, the complete set of possible values of ξ is $n = 0, 1, 2, \ldots$. The distribution in question, $P_\xi(n;\tau)$, gives the probability that in a single trial the number of zeros in the given interval τ is the given number n. Symbolically, we write

$$\mathcal{P}(\xi = n; \tau) = P_\xi(n;\tau) \qquad \text{for } n = 0, 1, 2, \ldots \tag{1}$$

We now proceed with the determination of $P_\xi(n;\tau)$.

A quantity required in the problem is the mean value of the number of zeros per unit time. Letting k be the mean, we write according to (8), Chapter 5, Sec. 1, for the mean per τ units of time

$$\begin{aligned}\bar{\xi} &= \sum_{n=0}^{\infty} nP_\xi(n;\tau) \\ &= k\tau \end{aligned}\tag{2}$$

Since the method of derivation depends upon the formulation of the problem in the form of a differential equation, the probability for a zero to occur, or not to occur, in the infinitesimal time $d\tau$ is the essential probability element. To determine this quantity, we consider a particular element $d\tau$ within the interval $\tau = t_2 - t_1$ as indicated in the figure. Inasmuch as the number of zeros in the element $d\tau$ is a random variable, we denote it by η. This variable has only two possible values, namely, 0 and 1. In making this statement, we have neglected the probability that the number of zeros in $d\tau$ exceeds 1, which can be shown to be an infinitesimal of higher order than the infinitesimal $d\tau$.* Hence the basic probability elements are $\mathcal{P}(\eta = 1; d\tau)$ and $\mathcal{P}(\eta = 0; d\tau)$ which are, respectively, the probabilities that η takes the values 1 and 0 in the particular $d\tau$ under consideration.

To find these probabilities, we associate with $d\tau$ the interval τ and the probability distribution (1). We divide interval τ into elements each of duration $d\tau$, thereby obtaining a total of $\tau/d\tau$ elements. Now, consider the event A, which is: the number of zeros in τ is any number n ($n = 1, 2, 3, \ldots$), and the number of zeros in the particular $d\tau$ is 1. Symbolically,

$$\mathcal{P}(A) = \mathcal{P}(\xi = n; \tau, \eta = 1; d\tau) \qquad \text{for } n = 1, 2, 3, \ldots \tag{3}$$

Clearly, this probability is the same as the probability of finding $\eta = 1$

* T. C. Fry, *Probability and Its Engineering Uses*, D. Van Nostrand Co., Princeton, 1928, pp. 221–224.

in the particular $d\tau$ independent of the number of zeros in τ if the number of zeros in τ is at least 1. Hence

$$\mathcal{P}(\xi = n; \tau, \eta = 1; d\tau) = \mathcal{P}(\eta = 1; d\tau) \qquad \text{for } n = 1, 2, 3, \ldots \tag{4}$$

Thus, if we evaluate (3), we have the desired probability $\mathcal{P}(\eta = 1; d\tau)$.

In the evaluation of (3) we apply the theorem of total probability and the theorem of compound probability. Event A has the mutually exclusive forms

$$\left.\begin{array}{l} A_1\colon \xi = 1 \text{ in } \tau, \text{ and } \eta = 1 \text{ in the particular } d\tau \\ A_2\colon \xi = 2 \text{ in } \tau, \text{ and } \eta = 1 \text{ in the particular } d\tau \\ \vdots \end{array}\right\} \tag{5}$$

and each of these events is a compound event. The probability expressions for (5) are

$$\left.\begin{array}{l} \mathcal{P}(A_1) = \mathcal{P}(\xi = 1; \tau, \eta = 1; d\tau) \\ \mathcal{P}(A_2) = \mathcal{P}(\xi = 2; \tau, \eta = 1; d\tau) \\ \vdots \end{array}\right\} \tag{6}$$

By the theorem of compound probability, equations (6) are

$$\left.\begin{array}{l} \mathcal{P}(A_1) = \mathcal{P}(\xi = 1; \tau)\mathcal{P}(\eta = 1; d\tau \,|\, \xi = 1; \tau) \\ \mathcal{P}(A_2) = \mathcal{P}(\xi = 2; \tau)\mathcal{P}(\eta = 1; d\tau \,|\, \xi = 2; \tau) \\ \vdots \end{array}\right\} \tag{7}$$

Consider $\mathcal{P}(A_1)$ of (7). The unconditional probability $\mathcal{P}(\xi = 1; \tau)$ is for the event that $\xi = 1$ in τ. Clearly, this is the probability distribution (1) when $n = 1$. Hence $\mathcal{P}(\xi = 1; \tau) = P_\xi(1; \tau)$. The conditional probability $\mathcal{P}(\eta = 1; d\tau \,|\, \xi = 1; \tau)$ is for the event that $\eta = 1$ in $d\tau$ when it is known that $\xi = 1$ in τ. Since τ is divided into $\tau/d\tau$ elements and since $\xi = 1$ in τ, the probability that it occurs in the particular differential $d\tau$ is, according to the definition of mathematical probability, $1/(\tau/d\tau)$. In making this statement, *we have made an important assumption, to the effect that, if a zero occurs in τ, it has the same probability of occurring in any one of the infinitesimal elements into which τ is divided.* We have, then, $\mathcal{P}(\eta = 1; d\tau \,|\, \xi = 1; \tau) = d\tau/\tau$, and

$$\mathcal{P}(A_1) = P_\xi(1; \tau) \frac{d\tau}{\tau} \tag{8}$$

The probability $\mathcal{P}(A_2)$ in (7) is determined in a similar manner. First, $\mathcal{P}(\xi = 2; \tau) = P_\xi(2; \tau)$. Then, for the conditional probability $\mathcal{P}(\eta = 1; d\tau | \xi = 2; \tau)$, we have the situation that $\xi = 2$ in τ. Making the same assumption as before, we have the probability that one of the zeros occurring in the particular element $d\tau$ is $1/(\tau/d\tau)$ and the probability that the other zero occurring in the same element is also $1/(\tau/d\tau)$. By the theorem of total probability, the probability that one of the two zeros occurs in $d\tau$ is $2/(\tau/d\tau)$. Thus

$$\mathcal{P}(A_2) = P_\xi(2; \tau) \frac{2\, d\tau}{\tau} \tag{9}$$

It is easy to find the general expression.

It should be pointed out that the zeros are not to be considered as having an ordered sequence so that the first zero after $t = t_1$ prevents other zeros from being located between this zero and $t = t_1$. Our mathematical theory is rather that the probability that a zero occurs in a given element $d\tau$ is the same for all the other elements in τ. The distribution of zeros in τ is determined without any reference to the direction of time, but only with reference to the duration. In actual experiment, as the zeros occur, the direction of time is associated with them in the formation of the time function.

In accordance with the theorem of total probability,

$$\begin{aligned} \mathcal{P}(A) &= \mathcal{P}(A_1) + \mathcal{P}(A_2) + \cdots \\ &= [P_\xi(1; \tau) + 2P_\xi(2; \tau) + \cdots] \frac{d\tau}{\tau} \\ &= \left[\sum_{n=1}^{\infty} n P_\xi(n; \tau) \right] \frac{d\tau}{\tau} \end{aligned} \tag{10}$$

Since the expression in brackets is equal to the mean of ξ as given by (2), we have

$$\begin{aligned} \mathcal{P}(A) &= k\tau \frac{d\tau}{\tau} \\ &= k\, d\tau \end{aligned} \tag{11}$$

that is, as stated in (3) and (4),

$$\mathcal{P}(\eta = 1; d\tau) = k\, d\tau \tag{12}$$

Inasmuch as η has only two possible values, 0 and 1, the other probability element is determined by the relation

$$\mathcal{P}(\eta = 1; d\tau) + \mathcal{P}(\eta = 0; d\tau) = 1 \tag{13}$$

which follows from the fact that either one or the other event must occur in a trial. Therefore

$$\mathcal{P}(\eta = 0; d\tau) = 1 - k\,d\tau \tag{14}$$

Because the ensemble is stationary, (12) and (14) hold for a $d\tau$ at any time t.

Having determined the probability distribution of the number of zeros in the infinitesimal time $d\tau$, we proceed to find the distribution of zeros in a finite time τ. For this problem we consider the interval τ between $t = t_1$ and $t = t_2$ and the element $d\tau$ which is adjacent to, and on the right side of, the point $t = t_2$ as indicated in Fig. 1. In this situation, $d\tau$ is outside of the interval τ, whereas previously it was within the interval. The event B which we now consider is: within the interval $\tau + d\tau$ there is no zero. Expressed symbolically, the probability of the event is

$$\mathcal{P}(B) = \mathcal{P}(\xi + \eta = 0; \tau + d\tau) \tag{15}$$

Concerning this probability we make the second important assumption in the problem. *We assume that the number of zeros in an interval is independent of that in another interval if one interval does not overlap the other partially or completely.* On this basis the events which constitute the compound event B are independent. The component events of B are: (B_1) $\xi = 0$ in τ, and (B_2) $\eta = 0$ in $d\tau$. Hence the theorem of compound probability leads to

$$\mathcal{P}(\xi + \eta = 0; \tau + d\tau) = \mathcal{P}(\xi = 0; \tau)\mathcal{P}(\eta = 0; d\tau) \tag{16}$$

In this expression $\mathcal{P}(\eta = 0; d\tau)$ has been found and is given by (14), so that (16) is

$$\mathcal{P}(\xi + \eta = 0; \tau + d\tau) = \mathcal{P}(\xi = 0; \tau)(1 - k\,d\tau) \tag{17}$$

By rearrangement the equation can be put into the form

$$\frac{\mathcal{P}(\xi + \eta = 0; \tau + d\tau) - \mathcal{P}(\xi = 0; \tau)}{d\tau} = -k\mathcal{P}(\xi = 0; \tau) \tag{18}$$

On the left-hand side of the equation we have an expression of the change in $\mathcal{P}(\xi = 0; \tau)$ (due to the increment $d\tau$ in the interval τ, with the number of zeros in the total interval remaining equal to zero) divided by the increment. In other words, by definition, this expression is the derivative of the probability distribution, for $\xi = 0$, with respect to the continuous parameter τ. The probability distribution in question is given by (1) when $n = 0$. Thus (18) is the differential equation

$$\frac{d}{d\tau} P_\xi(0; \tau) = -kP_\xi(0; \tau) \tag{19}$$

The solution of this equation is

$$P_\xi(0; \tau) = Ae^{-k\tau} \tag{20}$$

To complete the solution, we evaluate the constant A by setting $\tau = 0$ so that (20) becomes

$$P_\xi(0; 0) = A \tag{21}$$

Since $P_\xi(0; 0)$ is the probability of finding no zeros in zero time, we can say that the probability is 1. This brings us to the result that the probability of finding no zeros in the interval τ is

$$P_\xi(0; \tau) = e^{-k\tau} \tag{22}$$

The determination of the distribution (1) is now partially done; it is done for $n = 0$. The maximum value of the function (22) is 1 at $\tau = 0$, as it should be. Furthermore, the function agrees with our expectation that, as the interval τ increases, the probability of finding no zeros in it decreases. Theoretically, no matter how long the interval τ is, the probability of finding no zero in it is finite; and, only when τ tends to infinity, does the probability tend to zero.

For the complete determination of $P_\xi(n; \tau)$ we consider the event that within the interval $\tau + d\tau$ there are n zeros. The probability of this event, which we call event C, is

$$\mathcal{P}(C) = \mathcal{P}(\xi + \eta = n; \tau + d\tau) \tag{23}$$

Because η must either be 1 or 0, the event C has two mutually exclusive forms; namely, (C_1) $\xi = n - 1$ in τ, and $\eta = 1$ in $d\tau$, and (C_2) $\xi = n$ in τ, and $\eta = 0$ in $d\tau$. With the aid of the theorem of total probability we are able to express (23) as

$$\mathcal{P}(\xi + \eta = n; \tau + d\tau)$$
$$= \mathcal{P}(\xi = n - 1; \tau, \eta = 1; d\tau) + \mathcal{P}(\xi = n; \tau, \eta = 0; d\tau) \tag{24}$$

Inasmuch as C_1 is a compound event, being the joint occurrence of (C_1') $\xi = n - 1$ in τ and (C_1'') $\eta = 1$ in $d\tau$, and since, by our second assumption, C_1' and C_1'' are independent events, we have

$$\mathcal{P}(C_1) = \mathcal{P}(C_1')\mathcal{P}(C_1'') \tag{25}$$

or

$$\mathcal{P}(\xi = n-1; \tau, \eta = 1; d\tau) = \mathcal{P}(\xi = n-1; \tau)\mathcal{P}(\eta = 1; d\tau) \quad (26)$$

Similarly, the last term in (24) is expressible as

$$\mathcal{P}(\xi = n; \tau, \eta = 0; d\tau) = \mathcal{P}(\xi = n; \tau)\mathcal{P}(\eta = 0; d\tau) \quad (27)$$

Noting that $\mathcal{P}(\eta = 1; d\tau)$ and $\mathcal{P}(\eta = 0; d\tau)$ are given by (12) and (14) respectively, we now express (24) as

$$\mathcal{P}(\xi + \eta = n; \tau + d\tau) = \mathcal{P}(\xi = n-1; \tau)k\,d\tau + \mathcal{P}(\xi = n; \tau)(1 - k\,d\tau) \quad (28)$$

Upon rearrangement the equation is

$$\frac{\mathcal{P}(\xi + \eta = n; \tau + d\tau) - \mathcal{P}(\xi = n; \tau)}{d\tau} + k\mathcal{P}(\xi = n; \tau) = k\mathcal{P}(\xi = n-1; \tau) \quad (29)$$

On the left-hand side of (29), the first term is the change in $\mathcal{P}(\xi = n; \tau)$ (due to an increment $d\tau$ in τ, with the number of zeros in the total interval, $\tau + d\tau$, remaining equal to n) divided by the increment. By definition, this ratio is the derivative of the function $P_\xi(n; \tau)$ with respect to τ with parameter n held constant. On the same side, the second term is clearly $kP_\xi(n; \tau)$. The last term of the equation is $kP_\xi(n-1; \tau)$. Substitution of these terms into (29) leads to the differential equation

$$\frac{d}{d\tau} P_\xi(n; \tau) + kP_\xi(n; \tau) = kP_\xi(n-1; \tau) \quad (30)$$

from which we can determine the desired distribution function $P_\xi(n; \tau)$.

The differential equation is of the type

$$\frac{dy}{dx} + R(x)y = S(x) \quad (31)$$

whose solution is known to be

$$ye^{\int R(x)\,dx} = \int S(x)e^{\int R(x)\,dx}\,dx \quad (32)$$

By comparison of (30) and (31) we note that $y = P_\xi(n; \tau)$, $R(x) = k$, $S(x) = kP_\xi(n-1; \tau)$, and $x = \tau$. Hence, $\int R(x)\,dx = \int k\,d\tau = k\tau$, and the solution of (30) is, according to (32),

$$P_\xi(n; \tau)e^{k\tau} = \int kP_\xi(n-1; \tau)e^{k\tau}\,d\tau \quad (33)$$

We shall determine $P_\xi(n;\tau)$ in successive steps. For $n = 1$, the general solution (33) becomes

$$P_\xi(1;\tau)e^{k\tau} = \int kP_\xi(0;\tau)e^{k\tau}\,d\tau \tag{34}$$

in which $P_\xi(0;\tau)$ has been found to be (22). Therefore

$$P_\xi(1;\tau)e^{k\tau} = k\tau + C \tag{35}$$

in which the constant of integration C is evaluated by setting $\tau = 0$. Thus

$$P_\xi(1;0) = C \tag{36}$$

and we conclude that C is zero, because $P_\xi(1;0)$ is the probability of finding one zero in an interval of zero duration, which obviously is an event of zero probability. This reduces (35) to

$$P_\xi(1;\tau) = k\tau e^{-k\tau} \tag{37}$$

The general expression becomes evident when the procedure is repeated for successive values of n. We indicate the procedure for $n = 2$ briefly as follows:

$$\begin{aligned} P_\xi(2;\tau)e^{k\tau} &= \int kP_\xi(1;\tau)e^{k\tau}\,d\tau \\ &= \int k[k\tau e^{-k\tau}]e^{k\tau}\,d\tau \\ &= \frac{(k\tau)^2}{2} + C \end{aligned} \tag{38}$$

and it can be shown that $C = 0$, so that

$$P_\xi(2;\tau) = \frac{(k\tau)^2}{2}\,e^{-k\tau} \tag{39}$$

The general expression is

$$P_\xi(n;\tau) = \frac{(k\tau)^n}{n!}\,e^{-k\tau} \qquad \text{for } n = 0, 1, 2, \ldots \tag{40}$$

This is the probability distribution function for ξ, the number of zeros in a given interval τ, if the average number of zeros per unit of τ is k. It is known as the *Poisson distribution*. A graph of this distribution is

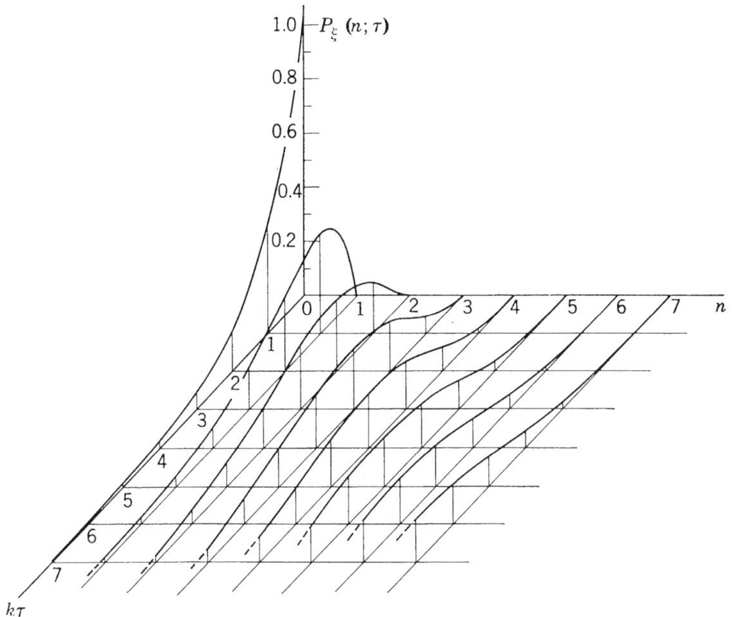

Fig. 2. The Poisson distribution.

given in Fig. 2. The function is discrete with respect to n but continuous with respect to τ.

To verify the fact that the average number of zeros per unit of time is k, we follow (9), Chapter 5, Sec. 1. Thus the mean of the number of zeros in a given interval τ is

$$\begin{aligned}\bar{\xi} &= \sum_{n=0}^{\infty} nP_\xi(n;\tau) = \sum_{n=0}^{\infty} n\frac{(k\tau)^n}{n!}e^{-k\tau} \\ &= k\tau e^{-k\tau}\left[1 + k\tau + \frac{(k\tau)^2}{2!} + \frac{(k\tau)^3}{3!} + \cdots\right] \\ &= k\tau e^{-k\tau}e^{k\tau} \\ &= k\tau \end{aligned} \tag{41}$$

This result tells us that the mean of ξ per unit of time is k.

To see if (40) satisfies the condition

$$\sum_{n=0}^{\infty} P_\xi(n;\tau) = 1 \tag{42}$$

as it should, we have

$$\begin{aligned}\sum_{n=0}^{\infty} P_\xi(n;\tau) &= \sum_{n=0}^{\infty} \frac{(k\tau)^n}{n!} e^{-k\tau} \\ &= e^{-k\tau} \sum_{n=0}^{\infty} \frac{(k\tau)^n}{n!} \\ &= e^{-k\tau} e^{k\tau} \\ &= 1 \end{aligned} \tag{43}$$

which confirms (42).

2. The Normal Density

By far the most important function in the theory of probability is the *normal density function.* The function is defined as

$$P_\xi(x) = \frac{1}{\sqrt{2\pi}\,\sigma} \exp\left[-\frac{(x-m)^2}{2\sigma^2}\right] \tag{44}$$

where ξ denotes the random variable, x its continuous range $(-\infty, \infty)$, m its mean, and σ^2 its variance. A form of this function has been given in several places as examples. The graphs of the normal density, for several values of σ, are given in Fig. 3. The function is maximum at $x = m$ where it has the value $1/(\sqrt{2\pi}\,\sigma)$, and it is an even function with respect to the line $x = m$.

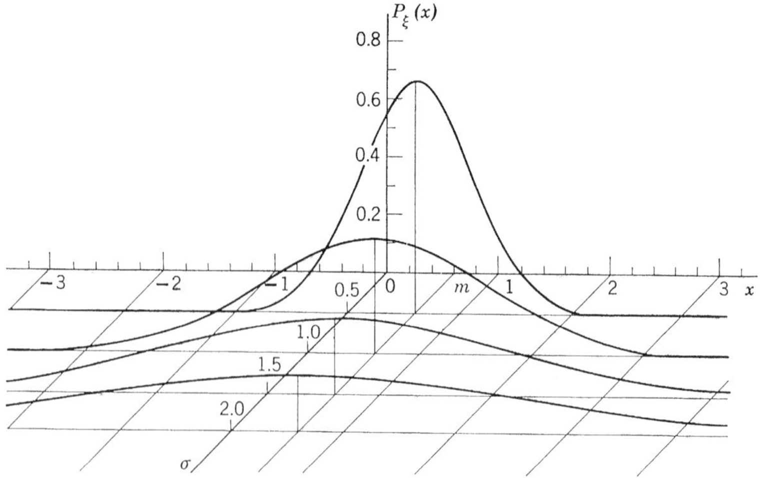

Fig. 3. The normal density.

Aside from its great theoretical importance, the normal density plays a central role in the theory of thermal noise and shot noise. It has been experimentally established that these noises obey the normal density function. Among a great variety of its applications outside the field of communications, we should mention the analysis of errors in physical measurements and the analysis of Brownian motion. Nevertheless, we should not regard the normal density as a universal density for random phenomena in which the variable has a continuous range. Many important random phenomena, such as speech waves and music waves, are not normally distributed.

To ascertain that the integral of (44) over $(-\infty, \infty)$ is 1, as it should be, we need only show that the factor A in

$$\int_{-\infty}^{\infty} A \exp\left[-\frac{(x-m)^2}{2\sigma^2}\right] dx = 1 \tag{45}$$

is $1/(\sqrt{2\pi}\,\sigma)$. Since the value of the integral in (45) is independent of m, we set $m = 0$ for convenience. Following the method given in most textbooks on advanced calculus, we simplify the evaluation of (45) by squaring both sides of it to obtain

$$\int_{-\infty}^{\infty} A \exp\left[-\frac{x^2}{2\sigma^2}\right] dx \int_{-\infty}^{\infty} A \exp\left[-\frac{y^2}{2\sigma^2}\right] dy = 1 \tag{46}$$

By a change of coordinates, from rectangular to polar, we get

$$A^2 \int_0^{2\pi} \int_0^{\infty} \exp\left[-\frac{r^2}{2\sigma^2}\right] r\, dr\, d\theta = 1 \tag{47}$$

The integral can be readily evaluated and (47) becomes

$$2\pi A^2 \sigma^2 = 1 \tag{48}$$

so that $A = 1/(\sqrt{2\pi}\,\sigma)$ which is the requirement for condition (45).

An important characteristic of the normal density is its characteristic function. According to definition (87), Chapter 5, Sec. 9, the function is

$$\begin{aligned} p_\xi(t) &= \int_{-\infty}^{\infty} \frac{1}{\sqrt{2\pi}\,\sigma} \exp\left[-\frac{(x-m)^2}{2\sigma^2}\right] e^{jtx}\, dx \\ &= \frac{1}{\sqrt{2\pi}\,\sigma} \int_{-\infty}^{\infty} \exp\left[-\frac{y^2}{2\sigma^2}\right] e^{jt(y+m)}\, dy \\ &= \frac{1}{\sqrt{2\pi}\,\sigma}\, e^{jmt} \int_{-\infty}^{\infty} \exp\left[-\frac{y^2}{2\sigma^2}\right] \cos ty\, dy \\ &= e^{jmt-\sigma^2t^2/2} \\ &= e^{-\sigma^2t^2/2}(\cos mt + j \sin mt) \end{aligned} \tag{49}$$

To show that the mean of ξ is m and the variance is σ^2 in the normal density, we first take the first two derivatives of the characteristic function (49). They are

$$p_\xi'(t) = (jm - \sigma^2 t)e^{jmt-\sigma^2t^2/2} \tag{50}$$

and

$$p_\xi''(t) = [(jm - \sigma^2 t)^2 - \sigma^2]e^{jmt-\sigma^2t^2/2} \tag{51}$$

By (91), Chapter 5, Sec. 9, the mean and mean square of ξ are

$$\bar{\xi} = \frac{1}{j} p_\xi'(0) = m \tag{52}$$

and

$$\begin{aligned}\overline{\xi^2} &= \frac{1}{j^2} p_\xi''(0) = \frac{1}{j^2}[(jm)^2 - \sigma^2] \\ &= m^2 + \sigma^2\end{aligned} \tag{53}$$

Hence the variance of ξ is

$$\begin{aligned}\sigma_\xi{}^2 &= \overline{\xi^2} - \bar{\xi}^2 = m^2 + \sigma^2 - m^2 \\ &= \sigma^2\end{aligned} \tag{54}$$

It should be remarked that the same results can be obtained by integration in accordance with the definitions of mean and variance without the aid of the characteristic function. Referring back to (44), we see that the normal density is completely determined by the mean and the variance of the random variable. Whereas the mean determines only the displacement of the curve from the origin and in no way affects the shape of the curve, the variance changes the shape of the curve by modifying both the horizontal and the vertical scales.

It is interesting to note that the distance between $x = m$ and the points of inflection of the normal density is the standard deviation σ. To show this, we equate the second derivative of the curve to zero in accordance with the behavior of the curve at points of inflection. Since

$$P_\xi''(x) = \frac{1}{\sqrt{2\pi}\,\sigma^3}\left[\frac{(x-m)^2}{\sigma^2} - 1\right]\exp\left[-\frac{(x-m)^2}{2\sigma^2}\right] \tag{55}$$

the points of inflection, from the equation $P_\xi''(x) = 0$, are

$$x = m \pm \sigma \tag{56}$$

Although the normal density is also known as a *Gaussian density*, the function originated in the work of De Moivre in 1733. It was many years later that the function appeared in the important work of Gauss and Laplace on the theory of probability.

3. The Sum of Normally Distributed Independent Variables

The sum of normally distributed independent random variables has some interesting properties. Physically, these random variables may represent thermal noise or shot noise from independent sources. Let the independent, normally distributed variables be $\xi_1, \xi_2, \ldots, \xi_n$, with the respective means $m_1, m_2, \ldots, m_n$ and variances $\sigma_1{}^2, \sigma_2{}^2, \ldots, \sigma_n{}^2$. Then according to (49) the respective characteristic functions are

$$\left.\begin{aligned} p_{\xi 1}(t) &= e^{jm_1t - \sigma_1{}^2t^2/2} \\ p_{\xi 2}(t) &= e^{jm_2t - \sigma_2{}^2t^2/2} \\ &\cdot \\ &\cdot \\ &\cdot \\ p_{\xi n}(t) &= e^{jm_nt - \sigma_n{}^2t^2/2} \end{aligned}\right\} \tag{57}$$

Now, consider the characteristic function $p_\xi(t)$ of the sum

$$\xi = \xi_1 + \xi_2 + \cdots + \xi_n \tag{58}$$

Since the variables are independent, we have by (102), Chapter 5, Sec. 9,

$$p_\xi(t) = \prod_{i=1}^{n} e^{jm_it - \sigma_i{}^2t^2/2} \tag{59}$$

According to (80), Chapter 5, Sec. 8, the mean of ξ, which we denote by m_o, is

$$m_o = m_1 + m_2 + \cdots + m_n \tag{60}$$

and as in (117), Chapter 5, Sec. 10, the variance of ξ, $\sigma_o{}^2$, is

$$\sigma_o{}^2 = \sigma_1{}^2 + \sigma_2{}^2 + \cdots + \sigma_n{}^2 \tag{61}$$

In terms of (60) and (61), (59) is

$$p_\xi(t) = e^{jm_ot - \sigma_o{}^2t^2/2} \tag{62}$$

Because (62) is in the form of the characteristic function of a normal density, we conclude that the probability density of the sum of normally distributed independent random variables is also normal and has the expression

$$P_\xi(x) = \frac{1}{\sqrt{2\pi}\,\sigma_o} \exp\left[-\frac{(x - m_o)^2}{2\sigma_o{}^2}\right] \tag{63}$$

This is an interesting result, for the normal form of the densities of the component random variables has not been changed. The only changes are in the displacement of the curve and its scales on both axes. If we consider the sum of two independent normally distributed random variables which have the same mean and variance, the mean and variance of the sum are both twice as large as the component mean and variance respectively. Figure 4 shows the graphs of this example.

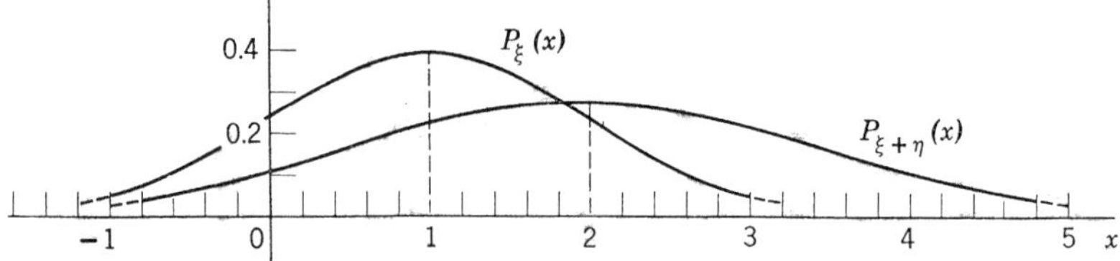

Fig. 4. Density $[P_\xi(x)]$ of normally distributed variable with mean = 1 and variance = 1, and density $[P_\xi + \eta(x)]$ of the sum of two independent normally distributed variables each with mean = 1 and variance = 1.

4. Transformation of Probability Densities

In studying the probability density of a random variable, we are often interested in the determination of the density after the random variable has gone through a transformation. Simple transformations which are of interest to us are: (1) multiplication of a random quantity by a constant a, which corresponds, for instance, to the amplification of a noise voltage by the factor a, and (2) translation of the random quantity by a constant b, which physically corresponds to the addition of a d-c component to the voltage. After such transformation, the random variables are other random variables. If ξ is the original random variable and η is ξ after a transformation, then for multiplication $\eta = a\xi$, and for translation $\eta = \xi + b$. The combination of both transformations leads obviously to $\eta = a\xi + b$. The process of squaring a random quantity is another important basic transformation. In general, the original random variable is related to the new variable by the equation

$$\eta = f(\xi) \tag{64}$$

Our problem is to determine the probability density of η in terms of the probability density of ξ and the given relation between the variables.

In the solution of this problem let us assume, for the moment, that the values of ξ and η have a one-to-one correspondence. This would be true if $\eta = a\xi$ or $\eta = a\xi + b$, but it would not be true if $\eta = \xi^2$, for then two values of ξ would correspond to only one value of η. Since every point on x (the range of ξ) has one and only one corresponding point on y (the range of η), we can state that

$$\mathcal{P}(x < \xi < x + dx) = \mathcal{P}(y < \eta < y + dy) \tag{65}$$

This equation is in agreement with the fact that the event of η occurring in $(y, y + dy)$ is dependent upon the event of ξ occurring in $(x, x + dx)$

and nowhere else, in accordance with the given relation (64). The non-occurrence of ξ results in the non-occurrence of η, and vice versa. Letting $P_\xi(x)$ and $P_\eta(y)$ be the probability densities of ξ and η respectively, as Fig. 5 shows, we express (65) in terms of the densities by the equation

$$P_\xi(x)\,dx = P_\eta(y)\,dy \tag{66}$$

which leads to the relation

$$P_\eta(y) = P_\xi(x)\left|\frac{dx}{dy}\right| \tag{67}$$

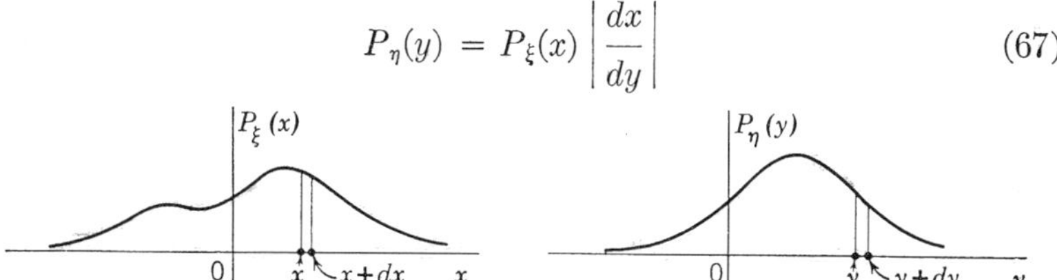

Fig. 5. Illustrating the transformation of probability densities.

Since each side of (66) is a probability, it must be a non-negative quantity. Furthermore, we know that the densities $P_\xi(x)$ and $P_\eta(y)$ are non-negative functions. Therefore, in expressing (66) in the form (67), we must take care to avoid the negative sign of dx/dy, if it arises. The relation between x and y follows from (64); that is,

$$y = f(x) \tag{68}$$

Assuming that the equation can be solved for x, we represent the result by

$$x = g(y) \tag{69}$$

Consequently, the solution to our problem is

$$P_\eta(y) = \frac{P_\xi(x)}{|f'(x)|}\bigg|_{x=g(y)} \tag{70}$$

in which $f'(x)$ denotes dy/dx. By this formula we can determine the probability density of the new random variable η in terms of the density of the original variable ξ and the given transformation. We should be reminded that the formula is invalid if a value of η has more than one corresponding value of ξ.

As an example of the application of (70), we consider the transformation

$$\eta = a\xi \tag{71}$$

with the given density $P_\xi(x)$. We have $y = ax$, so that $f'(x) = a$ and $x = y/a$. Substitution of these quantities into (70) yields

$$P_\eta(y) = \frac{1}{a}P_\xi\left(\frac{y}{a}\right) \tag{72}$$

Therefore, if ξ is multiplied by a constant a, the probability density of the new variable η should be the density of the original variable ξ with both scales changed by the factor $1/a$; that is, on the original density curve along both axes, the unit 1 should now read $1/a$.

Another simple example is the translation of a random variable. Clearly

$$\eta = \xi + b \tag{73}$$

so that $y = x + b$, and

$$P_\eta(y) = P_\xi(y - b) \tag{74}$$

When the relation between η and ξ is such that more than one value of ξ corresponds to one value of η, formula (70) does not hold, but it is possible to solve the problem by introducing the necessary modifications. For instance, if

$$\eta = \xi^2 \tag{75}$$

then the event that η occurs in $(y, y + dy)$ necessitates the occurrence of either the event ξ in $(x, x + dx)$ or ξ in $(-x, -[x + dx])$. Since the two events concerning ξ are mutually exclusive, the theorem of total probability is applicable, and we have

$$\begin{aligned}\mathcal{P}(y < \eta < y + dy) &= \mathcal{P}(x < \xi < x + dx) \\ &\quad + \mathcal{P}(-x < \xi < -[x + dx]) \qquad \text{for } \eta > 0\end{aligned} \tag{76}$$

In terms of the densities, (76) is

$$P_\eta(y)\,dy = [P_\xi(x) + P_\xi(-x)]\,dx \qquad \text{for } y > 0 \tag{77}$$

so that instead of (70) we have

$$P_\eta(y) = \begin{cases} \left.\dfrac{P_\xi(x) + P_\xi(-x)}{|f'(x)|}\right|_{x=g(y)} & \text{for } y > 0 \\ 0 & \text{for } y < 0 \end{cases} \tag{78}$$

Since $y = x^2$, $f'(x) = 2x$, and $x = \pm\sqrt{y}$,

$$P_\eta(y) = \begin{cases} \dfrac{P_\xi(\sqrt{y}) + P_\xi(-\sqrt{y})}{2\sqrt{y}} & \text{for } y > 0 \\ 0 & \text{for } y < 0 \end{cases} \tag{79}$$

To illustrate this result, consider a normally distributed thermal noise which passes through a squaring device. The probability density of the noise before squaring is

$$P_\xi(x) = \frac{1}{\sqrt{2\pi}\,\sigma} \exp\left[-\frac{(x - m)^2}{2\sigma^2}\right] \qquad \text{for } -\infty < x < \infty \tag{80}$$

After squaring, the density of the random noise is, by (79),

$$P_\eta(y) = \frac{1}{2\sqrt{2\pi}\,\sigma}\frac{1}{\sqrt{y}}\left(\exp\left[-\frac{(\sqrt{y}-m)^2}{2\sigma^2}\right] + \exp\left[-\frac{(-\sqrt{y}-m)^2}{2\sigma^2}\right]\right) \qquad \text{for } y > 0 \tag{81}$$

The expression for the desired density, when (81) is simplified, is

$$P_\eta(y) = \begin{cases} \dfrac{1}{\sqrt{2\pi}\,\sigma}\dfrac{1}{\sqrt{y}}\exp\left[-\dfrac{y+m^2}{2\sigma^2}\right]\cosh\dfrac{m}{\sigma^2}\sqrt{y} & \text{for } y > 0 \\ 0 & \text{for } y < 0 \end{cases} \tag{82}$$

If the noise has a zero mean value, (82) reduces to

$$P_\eta(y) = \begin{cases} \dfrac{1}{\sqrt{2\pi}\,\sigma}\dfrac{1}{\sqrt{y}}e^{-y/2\sigma^2} & \text{for } y > 0 \\ 0 & \text{for } y < 0 \end{cases} \tag{83}$$

The graph of this probability density is shown in Fig. 6.

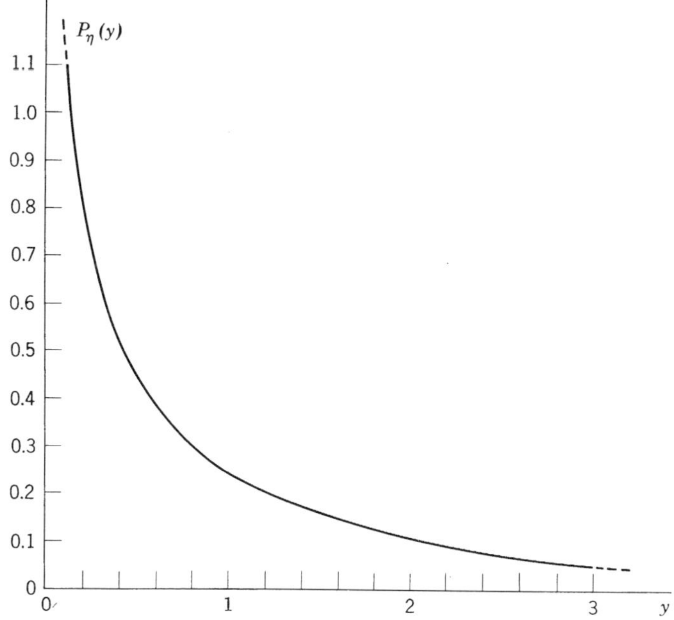

Fig. 6. Probability density of the square of a nor[illegible]y distributed variable. Drawn for $\sigma = 1$.

5. The Rectangular Density

If a random variable ξ is uniformly distributed over the continuous range (x_1, x_2), the probability density is rectangular (or uniform) as illustrated in Fig. 7. Because a probability density function has unit

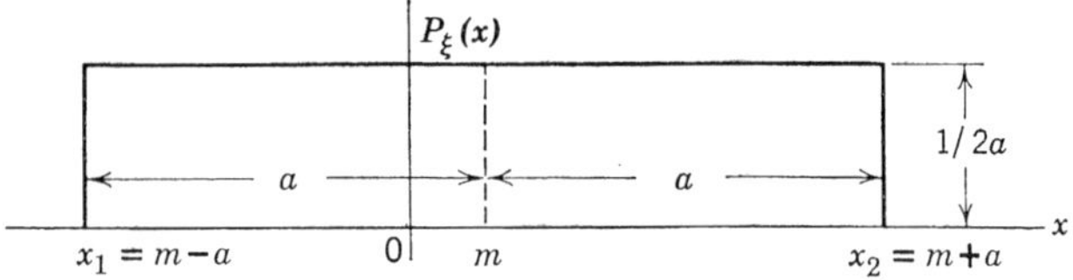

Fig. 7. The rectangular probability density.

area, the height of the rectangle is equal to $1/(x_2 - x_1)$, if $x_2 > x_1$. Thus the density is

$$P_\xi(x) = \begin{cases} \dfrac{1}{x_2 - x_1} & \text{for } x_1 < x < x_2 \\ 0 & \text{elsewhere} \end{cases} \tag{84}$$

Letting m be the distance between 0 and the midpoint of the interval (x_1, x_2) and letting $(x_2 - x_1)/2 = a$, as shown in the figure, we can readily show that

$$\bar{\xi} = m \tag{85}$$

and

$$\overline{\xi^2} = m^2 + \frac{a^2}{3} \tag{86}$$

It is an interesting fact that, by transformation, any probability density $P_\xi(x)$ can be reduced to the rectangular density

$$P_\eta(y) = \begin{cases} 1 & \text{for } 0 < y < 1 \\ 0 & \text{elsewhere} \end{cases} \tag{87}$$

The transformation that makes the reduction possible is the transformation

$$y = Q_\xi(x) \qquad \text{for } 0 < y < 1 \tag{88}$$

where $Q_\xi(x)$ is the cumulative distribution function of ξ; it is recalled that $Q_\xi(-\infty) = 0$ and $Q_\xi(\infty) = 1$ (see Chapter 4, Sec. 5). The transformation (88) is called the probability transformation.

To show that (87) results from transformation (88), we refer to (70) with the range of y restricted to $(0, 1)$ for the reason that this range of y has a one-to-one correspondence with the range $(-\infty, \infty)$ of x. Since

$f'(x)$ in (70) is the derivative of (88) with respect to x and since the derivative of the cumulative probability function is the probability density, we have

$$f'(x) = Q_\xi'(x)$$
$$= P_\xi(x) \tag{89}$$

We can easily see that substitution of (89) into (70) gives

$$P_\eta(y) = 1 \qquad \text{for } 0 < y < 1 \tag{90}$$

which is in agreement with (87). Outside the range (0, 1), y does not have corresponding points in x so that $P_\eta(y)$ vanishes as stated in (87).

6. The First Probability Density of an Ensemble of Periodic Functions

So far, the concept of an ensemble has been associated with random functions of time, and it has not been applied to the description of periodic functions with a random parameter. Inasmuch as periodic functions are frequently hidden in, or intermixed with, random waves, as in the case of a periodic signal in thermal noise, it is desirable to consider ensembles of periodic functions and to introduce a random parameter in these functions so that both random and periodic components are described in the same manner.

It is instructive to consider the ensemble of sine waves in Fig. 8. All the sine waves in the ensemble have the same maximum amplitude and the same frequency. The random parameter in the waves is the phase angle with respect to the time $t = 0$. We imagine that the sine waves are produced by identical oscillators.

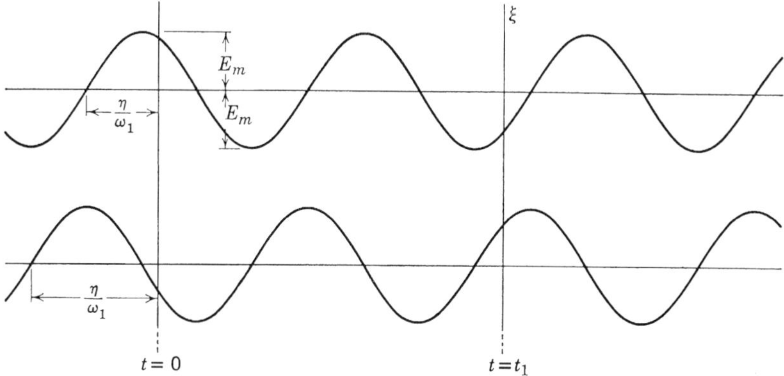

Fig. 8. An ensemble of sinusoids with random initial phase angles.

Letting η be the random phase angle of the sine wave whose angular frequency is ω_1 and whose maximum amplitude is E_m, we express the random amplitude, ξ, of the wave at $t = t_1$ by the equation

$$\xi = E_m \sin (\omega_1 t_1 + \eta) \tag{91}$$

The characteristic of interest to us is the probability density of the random variable ξ if we are given the probability density of η. In other words, we would like to know the probability density of the amplitude of the sine wave at $t = t_1$ from our knowledge of the probability density of the initial phase angle. This problem is recognized as a problem of transformation, and its solution can be obtained by the method discussed in Sec. 4. First of all, it is necessary that we specify the range of the phase angle η. Clearly, we need only consider the continuous range $(0, 2\pi)$. Let the probability density of η be $P_\eta(y)$ and that of ξ be $P_\xi(x)$. Now, an important point is that, for every value of ξ in its range $(-E_m, E_m)$, there are two possible values of η in its range $(0, 2\pi)$, except when $\xi = \pm E_m$.

If we put $\zeta = \omega_1 t_1 + \eta$, (91) is

$$\xi = E_m \sin \zeta \tag{92}$$

where the range of ζ is $(\omega_1 t_1, 2\pi + \omega_1 t_1)$ inasmuch as the range of η is $(0, 2\pi)$. The relation between the range variables x and z of the random variables ξ and ζ is

$$x = E_m \sin z \tag{93}$$

and z and y are related by

$$z = \omega_1 t_1 + y \tag{94}$$

Letting z_1 and z_2 be the two solutions of (93) for z in the interval $(\omega_1 t_1, 2\pi + \omega_1 t_1)$, we can state, in a manner similar to (76), that

$$\begin{aligned} &\mathcal{P}(x < \xi < x + dx) \\ &\qquad = \mathcal{P}(z_1 < \zeta < z_1 + dz) + \mathcal{P}(z_2 < \zeta < z_2 + dz) \end{aligned} \tag{95}$$

which leads to the equation

$$P_\xi(x)\, dx = P_\zeta(z_1)\, dz + P_\zeta(z_2)\, dz \tag{96}$$

Here $P_\zeta(z)$ is the probability density of ζ; and, by (74), $P_\zeta(z) = P_\eta(z - \omega_1 t_1)$. We now write (96) in the form

$$P_\xi(x) = \frac{P_\eta(z_1 - \omega_1 t_1) + P_\eta(z_2 - \omega_1 t_1)}{\left| \dfrac{dx}{dz} \right|} \qquad \text{for } -E_m < x < E_m \tag{97}$$

where

$$\left|\frac{dx}{dz}\right| = |E_m \cos z| = \sqrt{E_m{}^2 - x^2} \tag{98}$$

Hence the complete expression for $P_\xi(x)$ is

$$P_\xi(x) = \begin{cases} \dfrac{P_\eta(z_1 - \omega_1 t_1) + P_\eta(z_2 - \omega_1 t_1)}{\sqrt{E_m{}^2 - x^2}} & \text{for } -E_m < x < E_m \\ 0 & \text{elsewhere} \end{cases} \tag{99}$$

We see in (99) that $P_\xi(x)$ is generally a function of t_1 so that the ensemble is nonstationary. Since the stationary ensemble is of particular interest, we let $P_\eta(z - \omega_1 t_1)$ be a rectangular density and define it as

$$P_\eta(z - \omega_1 t_1) = \begin{cases} \dfrac{1}{2\pi} & \text{for } \omega_1 t_1 < z < 2\pi + \omega_1 t_1 \\ 0 & \text{elsewhere} \end{cases} \tag{100}$$

in order that (99) may be independent of t_1 and reduced to

$$P_\xi(x) = \begin{cases} \dfrac{1}{\pi\sqrt{E_m{}^2 - x^2}} & \text{for } -E_m < x < E_m \\ 0 & \text{elsewhere} \end{cases} \tag{101}$$

Because $z = \omega_1 t_1 + y$, condition (100) is equivalent to the condition

$$P_\eta(y) = \begin{cases} \dfrac{1}{2\pi} & \text{for } 0 < y < 2\pi \\ 0 & \text{elsewhere} \end{cases} \tag{102}$$

Therefore, we have the result that, if the probability density of the phase angle η is uniform, as given by (102) and shown in Fig. 9, the ensemble of sine waves is stationary and the probability density of the amplitude is given by (101). This density is plotted in Fig. 10.

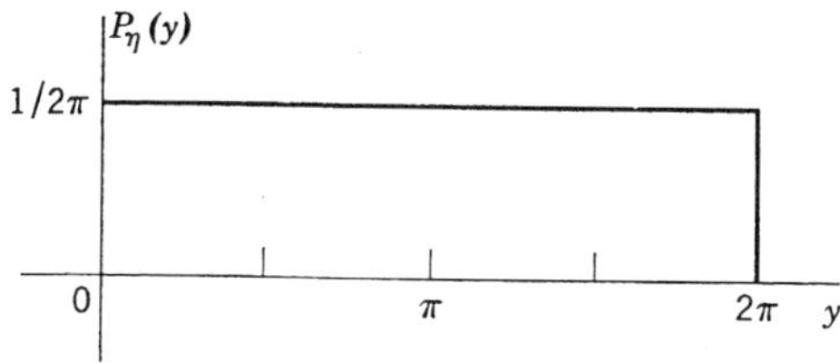

Fig. 9. Rectangular probability density for the initial phase angles of the sinusoids of Fig. 8.

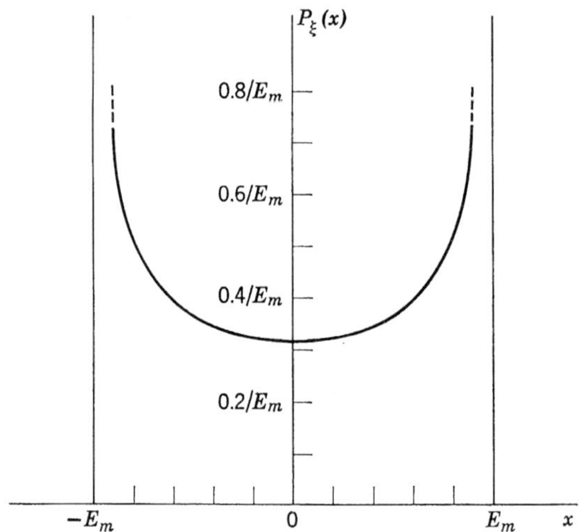

Fig. 10. First probability density of the ensemble of Fig. 8 with initial phase angles uniformly distributed.

It is clear that the method of transformation may be applied to ensembles of periodic functions of various waveforms for the determination of the first probability density.

PROBLEMS

1. We have given an ensemble of rectangular waves with zero crossings distributed according to the Poisson distribution. The average number of zero crossings per second is k. An observation is made on a member of the ensemble for the number of zero crossings in an interval τ from time $t = t_1$ to $t = t_1 + \tau$, where t_1 is arbitrarily chosen.

(a) Show whether the probability of finding an even number (2, 4, 6, ...) of zero crossings in τ is greater than, equal to, or less than, the probability of finding an odd number (1, 3, 5, ...) of zero crossings in τ. Note that the number zero is excluded.

(b) If the probabilities are unequal, determine the expression for the difference; if the probabilities are equal, determine the probability expression. The expression must be reduced to the simplest form.

2. (a) In Fig. P2a, $f_a(t)$ is a Poisson rectangular wave with the two possible values of amplitude $+E_m$ and $-E_m$ and with the average rate of k_1 zero crossings per second; $f_b(t)$ is another Poisson rectangular wave with the two possible values of amplitude $+E_m$ and $-E_m$ and the average rate of k_2 zero crossings per second. The two waves are from independent sources. The network A is a summing circuit, so that $f_1(t) = f_a(t) + f_b(t)$. Determine the probability distribution of the amplitude of $f_1(t)$.

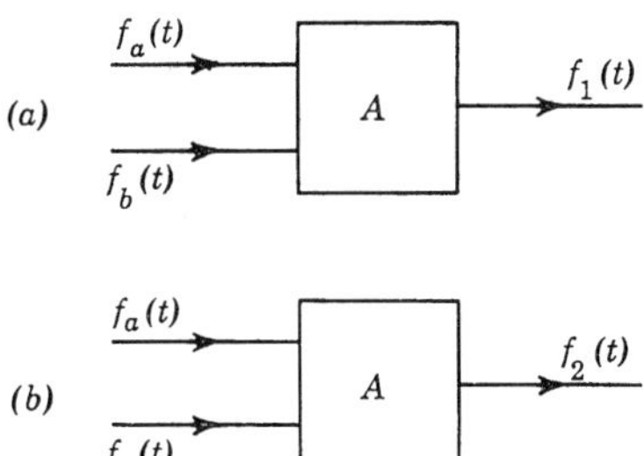

Fig. P2.

(*b*) In Fig. P2*b*, $f_a(t)$ is the same as before, but $f_c(t)$ is a Gaussian noise with the mean value m and the variance σ^2. The network A is a summing circuit so that $f_2(t) = f_a(t) + f_c(t)$. Determine the probability density of the amplitude of $f_2(t)$.

3. We have given a stationary ensemble with the amplitude probability density

$$P_\xi(x) = \frac{a}{2} e^{-2|x|} \qquad \text{for } -\infty < x < \infty$$

It is desired to form a normally distributed ensemble, with zero mean value and variance σ^2, by adding to the given ensemble an independent stationary ensemble with the probability density $P_\eta(y)$. If it is possible to do so, what is the density $P_\eta(y)$? If not, why not?

4. In the random function shown in Fig. P4 the probability density $P_\xi(x)$ of the pulse width ξ is linear from $x = 0$ to $x = \Delta$. The pulse widths are independent of one another. A stationary ensemble is now formed from functions such as the one described. If observations are made on an ensemble member at an arbitrary time $t = t_1$ and then at $t = t_1 + 2\Delta$, what is the probability that both observations give the value E? Each step in the solution of this problem should be clearly explained.

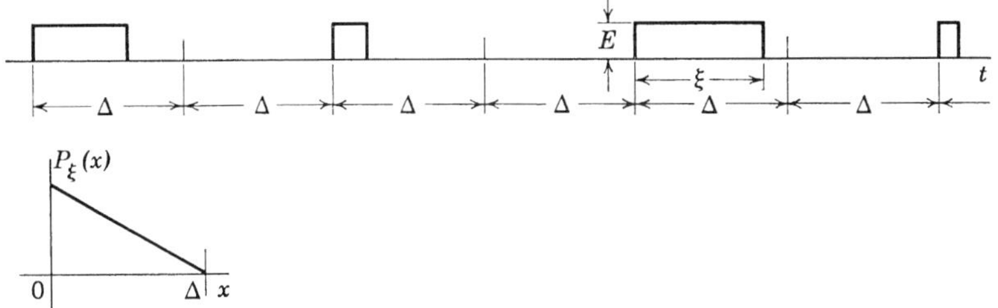

Fig. P4.

chapter 7

Time Averages and Ensemble Averages

The concept of an average is of basic importance in scientific analysis. This is particularly true in the study of random phenomena. In this connection we have pointed out that statistical regularity is a characteristic of the average value. Furthermore, we have emphasized the fact that correlation functions play a central role in the harmonic analysis of random processes. These functions are a special type of average.

In statistical communication theory there are two kinds of averages, namely, time averages and ensemble (or statistical) averages. Since a random process is described in terms of probability densities, the ensemble average can be evaluated if the necessary probability density is known. On the other hand, without an equivalent ensemble average, the time average of a random process can be estimated only from its past behavior because its future cannot be predicted exactly. This chapter considers in detail the expression of a time average as an ensemble average with particular attention on the correlation function.

1. The Aggregate of Elements of an Ensemble Member and the Aggregate of Elements of the Ensemble at a Given Time

Suppose that the past of each member of the ensemble $\{f(t)\}$ is divided into an infinitely large number of elements each of infinitesimal duration Δt as shown in Fig. 1. The amplitude of the ensemble is assumed to be bounded. We shall refer to this infinite aggregate of elements of a member function as the *member aggregate*. Since the ensemble members are generated by sources of identical nature, we make the assumption that all members of the ensemble have the same member aggregate although the order of the elements with respect to amplitude is different. In other words, our assumption is that, if the elements of each member aggregate were rearranged in an increasing (or decreasing) order of amplitude, all member aggregates would take an identical new form.

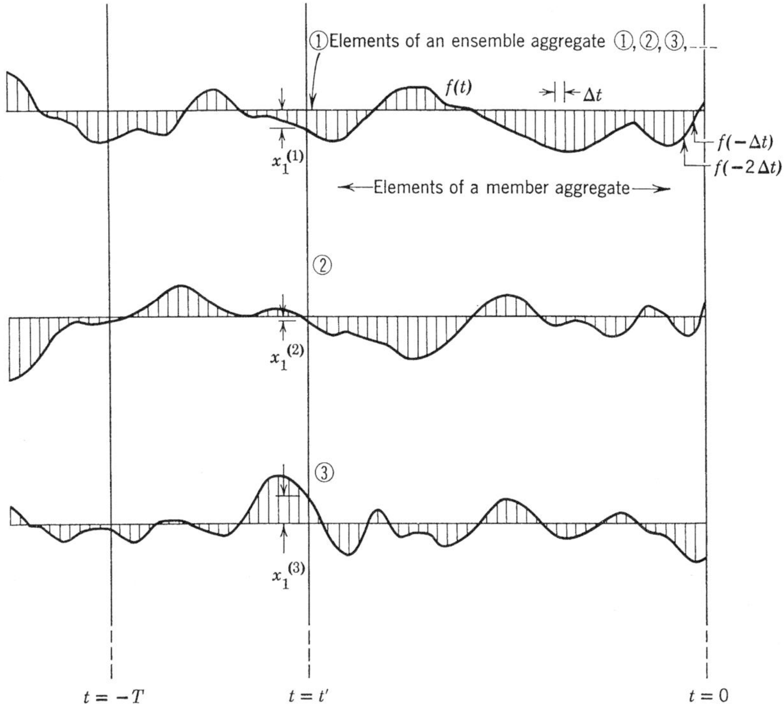

Fig. 1. Elements of a member aggregate and elements of an ensemble aggregate.

On the basis of this assumption, when the infinite past of the ensemble is considered, the time average of amplitude

$$\lim_{T\to\infty} \frac{1}{T} \int_{-T}^{0} f(t)\,dt \tag{1}$$

is the same for all ensemble members. To express this property of the ensemble members, as a consequence of the assumption, in a general form, we let $F(x)$ be an arbitrary function and state that

$$\lim_{T\to\infty} \frac{1}{T} \int_{-T}^{0} F[f(t)]\,dt \tag{2}$$

is the same for every ensemble member. An example of this time average is the mean square of $f(t)$

$$\lim_{T\to\infty} \frac{1}{T} \int_{-T}^{0} f^2(t)\,dt \tag{3}$$

or the mean nth power of $f(t)$

$$\lim_{T\to\infty} \frac{1}{T}\int_{-T}^{0} f^n(t)\,dt \tag{4}$$

With the ensemble members divided into infinitesimal elements, let us examine the ensemble from a different point of view. At time $t = t_1'$ in the past as indicated in Fig. 1, consider the aggregate of elements of the ensemble each from a member of the ensemble. This aggregate of elements shall be called the *ensemble aggregate*. Let us assume that the ensemble is such that the ensemble aggregate is invariant under a shift in time of observation although the order of the elements with respect to amplitude may be different. Now, because of the fact that the invariant ensemble aggregates are formed from functions each having the same aggregate of elements, the ensemble aggregate must be the same as the member aggregate except for the order in which the elements are arranged.

2. The Equivalence of the Time Average of an Ensemble Member and the Ensemble Average

We shall attempt to establish the relation between the average value of a member of the ensemble with respect to time t and the average value of amplitude of the ensemble at any time t. If, in the interval $(-T, 0)$, an ensemble member $f(t)$ is divided into n elements each of duration Δt so that $n = T/\Delta t$, the average of $f(t)$ with respect to t in the interval $(-\infty, 0)$ is

$$\overline{f(t)} = \lim_{\substack{T\to\infty \\ \Delta t\to 0}} \frac{f(-\Delta t) + f(-2\,\Delta t) + \cdots + f(-i\,\Delta t) + \cdots + f(-n\,\Delta t)}{\dfrac{T}{\Delta t}} \tag{5}$$

Clearly we can write (5) in the form

$$\overline{f(t)} = \lim_{\substack{T\to\infty \\ \Delta t\to 0}} \left[\frac{f(-\Delta t) + f(-2\,\Delta t) + \cdots + f(-i\,\Delta t) + \cdots + f(-n\,\Delta t)}{T}\right] \Delta t \tag{6}$$

which is

$$\overline{f(t)} = \lim_{T\to\infty} \frac{1}{T}\int_{-T}^{0} f(t)\,dt \tag{7}$$

This is the *time average* of a member of the ensemble over the interval $(-\infty, 0)$. A bar over a function that is expressed as a function of time indicates that the time average of the function is taken over an infinite interval.

In the average (5) the set of values

$$f(-\Delta t), f(-2\,\Delta t), \ldots, f(-i\,\Delta t), \ldots, f(-n\,\Delta t) \tag{8}$$

(as $T \to \infty$ and $\Delta t \to 0$) being the values of the amplitudes of the member aggregate, must correspond to the set of values

$$x_1^{(1)}, x_1^{(2)}, \ldots, x_1^{(i)}, \ldots, x_1^{(n)} \tag{9}$$

for the amplitudes of the ensemble aggregate, at $t = t_1'$, except for the order in which they appear. Here the superscript indicates the number of the trial in which the value $x_1^{(i)}$ is found at $t = t_1'$ (see Fig. 1). The correspondence follows from the argument that the member aggregate is the same as the ensemble aggregate if we allow a difference in the order of the elements. Assuming that the members of the ensemble have a continuous range, x, of amplitude in the interval $(-a, b)$, we can divide the interval into elements of length Δx for the purpose of obtaining an average value given by (5) on a statistical basis. Suppose that, with a sufficiently large number, M, of elements of the ensemble aggregate, we find that, of the set of values

$$x_1^{(1)}, \ldots, x_1^{(i)}, \ldots, x_1^{(M)} \tag{10}$$

for their amplitudes, there are

$$\left.\begin{array}{l} N_1 \text{ values falling in } (0, \Delta x) \\ N_2 \text{ values falling in } (\Delta x, 2\,\Delta x) \\ \vdots \end{array}\right\} \tag{11}$$

in the positive range and

$$\left.\begin{array}{l} N_{-1} \text{ values falling in } (0, -\Delta x) \\ N_{-2} \text{ values falling in } (-\Delta x, -2\,\Delta x) \\ \vdots \end{array}\right\} \tag{12}$$

in the negative range. Then, as in (13), Chapter 5, Sec. 2, the empirical average

$$\Delta x \frac{N_1}{M} + 2\,\Delta x \frac{N_2}{M} + \cdots - \Delta x \frac{N_{-1}}{M} - 2\,\Delta x \frac{N_{-2}}{M} - \cdots \tag{13}$$

is a close approximation of the average value of the set (8) and, therefore, is a close approximation of the average (5). Writing (13) in the form (15), Chapter 5, Sec. 2, we have

$$\sum_i i\,\Delta x\left(\frac{N_i}{M}\frac{1}{\Delta x}\right)\Delta x \tag{14}$$

where i ranges from $-a/\Delta x$ to $b/\Delta x$. As pointed out in the section referred to, the ratio $(N_i/M)/\Delta x$ is an estimate of the probability density $P_\xi(x)$, of the random variable ξ, which we assume to exist. The random variable is, of course, the amplitude of the ensemble. It should be noted that the assumption made earlier, to the effect that the ensemble aggregate is independent of the time of observation, means that the ensemble is stationary; that is to say, $P_\xi(x)$ is independent of the time of observation. For all practical purposes the empirical average (14), as M becomes sufficiently large and Δx sufficiently small, is

$$\int_{-a}^{b} xP_\xi(x)\,dx \tag{15}$$

or, in a more general form, it is

$$\int_{-\infty}^{\infty} xP_\xi(x)\,dx \tag{16}$$

This we know is the definition of the average value of amplitude of the ensemble at any time. Therefore, we have heuristically shown that the time average of $f(t)$ as given by (7) is, for all practical purposes, equivalent to the average (16) which is based upon the concept of the ensemble in which $f(t)$ is a member. The average (16) is called the *ensemble average* of $f(t)$.

In view of these considerations we make the assumption that, in a stationary ensemble whose members are produced by sources of identical nature, the time average of a member over the infinite past $(-\infty, 0)$ is exactly equal to the ensemble average; that is,

$$\lim_{T\to\infty}\frac{1}{T}\int_{-T}^{0} f(t)\,dt = \int_{-\infty}^{\infty} xP_\xi(x)\,dx \tag{17}$$

Concerning the infinite future of an ensemble member, we find it logical to assume that its mean over this interval of time is the same as that over its infinite past. Thus, we have

$$\lim_{T\to\infty}\frac{1}{T}\int_{-T}^{0} f(t)\,dt = \lim_{T\to\infty}\frac{1}{T}\int_{0}^{T} f(t)\,dt \tag{18}$$

Consequently, (17) becomes

$$\lim_{T\to\infty} \frac{1}{2T} \int_{-T}^{T} f(t)\,dt = \int_{-\infty}^{\infty} xP_\xi(x)\,dx \tag{19}$$

as an expression of the assumption that *the time average of an ensemble member, over the infinite past and future, is equal to the ensemble average.* The expression may be written simply as

$$\overline{f(t)} = \bar{\xi} \tag{20}$$

It follows from the equivalence of the ensemble aggregate and the member aggregate that we can put (19) into a general form by considering an arbitrary function of $f(t)$, $F[f(t)]$, instead of $f(t)$ itself. Thus, if

$$\lim_{T\to\infty} \frac{1}{2T} \int_{-T}^{T} F[f(t)]\,dt \tag{21}$$

exists, then, analogous to (19),

$$\lim_{T\to\infty} \frac{1}{2T} \int_{-T}^{T} F[f(t)]\,dt = \int_{-\infty}^{\infty} F(x)P_\xi(x)\,dx \tag{22}$$

or

$$\overline{F[f(t)]} = \overline{F(\xi)} \tag{23}$$

Clearly, a particular form of (22) is

$$\lim_{T\to\infty} \frac{1}{2T} \int_{-T}^{T} f^n(t)\,dt = \int_{-\infty}^{\infty} x^n P_\xi(x)\,dx \tag{24}$$

If the random process takes discrete values, expression (19) becomes

$$\lim_{T\to\infty} \frac{1}{2T} \int_{-T}^{T} f(t)\,dt = \sum_{i=-\infty}^{\infty} x_i P_\xi(x_i) \tag{25}$$

and (24) becomes

$$\lim_{T\to\infty} \frac{1}{2T} \int_{-T}^{T} F[f(t)]\,dt = \sum_{i=-\infty}^{\infty} F(x_i) P_\xi(x_i) \tag{26}$$

As an example, consider the ensemble of sinusoids shown in Fig. 8, Chapter 6. The first probability density of the stationary ensemble is given by (101), Chapter 6, Sec. 6, that is,

$$P_\xi(x) = \frac{1}{\pi\sqrt{E_m{}^2 - x^2}} \tag{27}$$

Let us consider the mean square value of the sinusoid. Obviously the time average of an ensemble member is

$$\overline{f^2(t)} = \lim_{T\to\infty} \frac{1}{2T} \int_{-T}^{T} E_m{}^2 \cos^2 \omega t \, dt$$

$$= \frac{E_m{}^2}{T_1} \int_0^{T_1} \cos^2 \omega t \, dt$$

$$= \frac{E_m{}^2}{2} \tag{28}$$

where T_1 is the period of the sinusoid; it is noticed that the phase angle is immaterial in the evaluation. The ensemble average of the square of the sinusoid is

$$\overline{\xi^2} = \int_{-E_m}^{E_m} x^2 \frac{1}{\pi\sqrt{E_m{}^2 - x^2}} \, dx$$

$$= \frac{1}{\pi}\left[-\frac{x\sqrt{E_m{}^2 - x^2}}{2} + \frac{E_m{}^2}{2} \sin^{-1} \frac{x}{E_m} \right]_{-E_m}^{E_m}$$

$$= \frac{E_m{}^2}{2} \tag{29}$$

Therefore, we have verified for the ensemble of sinusoids the assumption that

$$\overline{f^2(t)} = \overline{\xi^2} \tag{30}$$

in accordance with (23).

In this example we have been able to compute the time average from a member function because it is periodic and is specified for all values of time once the initial angle is given, but the average value does not depend upon the random initial angle. If the ensemble members are random, we cannot show for specific situations the equivalence of time and ensemble averages because the member functions cannot be expressed exactly for all values of time.

Clearly, when time averages, which are generally associated with physical measurements, are required, the equivalence of a time average and an ensemble average becomes important; for, although a random time function cannot be described with respect to time, its average can be replaced by the ensemble average.

3. Concerning the Justification for the Interchange of Time and Ensemble Averages

In heuristically showing the equivalence of time and ensemble averages, we note that the set of values (8), being the amplitudes of the elements of an enumerable ensemble, is enumerable. (Concerning enumerable and non-enumerable sets of points, reference is made to Chapter 4, Sec. 6.) On the other hand, the range of possible values of the amplitude, ξ, is a continuous variable, x. It has been pointed out that the number of points in a continuous range is a non-enumerable infinity which is of an order much higher than that of an enumerable infinity. Therefore, it is impossible for an enumerable set of points, such as the set (8), to fill all the points in a continuous line. We recall that the set (8) corresponds to the amplitudes of the infinite aggregate of elements of an ensemble member. Hence, however long in time the ensemble member may be extended, it is impossible to produce an aggregate great enough to occupy every point of the continuous range of possible values. In other words, in applying (13) for the mean value of set (8), we realize that there are large gaps in the continuous range, x, which can never be filled. But, since our interest is primarily in the practical equivalence of time and ensemble averages, we need not be concerned with the difference in the orders of infinity, for such a difference is a mathematical one which no physical measurement is able to reveal. As a mathematical problem, however, the equivalence deserves rigorous consideration.

Inasmuch as the concept of message and noise ensembles stems from statistical mechanics, the assumption of the equivalence of time and ensemble averages is found in the same source. In statistical mechanics, physical measurements such as the pressure and temperature of a gas are time averages of the motions of the particles in a system, but it is impossible to obtain a time average from particle motions whose instantaneous values cannot be completely predicted. Realizing that to be able to express a time average in terms of an average through integration over the ensemble is a fundamental requirement in the new mechanics, Gibbs originated the assumption that, in a closed system where the total energy remains constant, a time average over the motions of a system of particles has an equivalent average obtained by integration over a surface in phase space called the ergodic surface. To the physicist whose interest is in the application of Gibbs' mechanics to physical problems, the assumption is acceptable, but from the point of view of the mathematician it needs a rigorous demonstration.

We recall (Chapter 4, Sec. 7) that in statistical mechanics the state of a member at a given instant is represented by a point in the phase space and the point describes a path as time passes. In the attempt to justify Gibbs' assumption as questions concerning its validity arose, J. C. Maxwell proposed the hypothesis that ". . . the system, if left to itself in its actual state of motion, will, sooner or later, pass through every phase which is consistent with the equation of energy." * The phase under this condition is called an *ergodic surface*. This hypothesis is known as the *ergodic hypothesis*. The word "ergodic" is derived from Greek words which mean "work path." Systems which satisfy the hypothesis are called *ergodic systems*. Unfortunately, the belief in the existence of such systems is false, for it is now known that a differentiable path in space, no matter how long in length and how free its direction may be, cannot pass through every point of the space. This is the difficulty with which the mathematician is concerned.†

Owing to its contradictory nature the ergodic hypothesis has been replaced by the *quasi-ergodic hypothesis* which states in effect that there exists closed systems whose phase points pass infinitely close to every point of the ergodic surface, sooner or later, if they are allowed to follow their natural course of motion.

As the work of the mathematicians progressed, it became clear that the necessary tool in the justification of Gibbs' assumption was Lebesgue's theory of integration which appeared contemporaneously with Gibbs' statistical mechanics. That this fundamental work was difficult was borne out by the fact that the proof of the quasi-ergodic hypothesis was not established until some thirty years later. Since ergodic theorems belong to pure mathematics, they are beyond the scope of this book. Readers who are interested in this subject, which has come to be known as *ergodic theory*, ‡ are referred to the works of G. D. Birkhoff, J. von Neumann, and others.

Returning now to the ensemble of messages or noise, we find that the quasi-ergodic hypothesis of statistical mechanics would correspond to

* *The Scientific Papers of James Clark Maxwell*, Dover Publications, New York, 1952, p. 714.

† For a good account of this matter see *A Commentary on the Scientific Writings of Willard Gibbs*, Vol. II, Yale University Press, New Haven, 1936, p. 461.

‡ G. D. Birkhoff, "Proof of the Ergodic Theorem," *Proc. Nat. Acad. Sci.*, **17**, 656–660 (1931).

J. von Neumann, "Proof of the Quasi-Ergodic Hypothesis," *Proc. Nat. Acad. Sci.*, **18**, 70–82 (1932).

B. O. Koopman and J. von Neumann, "Dynamical Systems of Continuous Spectra," *Proc. Nat. Acad. Sci.*, **18**, 255–263 (1932).

J. von Neumann, "Physical Applications of the Ergodic Hypothesis," *Proc. Nat. Acad. Sci.*, **18**, 263–266 (1932).

N. Wiener, *Cybernetics*, John Wiley and Sons, New York, 1948, pp. 57–73.

the hypothesis that, *in a stationary ensemble of random functions produced by sources of identical nature and having a continuous range of possible values, the amplitudes of the aggregate of elements of an ensemble member come infinitely close to every point of the continuous range of possible values, sooner or later, if the ensemble member is allowed to take its natural course of fluctuation.* It is to be noted that the elements of the aggregate have the same infinitesimal duration dt and that the amplitude of each element is to be taken at an arbitrary point in the interval dt, such as either one of the end points, but the chosen point in dt is the same for all elements of the aggregate. Also, we note that the phase space for the ensemble with respect to its average value of amplitude is one dimensional, being the line which represents the variable x. Each element of the member aggregate is represented by a point in this phase space. The quasi-ergodic hypothesis is sufficient justification for the equivalence of time and ensemble averages in the statistical theory of communication involving physical systems.

When the range of possible values of an ensemble is discrete instead of continuous, the equivalence of time and ensemble averages is much more readily seen if we follow a heuristic proof similar to the one given for the continuous case. As we have seen, when the range of possible values is continuous, a difficulty in justifying the equivalence arises out of the discrepancy between the enumerable infinity of element amplitudes and the non-enumerable infinity of possible values in the continuous range. In the discrete case, however, both the element amplitudes and possible values are enumerable. Under this condition the quasi-ergodic hypothesis is unnecessary.

4. The Autocorrelation Function as a Time Average and as an Ensemble Average

Since the autocorrelation function occupies a central position in much of our work, the interpretation of its expression as an ensemble average in the light of its definition as a time average should be well understood. As we recall, the autocorrelation for a random function $f(t)$ is defined by (203), Chapter 2, Sec. C-2, which may be written as

$$\varphi_{ff}(\tau) = \lim_{T\to\infty} \frac{1}{2T} \int_{-T}^{T} f(t)f(t+\tau)\,dt \tag{31}$$

It is a time average. This expression may be abbreviated to

$$\varphi_{ff}(\tau) = \overline{f(t)f(t+\tau)} \tag{32}$$

Let us consider the time interval $(-\infty, 0)$ in which $f(t)$ is available for examination. The autocorrelation function for $f(t)$ on the basis of only its past, with $f(t) = 0$ for $t > 0$, is

$$\lim_{T\to\infty} \frac{1}{T} \int_{-T}^{0} f(t)f(t+\tau)\,dt \tag{33}$$

If, as it is done in Sec. 1, each member function is divided into infinitesimal elements of duration Δt, we can imagine that an aggregate of pairs of elements are formed, the elements of each pair being separated by the time displacement τ of (33), as indicated in Fig. 2. Thus

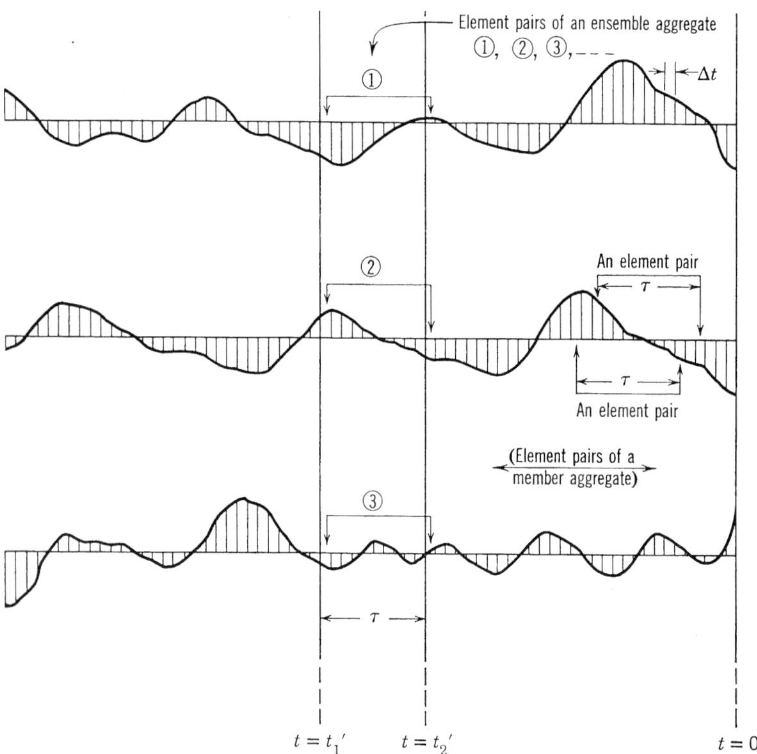

Fig. 2. Element pairs of a member aggregate and element pairs of an ensemble aggregate.

the elements of an ensemble member at $t = 0, -\Delta t, -2\,\Delta t, \ldots$ and those at $t = -\tau, -\tau - \Delta t, -\tau - 2\,\Delta t, \ldots$ respectively form an aggregate of pairs. We shall refer to this aggregate as a member aggregate of element pairs. Since the ensemble is generated by sources of identical nature, we assume that all the member aggregates of element pairs are the same except that the order in which the element pairs are arranged in accordance with the natural course of fluctuation, differs from member to member.

This assumption enables us to say that (33) is the same for all ensemble members and to assert further that, if $F(x)$ is an arbitrary integrable function, then

$$\lim_{T\to\infty} \frac{1}{T} \int_{-T}^{0} F[f(t)f(t+\tau)]\,dt \tag{34}$$

is the same for all members of the ensemble.

We now consider the element pairs with respect to the ensemble instead of the individual function. At times $t = t_1'$ and $t = t_2'$ in the past, which are separated by time τ (that is, $t_2' = t_1' + \tau$, as indicated in Fig. 2), let us form an aggregate of pairs of elements by taking an element at each of the two times from each member function. We shall call such an aggregate the ensemble aggregate of element pairs. Because we are concerned with only stationary ensembles, the ensemble aggregate of element pairs for a given value of τ is invariant for any value of t_1'. However, the order of the element pairs varies as the value of t_1' changes.

Now, since the invariant ensemble aggregates of element pairs are formed from member functions each of which has the same aggregate of pairs of elements, the ensemble aggregate must be the same as the member aggregate except for a difference in the order of the element pairs. Therefore, it is possible to say that the mean of the product of the two elements in a pair is the same no matter how it is expressed, whether as a mean with respect to the member aggregate or as a mean with respect to the ensemble aggregate. The mean with respect to the member aggregate is given by (33). Obviously this is a time average. On the other hand, the mean on the basis of the ensemble aggregate is expressed by (52), Chapter 5, Sec. 6, and it is an ensemble average. The equivalence of these two averages is expressed by

$$\lim_{T\to\infty} \frac{1}{T} \int_{-T}^{0} f(t)f(t+\tau)\,dt = \int_{-\infty}^{\infty}\int_{-\infty}^{\infty} x_1x_2P_{\xi_1\xi_2}(x_1, x_2; \tau)\,dx_1\,dx_2 \tag{35}$$

If we assume that the mean of $f(t)f(t+\tau)$ over the infinite past is equal to that over the infinite future, then

$$\lim_{T\to\infty} \frac{1}{T} \int_{-T}^{0} f(t)f(t+\tau)\,dt = \lim_{T\to\infty} \frac{1}{T} \int_{0}^{T} f(t)f(t+\tau)\,dt \tag{36}$$

(The right-hand integral is taken with $f(t) = 0$ for $t < 0$.) With this assumption we can show that (35) for the entire $f(t)$, both its past and its future, is

$$\lim_{T\to\infty}\frac{1}{2T}\int_{-T}^{T} f(t)f(t+\tau)\,dt = \int_{-\infty}^{\infty}\int_{-\infty}^{\infty} x_1x_2P_{\xi_1\xi_2}(x_1, x_2; \tau)\,dx_1\,dx_2 \tag{37}$$

which may be abbreviated to

$$\overline{f(t)f(t+\tau)} = \overline{\xi_1\xi_2} \tag{38}$$

as an expression of the conclusion that *the time average of* $f(t)f(t+\tau)$ *is equivalent to its ensemble average expressed as the average of* $\xi_1\xi_2$.

Inasmuch as the average under discussion is, by definition, the autocorrelation function as given by (31), the alternative expression for the autocorrelation function is

$$\varphi_{ff}(\tau) = \int_{-\infty}^{\infty}\int_{-\infty}^{\infty} x_1x_2P_{\xi_1\xi_2}(x_1, x_2; \tau)\,dx_1\,dx_2 \tag{39}$$

This is the autocorrelation function of $f(t)$ *as an ensemble average.* If $f(t)$ is a discrete process, (39) becomes

$$\varphi_{ff}(\tau) = \sum_{j=-\infty}^{\infty}\sum_{i=-\infty}^{\infty} x_{1i}x_{2j}P_{\xi_1\xi_2}(x_{1i}, x_{2j}; \tau) \tag{40}$$

Just as in statistical mechanics, the possibility of expressing a time average as an ensemble average in communication theory is of basic importance. It has been emphasized that messages and noise can be described adequately only by probability distributions, and this matter has been given thorough attention. When an average involving a message or a noise such as the autocorrelation function is required, its evaluation is theoretically possible only when the expression is reducible to one in terms of the probability functions which describe the random process concerned. Without the equivalent expression the average value can be estimated only on the basis of statistical data. It is to fulfill this need for autocorrelation that equations (39) and (40) have been developed.

5. General Forms of Equivalence of Time and Ensemble Averages

To generalize the equivalence of time and ensemble averages given by (37) for autocorrelation, we consider an arbitrary integrable function F. By a procedure much the same as that we have followed, it can be shown that

$$\lim_{T\to\infty}\frac{1}{2T}\int_{-T}^{T} F[f(t)f(t+\tau)]\,dt = \int_{-\infty}^{\infty}\int_{-\infty}^{\infty} F(x_1x_2)P_{\xi_1\xi_2}(x_1, x_2; \tau)\,dx_1\,dx_2 \tag{41}$$

or, in a shorter expression, the equivalence of the time and ensemble averages as expressed by (41) is

$$\overline{F[f(t)f(t+\tau)]} = \overline{F(\xi_1\xi_2)} \tag{42}$$

An example of (42) is

$$\overline{f^m(t)f^n(t+\tau)} = \overline{\xi_1{}^m\xi_2{}^n} \tag{43}$$

A further generalization of (37) is obtained by considering $F[f(t)f(t + \tau_1)f(t + \tau_1 + \tau_2) \ldots f(t + \tau_1 + \cdots + \tau_{n-1})]$. The extension is clearly the equivalence of time and ensemble averages expressed by

$$\overline{F[f(t)f(t+\tau_1)f(t+\tau_1+\tau_2) \ldots f(t+\tau_1+\cdots+\tau_{n-1})]} = \overline{F(\xi_1\xi_2\xi_3 \ldots \xi_n)} \tag{44}$$

A particular form of (44) is

$$\overline{f^{m_1}(t)f^{m_2}(t+\tau_1)f^{m_3}(t+\tau_1+\tau_2) \ldots f^{m_n}(t+\tau_1+\cdots+\tau_{n-1})} = \overline{\xi_1{}^{m_1}\xi_2{}^{m_2}\xi_3{}^{m_3} \ldots \xi_n{}^{m_n}} \tag{45}$$

In (44) and (45) the function f is a member of an ensemble so that the random variables $\xi_1, \xi_2, \ldots, \xi_n$ belong to the same ensemble at different times $t = t_1$, $t = t_2$, $\ldots$, $t = t_n$ respectively. However, as we have indicated in Chapter 5, Sec. 6, it is possible that an average value involves the product of the variables in different stationary ensembles of messages and noise. In such a circumstance we may let $f_1(t)$, $f_2(t + \tau_1)$, $f_3(t + \tau_1 + \tau_2)$, $\ldots$, $f_n(t + \tau_1 + \cdots + \tau_{n-1})$ represent the member functions of the ensembles $\{f_1(t)\}$, $\{f_2(t)\}$, $\{f_3(t)\}$, $\ldots$, $\{f_n(t)\}$, with displacements $\tau = 0$, $\tau = \tau_1$, $\tau = \tau_1 + \tau_2$, $\ldots$, $\tau = \tau_1 + \cdots + \tau_{n-1}$ respectively. Clearly, if some of the functions belong to the same ensemble, their subscripts then become identical. We shall let ξ_1, ξ_2, ξ_3, $\ldots$, ξ_n be the respective random variables at times $t = t_1$, $t = t_2$, $t = t_3$, $\ldots$, $t = t_n$ respectively. It should be noted that, although the symbols $\xi_1, \ldots, \xi_n$ are the same as those in (44) and (45), they now do not necessarily refer to the same ensemble. Analogous to (44), we have

$$\overline{F[f_1(t)f_2(t+\tau_1)f_3(t+\tau_1+\tau_2) \ldots f_n(t+\tau_1+\cdots+\tau_{n-1})]} = \overline{F(\xi_1\xi_2\xi_3 \ldots \xi_n)} \tag{46}$$

and, analogous to (45), we have

$$\overline{f_1{}^{m_1}(t)f_2{}^{m_2}(t+\tau_1)f_3{}^{m_3}(t+\tau_1+\tau_2) \ldots f_n{}^{m_n}(t+\tau_1+\cdots+\tau_{n-1})} = \overline{\xi_1{}^{m_1}\xi_2{}^{m_2}\xi_3{}^{m_3} \ldots \xi_n{}^{m_n}} \tag{47}$$

6. The Crosscorrelation Function as a Time Average and as an Ensemble Average

The special form

$$\overline{f_1(t)f_2(t+\tau_1)} = \overline{\xi_1\xi_2} \tag{48}$$

of (47) gives us the crosscorrelation function as an ensemble average; for, if the crosscorrelation is between the functions $f(t)$ and $g(t)$ of the ensembles $\{f(t)\}$ and $\{g(t)\}$, then $f_1(t) = f(t)$ and $f_2(t + \tau_1) = g(t + \tau)$, and $\xi_1\xi_2$ may be written as $\xi_1\eta_2$ so that (48) is

$$\overline{f(t)g(t+\tau)} = \overline{\xi_1\eta_2} \tag{49}$$

Since the crosscorrelation function of $f(t)$ and $g(t)$ is defined as the time average

$$\varphi_{fg}(\tau) = \lim_{T\to\infty} \frac{1}{2T} \int_{-T}^{T} f(t)g(t+\tau)\,dt \tag{50}$$

which is the left-hand member of (49) and since the right-hand member of (49) is given by (54), Chapter 5, Sec. 6, we have

$$\varphi_{fg}(\tau) = \int_{-\infty}^{\infty}\int_{-\infty}^{\infty} x_1y_2P_{\xi_1\eta_2}(x_1, y_2; \tau)\,dx_1\,dy_2 \tag{51}$$

This is the crosscorrelation function of $f(t)$ and $g(t)$ as an ensemble average if the range of possible values is continuous. The expression for discrete $f(t)$ and $g(t)$ is obviously

$$\varphi_{fg}(\tau) = \sum_{j=-\infty}^{\infty}\sum_{i=-\infty}^{\infty} x_{1i}y_{2j}P_{\xi_1\eta_2}(x_{1i}, y_{2j}; \tau) \tag{52}$$

7. The Quasi-Ergodic Hypothesis for Correlation and Other Averages

In heuristically showing the equivalence of time and ensemble averages in autocorrelation in Sec. 4, we see that element pairs of a member function are of basic importance. For a given time of displacement, τ, between the elements of a pair, each pair of elements in a member aggregate is represented by a phase point in the $(x_1, x_2; \tau)$ phase space. This is a three-dimensional phase space. In other words, the amplitudes of the elements in a pair determine a point in the (x_1, x_2) plane for a given value of τ which constitutes the phase space as far as autocorrelation is concerned. With respect to the element amplitudes let us be reminded that, as explained in Sec. 3, the amplitude of an element is taken at an arbitrary but fixed point in the interval dt of the element.

Now, inasmuch as the member aggregate of element pairs is an enumerable infinity, the points on the plane are an enumerable infinity.

On the other hand, if the random process in question is continuous, the phase space, insofar as autocorrelation is concerned, is the (x_1, x_2) plane for all values of displacement τ, and the number of points in the plane is a non-enumerable infinity. Therefore, however long the member function may be, it is impossible for the points representing the states of the member aggregate of element pairs to cover all points in the phase space. In fact the area filled by all the phase points of a member aggregate of element pairs is but a negligible portion of the available area. Such a discrepancy is, of course, due to the difference in the orders of infinities involved. In heuristically showing the equivalence of time and ensemble averages, we have overlooked this point and have tacitly assumed that the phase space is completely filled by the phase points.

Under this circumstance we introduce a quasi-ergodic hypothesis to strengthen the justification for the interchange of time and ensemble averages in autocorrelation, just as we do in the simple averaging of a random process. Analogous to the quasi-ergodic hypothesis in statistical mechanics, the hypothesis we need is that, *in a stationary ensemble produced by sources of identical nature with a continuous range of possible output values, the points representing the states of the element pairs of an ensemble member in autocorrelation will approach, sooner or later, infinitely close to every point in the continuous three-dimensional phase space for autocorrelation if the ensemble member is allowed to follow its natural course of fluctuation.*

In the general averages (44) through (47) the implied integrations involve not pairs of elements as in autocorrelation but groups of elements. In fact, for an average requiring the product of n time functions, or, alternatively, n random variables, the number of elements in an element group is n. Therefore, the member aggregate in a general average is an aggregate of such element groups. Clearly, the phase point of an element group is a point in the $(x_1, x_2, \ldots, x_n; \tau_1, \tau_2, \ldots, \tau_{n-1})$ phase space. This space is $(2n - 1)$ dimensional. The quasi-ergodic hypothesis for the general averages (44) through (47) is therefore that, *in a continuous stationary ensemble produced by sources of identical nature, the phase points of an ensemble member corresponding to such general averages will, if sufficient time is allowed, come as near as desired to all points in the* $(2n - 1)$*-dimensional phase space.*

PROBLEMS

1. Figure P1 shows an ensemble of periodic functions which are the same except for a displacement in time. Let the time from $t = 0$ to the first jump of each member function be ξ. If ξ has a uniform probability density, what is the

amplitude probability density of the ensemble? With this probability density show that the ensemble average of amplitude is equal to the time average of amplitude. Do the same for the square of the amplitude.

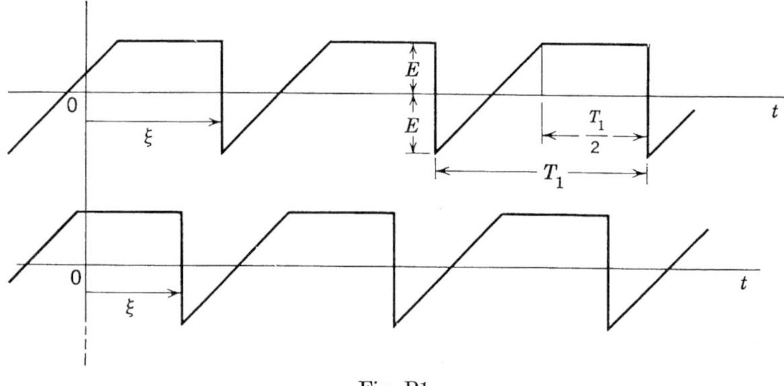

Fig. P1.

2. Figure P2 shows an ensemble of periodic functions which are the same except for a displacement in time. Let the time from $t = 0$ to the first positive peak of each member function be ξ. If ξ has a uniform probability density, what is the amplitude probability density of the ensemble? With this probability density show that the ensemble average of amplitude is equal to the time average of amplitude. Do the same for the square of the amplitude.

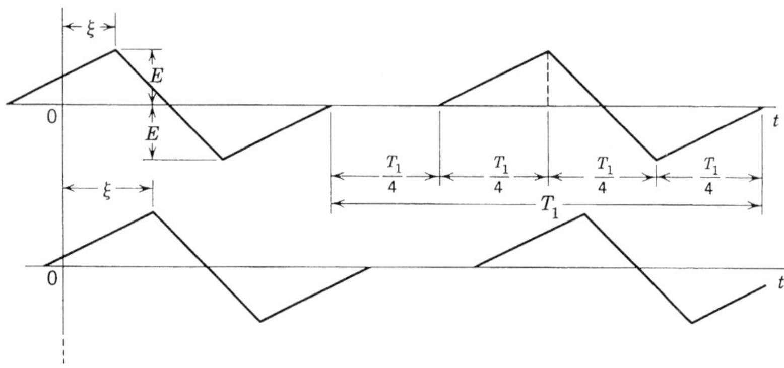

Fig. P2.

3. Figure P3 shows an ensemble of periodic functions which are identical except for a displacement in time. Let the time from $t = 0$ to the first jump from 0 to $-2E$ of each member function be ξ. If ξ has a uniform probability density, what is the amplitude probability density of the ensemble? With this probability density show that the ensemble average of amplitude is equal to the time average of amplitude. Do the same for the square of the amplitude.

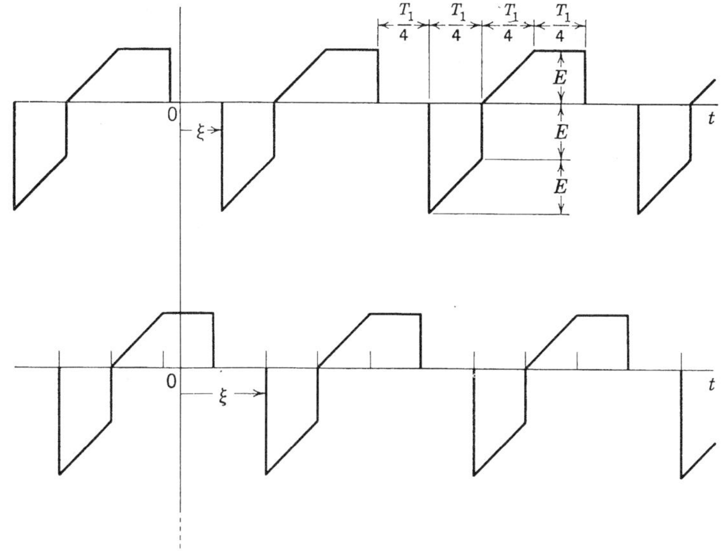

Fig. P3.

4. Figure P4 shows an ensemble of periodic functions which are identical except for a displacement in time. Let the time from $t = 0$ to the first positive jump of each member function be ξ. If ξ has a uniform probability density, what is the amplitude probability density of the ensemble? With this probability density show that the ensemble average of amplitude is equal to the time average of amplitude. Do the same for the square of the amplitude.

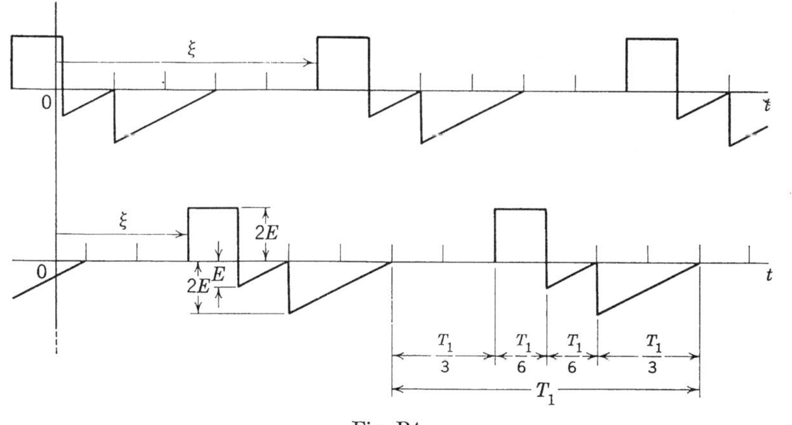

Fig. P4.

5. Figure P5 shows an ensemble of periodic functions which are identical except for a displacement in time. Let the time from $t = 0$ to the first initial

point of a pulse of each member function be ξ. If ξ has a uniform probability density, what is the amplitude probability density of the ensemble? With this probability density show that the ensemble average of amplitude is equal to the time average of amplitude. Do the same for the square of the amplitude.

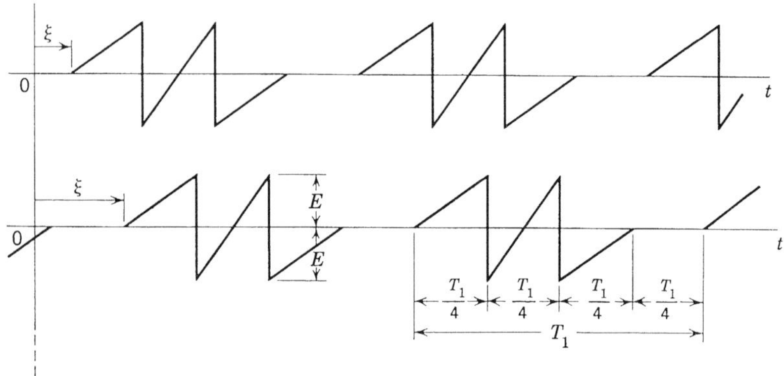

Fig. P5.

6. A stationary random noise containing no periodic components has the amplitude probability density

$$P_\xi(y) = \begin{cases} Cye^{-y} & \text{for } y \geq 0 \\ 0 & \text{elsewhere} \end{cases}$$

where C is to be determined. If to each member of the ensemble of periodic functions of Problem 1, 2, 3, 4, or 5 (assign one) a member of the noise ensemble is added, what is the amplitude probability density of the new ensemble? Sketch the density under the condition that both components have the same mean square value.

7. A stationary random process containing no periodic components has two possible values of amplitude $+E$ and $-E$ with probabilities $\frac{2}{3}$ and $\frac{1}{3}$ respectively. To each member function of the ensemble is now added a periodic function, which is shown in Fig. P7, thereby forming a new ensemble. The periodic functions added are identical except for a displacement in time. The time from $t = 0$ to the first jump is ξ, and the probability density of ξ is uniform (rectangular). Determine the waveform of a periodic function such that a stationary ensemble of periodic functions of this waveform has the same amplitude probability density as that of the new ensemble.

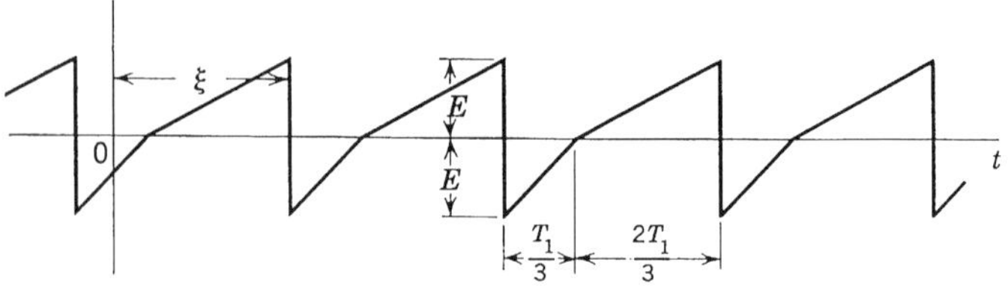

Fig. P7.

chapter 8

Analytical Determination of Correlation Functions and Power Density Spectrums of Random Processes

The determination of a correlation function is a statistical problem. It requires an adequate statistical description of the random process involved and an equivalent statistical expression for the definition of the function as a time average. Since both of the requirements have been considered in detail in several preceding chapters, we are ready to illustrate the methods and techniques of the problem by simple examples.

The concept of messages and noise in statistical communication theory is fundamentally different from that in classical communication theory. For their description in statistical theory we require probability theory and generalized harmonic analysis, whereas in classical theory we depend simply upon Fourier series and integrals. Because of this difference we expect that the methods and techniques in statistical communication theory are more complex and difficult. In this respect let us point out that a statistical description of an actual physical message, sufficient for the determination of the autocorrelation function, is in general not easy to achieve if at all possible. Although the first probability density, being only two dimensional, can be experimentally estimated without great effort, the second probability density, upon which the autocorrelation function depends, is a four-dimensional function which does not readily yield to complete measurement and simple representation. Moreover, aside from a few exceptional situations, the complexity of the causes of random fluctuation does not permit the theoretical determination of a valid second probability density for a physical system from simple assumptions. To avoid these difficulties, actual measurement of the autocorrelation function is made in place of its calculation based upon probability density when statistical theory is applied to a physical problem. Nevertheless, difficulty in obtaining the second probability density for practical application does not mean that expressing the autocorrelation function as an ensemble average is of no avail and unnecessary. The expression embodies basic concepts of the theory; it has great theoretical value. Furthermore, measurement of correlation functions on an ensemble basis should be guided by the physical interpretation of the expression.

Although the primary purpose of this chapter is to show the methods and techniques in finding the correlation function analytically for some relatively simple situations, certain properties of the function which are significant in communication problems are pointed out through examples.

1. Background of the Correlation Function

Correlation has long been an important concept in statistics. For certain problems pertaining to the strength of the relation between two random variables, the statistician introduces the *correlation coefficient.* The correlation coefficient, $\rho_{\xi\eta}$, of two random variables, ξ and η, is defined as

$$\rho_{\xi\eta} = \frac{\overline{(\xi - \bar{\xi})(\eta - \bar{\eta})}}{\sigma_\xi \sigma_\eta} \tag{1}$$

It is the mean of the product of the two random variables about their respective means, $\bar{\xi}$ and $\bar{\eta}$, normalized to the product of their standard deviations, σ_ξ and σ_η. When the coefficient is not normalized, it is called the *covariance*, or the *second-order product moment.* That is, the covariance, $\sigma_{\xi\eta}$, of ξ and η is

$$\sigma_{\xi\eta} = \overline{(\xi - \bar{\xi})(\eta - \bar{\eta})} \tag{2}$$

Clearly the correlation coefficient, or the covariance, of statistics and the correlation function of communication theory bear a close resemblance to each other because both of them are essentially mean values of the product of two random variables. Nevertheless, one important difference between them is that the correlation function, as the name implies, is a function of time whereas the correlation coefficient is a single mean value. Such a difference is to be expected, for in communication problems time is a variable of primary importance, but in statistical problems it is often not necessarily a relevant parameter.

The application of correlation to the spectral analysis of random phenomena is attributed to G. I. Taylor.* Before the incorporation of the concept in communication theory, Taylor had made a significant contribution to the theory of diffusion and turbulence with correlation analysis as a basic tool. However, in the general field of the analysis of random motion, the rudiment of correlation is to be found in the earlier studies of white light by A. Schuster, which led him to the development of periodogram analysis.

After the work of Schuster, Taylor, and others, rigorous mathematical development in the analysis of random phenomena began to appear. A most notable contribution in this development is that of N. Wiener. As we know, we have, in a heuristic manner, presented some of his work in Chapter 2, Secs. C and D, with particular emphasis on a theorem which bears his name.

2. Autocorrelation of a Rectangular Wave with Poisson Distributed Zeros

For an instructive illustration in the determination of autocorrelation functions, let us consider an ensemble of rectangular waves that alternate between the values E_m and $-E_m$. We have studied these waves in Chapter 6 in connection with the Poisson distribution; the ensemble is shown in Fig. 1 of that chapter. We shall assume that the distribution of zeros is Poisson; that is, if we denote the number of zeros of a member function in the duration τ by the random variable η, which has the possible values $n = 0, 1, 2, \ldots$, then the probability of finding n zeros in the duration τ of the member function is

$$P_\eta(n; \tau) = \frac{(k\tau)^n}{n!} e^{-k\tau} \tag{3}$$

where k is the average number of zeros per unit of time. This is the distribution (40), Chapter 6, Sec. 1. We shall refer to the waves as Poisson rectangular waves.

Since the process is discrete, we apply equation (40), Chapter 7, Sec. 4. Writing the expression for the condition that ξ_1 and ξ_2 have only two possible values, we have

$$\varphi_{ff}(\tau) = \sum_{j=-1}^{1} \sum_{i=-1}^{1} x_{1i}x_{2j}P_{\xi_1\xi_2}(x_{1i}, x_{2j}; \tau) \tag{4}$$

* G. I. Taylor, "Diffusion by Continuous Movements," *Proc. Lond. Math. Soc.*, **20,** 196–212 (1920).

G. I. Taylor, "The Spectrum of Turbulence," *Proc. Roy. Soc. Lond., Ser. A*, **164,** 476–490 (1938).

In the figure referred to, we have indicated that the amplitudes of an ensemble member, $f(t)$, at $t = t_1$ and $t = t_2 = t_1 + \tau$ are ξ_1 and ξ_2 respectively.

For computational purposes it is preferable to express the second probability distribution as the product of an unconditional distribution and a conditional distribution in accordance with (49), Chapter 5, Sec. 6. Thus

$$\varphi_{ff}(\tau) = \sum_{j=-1}^{1} \sum_{i=-1}^{1} x_{1i} x_{2j} P_{\xi 1}(x_{1i}) P_{\xi 2|\xi 1}(x_{2j}|x_{1i}; \tau) \tag{5}$$

In this expression

$$\left.\begin{aligned} x_{1,1} &= E_m \\ x_{1,-1} &= -E_m \\ x_{2,1} &= E_m \\ x_{2,-1} &= -E_m \end{aligned}\right\} \tag{6}$$

On the basis of the assumptions in the derivation of the Poisson distribution, we have

$$\begin{aligned} P_{\xi 1}(E_m) &= P_{\xi 1}(-E_m) \\ &= \tfrac{1}{2} \end{aligned} \tag{7}$$

In other words, a member function at $t = t_1$ has the same probability of taking the value E_m or $-E_m$.

With respect to the conditional distribution $P_{\xi 2|\xi 1}(x_{2j}|x_{1i}; \tau)$ in (5), let us first consider its value for $i = 1$ and $j = 1$; that is, let us consider

$$P_{\xi 2|\xi 1}(E_m|E_m; \tau) \tag{8}$$

This is the probability of the event that ξ_2, at $t = t_2 = t_1 + \tau$, takes the value E_m, given that ξ_1, at $t = t_1$, took the value E_m. We note that the event can occur in many ways. One way is to have no zeros in the duration τ, and another is to have two zeros in the interval so that the function which starts at $t = t_1$ with the value E_m will have returned to E_m τ seconds later. In fact, any even number of zeros in τ will satisfy the conditions. Therefore we have an event which can occur in a series of mutually exclusive forms. By the theorem of total probability we write

$$\mathcal{P}(\xi_2 = E_m | \xi_1 = E_m; \tau) = \sum_{n=0,2,4,\ldots} \mathcal{P}(\eta = n; \tau) \tag{9}$$

Hence the conditional probability (8) is

$$P_{\xi 2|\xi 1}(E_m|E_m; \tau) = \sum_{n=0,2,4,\ldots} P_\eta(n; \tau) \tag{10}$$

in which $P_\eta(n; \tau)$ is the Poisson distribution (3). In a similar manner we find that, for $i = -1$ and $j = -1$, the conditional probability has the same value; that is,

$$P_{\xi 2|\xi 1}(-E_m|-E_m;\tau) = \sum_{n=0,2,4,\ldots} P_\eta(n;\tau) \tag{11}$$

If we start with E_m at $t = t_1$ and wish to have $-E_m$ at $t = t_2$, it is quite clear that the number of zeros in τ must be an odd number. Conversely, if the initial value is $-E_m$ and the final value is E_m, the number of zeros in τ must also be odd. Accordingly, for $i = 1$ and $j = -1$, the conditional probability is

$$P_{\xi 2|\xi 1}(-E_m|E_m;\tau) = \sum_{n=1,3,5,\ldots} P_\eta(n;\tau) \tag{12}$$

and, for $i = -1$, $j = 1$, it is

$$P_{\xi 2|\xi 1}(E_m|-E_m;\tau) = \sum_{n=1,3,5,\ldots} P_\eta(n;\tau) \tag{13}$$

Summing in accordance with (5), we have the autocorrelation function

$$\varphi_{ff}(\tau) = 2 \times E_m \times E_m \times \tfrac{1}{2} \times \sum_{n=0,2,4,\ldots} P_\eta(n;\tau) - 2 \times E_m \times E_m \times \tfrac{1}{2} \times \sum_{n=1,3,5,\cdots} P_\eta(n;\tau) \tag{14}$$

Inasmuch as the variable τ in (3) is always positive whereas in autocorrelation the displacement τ ranges over the interval $(-\infty, \infty)$, it is necessary to write $P_\eta(n;\tau)$ in (14) with τ replaced by $|\tau|$. Hence applying (3) to (14) and simplifying, we have

$$\begin{aligned}\varphi_{ff}(\tau) &= E_m^2\left[\sum_{n=0,2,4,\ldots} \frac{(k|\tau|)^n}{n!} e^{-k|\tau|} - \sum_{n=1,3,5,\ldots} \frac{(k|\tau|)^n}{n!} e^{-k|\tau|}\right] \\ &= E_m^2 e^{-k|\tau|}\left[1 - \frac{k|\tau|}{1!} + \frac{(k|\tau|)^2}{2!} - \frac{(k|\tau|)^3}{3!} + \cdots\right] \\ &= E_m^2 e^{-2k|\tau|}\end{aligned} \tag{15}$$

This is the autocorrelation function of the Poisson rectangular wave. In connection with the discussion on generalized harmonic analysis, this function has served as an illustration in Sec. C-2, Chapter 2. Equation (15) above is (209) of the section referred to, with the graph shown in Fig. 29. As far as the author is aware, the autocorrelation of the Poisson rectangular wave first appeared in the work of G. W. Kenrick.*

* G. W. Kenrick, "The Analysis of Irregular Motions with Applications to the Energy-Frequency Spectrum of Static and of Telegraph Signals," *Phil. Mag., Ser.* 7, 7, 176–196 (1929).

The power density spectrum of the Poisson rectangular wave can now be readily evaluated by application of Wiener's theorem for autocorrelation which is expressed by (236) and (237), Chapter 2, Sec. C-3. In fact the spectrum appears as (241) at the end of that section, and it is

$$\Phi_{ff}(\omega) = \frac{E_m{}^2}{\pi} \frac{2k}{(2k)^2 + \omega^2} \tag{16}$$

with its graph shown in Fig. 30, Chapter 2.

The maximum value of the spectrum is at the origin and is $\Phi_{ff}(0) = E_m{}^2/2\pi k$. At the half-power-density point, that is, at $\Phi_{ff}(\omega) = \Phi_{ff}(0)/2$, the angular frequency is $\omega_o = 2k$ as indicated in Fig. 1. Clearly, the

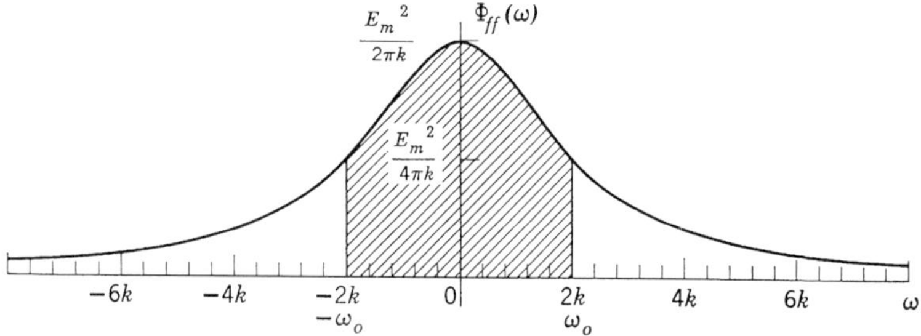

Fig. 1. Power density spectrum of Poisson rectangular wave.

greater the value of k, the greater is the angular frequency at which the power density has half the maximum value, but at the same time the maximum density is reduced in inverse proportion to k. Thus the spectrum is "flattened" and "lowered" at the peak as the average number of zeros per second is increased. Furthermore, the power in the frequency band $(-2k, 2k)$ is exactly half the total power of the random wave. We can easily see this because the total power is $\varphi_{ff}(0) = E_m{}^2$ whereas the power in the band $(-2k, 2k)$, which is represented by the shaded area of Fig. 1, is

$$\begin{aligned}\int_{-2k}^{2k} \Phi_{ff}(\omega)\,d\omega &= \frac{E_m{}^2}{\pi}\int_{-2k}^{2k} \frac{2k}{(2k)^2 + \omega^2}\,d\omega \\ &= \frac{E_m{}^2}{2}\end{aligned} \tag{17}$$

Since the half-power-density angular frequency $(\omega_o = 2k)$ divides the total power equally between the upper band $(\omega_o < \omega < \infty$ and $-\infty < \omega < -\omega_o)$ and the lower band $(-\omega_o < \omega < \omega_o)$, it is also the half-power angular frequency.

3. Autocorrelation of a Pulse-Width Modulated Wave

As a second illustration of the analytical determination of autocorrelation functions, we consider the pulse-width modulated wave $f_1(t)$ shown in Fig. 2. The wave is a series of rectangular pulses of height E with

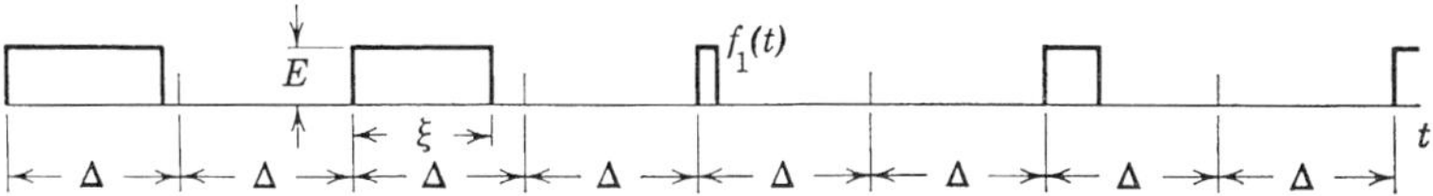

Fig. 2. A pulse-width modulated wave.

their leading edges at 2Δ apart. These pulses have random widths that are independent of one another. Let the pulse width be ξ and its probability density be $P_\xi(x)$. For the present we simplify the problem by assuming that the maximum pulse width is Δ. Later on we shall take account of the situation where a pulse may extend to the leading edge of the following pulse.

In this illustration we introduce a procedure which considerably simplifies the work that would be involved if the method for handling the rectangular Poisson wave were followed. We begin by noticing that the definition

$$\varphi_{11}(\tau) = \lim_{T\to\infty} \frac{1}{2T} \int_{-T}^{T} f_1(t) f_1(t+\tau)\, dt \tag{18}$$

when applied to the wave for $0 < \tau < \Delta$, as shown in Fig. 3, calls for

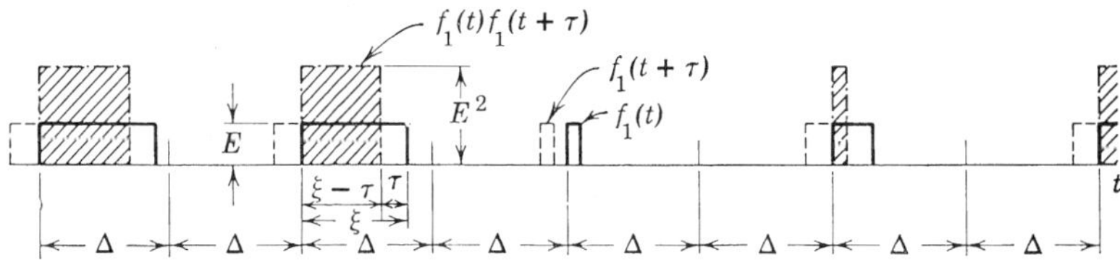

Fig. 3. Pertaining to the determination of the autocorrelation function of $f_1(t)$ in Fig. 2.

the displacement of $f_1(t)$, the multiplication of $f_1(t)$ by $f_1(t+\tau)$, and then the averaging, over the infinite interval, of the shaded rectangles each of height E^2 and width $\xi - \tau$. However, as far as the numerical value of the autocorrelation function is concerned, the rectangles of random width may be replaced by rectangles of the average width (per interval of length 2Δ, and as a function of τ). Accordingly, by first finding the average width, we replace the wave $f_1(t)f_1(t+\tau)$ in Fig. 3 by that in Fig. 4. Clearly the time average of $f_1(t)f_1(t+\tau)$ over the

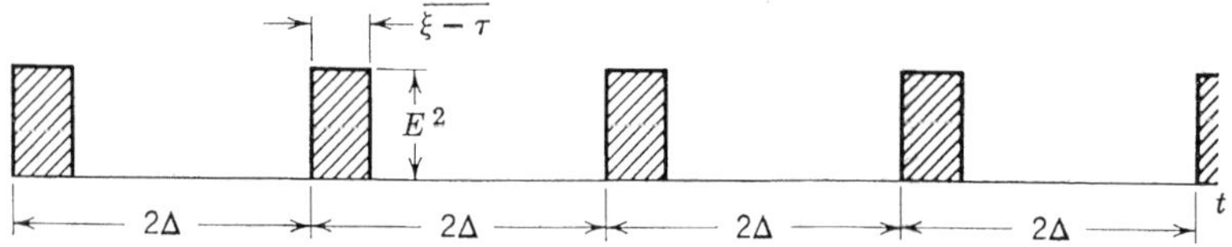

Fig. 4. Pertaining to the determination of the autocorrelation function of $f_1(t)$ in Fig. 2.

infinite interval is the same as the time average of the average rectangle over the interval 2Δ. Therefore if $a(\tau)$ is the area of the average rectangle (the rectangle of height E^2 and average width), then

$$\varphi_{11}(\tau) = \frac{a(\tau)}{2\Delta} \qquad \text{for } -\Delta < \tau < \Delta \tag{19}$$

To compute $a(\tau)$, let us consider a representative pulse of the given wave as shown in Fig. 5. We have, for $0 < \tau < \Delta$,

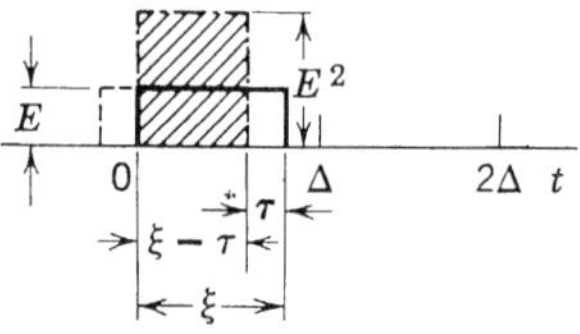

Fig. 5. A representative interval of the waves shown in Fig. 3.

(Integral of the product of the pulse and the pulse advanced by τ) $= E^2(\xi - \tau)$

and, for $-\Delta < \tau < 0$,

(Integral of the product of the pulse and the pulse delayed by $|\tau|$) $= E^2(\xi + \tau)$

This integral is represented by the shaded area for $0 < \tau < \Delta$. The average value of this integral is

$$a(\tau) = \int_{|\tau|}^{\Delta} E^2(x - |\tau|)P_\xi(x)\,dx \qquad \text{for } -\Delta < \tau < \Delta \tag{20}$$

Although we may carry the solution through with $P_\xi(x)$ in a general form, we assume, for simplicity, that

$$P_\xi(x) = \begin{cases} \dfrac{1}{\Delta} & \text{for } 0 < x < \Delta \\ 0 & \text{otherwise} \end{cases} \tag{21}$$

Then

$$a(\tau) = \frac{E^2}{2\Delta}(\Delta - |\tau|)^2 \qquad \text{for } -\Delta < \tau < \Delta \tag{22}$$

and

$$\varphi_{11}(\tau) = \frac{E^2}{4\Delta^2}(\Delta - |\tau|)^2 \qquad \text{for } -\Delta < \tau < \Delta \tag{23}$$

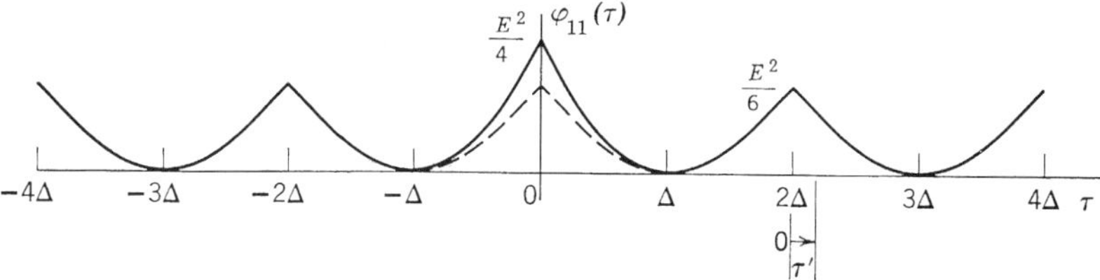

Fig. 6. Autocorrelation function of the pulse-width modulated wave shown in Fig. 2.

The solid curve in the interval $-\Delta < \tau < \Delta$ in Fig. 6 is a plot of this result. The remainder of the figure is explained in the following paragraphs.

We now proceed to determine the autocorrelation function for $2\Delta < \tau < 3\Delta$ and to show that the remainder of the function can be constructed from this portion and (23). When $2\Delta < \tau < 3\Delta$, the situation is as shown in Fig. 7. There are pulses that partially overlap, as those

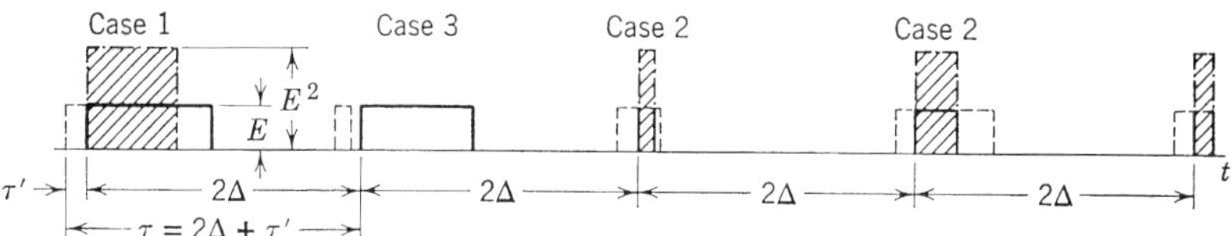

Fig. 7. Pertaining to the determination of the autocorrelation function shown in Fig. 6 for $2\Delta < \tau < 3\Delta$.

marked case 1; pulses of $f_1(t + \tau)$ that completely overlap those of $f_1(t)$, as those marked case 2; and pulses that do not overlap, as those marked case 3. There are no other forms of overlapping. Using the method for obtaining the first portion of the function, we find the average shaded rectangle of Fig. 7 as a function of τ and place this average rectangle in each 2Δ interval in a manner similar to that shown in Fig. 4. If $b(\tau)$ is the area of the average rectangle, then

$$\varphi_{11}(\tau) = \frac{b(\tau)}{2\Delta} \qquad \text{for } 2\Delta < \tau < 3\Delta \tag{24}$$

Since computations will be greatly simplified if $b(\tau)$ is expressed as the sum of three terms corresponding to the three cases of overlapping, we let

$b_1(\tau) =$ average value of areas resulting from partially overlapping pulses (case 1) for all the 2Δ intervals on the time axis. All areas resulting from overlapping pulses under case 2 are taken to be zero, and, of course, those from case 3 are actually zero.

$b_2(\tau)$ = average value of areas resulting from completely overlapping pulses (case 2) for all the 2Δ intervals on the time axis. All areas under case 1 are taken to be zero.

$b_3(\tau)$ = average value of areas resulting from non-overlapping pulses (case 3). Evidently $b_3(\tau) = 0$.

We then have

$$b(\tau) = b_1(\tau) + b_2(\tau) \tag{25}$$

To compute $b_1(\tau)$, let the width of a pulse of $f_1(t)$ be ξ and the width of the corresponding pulse of $f_1(t + \tau)$ be η as indicated in Fig. 8. For

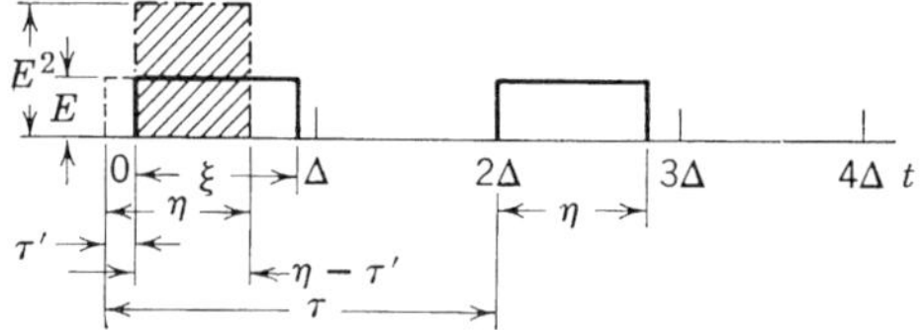

Fig. 8. Consideration of partially overlapping pulses in the waves of Fig. 7.

convenience, let $\tau = 2\Delta + \tau'$. We have

(Integral of the product of the two pulses, as represented by

$$\text{the shaded area}) = E^2(\eta - \tau') \tag{26}$$

This quantity is dependent upon the two random variables ξ and η. For its average value we refer to (61), Chapter 5, which states that

$$\overline{f(\xi_1, \eta_2)} = \int_{-\infty}^{\infty} \int_{-\infty}^{\infty} f(x_1, y_2) P_{\xi_1\eta_2}(x_1, y_2; \tau_1)\, dx_1\, dy_2 \tag{27}$$

in which subscripts 1 and 2 will be dropped for the present problem, a part of $f(\xi_1, \eta_2)$ is $E^2(\eta - \tau')$, and τ_1 is τ'. According to (64), Chapter 5, the joint probability element is

$$P_{\xi_1\eta_2}(x_1, y_2; \tau_1)\, dx_1\, dy_2 = P_\xi(x) P_{\eta|\xi}(y \,|\, x; \tau')\, dx\, dy \tag{28}$$

Since by assumption pulse widths are independent,

$$P_{\eta|\xi}(y \,|\, x; \tau')\, dy = P_\eta(y; \tau')\, dy \tag{29}$$

Before we find the average of (26) by means of (27), we must remember that the infinite limits in (27) indicate that the integrals are to be taken over all possible values of the random variables involved. Since we are writing the average value in two parts, we shall note carefully the corresponding limits of integration. For the average value $b_1(\tau)$ we find that the lower limit of ξ is $\xi = y - \tau'$ and the upper limit is $\xi = \Delta$; the lower limit of η is $\eta = \tau'$, and the upper limit is $\eta = \Delta$. Applying (27) through (29), we find

$$b_1(\tau') = \int_{x=y-\tau'}^{x=\Delta} \int_{y=\tau'}^{y=\Delta} E^2(y - \tau')P_\xi(x)P_\eta(y; \tau')\,dx\,dy \tag{30}$$

Since ξ is the width of one pulse and η is that of the following pulse, it is clear that the way these pulses are assumed to appear requires that

$$P_\xi(x) = P_\eta(y; \tau') \tag{31}$$

In expression (30) the probability densities are for all the pulses in the given random wave with none omitted although the integrations are confined to limits in which only partially overlapping pulses may occur. The use of the probability densities is in accord with the definition of $b_1(\tau)$ being an average value with the entire time interval as basis. If we evaluate (30) for the special simple density (21), we have

$$\begin{aligned} b_1(\tau') &= \frac{E^2}{\Delta^2} \int_{x=y-\tau'}^{x=\Delta} \int_{y=\tau'}^{y=\Delta} (y - \tau')\,dx\,dy \\ &= \frac{E^2}{6\Delta^2}[3\Delta(\Delta - \tau')^2 - 2(\Delta - \tau')^3] \qquad \text{for } 0 < \tau' < \Delta \end{aligned} \tag{32}$$

To calculate $b_2(\tau)$, we refer to Fig. 9. Here we consider all those

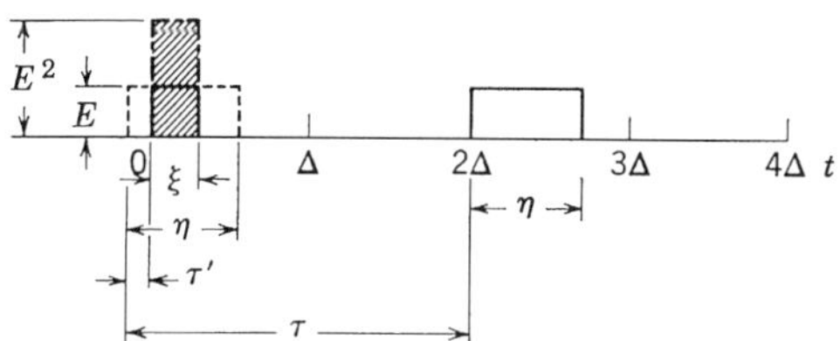

Fig. 9. Consideration of completely overlapping pulses in the waves of Fig. 7.

pulses of $f_1(t)$ that are completely overlapped by the corresponding ones of $f_1(t + \tau)$. With the same notations, we have

(Integral of the product of the two pulses, as represented by the shaded area) $= E^2\xi$ (33)

For all pulses under case 2 and no others, we find that the lower limit of ξ is $\xi = 0$ and the upper limit is $\xi = \Delta - \tau'$; the lower limit of η is $\eta = x + \tau'$, and the upper limit is $\eta = \Delta$. Hence

$$b_2(\tau') = \int_{x=0}^{x=\Delta-\tau'} \int_{y=x+\tau'}^{y=\Delta} E^2 x P_\xi(x)P_\eta(y; \tau')\,dx\,dy \tag{34}$$

Applying the rectangular density (21) to this result, we have

$$b_2(\tau') = \frac{E^2}{\Delta^2} \int_{x=0}^{x=\Delta-\tau'} \int_{y=x+\tau'}^{y=\Delta} x\,dx\,dy$$

$$= \frac{E^2}{6\Delta^2} (\Delta - \tau')^3 \qquad \text{for } 0 < \tau' < \Delta \tag{35}$$

The sum of (32) and (35) is

$$b(\tau') = \frac{E^2}{6\Delta^2} [3\Delta(\Delta - \tau')^2 - 2(\Delta - \tau')^3] + \frac{E^2}{6\Delta^2} (\Delta - \tau')^3$$

$$= \frac{E^2}{6\Delta^2} [3\Delta(\Delta - \tau')^2 - (\Delta - \tau')^3] \qquad \text{for } 0 < \tau' < \Delta \tag{36}$$

and it follows from (24) that

$$\varphi_{11}(\tau') = \frac{E^2}{12\Delta^3} [3\Delta(\Delta - \tau')^2 - (\Delta - \tau')^3] \qquad \text{for } 0 < \tau' < \Delta \tag{37}$$

This is the autocorrelation function of the random wave in the interval $(2\Delta < \tau < 3\Delta)$. Note that $\tau = 2\Delta + \tau'$. The curve in Fig. 6 in this interval is a plot of (37).

It is stated earlier that this portion of the autocorrelation function together with that in the interval $(0, \Delta)$ are sufficient to determine the entire autocorrelation function. To show this, let us consider first that portion of the function in the interval $(-2\Delta, -\Delta)$. This portion is obtained through displacing $f_1(t)$ to the right by $2\Delta - \tau'$, with $0 < \tau' < \Delta$, instead of to the left by $2\Delta + \tau'$, as shown in Fig. 7, when finding the function in the interval $(2\Delta, 3\Delta)$. Because the pulses start regularly at intervals of 2Δ and have independent widths and because the displacement concerned is greater than the maximum width of a pulse, it is not difficult to see that the average value of the integral of the product of the pulses at the displacement $\tau = 2\Delta + \tau'$ and that at $\tau = -(2\Delta - \tau')$ are the same. Therefore the curve in $(-2\Delta, -\Delta)$ is identical to that in $(2\Delta, 3\Delta)$, and the same reasoning leads to the fact that the curve is repeated in $(-4\Delta, -3\Delta)$, $(-6\Delta, -5\Delta)$, and so on, as shown in Fig. 6. Similarly, on the right-hand side of the τ-axis, the curve repeats in $(4\Delta, 5\Delta)$, $(6\Delta, 7\Delta)$, and so on. Finally, a property of the autocorrelation function that helps the completion of its determination is that the function is even. Thus, knowledge of its behavior in $(0, \Delta)$, $(-2\Delta, -\Delta)$, $(2\Delta, 3\Delta)$, $(-4\Delta, -3\Delta)$, $(4\Delta, 5\Delta)$, and so on, which we now have, determines its behavior in, respectively, $(0, -\Delta)$, $(\Delta, 2\Delta)$, $(-3\Delta, -2\Delta)$, $(3\Delta, 4\Delta)$, $(-5\Delta, -4\Delta)$, and so on. The complete curve is shown in Fig. 6.

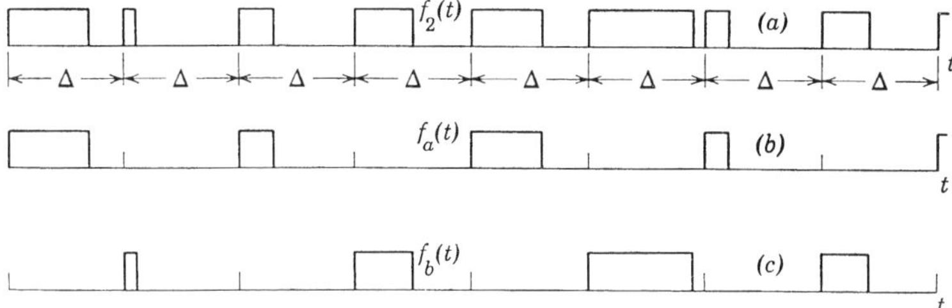

Fig. 10. Determination of the autocorrelation function of a pulse-width modulated wave by resolving the wave into two components.

Let us now consider the slightly more complex situation where a pulse is in every Δ interval as shown in Fig. 10*a*. The pulse widths are assumed to be independent as before. To determine the autocorrelation function $\varphi_{22}(\tau)$, we express the given function $f_2(t)$ as the sum of two components $f_a(t)$ and $f_b(t)$, that is,

$$f_2(t) = f_a(t) + f_b(t) \tag{38}$$

with the choice of the components in the manner shown in Figs. 10*b* and 10*c*. The function $f_a(t)$ is formed by taking only every other pulse in $f_2(t)$, and $f_b(t)$ is formed by taking all the other pulses. Since

$$\begin{aligned} \varphi_{22}(\tau) &= \overline{[f_a(t) + f_b(t)][f_a(t + \tau) + f_b(t + \tau)]} \\ &= \varphi_{aa}(\tau) + \varphi_{bb}(\tau) + \varphi_{ab}(\tau) + \varphi_{ba}(\tau) \end{aligned} \tag{39}$$

we shall consider the four components in connection with the results we have just obtained. First, the autocorrelation functions $\varphi_{aa}(\tau)$ and $\varphi_{bb}(\tau)$ should be exactly the same as $\varphi_{11}(\tau)$ just found, because $f_1(t)$ and $f_a(t)$ and $f_b(t)$ behave in exactly the same manner. The curve in Fig. 6 is repeated in Fig. 11*a* as $\varphi_{aa}(\tau)$ and $\varphi_{bb}(\tau)$. Next the crosscorrelation function $\varphi_{ab}(\tau)$ is obtained by displacing $f_b(t)$ by τ and performing the other operations in crosscorrelation. As we see in Figs. 10*a* and 10*b*, these operations involve the multiplication of pulses of independent widths in exactly the same manner as the calculation of $\varphi_{11}(\tau)$ for $|\tau| > \Delta$. Therefore, we can make use of the results just found. Accordingly, the function $\varphi_{ab}(\tau)$ in $(0, \Delta)$ should be the same as $\varphi_{11}(\tau)$ in $(\Delta, 2\Delta)$. Similarly, $\varphi_{ab}(\tau)$ in $(\Delta, 2\Delta)$ should be the same as $\varphi_{11}(\tau)$ in $(2\Delta, 3\Delta)$, and so on. It is clear from the manner in which $f_a(t)$ and $f_b(t)$ are formed that the crosscorrelation with $f_b(t)$ displaced either to the right or to the left, by the same amount, should be the same. Hence the curve is an even function as shown in Fig. 11*b*. Moreover, either from the fact that $\varphi_{ab}(\tau) = \varphi_{ba}(-\tau)$ or from considerations similar to those for $\varphi_{ab}(\tau)$, we find $\varphi_{ab}(\tau) = \varphi_{ba}(\tau)$. Thus we have completely determined the com-

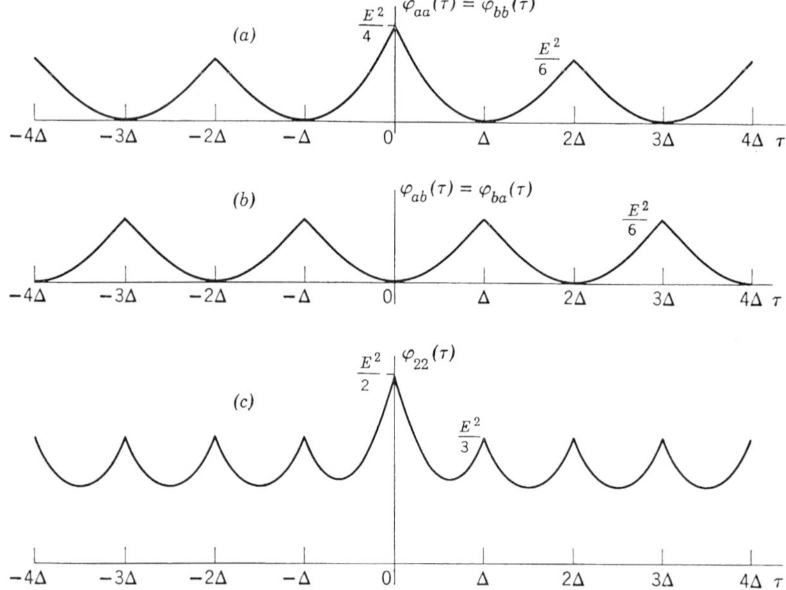

Fig. 11. Autocorrelation function of the wave shown at the top of Fig. 10 and its components.

ponents of $\varphi_{22}(\tau)$ in terms of the results obtained for $\varphi_{11}(\tau)$; the graph of $\varphi_{22}(\tau)$ is given in Fig. 11*c*.

4. Resolution of the Autocorrelation Function

In the autocorrelation function just determined, it is interesting to note that with the exception of the portion in the interval $(-\Delta, \Delta)$ the function is periodic. The excepted portion is particularly important and deserves further attention. As to the periodicity in the remainder of the function, it is readily seen that the periodic starting of the rectangular pulses is the cause.

To further investigate the nature of the random wave and its autocorrelation function, let us consider the central peak of the autocorrelation function in $(-\Delta, \Delta)$ as the sum of two parts, one equal to a peak of the periodic portion of the function and the other a remainder. This partition is indicated by a dotted curve in Fig. 6. Now, if we let

$[\varphi_{11}(\tau)]_p$ = periodic component of the autocorrelation function

$[\varphi_{11}(\tau)]_r$ = remaining nonperiodic component of the autocorrelation function we can express the autocorrelation function of the random wave as the sum of these components thus

$$\varphi_{11}(\tau) = [\varphi_{11}(\tau)]_p + [\varphi_{11}(\tau)]_r \tag{40}$$

The components are shown in Fig. 12.

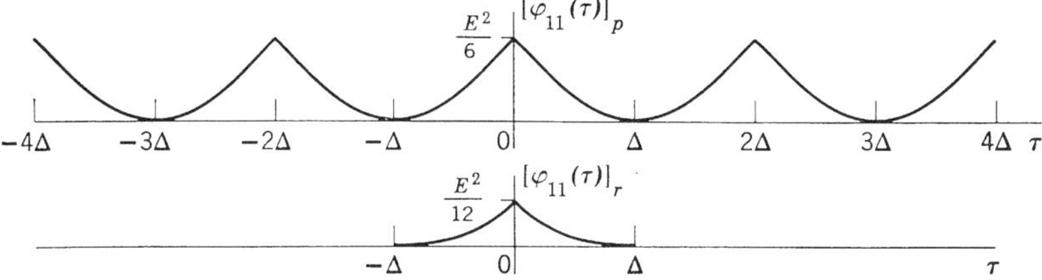

Fig. 12. Components of the autocorrelation function of Fig. 6.

For the complete specification of the periodic component, which has a period 2Δ, we need only its description in the interval $(0, \Delta)$. In accordance with the result (37) the expression for $[\varphi_{11}(\tau)]_p$ is

$$[\varphi_{11}(\tau)]_p = \frac{E^2}{12\Delta^3}[3\Delta(\Delta - \tau)^2 - (\Delta - \tau)^3] \qquad \text{for } 0 < \tau < \Delta \tag{41}$$

The other part of the autocorrelation function is

$$\begin{aligned}[\varphi_{11}(\tau)]_r &= \varphi_{11}(\tau) - [\varphi_{11}(\tau)]_p \qquad \text{for } 0 < \tau < \Delta \\ &= \frac{E^2}{4\Delta^2}(\Delta - \tau)^2 - \frac{E^2}{12\Delta^3}[3\Delta(\Delta - \tau)^2 - (\Delta - \tau)^3] \\ &\qquad\qquad \text{for } 0 < \tau < \Delta\end{aligned} \tag{42}$$

in which we have inserted the result (23). Simplifying this, and in accordance with the manner of resolution, we have

$$[\varphi_{11}(\tau)]_r = \begin{cases} \dfrac{E^2}{12\Delta^3}(\Delta - |\tau|)^3 & \text{for } -\Delta < \tau < \Delta \\ 0 & \text{elsewhere} \end{cases} \tag{43}$$

This method of resolving an autocorrelation function, which is periodic with the exception of the central region, suggests the resolution of the given random function into two components, one purely periodic and the other random. Thus let us put

$f_p(t)$ = periodic component of $f_1(t)$. This component includes the mean value of $f_1(t)$, that is, the d-c component, if $f_1(t)$ represents a voltage.

$f_r(t)$ = random component of $f_1(t)$. This component has no additive periodic term. The mean value of $f_r(t)$ is zero.

It is reasonable to assume that these components are independent. Next, we shall assume that

$$f_1(t) = f_p(t) + f_r(t) \tag{44}$$

To show that this assumption is in agreement with the results we have obtained, let us note that the autocorrelation of $f_1(t)$ in terms of $f_p(t)$ and $f_r(t)$ is

$$\begin{aligned}\varphi_{11}(\tau) &= \overline{[f_p(t) + f_r(t)][f_p(t+\tau) + f_r(t+\tau)]} \\ &= \overline{f_p(t)f_p(t+\tau)} + \overline{f_r(t)f_r(t+\tau)} + \overline{f_p(t)f_r(t+\tau)} \\ &\quad + \overline{f_r(t)f_p(t+\tau)} \end{aligned} \tag{45}$$

The evaluation of $f_p(t)f_r(t+\tau)$ involves ensembles of $f_p(t)$ and of $f_r(t)$ which are statistically independent. If ξ denotes the random amplitude of the ensemble of periodic functions $\{f_p(t)\}$, and η the random amplitude of the ensemble of random functions $\{f_r(t)\}$ with the displacement τ from ξ, we shall have

$$\overline{f_p(t)f_r(t+\tau)} = \overline{\xi\eta} = \bar{\xi}\bar{\eta} \tag{46}$$

in accordance with (68), Chapter 5, for independent variables. Since

$$\overline{f_r(t)} = \bar{\eta} = 0 \tag{47}$$

we conclude that

$$\overline{f_p(t)f_r(t+\tau)} = 0 \tag{48}$$

and, similarly, in (45)

$$\overline{f_r(t)f_p(t+\tau)} = 0 \tag{49}$$

Therefore (45) reduces to

$$\varphi_{11}(\tau) = \varphi_{pp}(\tau) + \varphi_{rr}(\tau) \tag{50}$$

which corresponds to (40) so that the autocorrelation of $f_p(t)$ is

$$\varphi_{pp}(\tau) = [\varphi_{11}(\tau)]_p \tag{51}$$

and the autocorrelation of $f_r(t)$ is

$$\varphi_{rr}(\tau) = [\varphi_{11}(\tau)]_r \tag{52}$$

Let us point out that the choice of $f_r(t)$ to have a zero mean value is purely a matter of convenience since we see that as a consequence (45) is reduced to two terms. However, the choice of $\overline{f_p(t)} = 0$ is just as good. Since we have taken $\overline{f_r(t)} = 0$, it is necessary that $[\varphi_{11}(\tau)]_r$ vanishes as $\tau \to \infty$, for we know that an autocorrelation function of a random wave without a hidden periodic component tends to the square of the mean value as $\tau \to \infty$ (Sec. 7, Chapter 5). This requirement is met in (43).

In going through the present problem for a better understanding of autocorrelation, we have brought out the interesting fact that under suitable conditions it is possible to resolve an autocorrelation function into two components, one periodic and the other nonperiodic. These are the autocorrelation functions of the periodic and random components of the given random wave. Although it is impossible to determine the waveform of the periodic component by the present method, the separation of the components in the form of autocorrelation functions can often be of importance in communication problems. From the autocorrelation of the periodic component, we can determine its power spectrum, and, from that of the random component, its power density spectrum. To illustrate this point, let us consider the random component of the pulse-width modulated wave. The component has the autocorrelation given by (43). By the Wiener theorem as expressed by (237), Chapter 2, the power density spectrum of the random component is

$$\begin{aligned}\Phi_{rr}(\omega) &= \frac{1}{2\pi}\int_{-\infty}^{\infty} \varphi_{rr}(\tau)\cos\omega\tau\,d\tau \\ &= \frac{E^2}{12\pi\Delta^3}\int_0^{\Delta}(\Delta-\tau)^3\cos\omega\tau\,d\tau \\ &= \frac{E^2\Delta}{4\pi}\frac{1}{(\Delta\omega)^2}\left[1-\left(\frac{\sin\frac{\Delta\omega}{2}}{\frac{\Delta\omega}{2}}\right)^2\right]\end{aligned} \tag{53}$$

A graph of this spectrum is shown in Fig. 13.

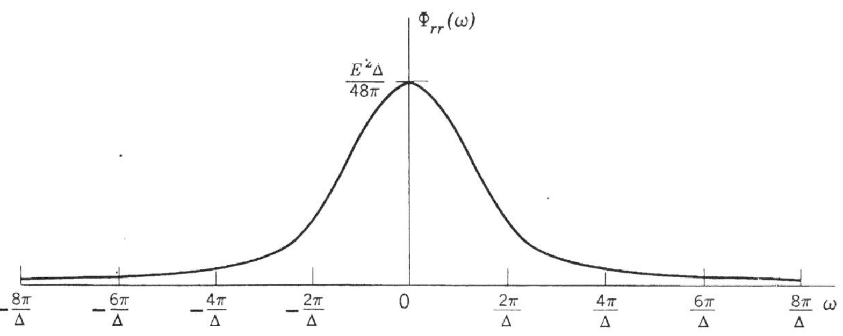

Fig. 13. Power density spectrum of the random component of the pulse-width modulated wave shown in Fig. 2.

5. Waveform Determination of Periodic Component in Random Wave

The method just presented enables us to determine the power spectrum of the periodic component in a random wave. However, having lost all phase angles of the harmonics through autocorrelation, we are unable to find the waveform. An interesting question at this point is whether it is possible to find the waveform by correlation techniques. We shall now show that it is possible to do so with the aid of crosscorrelation.

For this problem we refer to Sec. A-7, Chapter 2, with particular emphasis on the use of the periodic unit-impulse function. One important fact which we shall apply is that, except for a constant factor, the crosscorrelation of the periodic unit-impulse function and a periodic function of the same period restores the periodic function, as expressed by (120), Chapter 2. If the periodic function concerned is the periodic component $f_p(t)$ and the periodic unit-impulse function is denoted by $u_p(t)$, then

$$\frac{1}{T_1} f_p(t) = \varphi_{u_p f_p}(t) \tag{54}$$

or

$$\frac{1}{T_1} f_p(t) = \varphi_{f_p u_p}(-t) \tag{55}$$

where T_1 is the period of $f_p(t)$ and $u_p(t)$. We shall slightly modify the form of (55) to suit the present situation. First, we know that according to the definition of $\varphi_{u_p f_p}(t)$, (55) is

$$\frac{1}{T_1} f_p(t) = \frac{1}{T_1} \int_{-T_1/2}^{T_1/2} f_p(\tau) u_p(\tau - t)\, d\tau \tag{56}$$

Since $f_p(t)$ and $u_p(t)$ are periodic functions of the same period, (56) may be written as an average over the infinite interval instead of over a period. Thus

$$\frac{1}{T_1} f_p(t) = \lim_{T\to\infty} \frac{1}{2T} \int_{-T}^{T} f_p(\tau) u_p(\tau - t)\, d\tau \tag{57}$$

Interchanging the symbols t and τ so that later results will appear in familiar symbols, we have

$$\frac{1}{T_1} f_p(\tau) = \lim_{T\to\infty} \frac{1}{2T} \int_{-T}^{T} f_p(t) u_p(t - \tau)\, dt \tag{58}$$

This expression is the desired form of (55) for the problem at hand.

Now, let us write the crosscorrelation $\varphi_{f1up}(-\tau)$ between the given function $f_1(t)$ and $u_p(t)$ in terms of the periodic component $f_p(t)$ and the random component $f_r(t)$ of $f_1(t)$ as given by (44). The expression is

$$\begin{aligned}\varphi_{f1up}(-\tau) &= \lim_{T\to\infty} \frac{1}{2T} \int_{-T}^{T} [f_p(t) + f_r(t)]u_p(t-\tau)\,dt \\ &= \lim_{T\to\infty} \frac{1}{2T} \int_{-T}^{T} f_p(t)u_p(t-\tau)\,dt \\ &\quad + \lim_{T\to\infty} \frac{1}{2T} \int_{-T}^{T} f_r(t)u_p(t-\tau)\,dt \end{aligned} \tag{59}$$

In writing this equation, we have assumed that $f_1(t)$ has only one periodic component $f_p(t)$ and that it has the period T_1. This period is assumed to be known. In the pulse-width modulated wave under discussion, T_1 can be found from autocorrelation.

On the right-hand side of (59) the first term, according to (58), is $(1/T_1)f_p(\tau)$. For the interpretation of the last term of (59) we may first consider a particular impulse of $u_p(t)$, say the impulse at $t = 0$, with all other impulses reduced to zero. This would reduce $u_p(t - \tau)$ to a unit impulse displaced by τ, which we shall denote by $u(t - \tau)$. Let the integration be taken over the infinite interval as before, but without averaging. In other words, we consider

$$\lim_{T\to\infty} \int_{-T}^{T} f_r(t)u(t-\tau)\,dt \tag{60}$$

or, since $u(t)$ is an even function,

$$\lim_{T\to\infty} \int_{-T}^{T} f_r(t)u(\tau-t)\,dt \tag{61}$$

This we know is the convolution of a unit impulse and $f_r(t)$ so that it leaves $f_r(t)$ unchanged; that is,

$$f_r(\tau) = \lim_{T\to\infty} \int_{-T}^{T} f_r(t)u(\tau-t)\,dt \tag{62}$$

Thus the integral (60) corresponds to taking the value of $f_r(t)$ at $t = \tau$. Returning to (59), we see that the last term is the average of an infinite series of values of $f_r(t)$ that are taken at intervals of T_1 as illustrated in Fig. 14. This average value is the value per interval T_1. The random

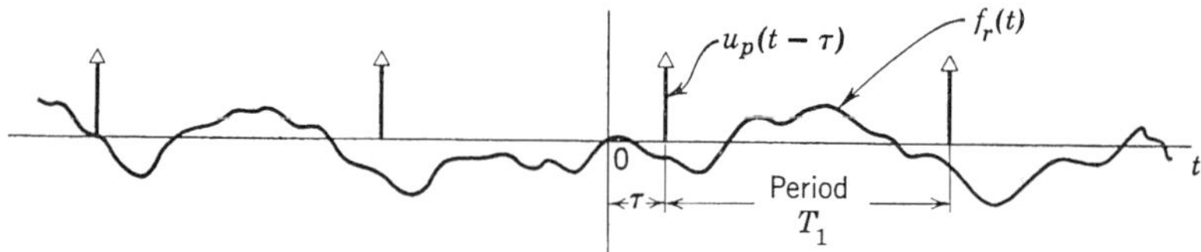

Fig. 14. Pertaining to the convolution of a random wave without any periodic components and a periodic unit-impulse function.

component $f_r(t)$ in the figure is drawn in a convenient manner for illustrating the method and is not the exact form of the random component of the pulse-width modulated wave. The last term of (59) is therefore

$$\lim_{T\to\infty} \frac{1}{2T} \int_{-T}^{T} f_r(t) u_p(t - \tau)\, dt = \lim_{N\to\infty} \frac{\sum_{k=-N}^{N} f_r(\tau + kT_1)}{(2N + 1)T_1} \tag{63}$$

Inasmuch as $f_r(t)$ is assumed to have a zero mean and is devoid of any periodic components, it is necessary that (63) have zero value for all values of τ. With

$$\lim_{T\to\infty} \frac{1}{2T} \int_{-T}^{T} f_r(t) u_p(t - \tau)\, dt = 0 \tag{64}$$

it follows that (59) is

$$\varphi_{f1up}(-\tau) = \frac{1}{T_1} f_p(\tau) \tag{65}$$

This result states that the crosscorrelation of $f_1(t)$, which is the sum of a random component and a periodic component of period T_1, and a periodic unit-impulse function of the same period yields the periodic component with the factor $1/T_1$. This is the basic equation for extracting the waveform of a periodic wave that has been mixed with a random wave.

In applying this method of extraction to the pulse-width modulated wave, we have the situation shown in Fig. 15. From autocorrelation we

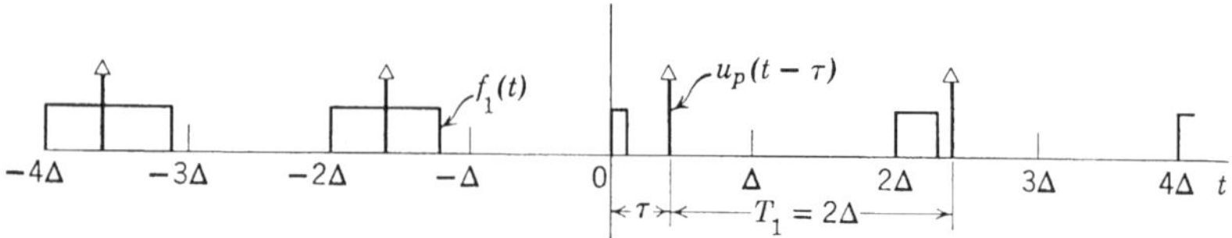

Fig. 15. Pertaining to the convolution of a pulse-width modulated wave and a periodic unit-impulse function.

know that $f_p(t)$ of the wave has the period 2Δ. Hence $u_p(t)$ should have this period. The origin of $u_p(t)$ has been placed at a leading edge of a pulse of $f_1(t)$ for convenience. Following the method presented in Sec. 3, we compute the crosscorrelation $\varphi_{f1up}(-\tau)$ by first finding the average of the integral of $f_1(t)u_p(t - \tau)$ and then dividing by 2Δ. For the interval $0 < \tau < \Delta$ in the crosscorrelation function we compute according to Fig. 16 which is a representative period of Fig. 15. Clearly

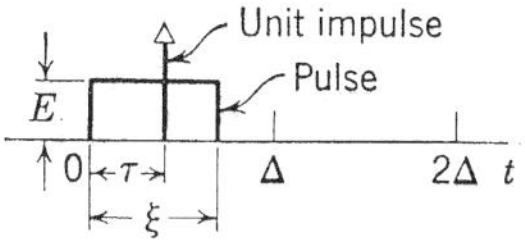

Fig. 16. A representative period of the waves shown in Fig. 15.

(Integral of product of the pulse and the unit impulse)

$$= \begin{cases} E & \text{for } \tau < \xi < \Delta \\ 0 & \text{for } 0 < \xi < \tau \end{cases}$$

The average value of this integral is

$$c(\tau) = \int_{\tau}^{\Delta} EP_{\xi}(x)\, dx$$

$$= \frac{E}{\Delta}(\Delta - \tau) \qquad \text{for } 0 < \tau < \Delta \tag{66}$$

Consequently

$$\varphi_{f1up}(-\tau) = \frac{c(\tau)}{2\Delta}$$

$$= \frac{E}{2\Delta^2}(\Delta - \tau) \qquad \text{for } 0 < \tau < \Delta \tag{67}$$

This crosscorrelation function yields the periodic component; for, in accordance with (65),

$$\frac{1}{T_1} f_p(\tau) = \frac{E}{2\Delta^2}(\Delta - \tau) \qquad \text{for } 0 < \tau < \Delta \tag{68}$$

and, since $T_1 = 2\Delta$,

$$f_p(\tau) = \frac{E}{\Delta}(\Delta - \tau) \qquad \text{for } 0 < \tau < \Delta \tag{69}$$

This expression represents a straight line indicating that the periodic component starts with the value E at $\tau = 0$ decreasing linearly to zero at $\tau = \Delta$. The component is zero from Δ to 2Δ. We have now determined a complete period (2Δ) of the periodic component, and it is shown in Fig. 17a as $f_p(t)$.

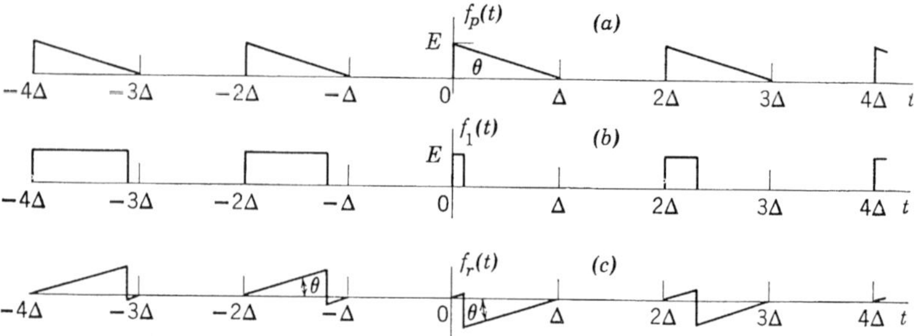

Fig. 17. The pulse-width modulated wave and its periodic and random components.

To find the waveform of the random component, it is merely necessary to take the difference between $f_1(t)$ and $f_p(t)$; that is,

$$f_r(t) = f_1(t) - f_p(t) \tag{70}$$

In Fig. 17*b* the graph of $f_1(t)$ is repeated for reference so that $f_r(t)$ may be drawn as in Fig. 17*c* by following (70). Since the slope of the triangles in $f_p(t)$ is $-\tan^{-1}(E/\Delta)$ [that is, the angle θ (in magnitude) in the figure is $\tan^{-1}(E/\Delta)$], the angles θ in Fig. 17*c* have the same value in magnitude. Thus $f_r(t)$ consists of a series of pulses starting at regular intervals of 2Δ; each pulse (with duration Δ and separated from the next by Δ) starts with zero value increasing linearly with the slope $\tan^{-1}(E/\Delta)$ until time ξ, when the corresponding rectangular pulse in $f_1(t)$ terminates, then the pulse drops instantly by the amount E and continues linearly at the same positive slope to reach zero value. The random component $f_r(t)$, although from its appearance it might be suspected that a further periodic component still remains, actually has no periodicity in the sense that it is uncorrelated to a periodic function of any period. For instance, the crosscorrelation between it and $f_p(t)$ is zero everywhere, as it should be.

6. Autocorrelation of Poisson-Distributed Unit Impulses *

Waves formed by unit impulses have interesting applications in communication theory. One such wave is shown in Fig. 18*a*. It is a series of unit impulses that occur with the equal probability of being positive or negative. We shall assume that the distribution of the impulses is Poisson. If all the impulses occur with the same sign, the wave would appear as in Fig. 18*b*. Another variation of the wave is shown in Fig.

* Y. W. Lee, "Autocorrelation Function of Poisson Waves," *Quarterly Progress Report*, Research Laboratory of Electronics, M.I.T., January 15, 1955, pp. 56–62.

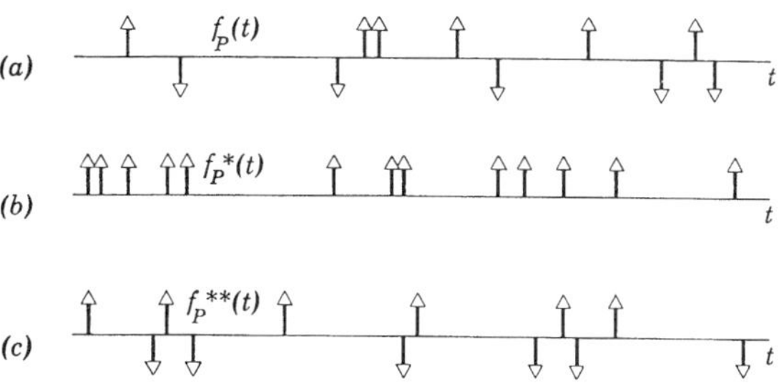

Fig. 18. Waves of Poisson-distributed unit impulses.

18*c*. Here the Poisson distributed impulses are alternately positive and negative. We shall study the autocorrelation of these waves.

(a) Wave of Equally Likely Positive and Negative Impulses. The present problem is closely related to the autocorrelation of a Poisson rectangular wave given in Sec. 2. Since the waves of unit impulses take discrete values, the autocorrelation function will be determined, as in Sec. 2, from (40), Chapter 7. First, consider the wave $f_P(t)$ of Fig. 18*a* and an ensemble $\{f_P(t)\}$ of such waves as shown in Fig. 19. The

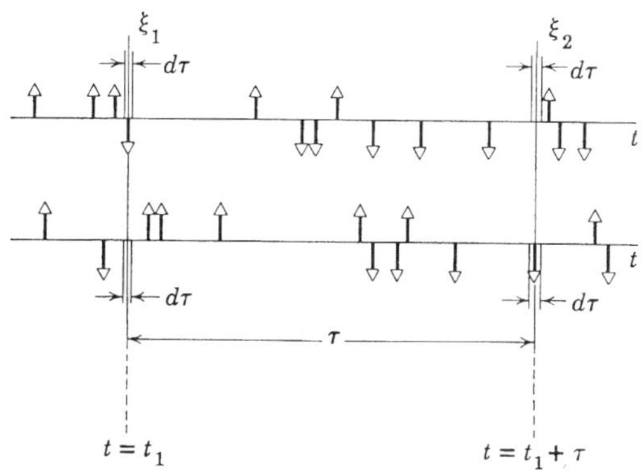

Fig. 19. An ensemble of Poisson waves of equally likely positive and negative unit impulses.

ensemble has the possible values ∞, 0, $-\infty$, but it is only necessary to consider the nonzero values. The autocorrelation $\varphi_{PP}(\tau)$ of this Poisson impulsive wave is

$$\varphi_{PP}(\tau) = \sum_{i=-1}^{1} \sum_{j=-1}^{1} x_{1i}x_{2j}P_{\xi_1\xi_2}(x_{1i}, x_{2j}; \tau)$$

$$= \sum_{i=-1}^{1} \sum_{j=-1}^{1} x_{1i}x_{2j}P_{\xi_1}(x_{1i})P_{\xi_2|\xi_1}(x_{2j}|x_{1i}; \tau) \tag{71}$$

We recall that a unit impulse is a pulse with an infinitesimal duration dt and a height A which approaches infinity and with a unit area under the pulse; that is, $A\,dt|_{A\to\infty} = 1$. It should be remarked that the height of the impulse approaches infinity in the entire infinitesimal interval dt, not at a point only.

In the ensemble at $t = t_1$, the amplitude ξ_1 has two possible nonzero values, namely, A and $-A$ if A represents the height of an impulse and $A \to \infty$. Note that, when we say at $t = t_1$, we refer to the infinitesimal time interval $d\tau$ with the line $t = t_1$ passing through its midpoint as shown. Similarly at $t = t_1 + \tau$ (the interval $d\tau$ with the line $t = t_1 + \tau$ passing through the $d\tau$), the amplitude ξ_2 has the same two possible nonzero values. We thus have

$$\left.\begin{aligned} x_{1,1} &= A \\ x_{1,-1} &= -A \\ x_{2,1} &= A \\ x_{2,-1} &= -A \end{aligned}\right\} \tag{72}$$

Since the impulses follow the Poisson distribution, the probability of finding an impulse in a particular $d\tau$ duration is, according to (12), Chapter 6,

$$\mathcal{P}(\eta = 1; d\tau) = k\,d\tau \tag{73}$$

in which η represents the number of impulses in a given time interval and k is the average number of impulses per second. Furthermore the fact that positive and negative impulses occur with equal probability leads to

$$\left.\begin{aligned} \mathcal{P}(\xi_1 = A) &= P_{\xi 1}(A) = \frac{k\,d\tau}{2} \\ \mathcal{P}(\xi_1 = -A) &= P_{\xi 1}(-A) = \frac{k\,d\tau}{2} \end{aligned}\right\} \tag{74}$$

For the autocorrelation function at $\tau = 0$, that is, in the interval $d\tau$ with the line $\tau = 0$ passing through its midpoint, we have

$$\begin{aligned} \varphi_{PP}(0) &= \sum_{i=-1}^{1} x_{1i}^2 P_{\xi 1}(x_{1i}) \\ &= \left.\frac{A^2 k\,d\tau}{2}\right|_{\substack{A\to\infty \\ A\,d\tau=1}} + \left.\frac{(-A)^2 k\,d\tau}{2}\right|_{\substack{A\to\infty \\ A\,d\tau=1}} \\ &= A^2 k\,d\tau\,|_{\substack{A\to\infty \\ A\,d\tau=1}} \end{aligned} \tag{75}$$

Since for a unit impulse $A\,d\tau|_{A\to\infty} = 1$,

$$\varphi_{PP}(0) = kA|_{A\to\infty} \tag{76}$$

Letting $u(t)$ be the unit-impulse function with the impulse at $t = 0$, we see that (76) is

$$\varphi_{PP}(0) = ku(0) \tag{77}$$

For $\tau \neq 0$ the impulses in the $d\tau$'s at $t = t_1$ and at $t = t_1 + \tau$ are independent, in accordance with the basic assumptions in the derivation of the Poisson distribution. Consequently

$$\begin{aligned} P_{\xi 2|\xi 1}(x_{2j}|x_{1i};\tau) &= P_{\xi 2}(x_{2j}) \\ &= P_{\xi 1}(x_{1i}) \end{aligned} \tag{78}$$

and

$$\begin{aligned} \varphi_{PP}(\tau) &= [(AA) + (-A)(-A) + A(-A) + (-A)A] \\ &\quad \times \left(\frac{k\,d\tau}{2}\right)\left(\frac{k\,d\tau}{2}\right)\bigg|_{\substack{A\to\infty \\ A\,d\tau=1}} \\ &= 0 \qquad \text{for } \tau \neq 0 \end{aligned} \tag{79}$$

Combining (77) and (79), we have

$$\varphi_{PP}(\tau) = ku(\tau) \tag{80}$$

as the autocorrelation of the impulsive Poisson wave $f_P(t)$. This function is shown in Fig. 20*a* as a k-unit impulse function, that is, a function consisting of an impulse with an area of k units.

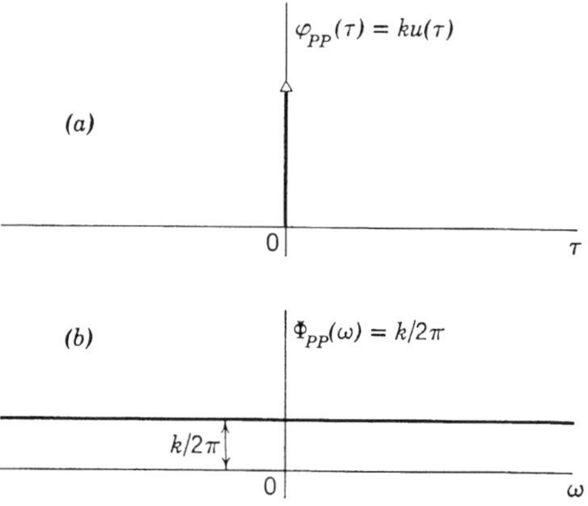

Fig. 20. Autocorrelation and power density spectrum of the ensemble of Poisson waves shown in Fig. 19.

To obtain the power density spectrum of the wave, we apply the Wiener theorem for autocorrelation. Thus

$$\begin{aligned}\Phi_{PP}(\omega) &= \frac{1}{2\pi}\int_{-\infty}^{\infty} ku(\tau)\cos\omega\tau\,d\tau \\ &= \frac{k}{2\pi}\end{aligned} \tag{81}$$

since the spectrum of a unit-impulse function, by (198), Chapter 2, is $1/2\pi$. Therefore the power density spectrum of the impulsive Poisson wave, with positive and negative impulses occurring with equal probability, is uniform as shown in Fig. 20*b*.

(b) Wave of Unit Impulses of the Same Sign. We now consider the wave $f_P{}^*(t)$ in Fig. 18*b*. The Poisson distributed unit impulses are of the same sign. At $t = t_1$, the amplitude ξ_1 of the ensemble $\{f_P{}^*(t)\}$, as shown in Fig. 21, is A if an impulse occurs in $d\tau$; it is zero if no impulse occurs. Therefore

$$\mathcal{P}(\xi_1 = A) = P_{\xi 1}(A) = k\,d\tau \tag{82}$$

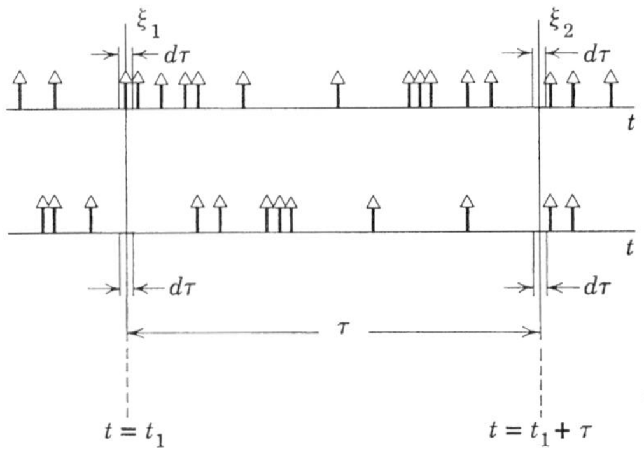

Fig. 21. An ensemble of Poisson waves of unit impulses of the same sign.

In the autocorrelation function $\overset{*}{\varphi}_{PP}(\tau)$ at $\tau = 0$, we have, analogous to (75),

$$\begin{aligned}\varphi^*_{PP}(0) &= x_{11}^2 P_{\xi 1}(x_{11}) \\ &= A^2 k\,d\tau\,\Big|_{\substack{A\to\infty \\ A\,d\tau=1}} \\ &= kA\,\Big|_{A\to\infty} \\ &= ku(0)\end{aligned} \tag{83}$$

This value is the same as (77). For $\tau \neq 0$, ξ_1 and ξ_2 are independent so that (78) holds and (71) becomes

$$\overset{*}{\varphi}_{PP}(\tau) = AAk\, d\tau\, k\, d\tau \,|_{\substack{A\to\infty \\ A\, d\tau=1}}$$

$$= k^2 \qquad \text{for } \tau \neq 0 \tag{84}$$

Taking the sum of (83) and (84), we have the autocorrelation function of $f_P{}^*(t)$,

$$\overset{*}{\varphi}_{PP}(\tau) = ku(\tau) + k^2 \tag{85}$$

The function consists of a k-unit impulse function and a constant as shown in Fig. 22*a*.

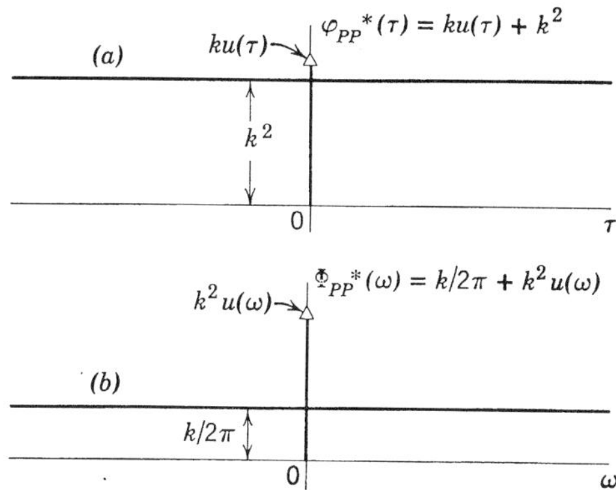

Fig. 22. Autocorrelation and power density spectrum of the ensemble of Poisson waves shown in Fig. 21.

The power density spectrum of $f_P{}^*(t)$ is, by Wiener's theorem,

$$\overset{*}{\Phi}_{PP}(\omega) = \frac{1}{2\pi}\int_{-\infty}^{\infty} [ku(\tau) + k^2] \cos \omega\tau\, d\tau$$

$$= \frac{k}{2\pi} + k^2u(\omega) \tag{86}$$

since the spectrum of an impulse is a constant, and, inversely, the spectrum of a constant is an impulse as given by (198) and (199), Chapter 2. The power density spectrum of the Poisson wave of positive (or negative) impulses is uniform except at $\omega = 0$ where we find a k^2-unit frequency impulse as shown in Fig. 22*b*.

(c) Wave of Alternately Positive and Negative Unit Impulses. In Fig. 18*c* the Poisson distributed unit impulses are alternately positive and negative. To determine the autocorrelation function $\overset{**}{\varphi}_{PP}(\tau)$ of the wave $f_P^{**}(t)$ at the point $\tau = 0$, we find that (75) through (77) may be applied to the ensemble $\{f_P^{**}(t)\}$ (Fig. 23). Without repeating the details, we state the result that

$$\overset{**}{\varphi}_{PP}(0) = ku(0) \tag{87}$$

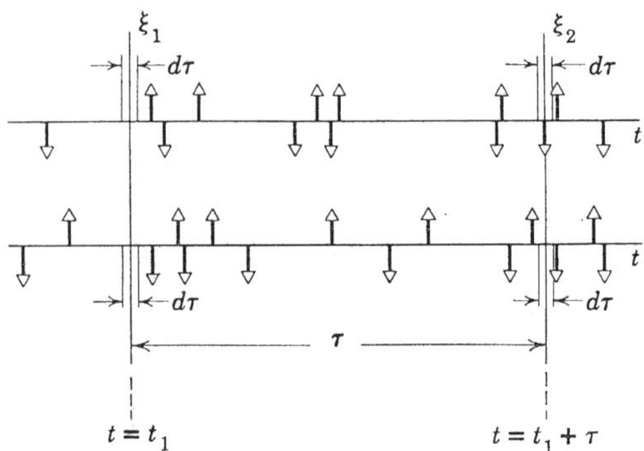

Fig. 23. An ensemble of Poisson waves of alternately positive and negative unit impulses.

For $\tau \neq 0$, we refer to the general expression (71) with $\varphi_{PP}(\tau)$ replaced by $\varphi_{PP}^{**}(\tau)$. As before, we have

$$\left.\begin{aligned} P_{\xi 1}(x_{1,1}) &= P_{\xi 1}(A) = \frac{k\,d\tau}{2} \\ P_{\xi 1}(x_{1,-1}) &= P_{\xi 1}(-A) = \frac{k\,d\tau}{2} \end{aligned}\right\} \tag{88}$$

In the conditional probability in (71), let us first consider $i = 1, j = 1$. We have

$$\mathcal{P}(\xi_2 = x_{2,1} \mid \xi_1 = x_{1,1}; \tau) = P_{\xi 2|\xi 1}(x_{2,1} \mid x_{1,1}; \tau) \tag{89}$$

or

$$\mathcal{P}(\xi_2 = A \mid \xi_1 = A; \tau) = P_{\xi 2|\xi 1}(A \mid A; \tau) \tag{90}$$

The event of finding a positive impulse ($x_{2,1}$) at $t = t_1 + \tau$, given that a positive impulse ($x_{1,1}$) occurred at $t = t_1$, is the result of the joint occurrence of event E_1, an odd number of impulses in the interval τ, and event E_2, an impulse at $t = t_1 + \tau$. All occurrences at the time specified are understood to be in the interval $d\tau$ in the manner already described. Applying the theorem of total probability, we have

$$\mathcal{P}(E_1) = \sum_{n=1,3,5,\ldots} \mathcal{P}(\eta = n; \tau) \tag{91}$$

where $\mathcal{P}(\eta = n; \tau)$ is the Poisson distribution (3). The probability of event E_2 is, as before

$$\mathcal{P}(E_2) = k\,d\tau \tag{92}$$

Since the impulses are Poisson distributed, events E_1 and E_2 are independent. Therefore

$$P_{\xi 2|\xi 1}(A \mid A; \tau) = \left[\sum_{n=1,3,5,\ldots} P_\eta(n; \tau)\right] k\,d\tau \tag{93}$$

By similar reasoning, for $i = -1$ and $j = -1$,

$$P_{\xi 2|\xi 1}(-A\,|\,-A\,;\tau) = \left[\sum_{n=1,3,5,\ldots} P_\eta(n;\tau)\right] k\,d\tau \tag{94}$$

If $i = 1$, $j = -1$, we consider the event of finding a negative impulse $(x_{2,-1})$ at $t = t_1 + \tau$, given that a positive impulse $(x_{1,1})$ occurred at $t = t_1$. This event is the joint occurrence of: (1) no impulse or an even number of impulses in τ, and (2) an impulse at $t = t_1 + \tau$. It is now clear that

$$P_{\xi 2|\xi 1}(A\,|\,-A\,;\tau) = \left[\sum_{n=0,2,4,\ldots} P_\eta(n;\tau)\right] k\,d\tau \tag{95}$$

and finally, for $i = -1$, $j = 1$,

$$P_{\xi 2|\xi 1}(-A\,|\,A\,;\tau) = \left[\sum_{n=0,2,4,\ldots} P_\eta(n;\tau)\right] k\,d\tau \tag{96}$$

Now, by substitution, (71) gives

$$\begin{aligned}
\varphi_{PP}^{**}(\tau) & \\
&= \left[\frac{AAk\,d\tau}{2} + \frac{(-A)(-A)k\,d\tau}{2}\right] \sum_{n=1,3,5,\ldots} P_\eta(n;\tau)k\,d\tau\,\Big|_{\substack{A\to\infty \\ A\,d\tau=1}} \\
&\quad + \left[\frac{A(-A)k\,d\tau}{2} + \frac{(-A)Ak\,d\tau}{2}\right] \sum_{n=0,2,4,\ldots} P_\eta(n;\tau)k\,d\tau\,\Big|_{\substack{A\to\infty \\ A\,d\tau=1}} \\
&= -k^2e^{-2k|\tau|} \qquad \text{for } \tau \neq 0
\end{aligned} \tag{97}$$

The sum of (87) and (97) is the complete expression for the autocorrelation of $f_P^{**}(t)$. It is

$$\varphi_{PP}^{**}(\tau) = ku(\tau) - k^2e^{-2k|\tau|} \tag{98}$$

The function is plotted in Fig. 24*a*. The power density spectrum of $f_P^{**}(t)$ is

$$\begin{aligned}
\Phi_{PP}^{**}(\omega) &= \frac{1}{2\pi}\int_{-\infty}^{\infty} [ku(\tau) - k^2e^{-2k|\tau|}] \cos\omega\tau\,d\tau \\
&= \frac{k}{2\pi} - \frac{k^2}{\pi}\frac{2k}{4k^2+\omega^2} \\
&= \frac{k}{2\pi}\frac{\omega^2}{4k^2+\omega^2}
\end{aligned} \tag{99}$$

A graph of this expression is given in Fig. 24*b*.

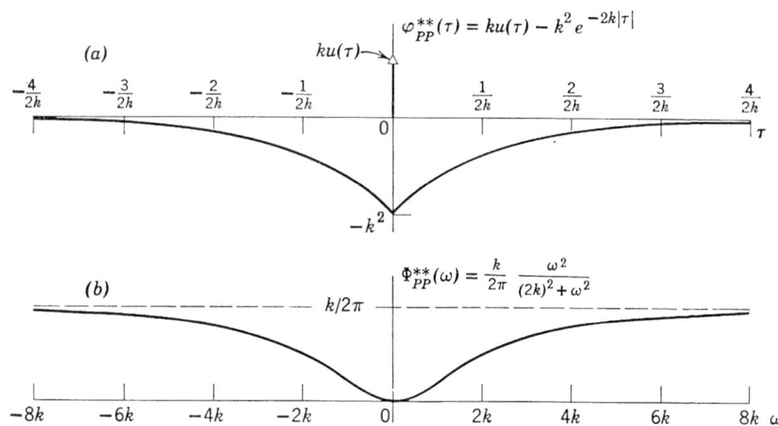

Fig. 24. Autocorrelation and power density spectrum of the ensemble of Poisson waves shown in Fig. 23.

Application of the results of this section to the determination of autocorrelation functions of other related random waves is presented in Chapter 13, Sec. 7.

PROBLEMS

1. The rectangular voltage pulses shown in Fig. P1 have random heights but a fixed duration, and they alternate between positive and negative values at the rate of 20,000 pulses per second. The pulse heights are completely independent of one another. The probability density $P_\xi(x)$ of the pulse height ξ is of the form

$$P_\xi(x) = Cxe^{-x}$$

where C is a constant to be determined. The probability density for negative values of ξ is the same as that for positive ξ. A 1-ohm load of pure resistance is assumed.

(a) Determine and plot the autocorrelation function.
(b) Determine and plot the power density spectrum.
(c) Determine and plot the integrated power spectrum.

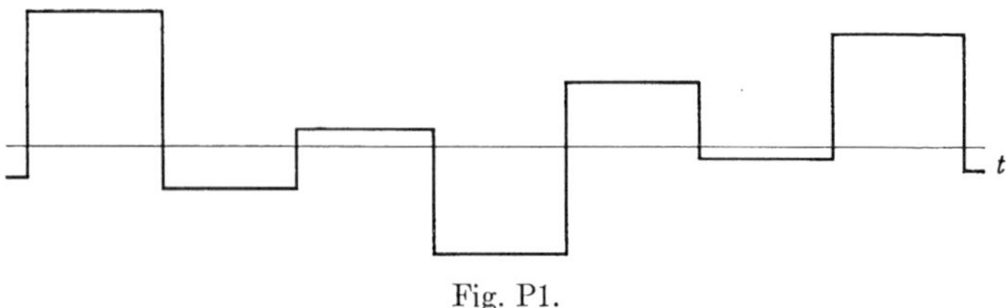

Fig. P1.

2. The rectangular voltage pulses of Fig. P2 have random heights but a fixed duration, and an equal probability of being positive or negative. All

pulse heights are independent of one another. The number of pulses per second is 20,000, and the load is a 1-ohm resistor. The pulse height probability density is the same as that given in Problem 1. Repeat parts a, b, and c of Problem 1.

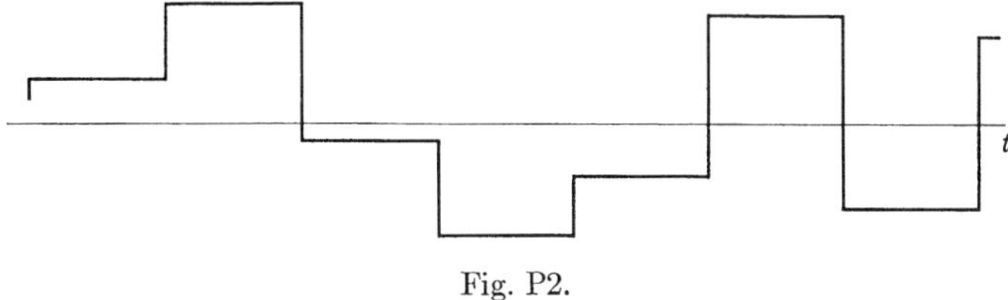

Fig. P2.

3. The pulses of Fig. P3 are the same as those of Fig. P1 except that half of each pulse has been removed. Repeat parts a, b, and c of Problem 1.

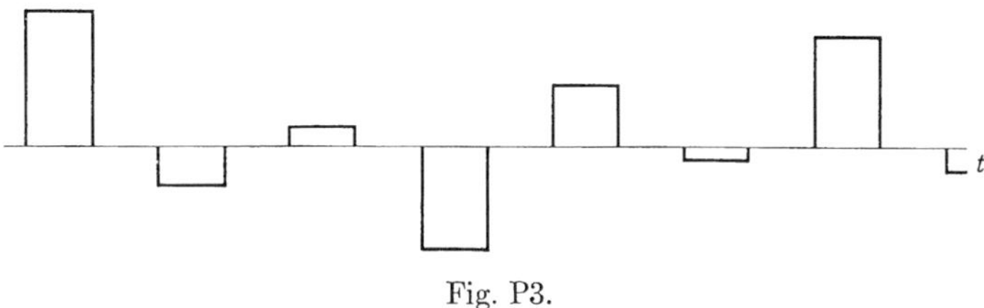

Fig. P3.

4. The pulses of Fig. P4 are the same as those of Fig. P2 except that half of each pulse has been removed. Repeat parts a, b, and c of Problem 1.

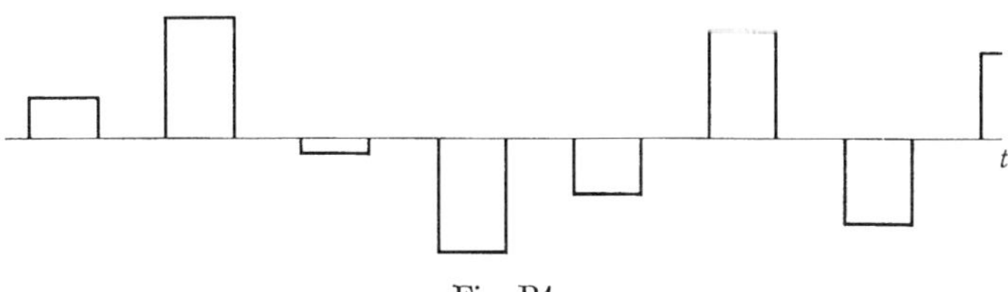

Fig. P4.

5. The series of rectangular pulses shown in Fig. P5 have a height of E volts and duration b seconds each and are separated from one another by b seconds. The pulses are on or off at random with equal probability. The state (on or off) of each pulse is independent of the state of other pulses.

(a) Determine and sketch the autocorrelation function of the series of pulses.

(b) Determine and sketch the power density spectrum of the random component of the series of pulses.

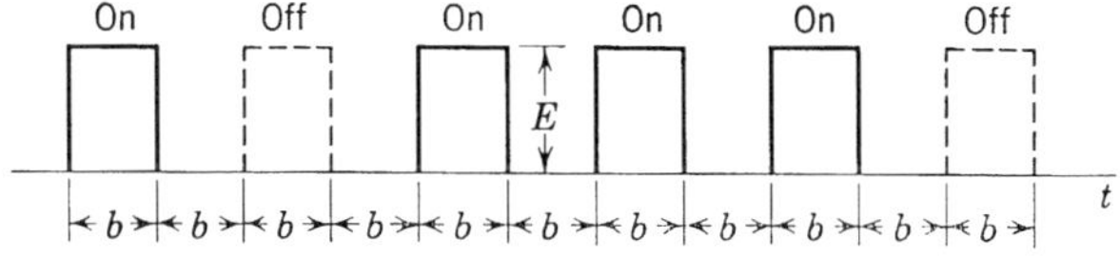

Fig. P5.

6. A series of rectangular pulses occurring at regular intervals of time is shown in Fig. P6. The height of a pulse is either E or $2E$ with equal probability of occurrence. The pulse duration is either one or two units of time, but, on the average, the longer pulses are three times as many as the others. The pulse height is independent of its duration. Determine and sketch the autocorrelation function.

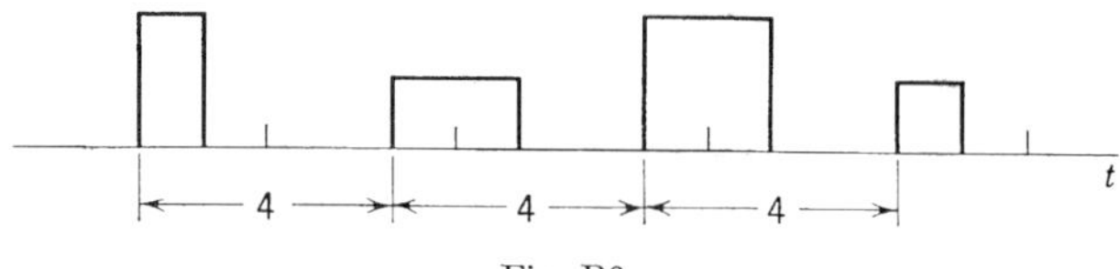

Fig. P6.

7. A random function consisting of groups of three rectangular pulses separated by regular intervals of duration Δ seconds is shown in Fig. P7. Each pulse may have a height of 0, 1, 2, or 3 volts with equal probability, and its duration is b seconds with a separation of the same duration between adjacent pulses in a group. All pulse heights are independent of each other. Determine and sketch the autocorrelation function of the random function.

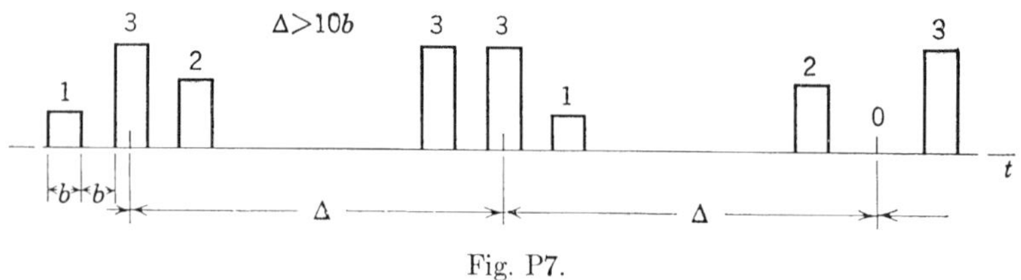

Fig. P7.

8. Figure P8 shows a series of pulses consisting of dots and dashes which occur at random at the average rate of 200 pulses per second of which there are, on the average, 100 dots and 100 dashes. The dashes are twice as long as the dots, and all pulses are separated by a space equal in length to that of a dash. Determine and sketch the autocorrelation function between $\tau = -5b$ and $\tau = 5b$.

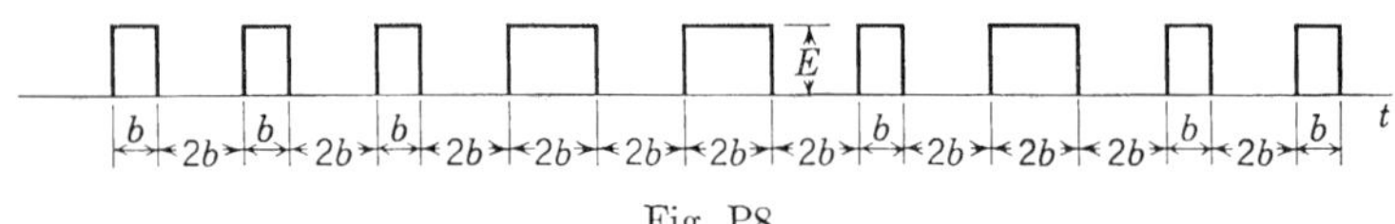

Fig. P8.

9. Figure P9 shows a series of pulses consisting of dots and dashes which occur at random at the average rate of 200 pulses per second of which there are, on the average, 100 dots and 100 dashes. The dashes are three times as long as the dots, and all pulses are separated by a space equal in length to that of a dot. Determine and sketch the autocorrelation function between $\tau = -4b$ and $\tau = 4b$.

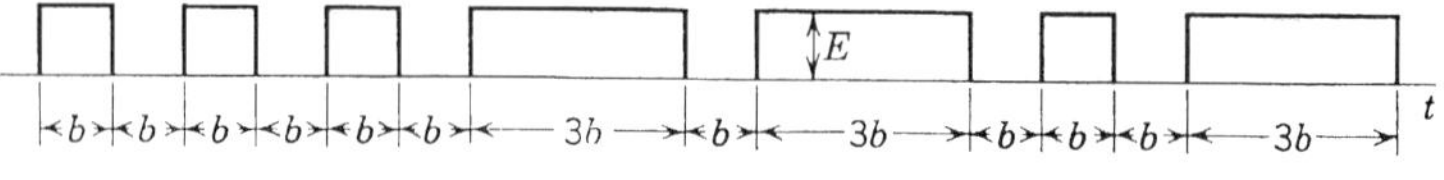

Fig. P9.

10. A random function formed by a series of rectangular pulses is shown in Fig. P10. The probability that pulse a is on is $\frac{2}{3}$. When this pulse is on, the probability that pulse b is also on is $\frac{3}{5}$. When a is off, the probability that b is on is $\frac{3}{4}$. The state of the pulses in one interval is independent of the state of those in any other interval. Determine and sketch the autocorrelation function giving its values at various points.

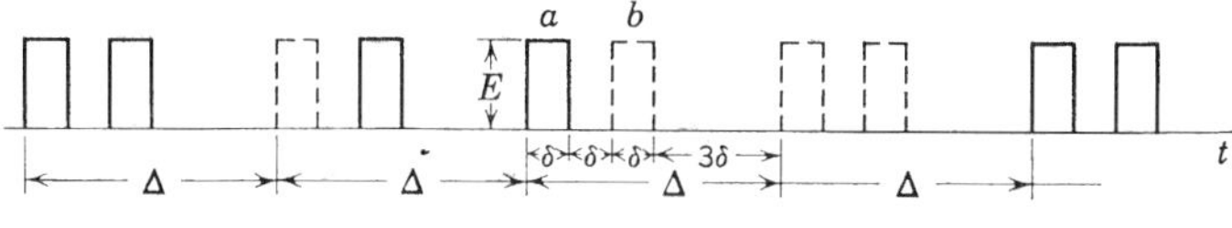

Fig. P10.

11. In a certain 12-channel pulse-code modulation system, one of the channels has a random wave as shown in Fig. P11. Each of the rectangular pulses goes on or off with equal probability, and the state (on or off) of each pulse is independent of the state of all other pulses. Each pulse has a duration of 0.50 μsec, and each space between two adjacent pulses in a group is 1.00 μsec. Other data are given in the figure. Determine the autocorrelation function of the wave in the form of a sketch with numerical values indicated.

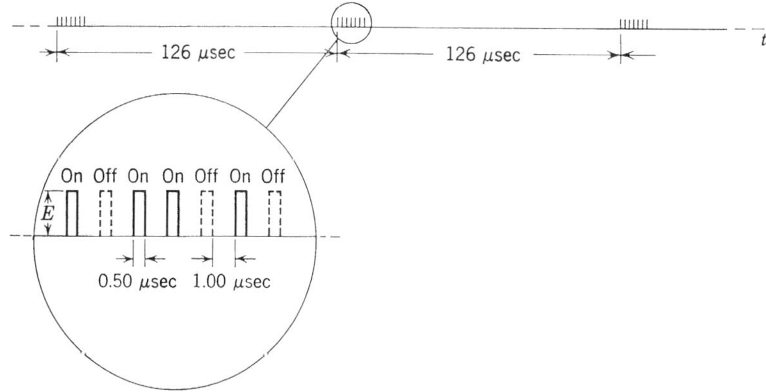

Fig. P11.

12. The numbers 1, 2, 3, and 4, each represented by the corresponding number of rectangular pulses, form a random wave as shown in Fig. P12a. The groups of pulses are separated by a regular interval of $10b$ seconds and the numbers have the same probability of independent occurrence. A second random wave is formed by a series of coded pulses following exactly the order of the random numbers in the first wave as Fig. P12b illustrates. The code is as follows:

Number	*Pulse Code*
1	Pulse in first position and pulse in second position
2	No pulse
3	Pulse in first position
4	Pulse in second position

Other data are given in the figures. Determine and sketch the crosscorrelation function $\varphi_{12}(\tau)$ of $f_1(t)$ and $f_2(t)$ and indicate values of the curve at various points.

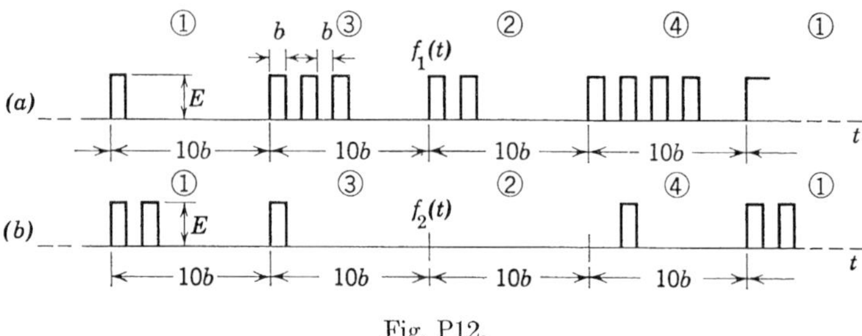

Fig. P12.

13. Figure P13 shows a series of rectangular pulses whose leading edges a occur at regular intervals of length Δ seconds. The trailing edges b terminate somewhere between the midpoint (marked by dotted line) of an interval and the beginning of the following pulse so that each pulse duration may have any value between $\Delta/2$ seconds and Δ seconds with equal probability. The duration of each pulse is independent of the durations of all other pulses. Determine and sketch the autocorrelation function.

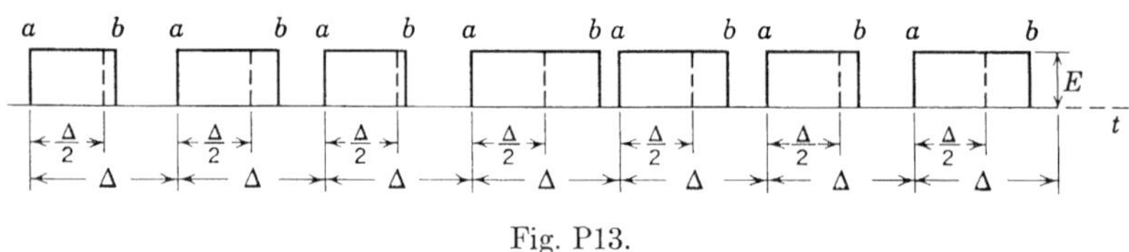

Fig. P13.

14. A series of rectangular pulses with a height of E volts and with midpoints a separated by a regular interval of Δ seconds is shown in Fig. P14. The duration of the pulses fluctuates between zero and Δ and has a uniform probability density. Assuming that the duration of each pulse is independent of the durations of all other pulses, determine and plot on graph paper the autocorrelation function of the series of pulses showing separately the periodic and aperiodic components.

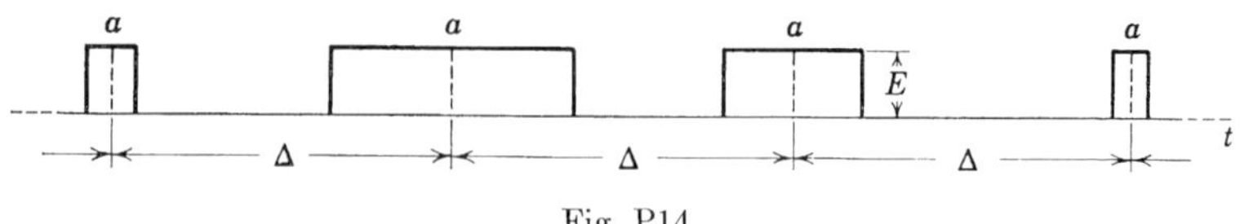

Fig. P14.

15. A series of rectangular pulses occurring at every other interval of Δ seconds is shown in Fig. P15. The pulse duration is b seconds and pulse height is E volts. A pulse takes its position in the interval in which it occurs at random, and the probability density of the position is uniform. The pulse edges do not extend beyond the interval. Assuming that the positions of the pulses are

independent of one another, determine and sketch the autocorrelation function indicating its values at various points.

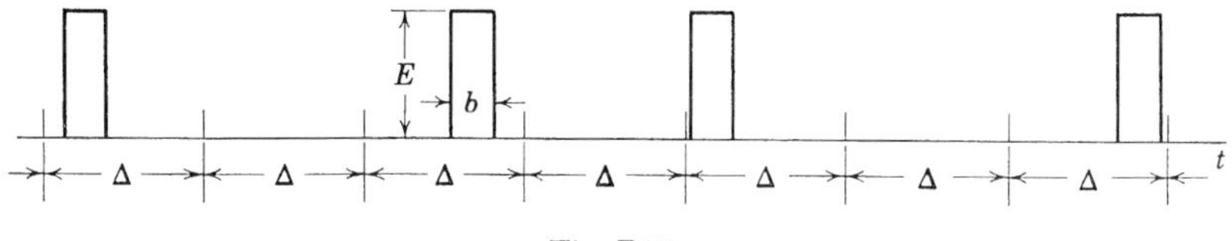

Fig. P15.

16. A series of triangular pulses with random heights is formed as shown in Fig. P16. The starting points of the pulses are separated by 2Δ seconds, and each pulse has a duration of Δ seconds. The maximum value of the random height ξ is E volts, and the minimum is zero. The probability density of ξ is a linear function, increasing from zero to maximum value. Assume that the pulse heights are independent of one another.

(a) Determine and sketch the autocorrelation function, indicating values at various points on the sketch.

(b) Determine and sketch the periodic component of the given function.

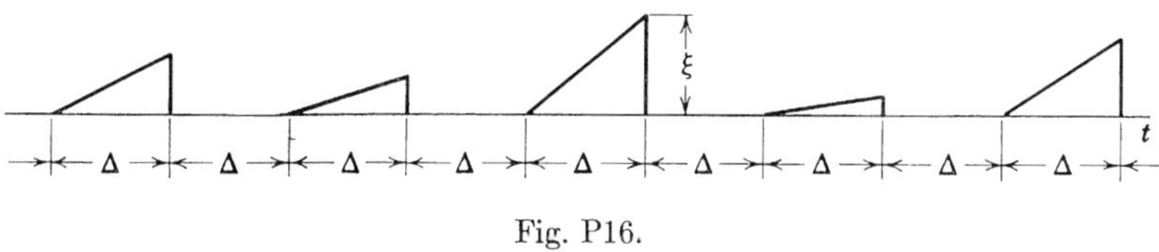

Fig. P16.

17. In the random wave $f_1(t)$ shown in Fig. P17, the triangular pulses start at regular intervals of length 2Δ. The pulses have the same slope, and the height ξ which is a random variable. When ξ is at its maximum value E, the duration of the pulse is Δ. The pulse heights are independent of one another. Assuming that the probability density $P_\xi(x)$ for the pulse height is linear as shown, determine:

(a) The waveform of the periodic component wave of $f_1(t)$.

(b) The autocorrelation function $\varphi_{11}(\tau)$ of $f_1(t)$ for $-\Delta < \tau < \Delta$.

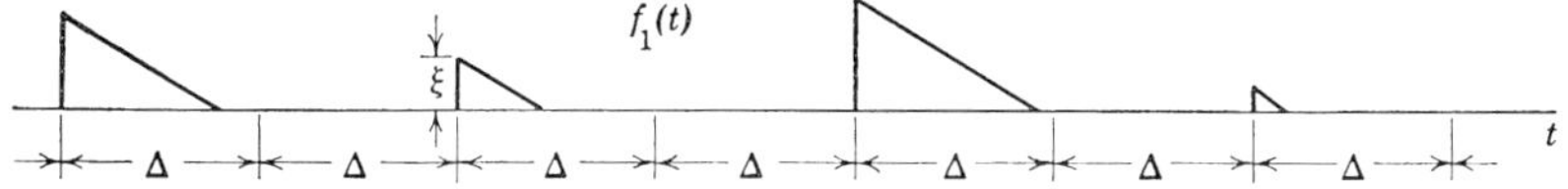

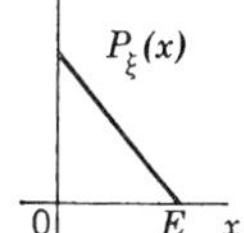

Fig. P17.

18. In the random function formed by a series of pulses as shown in Fig. P18*a*, the leading edges of the pulses occur at periodic intervals of duration 2Δ with the value $-E$. Each pulse has the duration $2b$ and a jump from $-E$ to $+E$ which is random and independent of all others. If the time from the midpoint of a pulse to the jump is ξ, the probability density of ξ is $P_\xi(x)$ and is shown in Fig. P18*b*. The function $P_\xi(x)$ is constant between $-b$ and 0 and linear between 0 and b. Determine and sketch the periodic component of the random function. Several numerical values should be given on the sketch for the periodic waveform.

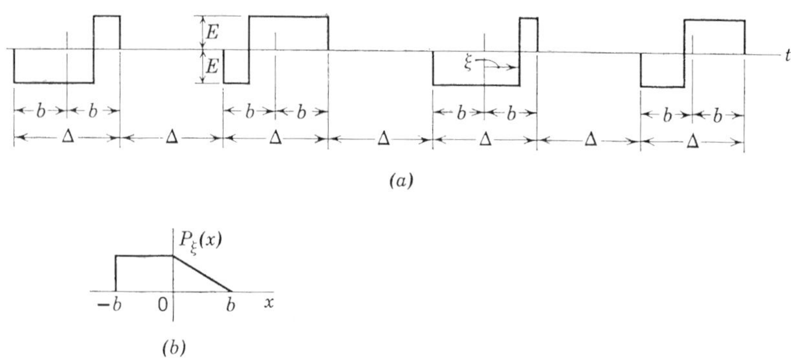

Fig. P18.

19. Figure P19 shows two random functions $f_1(t)$ and $f_2(t)$. Twenty-five per cent of the pulses are missed at random as indicated by dotted lines, and the disappearance of pulses occurs simultaneously in both functions. In each function the state (on or off) of each pulse is independent of the state of all other pulses in that function. Determine and sketch the crosscorrelation function of the two functions.

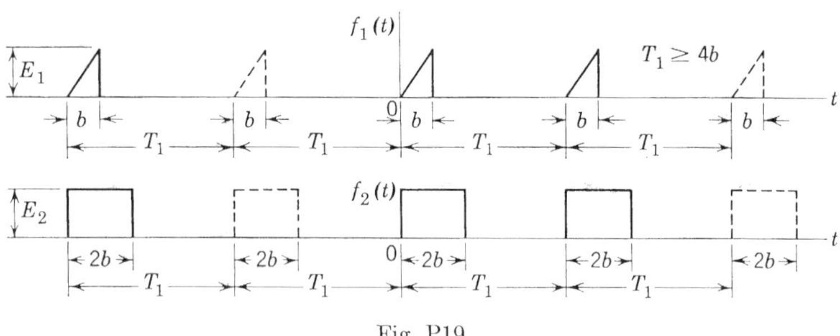

Fig. P19.

20. The wave shown in Fig. P20 is a series of jumps with a random height ξ and a random width ζ. The height has a continuous range of possible values in the interval $(-1, \infty)$, and its probability density $P_\xi(x)$ is as shown. In the interval $(0, \infty)$ the density has the form Ae^{-x}, where A is a constant to be determined; and in the interval $(-1, 0)$ the density is linear. All heights are independent of one another, in sign and in magnitude. The jumps occur at such

points in time, as indicated by the heavy dots in the figure, that they form a series of Poisson distributed points.

(a) Determine the autocorrelation function of the wave.

(b) Determine the probability density of the random width ζ of the jumps. All important steps in the solution of the problem must be clearly explained.

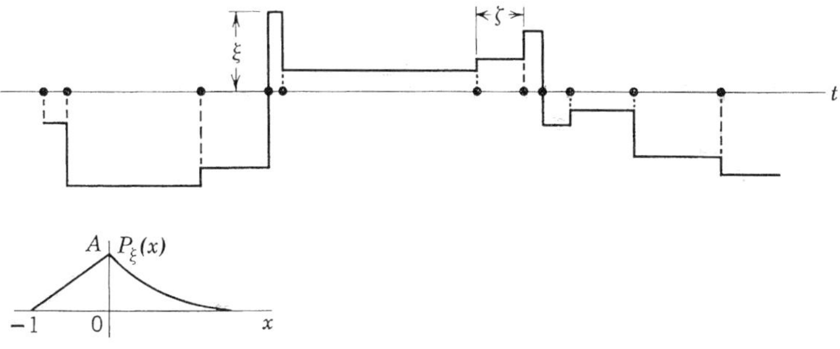

Fig. P20.

21. In Fig. P21, $f_a(t)$ is a Poisson rectangular wave with the average number of zero crossings per second equal to k; $f_b(t)$ is a wave of rectangular pulses each of duration b. At each point where $f_a(t)$ jumps from $-E$ to $+E$ a positive pulse of duration b starts in $f_b(t)$, and at each point where $f_a(t)$ drops from $+E$ to $-E$ a negative pulse of duration b starts in $f_b(t)$. In cases where a group of rectangular pulses of $f_b(t)$ overlap, the instantaneous values will be the algebraic sum of these pulses. Determine the cross-power density spectrum $\Phi_{ab}(\omega)$ of $f_a(t)$ and $f_b(t)$.

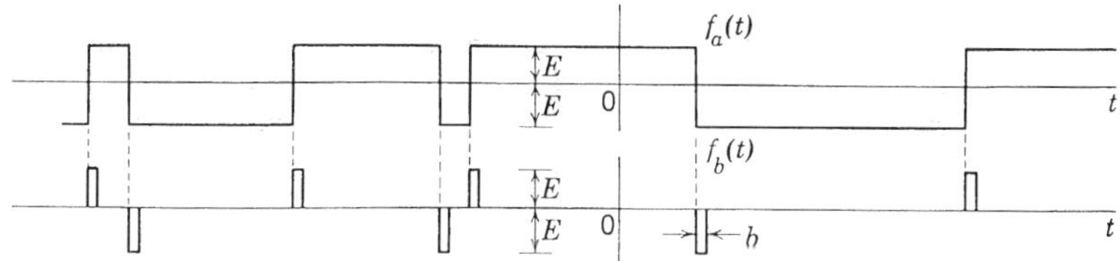

Fig. P21.

22. A random noise $f_1(t)$ has the autocorrelation function

$$\varphi_{11}(\tau) = A_1 e^{-|\tau|} + B_1$$

where the constant B_1 is positive. Another random noise $f_2(t)$ has the autocorrelation function

$$\varphi_{22}(\tau) = A_2 e^{-|\tau|} + B_2$$

where the constant B_2 is positive.

(a) If $f_1(t)$ and $f_2(t)$ come from independent sources, what is the autocorrelation function $\varphi_{oo}(\tau)$ for the sum $f_o(t) = f_1(t) + f_2(t)$?

(b) If $f_1(t)$ and $f_2(t)$ come from the same source, in the same manner, differing only in amplitude by a constant factor K, what is the autocorrelation function

$\varphi_{oo}(\tau)$ for the sum $f_o(t) = f_1(t) + f_2(t)$? [Do not assume the special case $f_1(t) = f_2(t)$.]

No constants other than those given should appear in the final expressions for both part a and part b.

23. In the circuit shown in Fig. P23 a random process $f_1(t)$ is applied to a multiplier to obtain $2f_1(t)$ at its output. This output is added to another random process $f_2(t)$ so that $f_o(t) = 2f_1(t) + f_2(t)$ is the output of the summing circuit.

(*a*) We have further given that $f_1(t)$ and $f_2(t)$ are Gaussian processes from independent sources, that they have the same amplitude probability density

$$P_{\xi 1}(x) = P_{\xi 2}(x) = \frac{1}{\sigma\sqrt{2\pi}} \exp\left[-\frac{(x-m)^2}{2\sigma^2}\right] \tag{1}$$

where m is the mean and σ^2 the variance, and that they have the same autocorrelation function

$$\varphi_{11}(\tau) = \varphi_{22}(\tau) = Ae^{-a|\tau|} + B \tag{2}$$

where A and B are expressible in terms of m and σ^2. Determine the amplitude probability density of $f_o(t)$ in terms of m and σ^2 and the autocorrelation function of $f_o(t)$ in terms of m, σ^2, and a.

(*b*) Now, suppose that $f_1(t)$ and $f_2(t)$ are from the same source so that $f_1(t) = f_2(t)$ and that the probability density and the autocorrelation function are the same as those given by equations (1) and (2) respectively. Determine the probability density of $f_o(t)$ in terms of m and σ^2, and the autocorrelation function of $f_o(t)$ in terms of m, σ^2, and a.

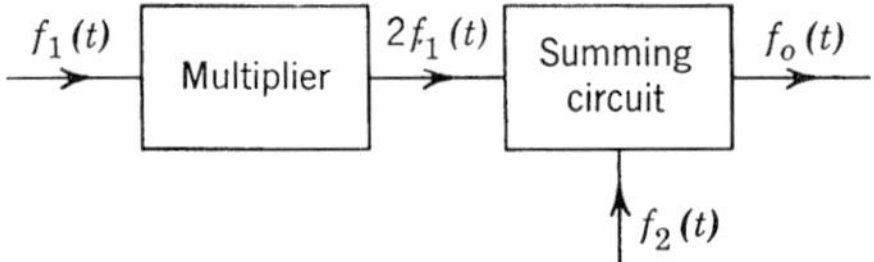

Fig. P23.

chapter 9

Summary of Basic Concepts and Tools

At this point it seems desirable to summarize the principal concepts and tools that we have discussed for the description and analysis of the various types of time functions in a communication or control system. It is desirable to tabulate the concepts and tools in relation to each other in order to bring out their significance and organization. The tabulation is shown in Table 1. Under the general heading of generalized harmonic analysis, we have the three types of functions, namely, periodic, aperiodic, and random. We assume that the periodic function has a finite mean square value, that the aperiodic function has a finite integral square value, and that the random function has a finite mean square value over the infinite time interval.

To describe a periodic function, we specify its values at all values of the argument over a complete period. Another way to describe the function is to give the amplitudes and phase angles of the harmonic components into which the function is resolved. These amplitudes and phase angles are contained in the complex spectrum which is obtained by an integration. Because this integration yields the information pertaining to the components, it is known as an analysis of the periodic function. There is a reverse process that recovers the periodic function. This process is a summation of the harmonics in accordance with the amplitudes and phase angles in the spectrum. We know that one reason for the great importance of Fourier series in the study of periodic phenomena is the fact that a periodic function can be resolved into simple harmonics by a simple integration and that the function can be reproduced by synthesis.

An aperiodic function is described by specifying its values for all values of the argument in the infinite interval. By the Fourier method, the function can also be described by an amplitude-phase density spectrum which is obtained by an integration. The spectral density is the result of an analysis. The aperiodic function is recoverable by synthesis

TABLE 1. GENERALIZED HARMONIC ANALYSIS

Periodic Functions	Aperiodic Functions	Stationary Random Processes: Without Hidden Periodic Component	Stationary Random Processes: With Hidden Periodic Component
$\frac{1}{T_1}\int_{-T_1/2}^{T_1/2} f^2(t)\,dt$ finite	$\int_{-\infty}^{\infty} f^2(t)\,dt$ finite	$\lim_{T\to\infty}\frac{1}{2T}\int_{-T}^{T} f^2(t)\,dt$ finite	
Analysis (amplitude and phase spectrums) $F(n) = \frac{1}{T_1}\int_{-T_1/2}^{T_1/2} f(t)e^{-jn\omega_1 t}\,dt$ *Synthesis* $f(t) = \sum_{n=-\infty}^{\infty} F(n)e^{jn\omega_1 t}$	*Analysis* (amplitude and phase density spectrums) $F(\omega) = \frac{1}{2\pi}\int_{-\infty}^{\infty} f(t)e^{-j\omega t}\,dt$ *Synthesis* $f(t) = \int_{-\infty}^{\infty} F(\omega)e^{j\omega t}\,d\omega$	*Probability densities* $P_{\xi 1}(x_1),\ P_{\xi 1\xi 2}(x_1, x_2; \tau_1),\ P_{\xi 1\xi 2\xi 3}(x_1, x_2, x_3; \tau_1, \tau_2), \ldots$	
Autocorrelation $\varphi_{11}(\tau) = \frac{1}{T_1}\int_{-T_1/2}^{T_1/2} f_1(t)f_1(t+\tau)\,dt$	*Autocorrelation* $\varphi_{11}(\tau) = \int_{-\infty}^{\infty} f_1(t)f_1(t+\tau)\,dt$	*Autocorrelation* $\varphi_{11}(\tau) = \lim_{T\to\infty}\frac{1}{2T}\int_{-T}^{T} f_1(t)f_1(t+\tau)\,dt$ (time average) $\varphi_{11}(\tau) = \int_{-\infty}^{\infty}\int_{-\infty}^{\infty} x_1x_2P_{\xi 1\xi 2}(x_1, x_2; \tau)\,dx_1\,dx_2$ (ensemble average)	
Analysis (power spectrum) $\Phi_{11}(n) = \frac{1}{T_1}\int_{-T_1/2}^{T_1/2} \varphi_{11}(\tau)\cos n\omega_1\tau\,d\tau$ *Synthesis* $\varphi_{11}(\tau) = \sum_{n=-\infty}^{\infty} \Phi_{11}(n)\cos n\omega_1\tau$	*Analysis* (energy density spectrum) $\Phi_{11}(\omega) = \frac{1}{2\pi}\int_{-\infty}^{\infty} \varphi_{11}(\tau)\cos\omega\tau\,d\tau$ *Synthesis* $\varphi_{11}(\tau) = \int_{-\infty}^{\infty} \Phi_{11}(\omega)\cos\omega\tau\,d\omega$	Time average=ensemble average (by ergodic hypothesis) Wiener theorem *Analysis* (power density spectrum) $\Phi_{11}(\omega) = \frac{1}{2\pi}\int_{-\infty}^{\infty} \varphi_{11}(\tau)\cos\omega\tau\,d\tau$ *Synthesis* $\varphi_{11}(\tau) = \int_{-\infty}^{\infty} \Phi_{11}(\omega)\cos\omega\tau\,d\omega$	Periodic Component: As under Periodic functions Aperiodic Component: Wiener theorem

which is another integration. The integration is a summation of sinusoids of all frequencies, each having an infinitesimal amplitude. The relative amplitudes and the phase angles are given by the amplitude-phase density spectrum. Again the importance of the Fourier transform lies, in part, in the simplicity of the components which are all sinusoids, in the simple method by which the spectrum is determined, and in the simple way by which the function is recovered.

A basic problem in the theory of linear systems is the determination of the output of a system whose input is a periodic function or a transient function. The idea of analysis and synthesis in Fourier theory has enabled us to develop a powerful method for solving the problem. As we know, the method is to resolve the input into sinusoidal components, then to determine the system output for each of the components, and finally to sum the system outputs for all the input components to obtain the total output. The techniques of the method have been so well developed that we frequently do not think of the fundamental ideas involved. However, for a clear understanding of the subject we should think of these ideas when we discuss harmonic analysis in communication problems.

Messages and noise are not completely predictable so that unique specification of these functions for all values of time is contrary to their nature. Only their recorded past is subject to definite instantaneous specification. However, our main interest is in their behavior in the future. Furthermore, a communication or control system must work with a class of messages and noise. Therefore the description of messages and noise must concern the future of ensembles of these messages and noise. A meaningful and useful method of description has been found to be the description by a set of probability densities (probability distributions if the ensemble is discrete). This is a fundamentally different concept in describing functions as compared with the concept of complete determinism and the concept of amplitude and phase spectrums for periodic and aperiodic functions.

Another important point indicated in the table is the fact that recovery of the original functions from the describing functions is not compatible with the nature of messages and noise. Concerning any particular member function in an ensemble at any given future time, we can only ask for the probability of the function to take a particular possible value (or the probability to be in a particular interval). Therefore, under random functions, there is no operation which corresponds to synthesis under periodic and aperiodic functions.

Since the concept of the spectrum is of such great importance in communication problems, it would be desirable if a meaningful and useful spectrum could be associated with a random function. Obviously this spectrum could not be the type that is associated with periodic and aperiodic functions from which the entire function can be reproduced. Fortunately, the work of Wiener shows that a spectrum that is meaningful and useful in the characterization of ensembles of messages and noise is the power density spectrum. To obtain this spectrum, an intermediate step is essential. This step is autocorrelation. Autocorrelation discards the information that is only characteristic of a particular ensemble member (information pertaining to phase angles) and at the same time completely preserves the information pertaining to the power density spectrum that is characteristic of all members of the ensemble. The Wiener theorem giving the reciprocal relations of the autocorrelation and the power density spectrum of an ensemble is the central theorem in many important communication and control problems.

Autocorrelation is applicable to periodic and aperiodic functions when appropriately defined. Similar to the Wiener theorem, we have reciprocal relations between autocorrelation and power spectrum for a periodic function and reciprocal relations between autocorrelation and energy density spectrum for an aperiodic function. We have introduced the concept of autocorrelation in these functions to complete the picture and for a better understanding of its properties. Whereas autocorrelation is unnecessary in the determination of the power spectrum and the energy density spectrum, it is the only means by which the power density spectrum of random functions is calculated.

Under suitable conditions autocorrelation of a random process with a hidden periodic function yields distinguishable periodic and aperiodic components. After the separation, the periodic part can be analyzed as an autocorrelation function of a periodic wave, and the aperiodic part can be treated by the Wiener theorem as indicated in the table.

In this table we see that the scope of communication and control theory has been considerably extended by the addition of a statistical description of messages and noise and some powerful tools for handling them. Instead of crudely representing messages and noise by periodic and aperiodic functions as in classical theory, we now can more precisely describe them in statistical language. The theory of harmonic analysis has been extended to include random functions so that the theory is effective for all three types of functions that are usually found in communication and control systems. As we have seen, the determination of the spectrum of a random function from an appropriate statistical description is possible. Here we find a remarkable combination of Fourier and statistical methods and techniques. This combination holds great power in the development of communication and control theory.

The table is not intended to include all concepts and tools in this field of work. For example, we have not included crosscorrelation and the expression of the Fourier integral in the complex domain. The purpose of the table is to give the reader an idea of the organization of the most important concepts and tools in statistical communication theory with particular emphasis on those coming from statistical theory.

chapter 10

Measurement of Correlation Functions and Probability Densities

With the rapid development of electronic devices and digital computers, it is not necessary in a book such as this to go into any of the methods that are now available for the measurement of correlation functions and probability densities in great detail. The principles of only one method for the measurement of correlation functions will be briefly described for the purpose of pointing out the application of statistical theory in the measurement of statistical characteristics. The theory for the measurement of the first and second probability densities will be discussed.

1. Measurement of Correlation Functions

The measurement of an autocorrelation or crosscorrelation function may be carried out either on the basis of a time average or on the basis of an ensemble average. We shall first consider the measurement of an autocorrelation function on the ensemble basis. Figure 1 shows a por-

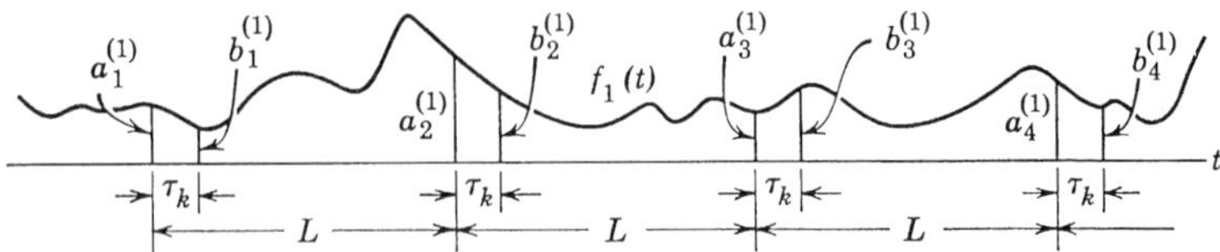

Fig. 1. Portion of a random function.

tion of a random function $f_1(t)$, whose autocorrelation is desired. We assume that the function is non-negative, as shown, for convenience in explaining the method of measurement. The assumption does not affect the generality of the theory. Let us divide the random function into sections each of duration L as indicated and further assume that the function contains no periodic components. When these sections are assembled in a vertical column in the manner shown in Fig. 2*a*, they form a finite ensemble of finite duration. We consider that L is sufficiently long if the values $a_1^{(1)}$, $a_2^{(1)}$, . . . of $f_1(t)$ at the beginning of the sections are independent. This statement implies that the values $b_1^{(1)}$,

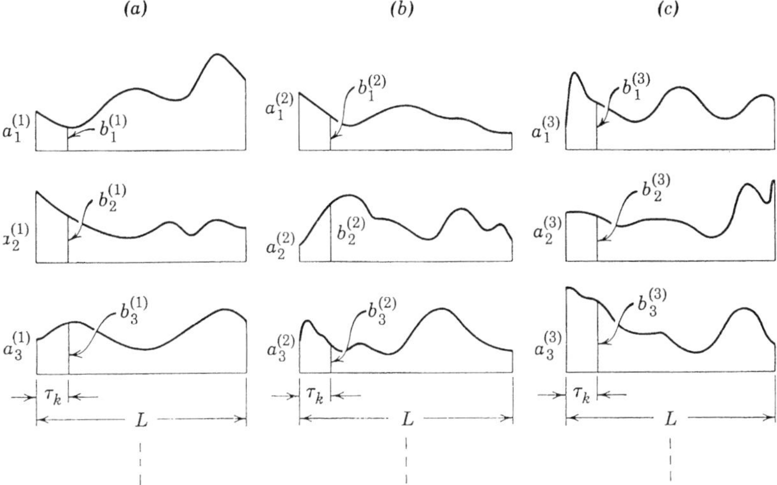

Fig. 2. Ensembles of random functions formed from sections of the function shown in Fig. 1.

$b_2^{(1)}$, . . . at time τ_k after the a-values are also independent. Forming an ensemble from a single function in this manner is a simple practical way of approximating a theoretical ensemble in which an infinite number of member functions are generated by identical but independent sources over an infinite duration. Clearly, the source of the random function $f_1(t)$ should be under normal steady-state conditions so that the ensemble formed from its sections may be considered stationary. It is reasonable to assume that the joint probability density $P_{\xi_1\xi_2}(x_1, x_2; \tau)$ of the ensemble in Fig. 2*a* when extended to an infinite ensemble is the same as that of the theoretical ensemble $\{f(t)\}$ of which $f_1(t)$ is a member. With these considerations, we see that, in accordance with the ensemble average for autocorrelation

$$\varphi_{11}(\tau) = \int_{-\infty}^{\infty}\int_{-\infty}^{\infty} x_1 x_2 P_{\xi_1\xi_2}(x_1, x_2; \tau)\, dx_1\, dx_2 \tag{1}$$

the autocorrelation function of $f_1(t)$ can be measured from the ensemble of Fig. 2a. At $\tau = \tau_k$, the function has the approximate value

$$\varphi_{11}(\tau_k) \cong \frac{1}{N} \sum_{n=1}^{N} a_n^{(1)} b_n^{(1)} \tag{2}$$

The superscript (1) indicates that the first set of N values of a_n and b_n are involved. We may refer to $a_n^{(1)}$ and $b_n^{(1)}$ as the values from $f_1(t)$ on the first measurement. A repeated measurement may be made from a second ensemble, shown in Fig. 2b, which has been formed from sections of the random wave that follow those in the first measurement. We assume that the joint probability density $P_{\xi_1\xi_2}(x_1, x_2; \tau)$ in the second ensemble is the same as that of the first ensemble in spite of the fact that entirely different sections of $f_1(t)$ have been taken to form the second ensemble. On the second measurement based upon the second ensemble, the approximate value of the autocorrelation function at $\tau = \tau_k$ is

$$\varphi_{11}(\tau_k) \cong \frac{1}{N} \sum_{n=1}^{N} a_n^{(2)} b_n^{(2)} \tag{3}$$

Theoretically (2) and (3) are practically the same as $N \to \infty$. Inasmuch as N must be finite in a physical measurement, these values vary in a manner that depends upon N. The variation is an error in measurement which will be considered in another chapter. Additional ensembles such as that in Fig. 2c may be formed with the same joint probability density. Since the displacement τ_k may be assigned various values such as $\tau_k = \tau_1, \tau_2, \tau_3, \ldots$ in the interval $(0, L)$, an autocorrelation function may be measured by taking sets of sample values of $f_1(t)$, each set for a particular value of τ_k.

Electronic correlators have been designed upon these principles to carry out the operations automatically. Figure 3 shows the characteristic waveforms of an electronic correlator (reference 1). For the computation of the autocorrelation curve for the random wave (A) two sets of timing pulses (B) of period L are derived from a master oscillator with the second set delayed by an adjustable time τ_k. By means of the first set of timing pulses, the values $a_1, a_2, a_3, \ldots$ are selected, and a boxcar wave (C) is generated so that the discretely varying heights are proportional to these values. A second boxcar wave (D) is similarly generated for the values $b_1, b_2, b_3, \ldots$ that have been obtained from (A) by the second set of timing pulses in (B). A sawtooth sampling wave converts the b-values as shown in (D) into a train of pulse-duration modulated pulses (E). A gating circuit multiplies the waveforms (C) and (E), resulting in the waveform (F), which is a series of pulses whose heights are proportional to the a-values and whose durations are proportional to the b-values. The summation of a large number N of these

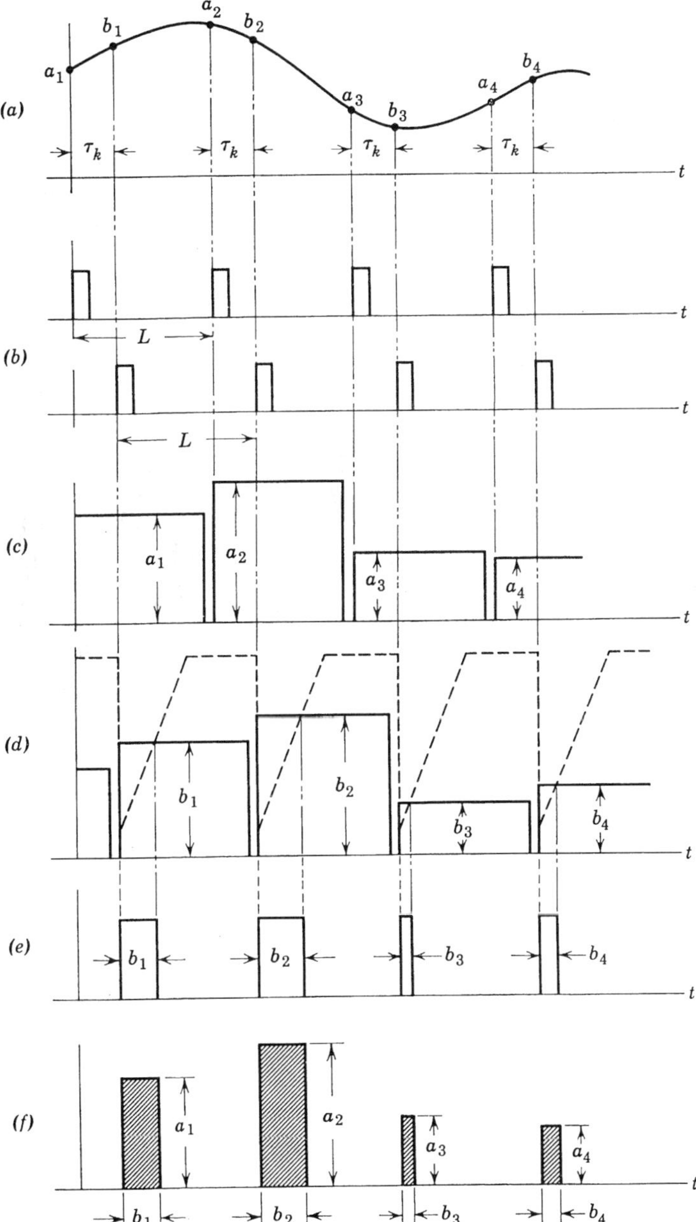

Fig. 3. Characteristic waveforms of an electronic correlator.

pulses by an integrating circuit is, except for a constant factor, the value $\varphi_{11}(\tau_k)$ as expressed by (2). The displacement τ_k, starting with $\tau_k = 0$, is increased in discrete steps each time after N pairs of samples have been processed. The number N is preset. The circuit that stores the sum of the products $a_n b_n$ is reset to zero after each N products have been summed. The correlator is entirely automatic, and the result is presented on an automatic recorder. A sample autocorrelation curve for filtered random noise from a Type 884 gas tube is shown in Fig. 4.

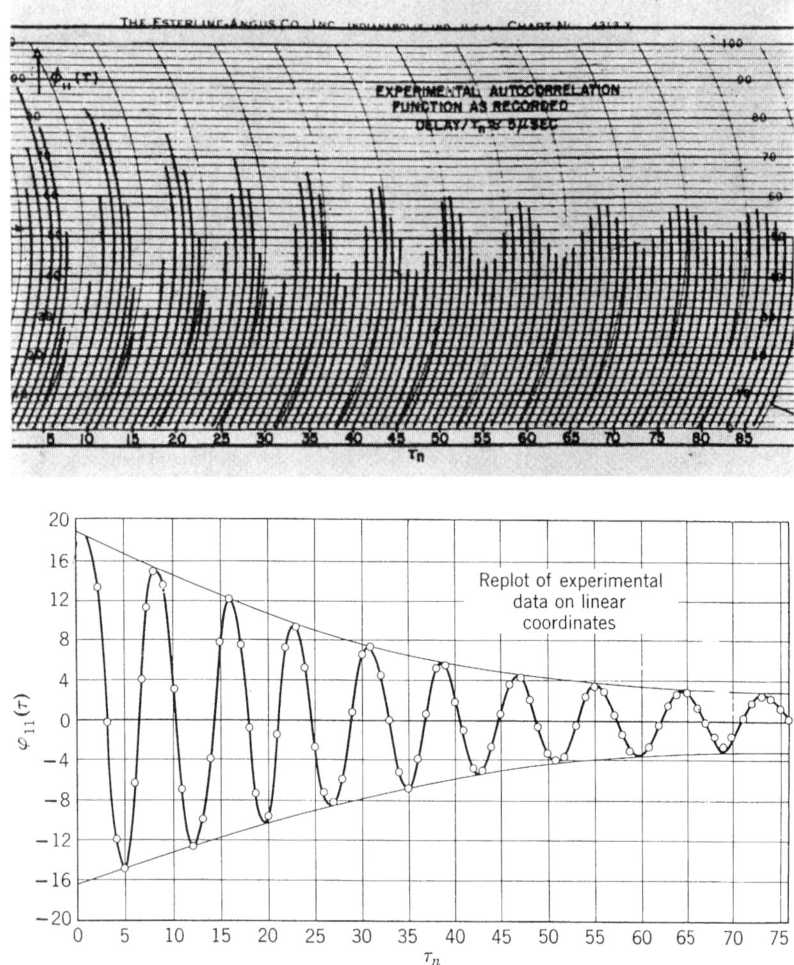

Fig. 4. Autocorrelation curve for filtered noise from Type 884 gas tube. From *Application of Statistical Methods to Communication Problems*, Y. W. Lee, Technical Report No. 181, 1950, Research Laboratory of Electronics, M.I.T.

Since the correlator does not take the mean value of the N products and there are other proportionality factors involved in the circuits, the final result is proportional to $\varphi_{11}(\tau_k)$. The proportionality factor must be determined by additional measurement or by calibration. The additional measurement could be the mean square value of the random wave because this value is represented by the autocorrelation curve at the origin.

In the method just described, the random wave is assumed available from its source directly or from its record on magnetic tape during the entire measurement. However, in many actual situations the time of observation is limited to a short duration due to various reasons. A simple way to meet these situations is to record the data on magnetic tape and to run the tape in a single loop continuously for the entire measurement. Obviously, a multichannel machine will save considerable time.

In measuring a crosscorrelation function, we apply the two sets of timing pulses in (B) to the two random functions to be crosscorrelated. The a-values are taken from one function and the b-values from the other with the difference in time τ_k. The rest of the procedure is the same as before. Thus the crosscorrelation function $\varphi_{12}(\tau_k)$ of the functions $f_1(t)$ and $f_2(t)$ is

$$\varphi_{12}(\tau_k) \cong \frac{1}{N} \sum_{n=1}^{N} a_n^{(1)} b_n^{(1)} \tag{4}$$

in which τ_k are positive values if $b_n^{(1)}$ are the values taken from $f_2(t)$ and $a_n^{(1)}$ are from $f_1(t)$ at times τ_k later, and τ_k are negative values if $a_n^{(1)}$ are from $f_1(t)$ and $b_n^{(1)}$ are from $f_2(t)$ at times $|\tau_k|$ later. This method of identifying the function for positive and negative τ_k follows from the definition

$$\varphi_{12}(\tau) = \lim_{T\to\infty} \frac{1}{2T} \int_{-T}^{T} f_1(t) f_2(t+\tau)\, dt \tag{5}$$

where τ is positive if $f_2(t)$ is displaced in the negative direction on the t-axis by the amount τ, and τ is negative if $f_2(t)$ is displaced in the positive direction on the t-axis by τ.

A point of some importance in the method described for measuring autocorrelation and crosscorrelation is the fact that samples are taken periodically and not at random. If the random function contains a periodic component, a difficulty may arise because the sampling period may happen to be the same as (or an integral multiple of) that of the periodic component. The sections of length L of the random wave will no longer form a stationary ensemble. Thus, if the samples happen to be taken near the peak of the periodic component, the result would be much greater than that from samples taken near small values of the periodic component. In the laboratory this difficulty can be easily remedied by changing the sampling period slightly. Although theoretically in the presence of a periodic component, the sampling period should be random with a probability density that will produce a stationary ensemble when the sections of random durations are assembled as in Fig. 2*a*, periodic sampling handled judiciously is sufficient for most purposes. Measurement error computations for periodic sampling when a periodic component is present are much more complex than those for random sampling, but the design of apparatus for measurement based upon periodic sampling is considerably simpler.

The computation of correlation functions from experimental data on the basis of time averaging has been discussed in Chapter 2, Sec. C-5. Here modern digital computers are a valuable aid. Electronic correlators for measurement on the time-average basis have been built. However, at the present stage of development of electronic techniques, discrete multiplication, upon which the ensemble-average (or sampling) correlator depends, is more satisfactory than continuous multiplication, upon which the time-average correlator depends. The advantage of discrete multiplication lies in greater possible accuracy and the comparatively greater range of the spectrum that can be adequately covered.

After an autocorrelation function is obtained, the transformation required for producing the power density spectrum may be performed by machine. As an illustration, an autocorrelation curve of filtered noise from a Type 884 gas tube which has been measured by an electronic correlator is shown in Fig. 5*a*. The power density spectrum of the noise, shown in Fig. 5*b*, has been obtained by a cosine transformation by means of the high-speed differential analyzer * developed at the Research Laboratory of Electronics, M.I.T.

2. Measurement of Probability Densities

For the measurement of the amplitude probability density of a random wave, we refer to Chapter 4, Sec. 2. As stated in that section, if the random amplitude ξ of the wave has the continuous range $(-a, b)$ and the probability density $P_\xi(x)$, we divide the range into small intervals each of length Δx. If we select an interval $(x_i, x_i + \Delta x)$ and if in a series of M independent trials there are N_i trials in which ξ is found in the interval, then $(N_i/M)/\Delta x$ is an estimate of $P_\xi(x)$ at the value of x selected, as illustrated by Figs. 3 and 4 in the section referred to. By

* A. B. Macnee, *An Electronic Differential Analyzer*, Technical Report No. 90, Research Laboratory of Electronics, M.I.T., 1948.

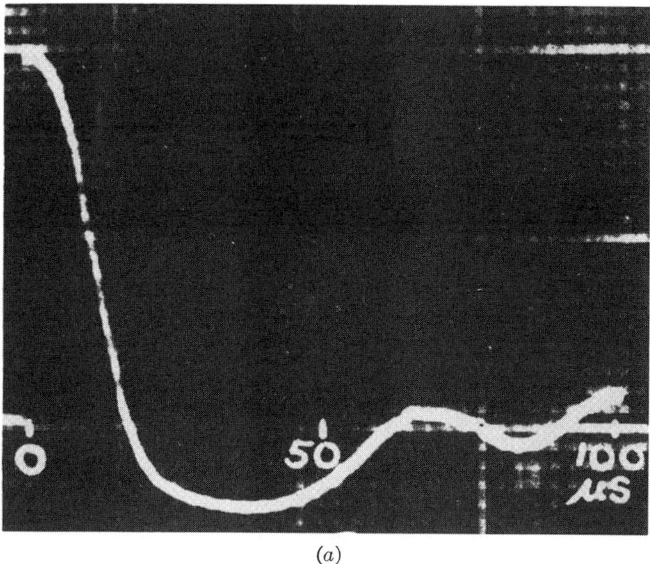

(a)

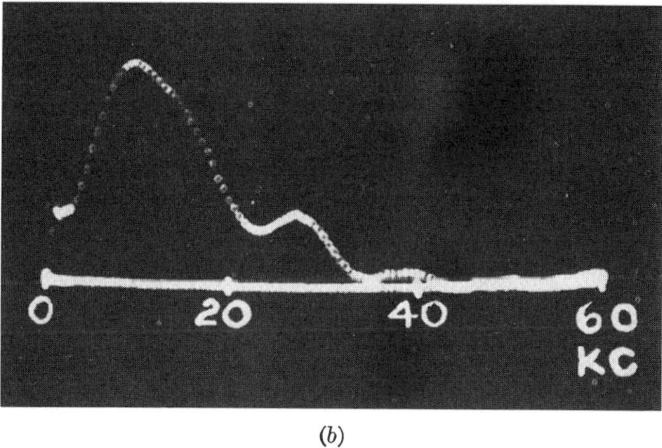

(b)

Fig. 5. (a) Autocorrelation curve of filtered noise from Type 884 gas tube measured by correlator. (b) Power density spectrum of the noise determined by differential analyzer. From *Application of Statistical Methods to Communication Problems*, Y. W. Lee, Technical Report No. 181, 1950, Research Laboratory of Electronics, M.I.T.

repeating the experiment for various values of x, an estimate of the entire probability density is obtained. In these statements we actually refer to a stationary ensemble $\{f(t)\}$, as shown in Fig. 6, whose amplitude

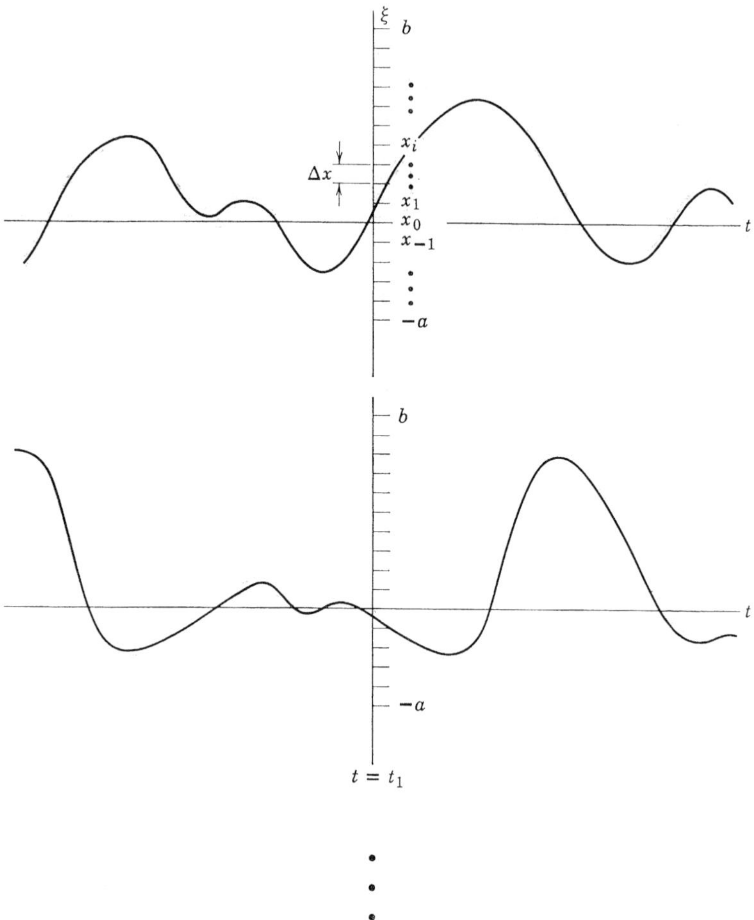

Fig. 6. Pertaining to the measurement of amplitude probability density of an ensemble.

at $t = t_1$ is ξ. The values of the series of trials are the values of the member functions corresponding to the trials at $t = t_1$.

Since actual measurement of a probability density is simpler on the basis of time averaging instead of ensemble averaging, we shall consider the relation between these two methods. At the point $x = x_i$, the probability that we wish to determine is

$$\mathcal{P}(x_i < \xi < x_i + \Delta x) \tag{6}$$

so that the probability density at this point is

$$\frac{1}{\Delta x}\,\mathcal{P}(x_i < \xi < x_i + \Delta x) \tag{7}$$

and the approximate probability density function is

$$P_\xi(x_i) \cong \frac{1}{\Delta x}\,\mathcal{P}(x_i < \xi < x_i + \Delta x) \quad i = \ldots, -2, -1, 0, 1, 2, \ldots \tag{8}$$

The smaller the interval Δx, the better is the approximation, but in practice Δx may be $1/50$ or $1/100$ of the total range of ξ.

Now let us consider a function v of ξ such that

$$v(\xi) = \begin{cases} 1 & \text{for } x_i < \xi < x_i + \Delta x \\ 0 & \text{elsewhere} \end{cases} \tag{9}$$

The mean value of $v(\xi)$ is

$$\begin{aligned} \overline{v(\xi)} &= \int_{-\infty}^{\infty} v(x) P_\xi(x)\,dx \\ &= \int_{x_i}^{x_i+\Delta x} P_\xi(x)\,dx = \mathcal{P}(x_i < \xi < x_i + \Delta x) \end{aligned} \tag{10}$$

Hence the approximate probability density function (8) is

$$P_\xi(x_i) \cong \frac{1}{\Delta x}\,\overline{v(\xi)} \tag{11}$$

This result states that, if a unit pulse $v(\xi)$ is generated whenever ξ falls in the interval $(x_i, x_i + \Delta x)$, the ensemble average of these unit pulses divided by Δx is the approximate value of the probability density of ξ at $x = x_i$.

By the ergodic hypothesis, the ensemble average in (11) may be replaced by a time average, as given by (23), Chapter 7. Hence

$$\overline{v(\xi)} = \overline{v[f_1(t)]} \tag{12}$$

and

$$P_\xi(x_i) \cong \frac{1}{\Delta x}\,\overline{v[f_1(t)]} \tag{13}$$

Here $f_1(t)$ is a member function of the ensemble, and

$$v[f_1(t)] = \begin{cases} 1 & \text{for } x_i < f_1(t) < x_i + \Delta x \\ 0 & \text{elsewhere} \end{cases} \tag{14}$$

The function $v[f_1(t)]$ can be generated with the aid of a level selector circuit. This circuit can be designed to give a constant output whenever the input amplitude is within a preset interval and a zero output otherwise. By integrating the output of the level selector circuit, we obtain the time average required in (13). Repeating the procedure for various values of x_i, we find that the measured probability density function takes the form shown in Fig. 7. Machines have been designed to record

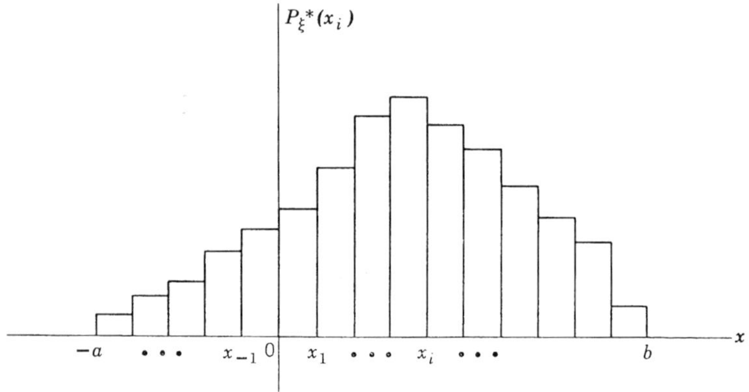

Fig. 7. Form of an experimental probability density.

the measurement automatically. It is to be noted that the time average in (13) is on the basis of the infinite time interval whereas physical measurements are necessarily of finite duration.

In the measurement of the second probability density, we consider the ensemble in Fig. 8 in which ξ_1 is the amplitude of the ensemble members at $t = t_1$ and ξ_2 is the amplitude of the corresponding ensemble members at $t = t_1 + \tau$. Assuming that the amplitude has a continuous range $(-a, b)$, we divide the range into small intervals of length Δx. The points of division are $\ldots, x_{1,-1}, 0, x_{11}, \ldots, x_{1i}, \ldots$ on the x_1-axis for ξ_1 and $\ldots, x_{2,-1}, 0, x_{22}, \ldots, x_{2j}, \ldots$ on the x_2-axis for ξ_2. The joint probability density to be measured is $P_{\xi_1\xi_2}(x_1, x_2; \tau)$; but, since

$$P_{\xi_1\xi_2}(x_1, x_2; \tau) = P_{\xi_1}(x_1)P_{\xi_2|\xi_1}(x_2|x_1; \tau) \tag{15}$$

and the measurement of $P_{\xi_1}(x_1)$ has been discussed, we need consider

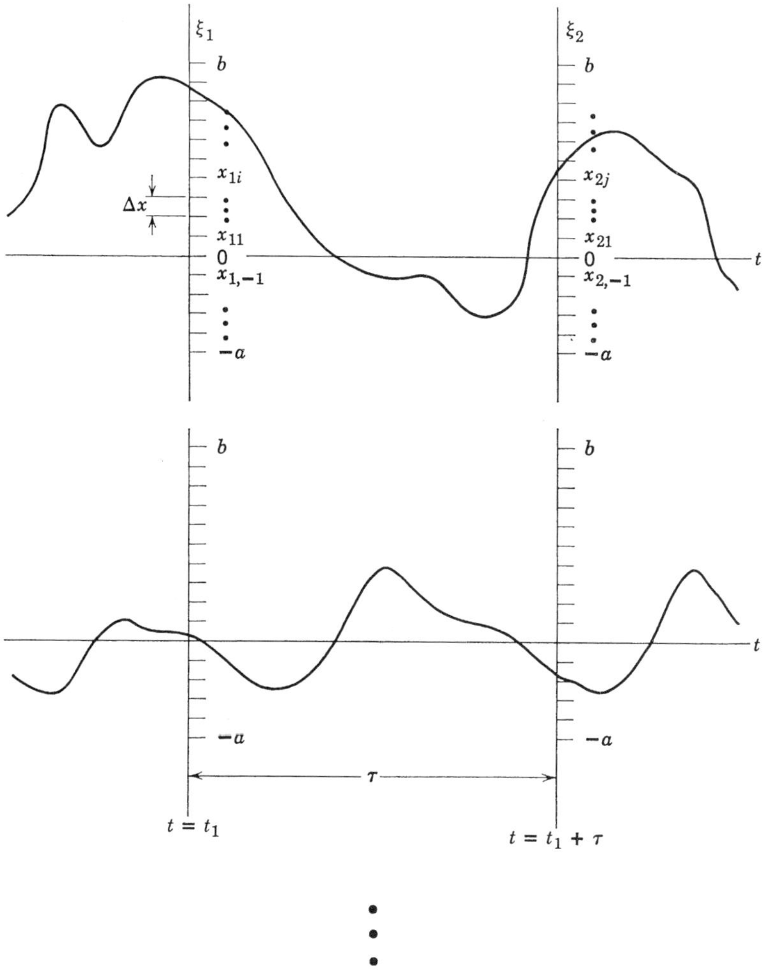

Fig. 8. Pertaining to the measurement of conditional amplitude probability density of an ensemble.

only the conditional probability density

$$P_{\xi_2|\xi_1}(x_2|x_1;\tau) \tag{16}$$

In measurement we select an interval $(x_{1i}, x_{1i} + \Delta x)$ for ξ_1 and an interval $(x_{2j}, x_{2j} + \Delta x)$ for ξ_2. For these intervals we wish to determine the conditional probability

$$\mathcal{P}(x_{2j} < \xi_2 < x_{2j} + \Delta x | x_{1i} < \xi_1 < x_{1i} + \Delta x; \tau) \tag{17}$$

so that the conditional probability density for $x_1 = x_{1i}$ and $x_2 = x_{2j}$ is

$$\frac{1}{\Delta x}\,\mathcal{P}(x_{2j} < \xi_2 < x_{2j} + \Delta x \,|\, x_{1i} < \xi_1 < x_{1i} + \Delta x;\, \tau) \tag{18}$$

and the approximate conditional probability density function is

$$P_{\xi 2|\xi 1}(x_{2j}|x_{1i};\,\tau) \cong \frac{1}{\Delta x}\,\mathcal{P}(x_{2j} < \xi_2 < x_{2j} + \Delta x \,|\, x_{1i} < \xi_1 < x_{1i} + \Delta x;\, \tau) \tag{19}$$

$$\text{for } i = \ldots, -1, 0, 1, \ldots$$
$$j = \ldots, -1, 0, 1, \ldots$$

For the method of measurement we introduce the random variable

$$v(\xi_2) = \begin{cases} 1 & \text{for } x_{2j} < \xi_2 < x_{2j} + \Delta x \\ 0 & \text{elsewhere} \end{cases} \tag{20}$$

We are concerned with the situation where, for any particular member function, ξ_1 is known to have occurred in the selected interval $(x_{1i}, x_{1i} + \Delta x)$, and we observe ξ_2 (the amplitude of the same function at time τ later) for its occurrence in the other selected interval $(x_{2j}, x_{2j} + \Delta x)$. Now we write

$$v(\xi_2)\,|\,x_{1i} < \xi_1 < x_{1i} + \Delta x;\, \tau \tag{21}$$

as a random variable that, given $x_{1i} < \xi_1 < x_{1i} + \Delta x$, has the value 1 if, at time τ later, $x_{2j} < \xi_2 < x_{2j} + \Delta x$, and the value zero otherwise. We are interested in the mean of this variable. For its calculation we need

$$\mathcal{P}(x_2 < \xi_2 < x_2 + dx_2 \,|\, x_{1i} < \xi_1 < x_{1i} + \Delta x;\, \tau) \tag{22}$$

which can be expressed as

$$\mathcal{P}(x_2 < \xi_2 < x_2 + dx_2 \,|\, x_{1i} < \xi_1 < x_{1i} + \Delta x;\, \tau)$$
$$= \frac{\mathcal{P}(x_{1i} < \xi_1 < x_{1i} + \Delta x,\, x_2 < \xi_2 < x_2 + dx_2;\, \tau)}{\mathcal{P}(x_{1i} < \xi_1 < x_{1i} + \Delta x)}$$
$$= \frac{dx_2 \displaystyle\int_{x_{1t}}^{x_{1t}+\Delta x} P_{\xi 1 \xi 2}(x_1, x_2;\, \tau)\, dx_1}{\displaystyle\int_{x_{1t}}^{x_{1t}+\Delta x} P_{\xi 1}(x_1)\, dx_1} \tag{23}$$

The mean of (21) is therefore

$$\overline{v(\xi_2)}\,|\,x_{1i} < \xi_1 < x_{1i} + \Delta x; \tau$$

$$= \frac{\int_{-\infty}^{\infty} v(x_2)\,dx_2 \int_{x_{1i}}^{x_{1i}+\Delta x} P_{\xi_1\xi_2}(x_1, x_2; \tau)\,dx_1}{\int_{x_{1i}}^{x_{1i}+\Delta x} P_{\xi_1}(x_1)\,dx_1}$$

$$= \frac{\int_{x_{2j}}^{x_{2j}+\Delta x} dx_2 \int_{x_{1i}}^{x_{1i}+\Delta x} P_{\xi_1\xi_2}(x_1, x_2; \tau)\,dx_1}{\int_{x_{1i}}^{x_{1i}+\Delta x} P_{\xi_1}(x_1)\,dx_1}$$

$$= \frac{\mathcal{P}(x_{1i} < \xi_1 < x_{1i} + \Delta x,\ x_{2j} < \xi_2 < x_{2j} + \Delta x; \tau)}{\mathcal{P}(x_{1i} < \xi_1 < x_{1i} + \Delta x)}$$

$$= \mathcal{P}(x_{2j} < \xi_2 < x_{2j} + \Delta x\,|\,x_{1i} < \xi_1 < x_{1i} + \Delta x; \tau) \qquad (24)$$

Referring back to (19), we find that, with the result just obtained,

$$P_{\xi_2|\xi_1}(x_{2j}|x_{1i}; \tau) \cong \frac{1}{\Delta x}\,\overline{v(\xi_2)}\,|\,x_{1i} < \xi_1 < x_{1i} + \Delta x; \tau \qquad (25)$$

This expression states that the approximate conditional probability density for $\xi_2 = x_{2j}$, given that $\xi_1 = x_{1i}$ at time τ earlier, is $1/\Delta x$ times the ensemble average of the unit pulses that are generated by ξ_2, given that $x_{1i} < \xi_1 < x_{1i} + \Delta x$ at time τ earlier.

To express (25) as a time average by applying the ergodic hypothesis, we write

$$P_{\xi_2|\xi_1}(x_{2j}|x_{1i}; \tau) \cong \frac{1}{\Delta x}\,\overline{v[f_1(t + \tau)]}\,|\,x_{1i} < f_1(t) < x_{1i} + \Delta x \qquad (26)$$

in which $f_1(t)$ is a member of the ensemble, and

$$v[f_1(t + \tau)]\,|\,x_{1i} < f_1(t) < x_{1i} + \Delta x$$

$$= \begin{cases} 1 & \text{for } x_{2j} < f_1(t + \tau) < x_{2j} + \Delta x \\ & \quad \text{given that } x_{1i} < f_1(t) < x_{1i} + \Delta x \\ 0 & \text{otherwise} \end{cases} \qquad (27)$$

Various types of electronic circuits can be designed to yield a unity output whenever the first input $f_1(t)$ lies in the preset level $(x_{1i}, x_{1i} + \Delta x)$, and the second input $f_1(t + \tau)$, which is the first input at time τ later, lies in the preset interval $(x_{2j}, x_{2j} + \Delta x)$. The integral of the output over those times in which $x_{1i} < f_1(t) < x_{1i} + \Delta x$ is proportional to the time average required in (26). As noted before, the theoretical average is over an infinite interval, but the actual integration time depends upon the required accuracy and physical limitations. Repetition of the measurement for a series of values for each of the variables $(x_{1i}, x_{2j}; \tau)$ gives the experimental conditional probability density in the form of a family of surfaces.

REFERENCES

1. Lee, Y. W., T. P. Cheatham, Jr., and J. B. Wiesner, "Application of Correlation Analysis to the Detection of Periodic Signals in Noise," *Proc. IRE*, **38,** 1165–1171 (1950).
2. Singleton, H. E., "A Digital Electronic Correlator," *Proc. IRE*, **38,** No. 12, December 1950.
3. Reintjes, J. F., "An Analogue Electronic Correlator," *Proc. NEC*, **7,** 390–400 (1951).
4. Cowley, P., R. Fano, and B. Basore, *Short-Time Correlator for Speech Waves*, Technical Report No. 174, Research Laboratory of Electronics, M.I.T., 1951.
5. Bell, H., and V. Rideout, "A High-Speed Correlator," *Proc. IRE*, **EC-3,** No. 2, 30–36, June 1954.
6. Peboy-Peyroula, J. C., and P. Simon, "Direct-Reading Electronic Correlator," *Jour. Phys. Radium*, **15,** suppl. to No. 5, 101A–108A, May 1954.
7. Revesz, G., "An Autocorrelogram Computer," *Jour. Sci. Instr.*, **31,** 406–410, November 1954.
8. Barlow, J. S., and R. M. Brown, *An Analog Correlator System for Brain Potentials*, Technical Report No. 300, Research Laboratory of Electronics, M.I.T., July 1955.
9. Furth, R., and D. K. C. MacDonald, "Statistical Analysis of Spontaneous Electrical Fluctuations," *Proc. Phys. Soc.*, **59,** 388–403 (1947).
10. Winter, D., "Investigation of Electro-Encephalic Components between 200 and 14,000 cps," S.M. Thesis, Department of Electrical Engineering, M.I.T., 1948.
11. Knudtzon, N., *Experimental Study of Statistical Characteristics of Filtered Random Noise*, Technical Report No. 115, Research Laboratory of Electronics, M.I.T., 1949.
12. Davenport, W. B., *A Study of Speech Probability Distributions*, Technical Report No. 148, Research Laboratory of Electronics, M.I.T., 1950.
13. Stewart, R., and D. Stebbins, "The Application of Electrical Counting to the Compilation of Frequency Distributions and Correlation Tables," *Trans. Amer. Geophys. Union*, **32,** 341–346 (1951).
14. King, A., *An Amplitude-Distribution Analyzer*, Report 3890, Naval Research Laboratory, Washington, D. C., 1951.
15. Stewart, R., and A. Kassander, "An Electronic Statistical Tabulator," *Proc. NEC*, **8,** 657–667 (1952).
16. Schrieber, W., "Probability Distributions of Television Signals," Ph.D. Thesis, Harvard University, December 1952.
17. Kretzmer, E., "Statistics of Television Signals," *Bell System Technical Jour.*, **31,** 751–763, July 1952.

18. Easter, B., "An Electronic Amplitude Probability Distribution Analyzer," S.M. Thesis, Department of Electrical Engineering, M.I.T., 1953.
19. Davis, H., and G. Cooper, "Direct-Reading Probability Distribution Meter," *Proc. NEC*, **10**, 358–366 (1954).
20. Capon, J., "Bounds to the Entropy of Television Signals," S.M. Thesis, Department of Electrical Engineering, M.I.T., 1955.
21. Stoddard, J., "The Measurement of Second Order Probability Distributions by Digital Means," S.M. Thesis, Department of Electrical Engineering, M.I.T., 1955.
22. Jordan, K. L., Jr., "A Digital Probability Density Analyzer," S.M. Thesis, Department of Electrical Engineering, M.I.T., 1956.
23. White, H. E., "An Analog Probability Density Analyzer," S.M. Thesis, Department of Electrical Engineering, M.I.T., 1957.

chapter 11

Sampling Theory

For the analysis of errors in the measurement of correlation functions and probability densities, it is necessary to introduce some elementary principles of sampling. In this chapter we shall consider independent discrete samples, dependent discrete samples, and continuous samples.

1. Variance of Sample Mean of Independent Discrete Samples

Let us consider a stationary random wave $f(t)$ shown in Fig. 1 where $z_1^{(1)}, z_2^{(1)}, z_3^{(1)}, \ldots, z_n^{(1)}$ are samples of the wave taken at intervals of length L.

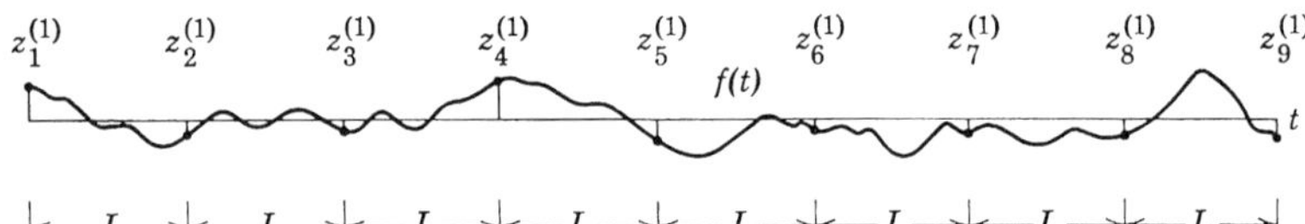

Fig. 1. Periodic sampling of a random function.

It is assumed that $f(t)$ contains no periodic components and that L is sufficiently long to ensure the independence of the samples. Let the mean of $f(t)$ be denoted by m and its variance by $\sigma_f{}^2$; that is,

$$\overline{f(t)} = m \tag{1}$$

$$\overline{f^2(t)} - m^2 = \sigma_f{}^2 \tag{2}$$

If n samples are taken as indicated and the mean value of the n samples is taken, we shall have a number that is known as a *sample mean of size n*. Thus, if the sample mean of size n for the first n samples is denoted by $\dot{z}^{(1)}$, we have

$$\dot{z}^{(1)} = \frac{1}{n}\,(z_1^{(1)} + z_2^{(1)} + \cdots + z_i^{(1)} + \cdots + z_n^{(1)}) \tag{3}$$

The superscript in brackets indicates the number of the trial. Additional sets of n samples each may be taken to form subsequent sample means. for example,

$$\dot{z}^{(2)} = \frac{1}{n}\,(z_1^{(2)} + z_2^{(2)} + \cdots + z_i^{(2)} + \cdots + z_n^{(2)}) \tag{4}$$

$$\dot{z}^{(3)} = \frac{1}{n}\,(z_1^{(3)} + z_2^{(3)} + \cdots + z_i^{(3)} + \cdots + z_n^{(3)}) \tag{5}$$

$$\vdots$$

As explained under measurement of correlation functions, the sections of the random wave may be grouped to form ensembles as shown in Fig. 2. The first n sections with the samples $z_1^{(1)}$, $z_2^{(1)}$, ..., $z_n^{(1)}$ and the

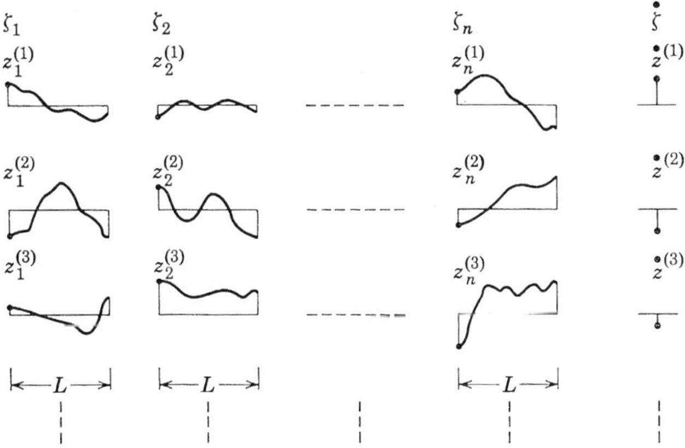

Fig. 2. Ensembles of samples and sample mean from the random wave of Fig. 1.

sample mean $\dot{z}^{(1)}$ are placed on the first line; the second n sections with the samples $z_1^{(2)}$, $z_2^{(2)}$, ..., $z_n^{(2)}$ and the sample mean $\dot{z}^{(2)}$ are placed on the second line; and so on. If the procedure is repeated indefinitely, we see that each column of sections of $f(t)$ is an ensemble. As long as the samples are within the duration L, our analysis is not in any way affected by the fact that the ensemble is of finite duration. If $f(t)$ is from a source under steady-state conditions, it is reasonable to assume that the ensembles we form are stationary and the probability density is the same for all the ensembles. Let us denote the amplitudes of the ensembles by $\zeta_1, \zeta_2, \ldots, \zeta_i, \ldots, \zeta_n$ as indicated, of which $z_1^{(1)}, z_1^{(2)}, \ldots; z_2^{(1)}, z_2^{(2)}, \ldots; z_i^{(1)}, z_i^{(2)}, \ldots; z_n^{(1)}, z_n^{(2)}, \ldots$ are their respective sample values. The sample mean of size n as a random variable is denoted by $\dot{\zeta}$, with $\dot{z}^{(1)}$,

$\dot{z}^{(2)}, \ldots$ as its corresponding values for the samples just given. Thus we have

$$\dot{\zeta} = \frac{1}{n}(\zeta_1 + \zeta_2 + \cdots + \zeta_i + \cdots + \zeta_n) \tag{6}$$

as the relation between the sample mean as a random variable and the samples as random variables.

By the ergodic hypothesis the mean of each sample as a random variable is equal to the mean of the sampled time function, and a similar relation holds for the mean square value; that is,

$$\overline{\zeta_i} = \overline{f(t)} = m \quad i = 1, 2, \ldots, n \tag{7}$$

and

$$\overline{\zeta_i^2} = \overline{f^2(t)} \qquad i = 1, 2, \ldots, n \tag{8}$$

Of particular interest in the study of sampling errors is the mean value of the sample mean, that is,

$$\bar{\dot{\zeta}} = \frac{1}{n}\overline{(\zeta_1 + \zeta_2 + \cdots + \zeta_i + \cdots + \zeta_n)} \tag{9}$$

Since this mean of a sum can be replaced by a sum of the individual means and since each mean has the value (7), we have

$$\begin{aligned} \bar{\dot{\zeta}} &= \frac{1}{n}(\bar{\zeta}_1 + \bar{\zeta}_2 + \cdots + \bar{\zeta}_i + \cdots + \bar{\zeta}_n) \\ &= \overline{f(t)} \end{aligned} \tag{10}$$

This result states that the mean of the sample mean of size n is equal to the mean of the sampled time function. The size does not affect the equality.

Another important quantity in the study of sampling errors is the variance of the sample mean. This variance is, by definition,

$$\sigma_{\dot{\zeta}}^{\ 2} = \overline{\dot{\zeta}^2} - \bar{\dot{\zeta}}^2 \tag{11}$$

Since $\bar{\dot{\zeta}}^2 = m^2$ according to (10), we proceed to evaluate $\overline{\dot{\zeta}^2}$. Taking the mean square of (6), we have

$$\begin{aligned} \overline{\dot{\zeta}^2} &= \frac{1}{n^2}\overline{(\zeta_1 + \zeta_2 + \cdots + \zeta_i + \cdots + \zeta_n)^2} \\ &= \frac{1}{n^2}\left[\sum_{i=1}^{n} \overline{\zeta_i^2} + \sum_{i=1}^{n} \sum_{\substack{j=1 \\ i \neq j}}^{n} \overline{\zeta_i \zeta_j}\right] \end{aligned} \tag{12}$$

In this expression

$$\overline{\zeta_i\zeta_j} = \bar{\zeta}_i\bar{\zeta}_j \qquad \text{for } i \neq j \tag{13}$$

since the variables involved are independent. By (7) we then have

$$\overline{\zeta_i\zeta_j} = m^2 \qquad \text{for } i \neq j \tag{14}$$

Substitution of (14), (8), and (2) into (12) gives

$$\begin{aligned}\overline{\dot{\zeta}^2} &= \frac{1}{n^2}[n\overline{f^2(t)} + (n^2 - n)m^2] \\ &= \frac{1}{n}\sigma_f^2 + m^2 \end{aligned} \tag{15}$$

Finally, in accordance with (11)

$$\sigma_{\dot{\zeta}}^2 = \frac{1}{n}\sigma_f^2 \tag{16}$$

This result states that the variance of the sample mean is equal to the variance of the sampled function divided by the sample size.

As a function of sample size, the variance of the sample mean is maximum when $n = 1$. Since it is convenient to compare the variance of the sample mean to its maximum value, we define

$$\sigma_{\dot{\zeta}}^{2*} = \frac{\sigma_{\dot{\zeta}}^2}{\sigma_f^2} \tag{17}$$

as the normalized variance of the sample mean. The normalized variance of the sample mean of independent samples is therefore

$$\sigma_{\dot{\zeta}}^{2*} = \frac{1}{n} \tag{18}$$

Returning to Figs. 1 and 2, we see that, if $f(t)$ has a periodic component or if $f(t)$ is a series of random pulses that occur at periodic intervals even though there is no periodic component, the periodic sampling procedure will not give us independent samples in each ensemble. Similarly, the variables $\zeta_1, \zeta_2, \ldots, \zeta_n$ representing the amplitudes of the ensembles will not be independent. In order that the preceding results be applicable to random waves with periodic components and random pulses that occur periodically, we must modify the sampling procedure.

If we select samples not at regular intervals but at such random intervals that all the ensembles formed from sections of $f(t)$ are stationary, we shall have the desired independence among the samples and among the ensembles. [Concerning stationary ensembles of periodic functions, see Chapter 6, Sec. 6.] To change the sampling interval L to a random variable, we might introduce a noise in the timing circuit to approximate the requirement. For theoretical discussion we shall include the method of random sampling so that for waves with periodic components and waves of periodic random pulses the results of this section are valid.

2. Variance of Sample Mean of Dependent Discrete Samples

The preceding discussion has the strong restriction that the samples taken from the random time function are independent. J. P. Costas * has extended the analysis to include sample means of dependent samples. The samples are obtained periodically as before. Since the samples need not be independent, we shall allow the existence of a periodic component in the random wave or the wave to be a series of periodic random pulses. We shall form ensembles with sections from the random wave $f(t)$ in the manner shown in Fig. 2 but relax the restriction that $\zeta_1, \zeta_2, \ldots, \zeta_n$ be independent so that the section length L can be any finite length. However, the members of each ensemble are assumed to be from independent sources as before so that in the formation of each ensemble, if $f(t)$ does not contain a periodic component and is not a series of periodic random pulses, we must have a sufficiently long interval between the member functions. In other words, the $n - 1$ sections between members are long enough to ensure the independence of the samples $z_1^{(1)}, z_1^{(2)}, \ldots$ in the first ensemble, the independence of $z_2^{(1)}, z_2^{(2)}, \ldots$ in the second ensemble, and so on. On the other hand, if $f(t)$ contains a periodic component or is a series of periodic random pulses, we must have a uniform probability density in the ensemble for the starting phase angles of the periodic component or the periodic reference points of the random pulses.

Equations (1) to (10) are applicable to the present situation, and we particularly note that the mean of the sample mean as given by (10) is the same whether or not $\zeta_1, \zeta_2, \ldots, \zeta_n$ are independent.

The dependence of the samples has an important effect in the evaluation of the mean square of the sample mean. For this evaluation we refer back to (12). In this expression the products $\overline{\zeta_i\zeta_j}$ are now products of dependent variables. Let us first note that $\overline{\zeta_i\zeta_j}$ are correlation functions. For instance

$$\left.\begin{aligned} \overline{\zeta_1\zeta_2} &= \varphi_{ff}(L) \\ \overline{\zeta_1\zeta_3} &= \varphi_{ff}(2L) \end{aligned}\right\} \qquad (19)$$

* J. P. Costas, *Periodic Sampling of Stationary Time Series*, Technical Report No. 156, Research Laboratory of Electronics, M.I.T., 1950.

and so on. They are crosscorrelation functions between the different ensembles with displacements in integral multiples of the length L. Since the ensembles are formed from the same function, crosscorrelation is in fact autocorrelation. We have as a general expression of (19)

$$\begin{aligned}\overline{\zeta_i\zeta_j} &= \varphi_{ff}[(i-j)L] \\ &= \varphi_{ff}(kL) \qquad \text{for } i \neq j \end{aligned} \tag{20}$$

In the sum

$$\sum_{\substack{i=1 \\ i\neq j}}^{n} \sum_{j=1}^{n} \overline{\zeta_i\zeta_j} \tag{21}$$

there are $n(n-1)$ terms; two terms of $\varphi_{ff}[(n-1)L]$, four terms of $\varphi_{ff}[(n-2)L]$, . . . ; and $2(n-1)$ terms of $\varphi_{ff}(L)$. Hence

$$\sum_{\substack{i=1 \\ i\neq j}}^{n} \sum_{j=1}^{n} \overline{\zeta_i\zeta_j} = \sum_{k=1}^{n-1} 2(n-k)\varphi_{ff}(kL) \tag{22}$$

and we know that

$$\sum_{k=1}^{n-1} 2(n-k) = n(n-1) \tag{23}$$

Substitution of (22) and

$$\overline{\zeta_i^{\,2}} = \sigma_f^{\,2} + m^2 \tag{24}$$

in (12) yields the mean square of the sample mean

$$\overline{\dot{\zeta}^2} = \frac{\sigma_f^{\,2} + m^2}{n} + \frac{1}{n^2}\sum_{k=1}^{n-1} 2(n-k)\varphi_{ff}(kL) \tag{25}$$

It follows immediately from (11) that the variance of the sample mean is

$$\sigma_{\dot{\zeta}}^{\,2} = \frac{\sigma_f^{\,2}}{n} + \frac{(1-n)m^2}{n} + \frac{2}{n^2}\sum_{k=1}^{n-1} (n-k)\varphi_{ff}(kL) \tag{26}$$

Application of (23) in (26) simplifies the variance to

$$\sigma_{\dot{\zeta}}^{\,2} = \frac{\sigma_f^{\,2}}{n} + \frac{2}{n^2}\sum_{k=1}^{n-1} (n-k)[\varphi_{ff}(kL) - m^2] \tag{27}$$

In this result we note that, if $f(t)$ is free of periodic components and L is sufficiently long to ensure the independence of the samples, then $\varphi_{ff}(kL) = m^2$, and (27) reduces to (16).

Expression (27) may be reduced to a simpler form. Since

$$\begin{aligned} \varphi_{ff}(0) &= \overline{f^2(t)} \\ &= \sigma_f^{\,2} + m^2 \end{aligned} \tag{28}$$

we have

$$\sigma_f^{\,2} = \varphi_{ff}(0) - m^2 \tag{29}$$

Furthermore, $\varphi_{ff}(kL) - m^2$ is an even function. Therefore we may reduce (27) to the following form in which k includes 0.

$$\sigma_{\dot{\zeta}}^{\,2} = \sum_{k=-(n-1)}^{n-1} \frac{n - |k|}{n^2} [\varphi_{ff}(kL) - m^2] \tag{30}$$

The maximum value of the variance is $\varphi_{ff}(0) - m^2$ when $n = 1$. As indicated by (29) the maximum value is the variance of the sampled function. Let us define a normalized autocorrelation function by

$$\begin{aligned} \overset{*}{\varphi}_{ff}(kL) &= \frac{\varphi_{ff}(kL) - m^2}{\varphi_{ff}(0) - m^2} \\ &= \frac{\varphi_{ff}(kL) - m^2}{\sigma_f^{\,2}} \end{aligned} \tag{31}$$

and a normalized variance of the sample mean of dependent samples by

$$\sigma_{\dot{\zeta}}^{2*} = \frac{\sigma_{\dot{\zeta}}^{\,2}}{\sigma_f^{\,2}} \tag{32}$$

It follows immediately that (30) in the normalized form is

$$\sigma_{\dot{\zeta}}^{2*} = \sum_{k=-(n-1)}^{n-1} \frac{n - |k|}{n^2} \overset{*}{\varphi}_{ff}(kL) \tag{33}$$

It is interesting to note that the autocorrelation function of the sampled function determines the variance of the sample mean. To illustrate the significance of this expression, a representative autocorrelation function for a random wave without a periodic component is shown in Fig. 3 with the various quantities involved in (33) indicated. The normalized variance of the sampled mean depends upon the values of the normalized autocorrelation function at discrete values of the argument 0, $\pm L$, $\pm 2L$, $\pm 3L$, $\ldots$, $\pm(n - 1)L$. These values of the autocorrelation function are given a series of decreasing weights from the origin to $kL = \pm(n - 1)L$. The weights are n/n^2, $(n - 1)/n^2$, $(n - 2)/n^2$, $\ldots$, $3/n^2$, $2/n^2$, $1/n^2$. The sum of the weighted ordinates of the normalized autocorrelation function at the discrete points is the variance of the sample mean. The weighting function $(n - |k|)/n^2$ as shown in the figure is an even discrete function which falls on straight lines running from n/n^2 at $k = 0$ to zero at $k = \pm n$.

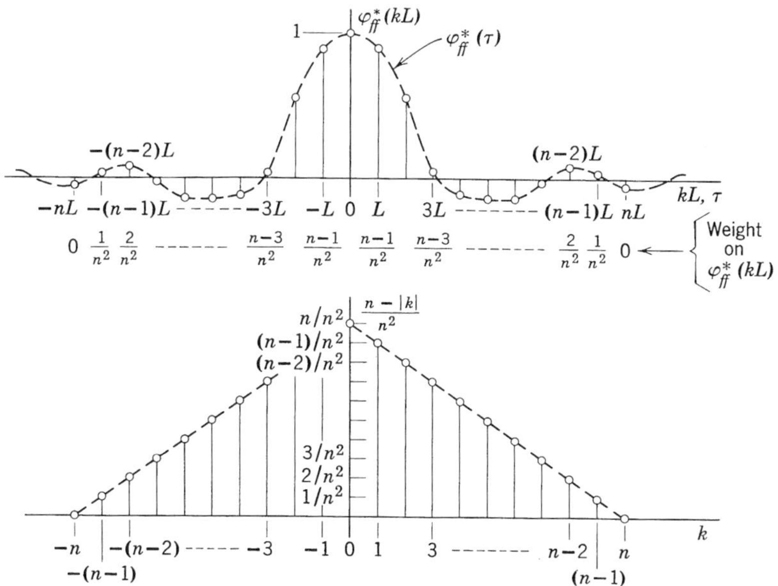

Fig. 3. Concerning the interpretation of equation (33).

It is clear from this figure that, if L is extended to a point where the autocorrelation has practically reached its final value, the variance of the sample mean will depend only on the autocorrelation function at the origin which is the mean square value of the sampled function. When L has this length, the samples are independent since the autocorrelation function at this point is the mean of the product of independent variables. We repeat that the sampled function is assumed to be free of periodic components. The variance of the sample mean under these conditions is in agreement with that obtained in the preceding section.

3. Variance of Sample Mean of Continuous Samples

We now consider the sample mean obtained by an integration over a finite time interval of the random wave. This sample mean is actually the limit of the sample mean of discrete samples as the interval between samples tends to zero. Hence the variance of the sample mean of continuous samples is the limiting form of (33). As we approach this form

$$\left.\begin{aligned} kL &\to \tau \qquad \text{a continuous variable} \\ \varphi_{ff}^*(kL) &\to \varphi_{ff}^*(\tau) \\ L &\to d\tau \\ nL &= T \end{aligned}\right\} \tag{34}$$

Here T is the total time for obtaining the samples for a sample mean. For continuous sampling, T is the total time of integration. We note further that the summation limits $k = -(n-1)$, $k = n - 1$ will tend to the integration limits $\tau = -T$, $\tau = T$. Substituting these values in (33) and changing it to an integral, we have

$$\sigma_{\zeta}^{2*} = \int_{-T}^{T} \frac{T - |\tau|}{T^2} \overset{*}{\varphi}_{ff}(\tau)\, d\tau \tag{35}$$

This is the normalized variance of the sample mean of continuous samples in an interval T. For comparison with discrete sampling, the quantities involved in (35) are shown in Fig. 4. The weighting function $(T - |\tau|)/T^2$ is an isosceles triangle in the interval $(-T, T)$ and zero elsewhere.

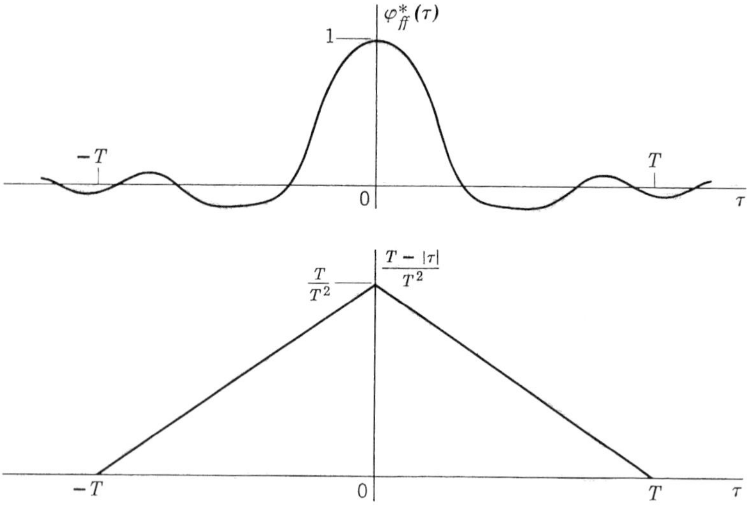

Fig. 4. Concerning the interpretation of equation (35).

4. An Example

As an illustrative example, let us consider the rectangular Poisson wave with zero mean. Suppose that by means of an ideal integrator the mean value of the wave is to be measured experimentally. If the Poisson wave is $f(t)$ and the integration interval is T, the integrator is to perform the integral

$$\frac{1}{T}\int_{0}^{T} f(t)\, dt \tag{36}$$

to give the mean value of $f(t)$ in the interval T. Repeated experiments with sections of $f(t)$, each of duration T, will yield various values in the neighborhood of the correct result, zero. We are interested in the variance of these values as a function of the integration time T.

The autocorrelation function of the sampled function is known to be

$$\varphi_{ff}(\tau) = E_m{}^2 e^{-2k|\tau|} \tag{37}$$

in which E_m is the amplitude of the wave and k is the average number of zero crossings per second. Accordingly, the normalized autocorrelation function is

$$\varphi_{ff}^{*}(\tau) = e^{-2k|\tau|} \tag{38}$$

By (35) we have

$$\sigma_{\zeta}^{2*} = \int_{-T}^{T} \frac{T - |\tau|}{T^2} e^{-2k|\tau|}\, d\tau$$

$$= \frac{2}{(2kT)^2}\left[2kT - 1 + e^{-2kT}\right] \tag{39}$$

The graph of this function is given in Fig. 5. For a particular numerical

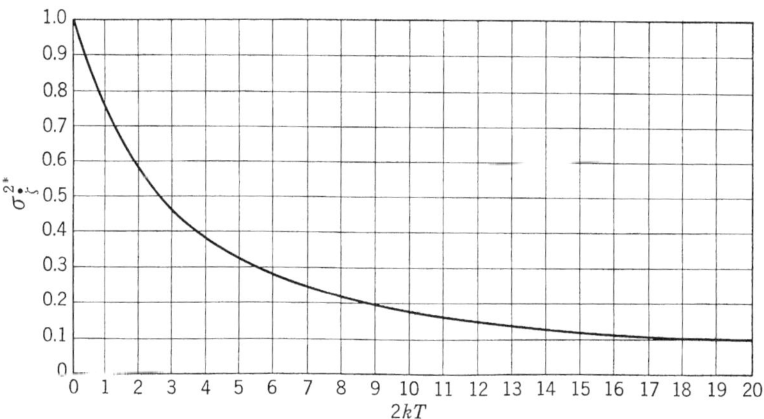

Fig. 5. Normalized variance of the sample mean in the measurement of the mean of a rectangular Poisson wave.

value, let us take $\sigma_{\zeta}^{2*} = 0.1$. To measure the mean value of $f(t)$ to this degree of accuracy statistically, the ideal integrator must have an integration time

$$T \cong \frac{20}{2k} = \frac{10}{k} \text{ seconds} \tag{40}$$

If $2/k$ is the average "period" of the Poisson wave, the integration time will be approximately 5 average periods.

chapter 12

Detection of a Periodic Signal in Noise by Correlation*

The extraction of a periodic wave which is masked by random noise is an important basic communication problem. The detection of a radar signal in the presence of noise is one such problem. This basic problem is found in other fields of scientific work. For example, in the study of brain waves one important problem is the search for periodic oscillations. In ocean wave analysis the separation of the cyclical component and the random component is the separation of the sea surface wave pattern into "sea swell" and "local sea." In many areas of geophysics including meteorology the problem of separation of periodic and random fluctuations is a major one. Correlation theory and techniques have been of considerable value in the solution of these problems. This chapter considers some elementary principles in the application of both autocorrelation and crosscorrelation to the detection of a periodic wave in noise.

1. Detection by Autocorrelation

Let $S(t)$ be a periodic signal with zero mean and $N(t)$ be a random noise without any periodic component and with zero mean. Let $f(t)$ be their sum; that is,

$$f(t) = S(t) + N(t) \tag{1}$$

The specification of zero mean for both components is a matter of convenience; it does not affect the theory. The autocorrelation function of $f(t)$ is

$$\begin{aligned}\varphi_{ff}(\tau) &= \overline{[S(t) + N(t)][S(t + \tau) + N(t + \tau)]} \\ &= \varphi_{SS}(\tau) + \varphi_{NN}(\tau) + \varphi_{SN}(\tau) + \varphi_{NS}(\tau)\end{aligned} \tag{2}$$

* Sections 1 through 4 of this chapter are based upon reference 1, Chapter 10.

We shall first consider the crosscorrelation functions

$$\left.\begin{aligned}\varphi_{SN}(\tau) &= \overline{S(t)N(t+\tau)}\\ \varphi_{NS}(\tau) &= \overline{N(t)S(t+\tau)}\end{aligned}\right\} \quad (3)$$

In considering these averages, we go from the time average, say $\overline{S(t)N(t+\tau)}$, to the equivalent ensemble average by the ergodic hypothesis so that

$$\overline{S(t)N(t+\tau)} = \overline{\xi\eta} \quad (4)$$

in which ξ is the amplitude of the stationary ensemble $\{S(t)\}$ and η is the amplitude of the stationary ensemble $\{N(t+\tau)\}$. Inasmuch as ξ and η are assumed independent,

$$\overline{\xi\eta} = \bar{\xi}\bar{\eta} \quad (5)$$

Now, going back to the equivalent time averages for ξ and η, we find

$$\left.\begin{aligned}\bar{\xi} &= \overline{S(t)}\\ \text{and} \qquad \bar{\eta} &= \overline{N(t+\tau)}\end{aligned}\right\} \quad (6)$$

Since $S(t)$ and $N(t)$ have zero means,

$$\left.\begin{aligned}\bar{\xi} &= 0\\ \text{and} \qquad \bar{\eta} &= 0\end{aligned}\right\} \quad (7)$$

Hence we write

$$\left.\begin{aligned}\varphi_{SN}(\tau) &= \overline{S(t)N(t+\tau)} = \overline{S(t)}\,\overline{N(t+\tau)}\\ &= 0\\ \text{and} \qquad \varphi_{NS}(\tau) &= \overline{N(t)S(t+\tau)} = \overline{N(t)}\,\overline{S(t+\tau)}\\ &= 0\end{aligned}\right\} \quad (8)$$

Consequently (2) is reduced to

$$\varphi_{ff}(\tau) = \varphi_{SS}(\tau) + \varphi_{NN}(\tau) \quad (9)$$

In Fig. 1 the components of this equation are shown in a representative manner. Theoretically $\varphi_{NN}(\tau)$ tends to zero as τ tends to infinity. However, in a practical situation it reaches zero at some sufficiently large value of τ, say τ_o. Beyond τ_o, $\varphi_{ff}(\tau)$ is practically the same as $\varphi_{SS}(\tau)$. The problem of detection may require the determination of whether or not a periodic component exists in $f(t)$. Frequently the answer to this question, as provided by the presence or absence of a

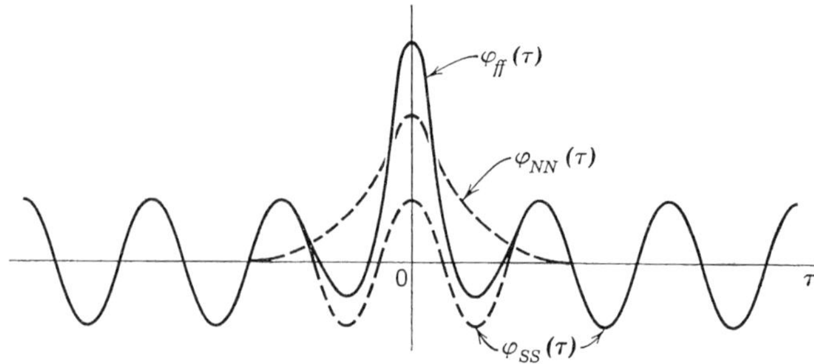

Fig. 1. Autocorrelation function of a periodic signal plus random noise.

periodic variation in $\varphi_{ff}(\tau)$ after $\tau = \tau_o$, is of great importance. On the other hand the entire autocorrelation may be wanted from which the components $\varphi_{SS}(\tau)$ and $\varphi_{NN}(\tau)$ are obtained. We shall apply sampling theory to calculate the errors in the solution of these problems with the aid of the electronic correlator; but, before going into this matter, we consider another aspect of the problem.

2. Detection by Crosscorrelation

Suppose that the period of the periodic signal is known or that it has been determined by autocorrelation. Then it would be desirable to apply crosscorrelation. Aside from other advantages, crosscorrelation is much more effective than autocorrelation in the extraction of a periodic signal in noise.

Let the disturbed signal $f(t)$ be the same as before, and let a local signal $g(t)$ be generated for crosscorrelation. The local signal is a periodic wave $C(t)$ with its period the same as that of $S(t)$. For the present the specific waveform of the local signal need not be given. We write for the local signal

$$g(t) = C(t) \tag{10}$$

and assume that $\overline{C(t)} = 0$.

The crosscorrelation of the disturbed signal and the local signal is

$$\begin{aligned} \varphi_{fg}(\tau) &= \overline{[S(t) + N(t)]C(t + \tau)} \\ &= \varphi_{SC}(\tau) + \varphi_{NC}(\tau) \end{aligned} \tag{11}$$

We have, similar to (8),

$$\varphi_{NC}(\tau) = \overline{N(t)C(t + \tau)} = \overline{N(t)}\ \overline{C(t + \tau)} = 0 \tag{12}$$

so that (11) is reduced to

$$\varphi_{fg}(\tau) = \varphi_{SC}(\tau) \tag{13}$$

This result states that the crosscorrelation between the disturbed input periodic signal and the local noise-free periodic signal with the same period is the crosscorrelation between the two signals. As we know, this crosscorrelation function is periodic. Figure 2 shows a represent-

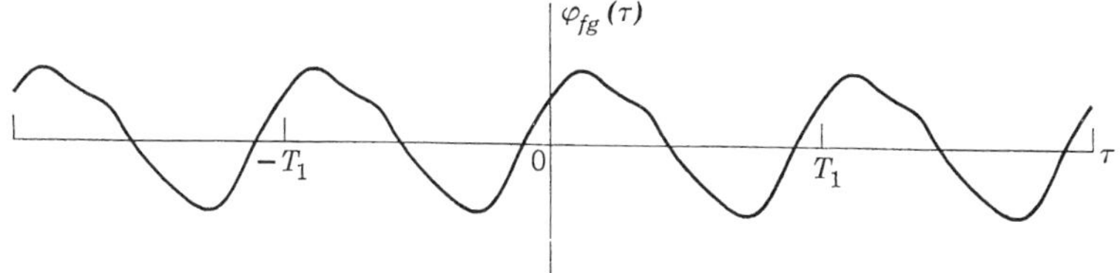

Fig. 2. Crosscorrelation function of a periodic signal plus random noise and another periodic signal of the same period.

ative result. Whereas in autocorrelation the wanted signal appears as an autocorrelation function with the noise autocorrelation as an additional term, in crosscorrelation the wanted signal appears as a crosscorrelation with the local known signal with no other terms. Because of the noise component in autocorrelation, measurement of the signal component starts at $\tau = \tau_o$ where it is practically free from the noise component. The absence of a noise component in crosscorrelation makes it possible to start measurement at any value of τ.

The greater effectiveness in detection by crosscorrelation as compared with autocorrelation is due to the fact that, whereas crosscorrelation has two terms in the general expression (11), autocorrelation has four terms in the corresponding expression (2). This point will be made clear in the calculations that follow.

3. Calculation of Gain in Detection by Autocorrelation

In the measurement of $\varphi_{ff}(\tau)$ given by (9), the input to the correlator is $f(t)$ as given by (1). The correlator selects pairs of samples $x_1^{(1)}, y_1^{(1)}$; $x_2^{(1)}, y_2^{(1)}; \ldots; x_n^{(1)}, y_n^{(1)}$ from $f(t)$ at intervals $L_1, L_2, \ldots, L_n$ as shown in Fig. 3. Here we assume that the samples are taken at random in-

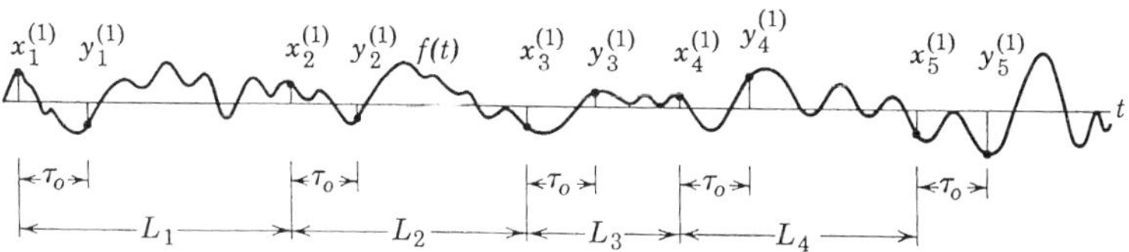

Fig. 3. Pertaining to the measurement of the autocorrelation function of Fig. 1 on the ensemble basis.

tervals instead of periodic intervals. In the manner described in Chapter 10 these pairs of values are multiplied and summed to determine a point on the correlation curve at the preset value of τ. The procedure is repeated for various values of τ. The first n pairs of values for $\tau = \tau_o$ result in the point $\dot{z}^{(1)}$ which is a sample mean of size n. The superscript in brackets denotes the number of the trial in the experiment. This point is indicated in Fig. 4 which shows the experimental correlation curve plotted by the electronic correlator. Specifically the point $\dot{z}^{(1)}$ is

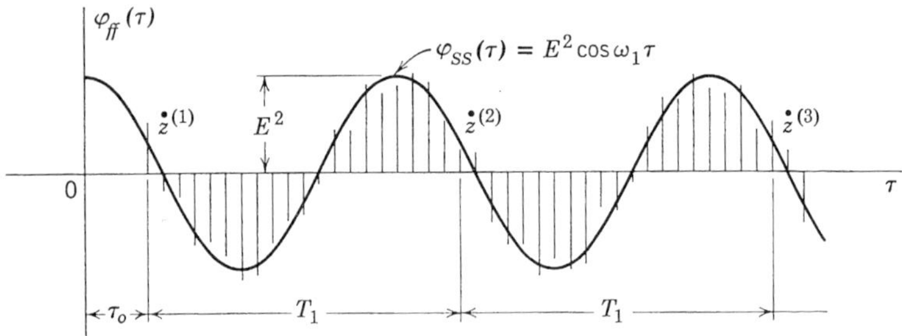

Fig. 4. Form of experimental autocorrelation curve of Fig. 1 plotted by the electronic correlator.

$$\dot{z}^{(1)} = \frac{1}{n} (x_1^{(1)}y_1^{(1)} + x_2^{(1)}y_2^{(1)} + \cdots + x_n^{(1)}y_n^{(1)}) \tag{14}$$

With $z = xy$, this equation is

$$\dot{z}^{(1)} = \frac{1}{n} (z_1^{(1)} + z_2^{(1)} + \cdots + z_n^{(1)}) \tag{15}$$

Let us now consider the sampled function. In (14) we have the x-samples from the function $S(t) + N(t)$ and the y-samples from the same function delayed by τ_o, that is, from $S(t - \tau_o) + N(t - \tau_o)$ with each pair of values separated by τ_o. The products are then taken and summed. As far as the sample mean is concerned, these steps are equivalent to those involved in (15). In this expression we imagine that $S(t) + N(t)$ and $S(t - \tau_o) + N(t - \tau_o)$ are first multiplied and then samples are taken from the product wave. These samples $z_1^{(1)}, z_2^{(1)}, \ldots, z_n^{(1)}$ are single samples instead of pairs. The timing of sampling is the same as that for the y-samples. Hence we say that the sampled function for the sample mean $\dot{z}^{(1)}$ is

$$s(t, \tau_o) = [S(t) + N(t)][S(t - \tau_o) + N(t - \tau_o)] \tag{16}$$

The sample mean for the next trial may be obtained by repeating the experiment with a different portion of the sampled function, or it may be obtained at $\tau = \tau_o + T_1$ instead of the same value of τ. Here T_1 is the period of the periodic component. To see this, let us consider the point $\dot{z}^{(2)}$ in Fig. 4. This point is separated from $\dot{z}^{(1)}$ by T_1. In accordance with the discussion on $\dot{z}^{(1)}$, the sample mean $\dot{z}^{(2)}$ is

$$\begin{aligned} \dot{z}^{(2)} &= \frac{1}{n}\,(x_1^{(2)}y_1^{(2)} + x_2^{(2)}y_2^{(2)} + \cdots + x_n^{(2)}y_n^{(2)}) \\ &= \frac{1}{n}\,(z_1^{(2)} + z_2^{(2)} + \cdots + z_n^{(2)}) \end{aligned} \tag{17}$$

and the sampled function for this sample mean is

$$s(t, T_1 + \tau_o) = [S(t) + N(t)][S(t - T_1 - \tau_o) + N(t - T_1 - \tau_o)] \tag{18}$$

Since τ_o is increased by exactly a period of $S(t)$, then as far as the periodic component is concerned there is no difference in amplitude between the $S(t - \tau_o)$ in (16) and $S(t - T_1 - \tau_o)$ in (18). Furthermore, the additional delay T_1 on $N(t)$ does not change the mean or the mean square of (16). We recall that τ_o is already a sufficiently long delay to ensure independence of $N(t)$ and $N(t - \tau_o)$. Hence an additional delay T_1 on $N(t)$ has no effect on the independence. In other words the variances of (16) and (18) are the same. By similar reasoning the points $\dot{z}^{(3)}$, $\dot{z}^{(4)}$, . . ., each a period T_1 from the other, are sample means from functions having the same variance. Inasmuch as the variance of the sample mean depends upon only the variance of the sampled function, we may proceed to calculate the variance of the sample means $\dot{z}^{(1)}$, $\dot{z}^{(2)}$, $\dot{z}^{(3)}$, . . . by finding the variance of (16).

As an illustration, let the periodic signal be a sinusoid; that is, let

$$S(t) = E_m \sin(\omega_1 t + \theta) \tag{19}$$

with the rms value $E = E_m/\sqrt{2}$, and let the variance of the noise $N(t)$ be $\sigma_N{}^2$. Since $\overline{N(t)} = 0$, we have $\overline{N^2(t)} = \sigma_N{}^2$.

If $\dot{\zeta}$ is the sample mean as a random variable, which is the measured value of the autocorrelation $\varphi_{ff}(\tau)$ at $\tau = \tau_o$, or at $\tau = \tau_o + kT_1$ ($k = 1, 2, \ldots$), on the basis of random sampling, then its variance is

$$\sigma_{\dot{\zeta}}{}^2 = \frac{1}{n}\,\sigma_s{}^2 \tag{20}$$

where $\sigma_s{}^2$ is the variance of the sampled function $s(t, \tau_o)$. Since

$$\sigma_s{}^2 = \overline{s^2(t, \tau_o)} - \overline{s(t, \tau_o)}^2 \tag{21}$$

we shall evaluate $\overline{s(t, \tau_o)}$ and then $\overline{s^2(t, \tau_o)}$.

We have, from (16),

$$\overline{s(t, \tau_o)} = \overline{[S(t) + N(t)][S(t - \tau_o) + N(t - \tau_o)]}$$
$$= \varphi_{SS}(\tau_o) + \varphi_{NN}(\tau_o) + \varphi_{NS}(\tau_o) + \varphi_{SN}(\tau_o) \qquad (22)$$

As in (8), $\varphi_{NS}(\tau_o) = 0$, $\varphi_{SN}(\tau_o) = 0$. Since $\overline{N(t)} = 0$, and $\overline{N(t)N(t - \tau_o)} = \overline{N(t)}\,\overline{N(t - \tau_o)}$ (practically), we have $\varphi_{NN}(\tau_o) = 0$ (practically). Therefore

$$\overline{s(t, \tau_o)} = \varphi_{SS}(\tau_o) \qquad (23)$$

Taking the sinusoid as the signal, we have

$$\overline{s(t, \tau_o)} = E^2 \cos \omega_1 \tau_o \qquad (24)$$

For the mean square of $s(t, \tau_o)$ we have

$$\begin{aligned}\overline{s^2(t, \tau_o)} &= \overline{[S(t) + N(t)]^2[S(t - \tau_o) + N(t - \tau_o)]^2} \\ &= \overline{S^2(t)S^2(t - \tau_o)} + \overline{N^2(t)N^2(t - \tau_o)} + \overline{S^2(t)N^2(t - \tau_o)} \\ &\quad + \overline{N^2(t)S^2(t - \tau_o)} + \overline{2S^2(t)S(t - \tau_o)N(t - \tau_o)} \\ &\quad + \overline{2N^2(t)S(t - \tau_o)N(t - \tau_o)} + \overline{4S(t)N(t)S(t - \tau_o)N(t - \tau_o)} \\ &\quad + \overline{2S^2(t - \tau_o)S(t)N(t)} + \overline{2N^2(t - \tau_o)S(t)N(t)} \qquad (25)\end{aligned}$$

The first term on the right-hand side is

$$\begin{aligned}\overline{S^2(t)S^2(t - \tau_o)} &= \frac{1}{T_1}\int_0^{T_1} E_m{}^2 \sin^2 \omega_1 t\, E_m{}^2 \sin^2 \omega_1(t - \tau_o)\, dt \\ &= \frac{E^4}{2} + E^4 \cos^2 \omega_1 \tau_o \qquad (26)\end{aligned}$$

the second term is

$$\begin{aligned}\overline{N^2(t)N^2(t - \tau_o)} &= \overline{N^2(t)}\;\overline{N^2(t - \tau_o)} = \sigma_N{}^2\sigma_N{}^2 \\ &= \sigma_N{}^4 \qquad (27)\end{aligned}$$

the third term is

$$\overline{S^2(t)N^2(t - \tau_o)} = \overline{S^2(t)}\;\overline{N^2(t - \tau_o)} = E^2\sigma_N{}^2 \qquad (28)$$

and the fourth term is

$$\overline{N^2(t)S^2(t - \tau_o)} = \overline{N^2(t)}\,\overline{S^2(t - \tau_o)} = \sigma_N{}^2E^2 \qquad (29)$$

The remaining terms vanish. For example,

$$\overline{S^2(t)S(t - \tau_o)N(t - \tau_o)} = \overline{S^2(t)S(t - \tau_o)}\ \overline{N(t - \tau_o)} = 0 \qquad (30)$$

Collecting terms, we have

$$\overline{s^2(t, \tau_o)} = \frac{E^4}{2} + E^4 \cos^2 \omega_1\tau_o + \sigma_N{}^4 + 2E^2\sigma_N{}^2 \qquad (31)$$

and (21) becomes

$$\sigma_s{}^2 = \frac{E^4}{2} + 2E^2\sigma_N{}^2 + \sigma_N{}^4 \qquad (32)$$

It is interesting to note that the variance of the sampled function is independent of τ_o. We now have, in accordance with (20),

$$\sigma_{\xi}{}^2 = \frac{1}{n}\left(\frac{E^4}{2} + 2E^2\sigma_N{}^2 + \sigma_N{}^4\right) \qquad (33)$$

The theoretical autocorrelation function for $\tau > \tau_o$ is practically the same as the autocorrelation function of the signal, which is

$$\varphi_{SS}(\tau) = E^2 \cos \omega_1\tau \qquad (34)$$

as indicated by the solid curve in Fig. 4. Clearly the variance of the sample mean as given by (33) is a measure of the amount by which the correlator fails to give the theoretical result because only a finite sample size is allowed in physical measurement. If the sample size n tends to infinity, of course $\sigma_{\xi}{}^2$ tends to zero, and the theoretical curve is obtained. We may consider this imperfection in measurement at the output of the correlator to be a "noise." Denoting the autocorrelator output noise by N_{oa}, in rms value, we have

$$N_{oa} = \sigma_{\xi} = \sqrt{\frac{1}{n}\left(\frac{E^4}{2} + 2E^2\sigma_N{}^2 + \sigma_N{}^4\right)} \qquad (35)$$

Since the ideal output signal is (34), the rms value of the signal is

$$S_{oa} = \frac{E^2}{\sqrt{2}} \qquad (36)$$

At the input of the autocorrelator, let N_i be the input noise in rms value and S_i be the input signal in rms value, and let

$$\rho_i = \frac{N_i}{S_i} \qquad (37)$$

be the input noise-to-signal ratio. With the sinusoid as the input signal and with the variance of the noise with a zero mean being σ_N^2, we have

$$\rho_i = \frac{\sigma_N}{E} \tag{38}$$

In addition, let R_i be the input signal-to-noise ratio in decibels so that

$$R_i = 20 \log_{10} \frac{1}{\rho_i} \quad \text{db} \tag{39}$$

At the output of the autocorrelator, let R_{oa} be the output signal-to-noise ratio in decibels; that is,

$$R_{oa} = 20 \log_{10} \frac{S_{oa}}{N_{oa}} \quad \text{db} \tag{40}$$

Substitution of (35) and (36) into (40) gives

$$R_{oa} = 10 \log_{10} \frac{nE^4}{E^4 + 4E^2\sigma_N^2 + 2\sigma_N^4} \quad \text{db} \tag{41}$$

In terms of the input noise-to-signal ratio, (41) is

$$R_{oa} = 10 \log_{10} \frac{n}{1 + 4\rho_i^2 + 2\rho_i^4} \quad \text{db} \tag{42}$$

To put this output signal-to-noise ratio in still another form, let n_a be the sample size in autocorrelation necessary for the output signal (in rms value) to be equal to the output noise (in rms value), at a given value of ρ_i. In other words, n_a is the sample size for $S_{oa} = N_{oa}$. This amounts to saying that

$$n_a = 1 + 4\rho_i^2 + 2\rho_i^4 \tag{43}$$

We now write (42) in terms of n_a to get

$$R_{oa} = 10 \log_{10} \frac{n}{n_a} \quad \text{db} \tag{44}$$

It seems desirable to state briefly that this expression, giving the autocorrelator output signal-to-noise ratio in decibels for $\tau > \tau_o$ as a function of the input noise-to-signal ratio and the sample size, has been established for an input consisting of a random noise plus a sinusoid as the signal, under the conditions of random sampling.

Formula (44) has been plotted in Fig. 5 as R_{oa} versus n for various values of n_a in the range 10 to 100,000. Since n_a is a function of the input noise-to-signal ratio ρ_i, we have plotted the relationship as given by (43) in Fig. 6. In both figures the symbols R_{oc} and n_c will be referred

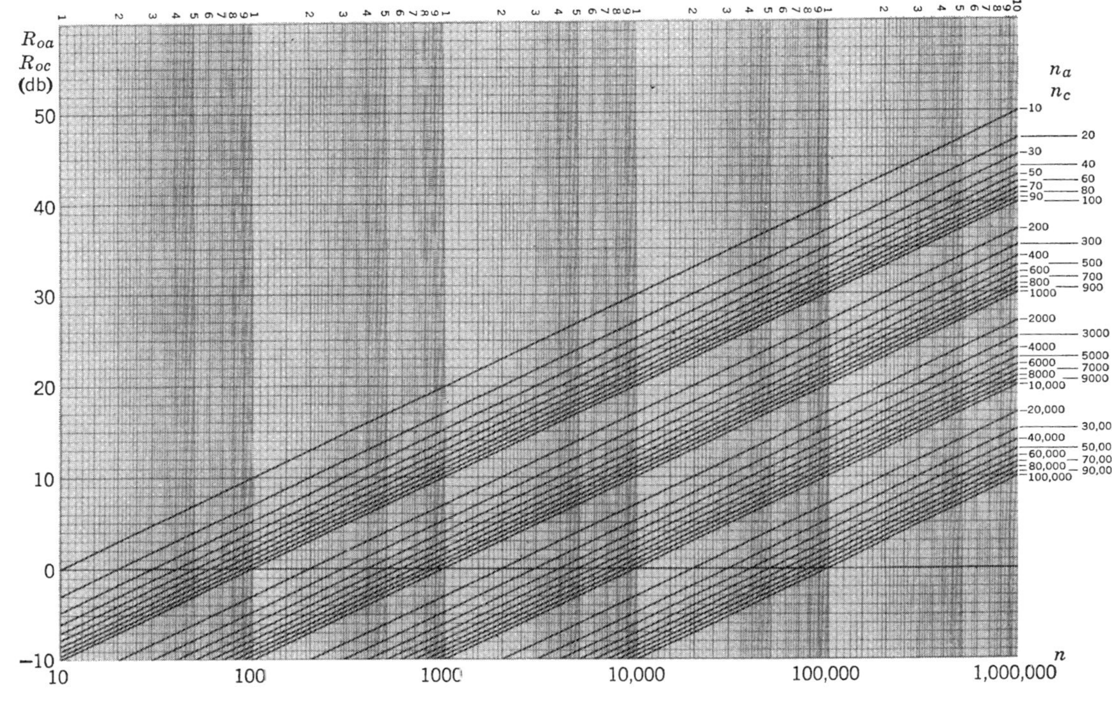

Fig. 5. Plot of equations (44) and (69).

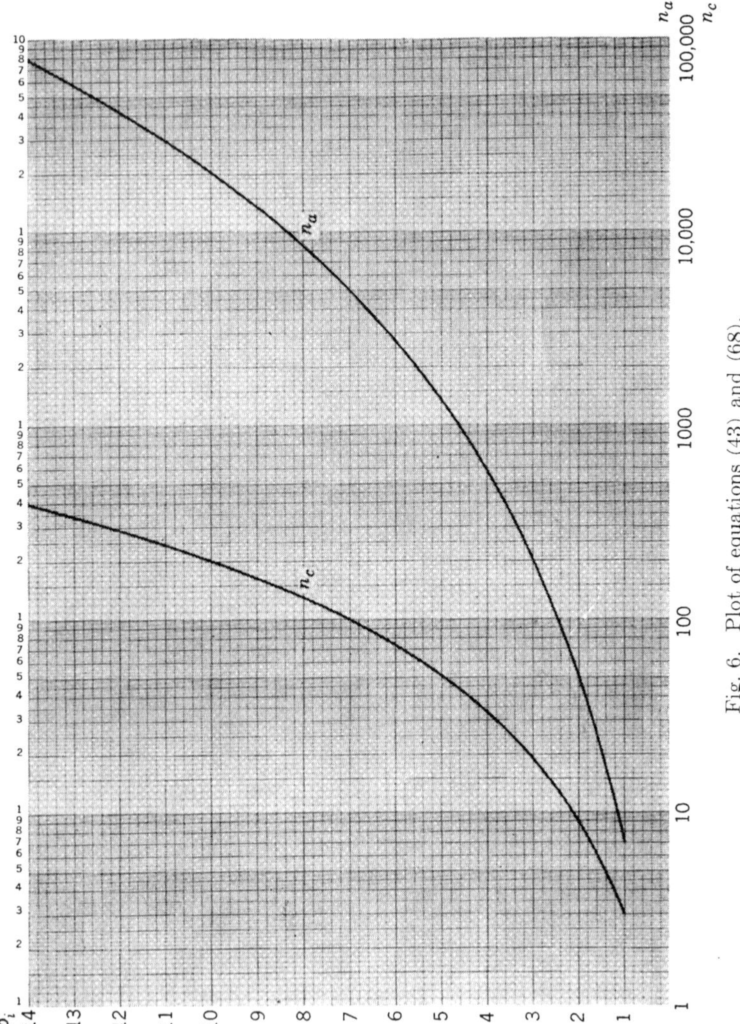

Fig. 6. Plot of equations (43) and (68).

to in the discussion on detection by crosscorrelation. As a numerical illustration, suppose that the disturbed sinusoid to be autocorrelated has a noise-to-signal ratio $\rho_i = 10$ which corresponds to an input signal-to-noise ratio of $R_i = -20$ db. From Fig. 6, or (43), we find $n_a = 20{,}400$. With this number for n_a we locate the R_{oa} versus n curve in Fig. 5. At $n = 20{,}400$, R_{oa} is zero; for $n > 20{,}400$, R_{oa} is positive; and, at $n = 50{,}000$, for example, $R_{oa} = 4$ db. Hence at this sample size the total gain obtained by autocorrelation is $20 + 4$ db or 24 db. This is a substantial improvement in signal-to-noise ratio. Although further gain can be obtained by increasing the sample size and theoretically the gain may be made as large as we please, errors in the electronic circuits, which have not been accounted for in the analysis, will play an important role as the sample size becomes sufficiently large.

It is interesting to note that in this illustration, if $n < 204$, $R_{oa} < -20$ db. This simply means that, if insufficient number of samples are taken, the output will be worse than the input in signal-to-noise ratio.

Solving (44) for n, we have

$$n = n_a e^{0.2303 R_{oa}} \tag{45}$$

which is convenient for determining the necessary sample size for given values of R_{oa} and n_a. The sample size may also be found from the curves of Fig. 5.

4. Calculation of Gain in Detection by Crosscorrelation

In the detection of a periodic signal in noise by crosscorrelation, $f(t)$ as shown in Fig. 7 is the same as before, being the sum of a random noise

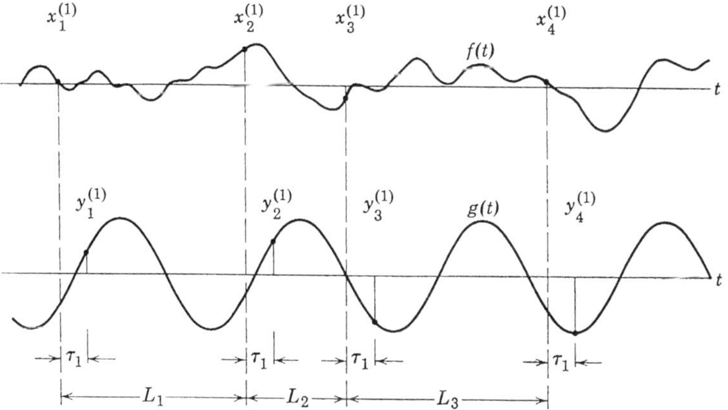

Fig. 7. Pertaining to the measurement of the crosscorrelation function of Fig. 2 on the ensemble basis.

and a periodic signal. The period T_1 of the signal is assumed to be given or determined by autocorrelation. The local periodic signal $g(t)$ has the period T_1. The correlator selects pairs of samples $x_1^{(1)}, y_1^{(1)}; x_2^{(1)}, y_2^{(1)}; \ldots; x_n^{(n)}, y_n^{(n)}$ to determine the point $\dot{z}^{(1)}$ on the crosscorrelation function $\varphi_{gf}(\tau)$ as shown in Fig. 8. This point is the sample mean

$$\begin{aligned} \dot{z}^{(1)} &= \frac{1}{n}\,(x_1^{(1)}y_1^{(1)} + x_2^{(1)}y_2^{(1)} + \cdots + x_n^{(1)}y_n^{(1)}) \\ &= \frac{1}{n}\,(z_1^{(1)} + z_2^{(1)} + \cdots + z_n^{(1)}) \end{aligned} \tag{46}$$

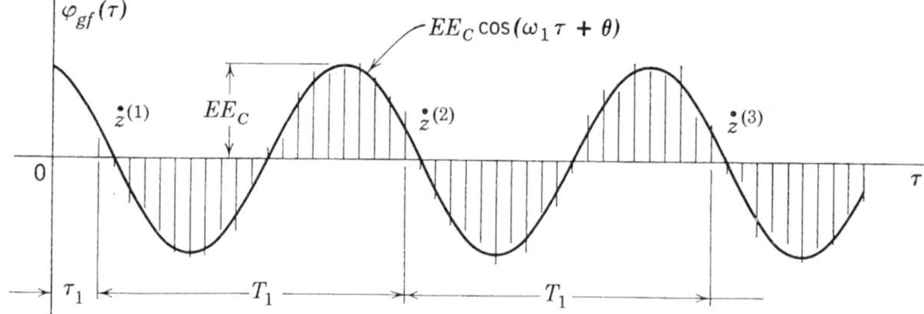

Fig. 8. Form of experimental crosscorrelation curve of Fig. 2 plotted by the electronic correlator.

The samples are taken at random intervals $L_1, L_2, \ldots$ in the manner described before. Similar to (16) for autocorrelation, the sampled function for the sample mean (46) is

$$s(t, \tau_1) = [S(t) + N(t)]C(t - \tau_1) \tag{47}$$

Other measured points on $\varphi_{gf}(\tau)$ at intervals of T_1 apart are sample means from

$$s(t, kT_1 + \tau_1) = [S(t) + N(t)]C(t - kT_1 - \tau_1) \qquad k = 1, 2, \ldots \tag{48}$$

Since

$$\overline{s(t, \tau_1)} = \overline{s(t, kT_1 + \tau_1)} \qquad k = 1, 2, \ldots \tag{49}$$

and

$$\overline{s^2(t, \tau_1)} = \overline{s^2(t, kT_1 + \tau_1)} \qquad k = 1, 2, \ldots \tag{50}$$

the variance of the sample mean $\dot{\zeta}$, of which $\dot{z}^{(1)}, \dot{z}^{(2)}, \ldots$ are some particular values, can be found from (47).

We shall assume that $S(t)$ and $C(t)$ are both sinusoids and let

$$S(t) = E_{m1} \sin(\omega_1 t + \theta_1) \tag{51}$$

$$C(t) = E_{m2} \sin(\omega_1 t + \theta_2) \tag{52}$$

whose rms values are $E = E_{m1}/\sqrt{2}$ and $E_C = E_{m2}/\sqrt{2}$ respectively. The noise $N(t)$ is the same as in autocorrelation. We have

$$\overline{s(t, \tau_1)} = \overline{S(t)C(t - \tau_1)} + \overline{N(t)C(t - \tau_1)} \tag{53}$$

Since $\overline{N(t)C(t - \tau_1)} = 0$,

$$\begin{aligned}\overline{s(t, \tau_1)} &= \frac{1}{T_1}\int_0^{T_1} E_{m1} \sin (\omega_1 t + \theta_1)\, E_{m2} \sin [\omega_1(t - \tau_1) + \theta_2]\, dt \\ &= EE_C \cos (\omega_1\tau_1 + \theta)\end{aligned} \tag{54}$$

where $\theta = \theta_1 - \theta_2$, and

$$\overline{s(t, \tau_1)}^2 = \tfrac{1}{2}E^2E_C{}^2[1 + \cos 2(\omega_1\tau_1 + \theta)] \tag{55}$$

The mean square of (47) is

$$\overline{s^2(t, \tau_1)} = \overline{S^2(t)C^2(t - \tau_1)} + \overline{N^2(t)C^2(t - \tau_1)} + \overline{2S(t)N(t)C^2(t - \tau_1)} \tag{56}$$

in which

$$\begin{aligned}\overline{S^2(t)C^2(t - \tau_1)} &= \frac{1}{T_1}\int_0^{T_1} E_{m1}^2 \sin^2 (\omega_1 t + \theta_1)\, E_{m2}^2 \\ &\qquad \times \sin^2 [\omega_1(t - \tau_1) + \theta_2]\, dt \\ &= E^2E_C{}^2[\tfrac{1}{2} \cos 2(\omega_1\tau_1 + \theta) + 1]\end{aligned} \tag{57}$$

$$\begin{aligned}\overline{N^2(t)C^2(t - \tau_1)} &= \overline{N^2(t)}\,\overline{C^2(t - \tau_1)} \\ &= \sigma_N{}^2E_C{}^2\end{aligned} \tag{58}$$

$$\overline{S(t)N(t)C^2(t - \tau_1)} = 0 \tag{59}$$

Hence

$$\overline{s^2(t, \tau_1)} = E^2E_C{}^2[\tfrac{1}{2} \cos 2(\omega_1\tau_1 + \theta) + 1] + \sigma_N{}^2E_C{}^2 \tag{60}$$

and the combination of (55) and (60) gives

$$\sigma_s{}^2 = E_C{}^2(\tfrac{1}{2}E^2 + \sigma_N{}^2) \tag{61}$$

It follows immediately that the variance of the sample mean ζ in detection by crosscorrelation is

$$\sigma_\zeta{}^2 = \frac{E_C{}^2}{n}\,(\tfrac{1}{2}E^2 + \sigma_N{}^2) \tag{62}$$

and it is the output "noise" of the crosscorrelator, as a mean square value. The rms value of the output noise is therefore

$$N_{oc} = \sigma_{\dot{\zeta}} = \frac{1}{\sqrt{n}} E_C \sqrt{\tfrac{1}{2}E^2 + \sigma_N{}^2} \tag{63}$$

Inasmuch as the ideal output signal of the crosscorrelator is

$$\begin{aligned} \varphi_{gf}(\tau) &= \overline{S(t)C(t-\tau)} \\ &= EE_C \cos(\omega_1\tau + \theta) \end{aligned} \tag{64}$$

the rms value of the ideal output signal is

$$S_{oc} = \frac{EE_C}{\sqrt{2}} \tag{65}$$

Denoting the output signal-to-noise ratio in decibels in crosscorrelation by R_{oc}, we have

$$\begin{aligned} R_{oc} &= 20 \log_{10} \frac{S_{oc}}{N_{oc}} \quad \text{db} \\ &= 10 \log_{10} \frac{nE^2}{E^2 + 2\sigma_N{}^2} \quad \text{db} \end{aligned} \tag{66}$$

In terms of the input noise-to-signal ratio ρ_i, (66) is

$$R_{oc} = 10 \log_{10} \frac{n}{1 + 2\rho_i{}^2} \quad \text{db} \tag{67}$$

This result, as we know, holds only for random sampling and for sinusoidal input and local signals.

In obtaining (66), we note that the term E_C drops out of the ratio by cancellation, showing that R_{oc} is independent of the strength of the local signal.

Let n_c be the sample size for $S_{oc} = N_{oc}$. We then have

$$n_c = 1 + 2\rho_i{}^2 \tag{68}$$

and (67) reads

$$R_{oc} = 10 \log_{10} \frac{n}{n_c} \quad \text{db} \tag{69}$$

from which we find that

$$n = n_c e^{0.2303 R_{oc}} \tag{70}$$

In Fig. 5 are a set of curves plotted in accordance with (69) as a function of n and n_c. The relation between n_c and ρ_i as given by (68) is plotted in Fig. 6. For comparison with detection by autocorrelation, we take $\rho_i = 10$ and $n = 50{,}000$ as before. From Fig. 6, or (68), we find $n_c = 201$; and, entering Fig. 5 with these values, we read $R_{oc} = 24$ db. The total gain is therefore $20 + 24$ db, or 44 db. For $\rho_i = 10$ and $n =$

50,000, crosscorrelation provides an additional 20 db gain over autocorrelation. This is a substantial difference.

The additional gain provided by crosscorrelation over autocorrelation for the same input noise-to-signal ratio and the same sample size is expressed by the difference of (42) and (67). Let G be the additional gain so that

$$G = R_{oc} - R_{oa} = 10 \log_{10} \frac{1 + 4\rho_i^2 + 2\rho_i^4}{1 + 2\rho_i^2} \text{ db} \tag{71}$$

This expression is plotted in Fig. 9. The additional gain may be traced

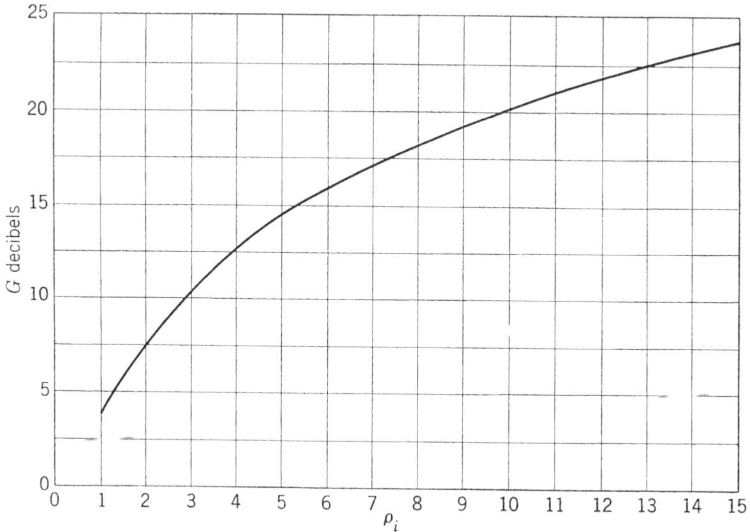

Fig. 9. Plot of equation (71).

to the difference between the sampled functions. The output noise is measured by the variance of the sample mean, which is the variance of the sampled function divided by the sample size. Now, the sampled function (16) in autocorrelation has a variance greater than that of (47) in crosscorrelation. Hence the output in autocorrelation has more noise.

5. Detection of Periodic Signal Waveform by Crosscorrelation with Periodic Impulses

In Sec. 5, Chapter 8 we have shown that the crosscorrelation between a function $f(t)$, having a random component and a periodic component, and a periodic unit-impulse function $u_p(t)$ with the same period as that of the periodic component yields, except for a constant factor, the periodic component without distortion in the waveform. Explicitly,*

$$\varphi_{uf}(\tau) = \frac{1}{T_1} f_p(\tau) \tag{72}$$

In this chapter $f_p(t) = S(t)$ so that

$$\varphi_{uf}(\tau) = \frac{1}{T_1} S(\tau) \tag{73}$$

The electronic correlator can be adjusted to perform the operations involved in $\varphi_{uf}(\tau)$ in a simple manner. In fact, since the crosscorrelation of a periodic unit-impulse function and $f(t)$ amounts to taking values of $f(t)$ at the moments of occurrence of the periodic unit impulses, over a sufficiently long duration, and then averaging these values for each value of the displacement τ, the electronic circuits are much simplified.

Figure 10 shows the input to the correlator $f(t) = S(t) + N(t)$ and a

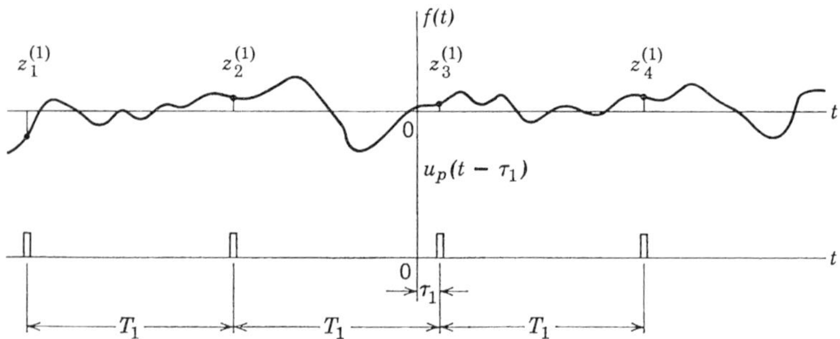

Fig. 10. Sampling a periodic signal plus noise.

series of periodic timing pulses that correspond to $u_p(t - \tau_1)$ in theory. The correlator selects values of $f(t)$ in accordance with the timing pulses, and these values are $z_1^{(1)}, z_2^{(1)}, z_3^{(1)}, \ldots, z_n^{(1)}$. The correlator sums these values, and the point $\dot{z}^{(1)}$ on the output graph is proportional to the crosscorrelation function at $\tau = \tau_1$. The procedure is repeated on subsequent sections of $f(t)$ with additional delays on the timing pulses in successive steps so that the output graph shows a series of points that outline the periodic wave $S(t)$ with the constant factor $1/T_1$ as shown in Fig. 11. Note that in this method of detection we allow $S(t)$ to have a general waveform, not necessarily sinusoidal. Furthermore, the result will be the original waveform distorted only by error in measurement.

* For simplicity we shall drop the letter p in the subscript of φ in equation (54), Chapter 8, Sec. 5.

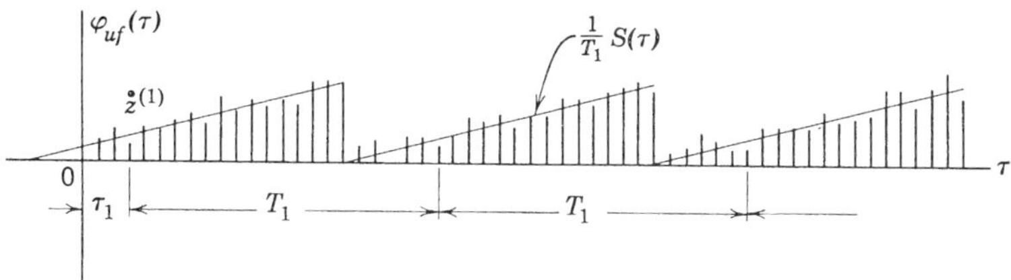

Fig. 11. Form of experimental curve in the detection of a periodic waveform in noise by sampling.

Of course, as in the preceding section, the period of the periodic component is assumed to be known.

For the determination of the signal-to-noise ratio it is only necessary to consider the repeated experiments at one point of the crosscorrelation function. Thus at $\tau = \tau_1$, repeated experiments will result in sample means $\dot{z}^{(1)}, \dot{z}^{(2)}, \dot{z}^{(3)}, \ldots$. We denote the sample mean as a random variable by $\dot{\zeta}$ as before and proceed to determine its variance. As mentioned before the variance of the sample mean at a fixed value of τ, which is what we shall consider here, is the same as the variance of the series of points at integral multiples of the period T_1.

When we consider the variance of the sample mean, one point of particular importance is that the samples of $f(t)$ are now taken at periodic intervals. The sampled function for the crosscorrelation is $f(t)$ itself. With respect to the random component alone in $f(t)$, we shall assume that T_1 is a sufficiently long duration to ensure the independence of the samples. With respect to the periodic component alone in $f(t)$, we see clearly that the samples must all have the same value since they are selected at integral multiples of the period. Therefore, the periodic component contributes nothing to the variance of the sample mean. In other words, if the random component is zero and there is only a periodic component, the sample mean that we consider must have zero variance. We conclude that we need only consider the variance of the random component $N(t)$. Therefore

$$\begin{aligned} \sigma_{\dot{\zeta}}^2 &= \frac{1}{n} \sigma_f^2 \\ &= \frac{1}{n} \sigma_N^2 \end{aligned} \tag{74}$$

which means that the output noise in rms value is $\sigma_N/\sqrt{n}$.

Let R_o be the output signal-to-noise ratio in decibels and E be the rms value of the input signal. Since in this method the ideal output is identical with the input signal, E is also the rms value of the ideal output.

Hence

$$R_o = 20 \log_{10} \frac{E}{\left(\dfrac{\sigma_N}{\sqrt{n}}\right)}$$

$$= 10 \log_{10} \frac{n}{\rho_i^2} \quad \text{db} \tag{75}$$

As before, the input signal-to-noise ratio is

$$R_i = 10 \log_{10} \frac{1}{\rho_i^2} \quad \text{db} \tag{76}$$

From (75) and (76) we find that the total gain in signal-to-noise ratio is

$$G_T = R_o - R_i$$

$$= 10 \log_{10} n \quad \text{db} \tag{77}$$

By comparison with (67) we see that the detection of a sinusoid by the method of periodic sampling at the period of the sinusoid yields a higher gain than that obtainable from crosscorrelating the input with a sinusoid. For example, if $n = 50{,}000$, $G_T = 47$ db which is slightly higher than the gain of 44 db in the other method.

To check the results, a periodic signal and a random noise in the relative magnitudes as shown in (*a*) and (*b*) of Fig. 12 were mixed, and a single sweep picture of the mixture is shown in (*c*). This mixture was fed into the correlator whose sampling rate was synchronized with the repetition rate of the signal. The output of the correlator (*d*) being presented by a recording meter is in the form of a set of lines whose envelope is the signal. For comparison a much enlarged input signal (*e*) is shown. The sample size was 14,000 so that according to (77) the total gain is 41 db. Figure 13 illustrates the detection of a triangular waveform in the same manner.

The method and technique that we have developed, when applied to a small radar set as a test for possibility of practical application, resulted in the presentations shown in Figs. 14 and 15. In these figures the A-scope presentations are compared with the corresponding presentations from the correlator which performed a crosscorrelation of the radar signal with unit impulses. These results show that the increase in clarity in the correlated output is undoubtedly appreciable. The sample size for both experiments was 7000. In both experiments the radar was directed toward the same object, but in the second of these experiments (Fig. 15) the radar signal power was reduced to such an extent that the A-scope showed no sign of an echo.

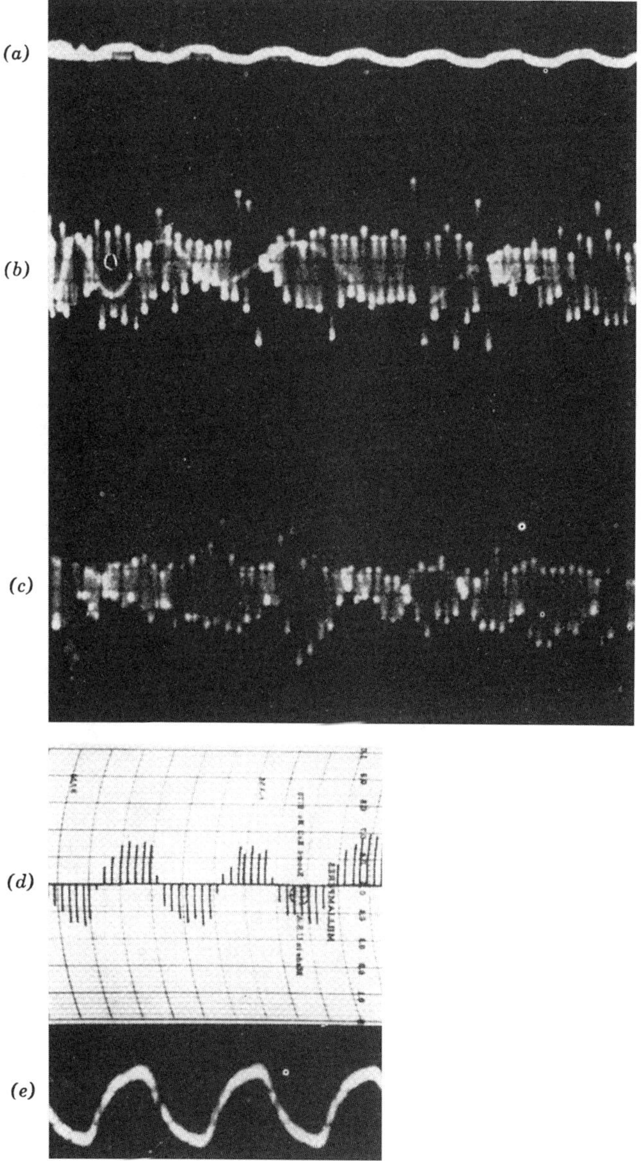

Fig. 12. Waveform detection by correlation. (*a*) Signal, (*b*) noise, (*c*) signal plus noise, (*d*) correlator output, (*e*) enlarged signal for comparison with output. From *Application of Statistical Methods to Communication Problems*, Y. W. Lee, Technical Report No. 181, 1950, Research Laboratory of Electronics, M.I.T.

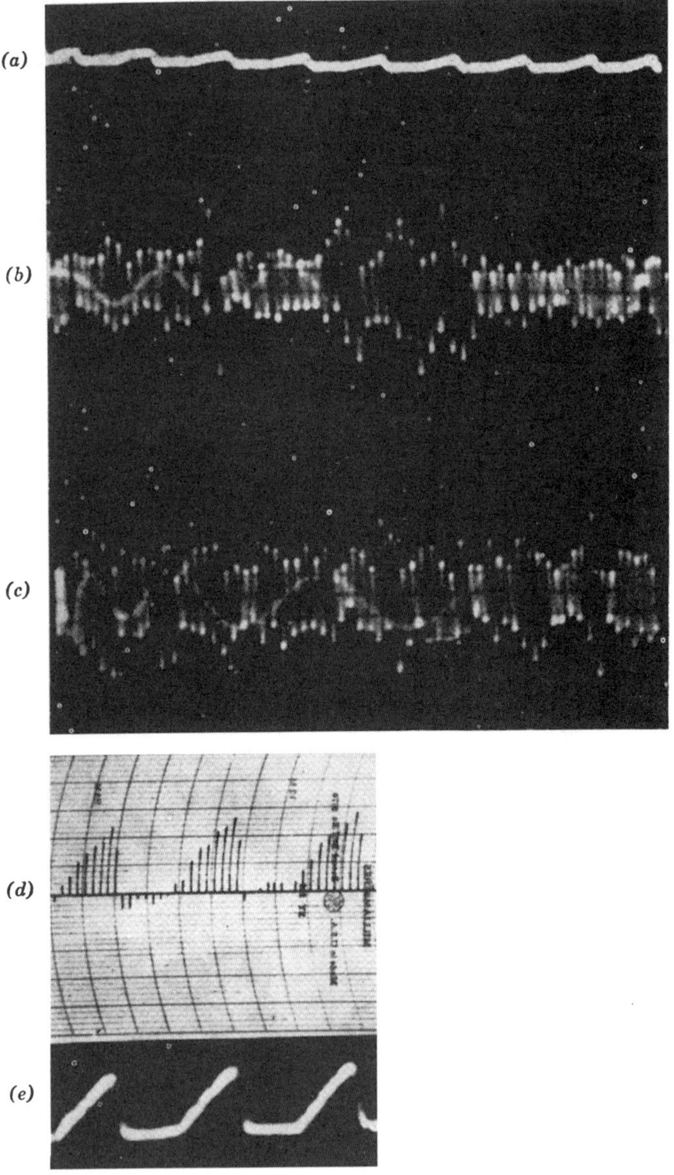

Fig. 13. Waveform detection by correlation. (*a*) Signal, (*b*) noise, (*c*) signal plus noise, (*d*) correlator output, (*e*) enlarged signal for comparison with output. From *Application of Statistical Methods to Communication Problems*, Y. W. Lee, Technical Report No. 181, 1950, Research Laboratory of Electronics, M.I.T.

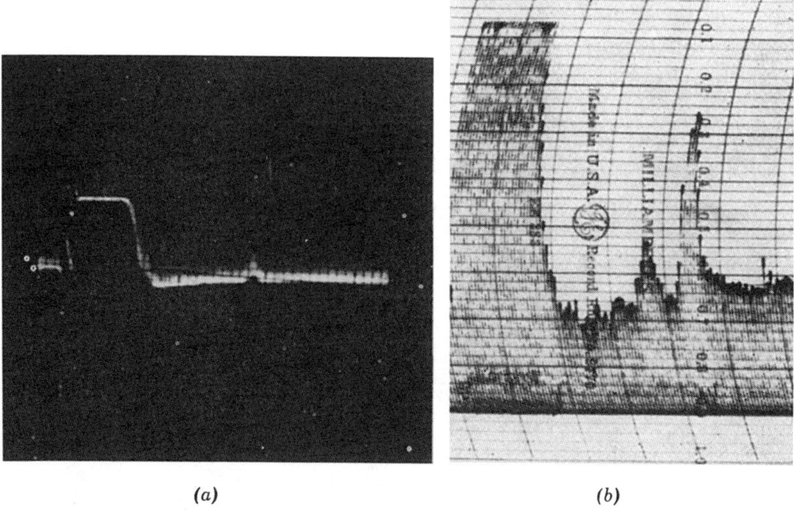

(a) (b)

Fig. 14. (a) Radar A-scope presentation, (b) correlated A-scope presentation. From *Application of Statistical Methods to Communication Problems*, Y. W. Lee, Technical Report No. 181, 1950, Research Laboratory of Electronics, M.I.T.

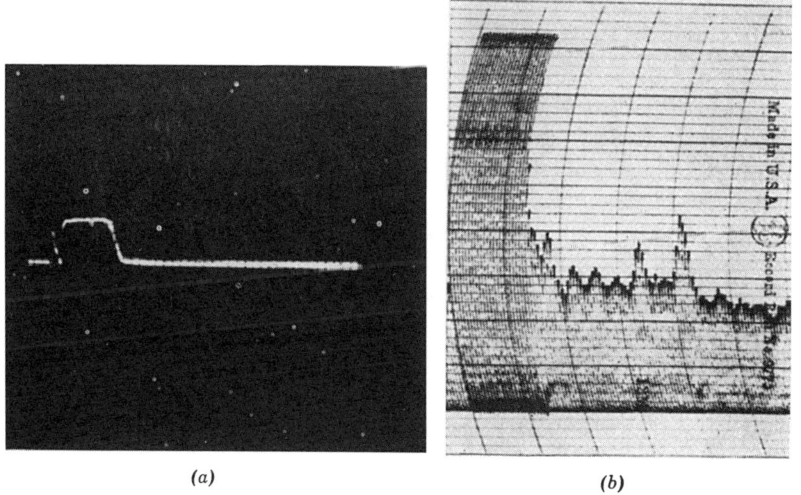

(a) (b)

Fig. 15. (a) Radar A-scope presentation with gain greatly reduced, (b) correlated A-scope presentation. From *Application of Statistical Methods to Communication Problems*, Y. W. Lee, Technical Report No. 181, 1950, Research Laboratory of Electronics, M.I.T.

As a further test of the power of the method and technique, a periodic sawtooth wave and a periodic rectangular wave were added to a random noise in an experiment to show that each of the periodic waveforms could be extracted from the mixture. As shown in Fig. 16,* the periodic

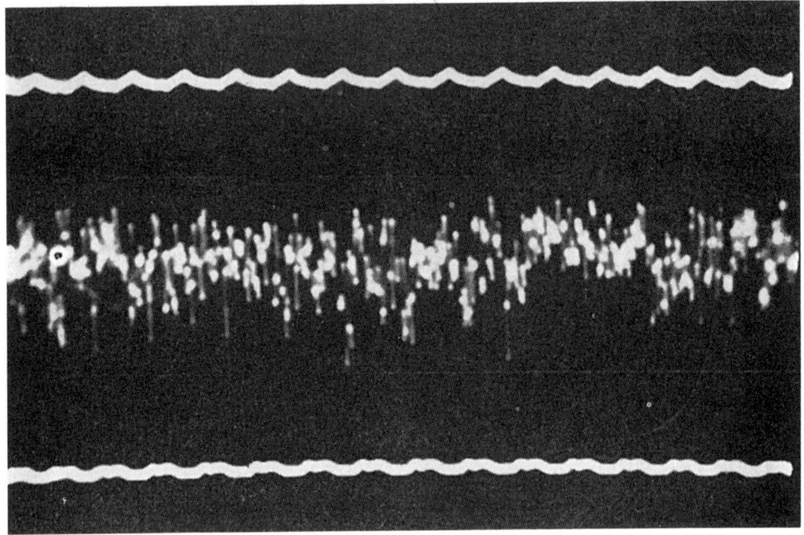

Fig. 16. Periodic components and random noise in an experiment on waveform detection by correlation.

components are small in comparison with the random noise. The actual relative amplitudes are as shown. To show the effectiveness of the method, the two periodic waves were generated with an extremely small difference in the fundamental frequencies. In fact, one wave had the fundamental frequency 30,000 cps, and the other 30,005 cps. In Fig. 17 * one of the waves is the sum of the three components, and the other is the noise without periodic components. By the method of crosscorrelation with periodic unit impulses, we first extracted the sawtooth wave. The synchronization of the impulses with the sawtooth wave was a simple matter in the laboratory since they were controlled by the same oscillator. The result is shown in Fig. 18 * in comparison with the actual wave after amplification. To extract the rectangular wave, the impulses were synchronized with the fundamental frequency of the wave. Figure 19 * shows the result of this experiment. The sample size was 60,000 for both experiments.

* From *Detection of Repetitive Signals in Noise by Correlation*, Y. W. Lee and L. G. Kraft, a paper presented at IRE National Convention, New York, 1951 (unpublished).

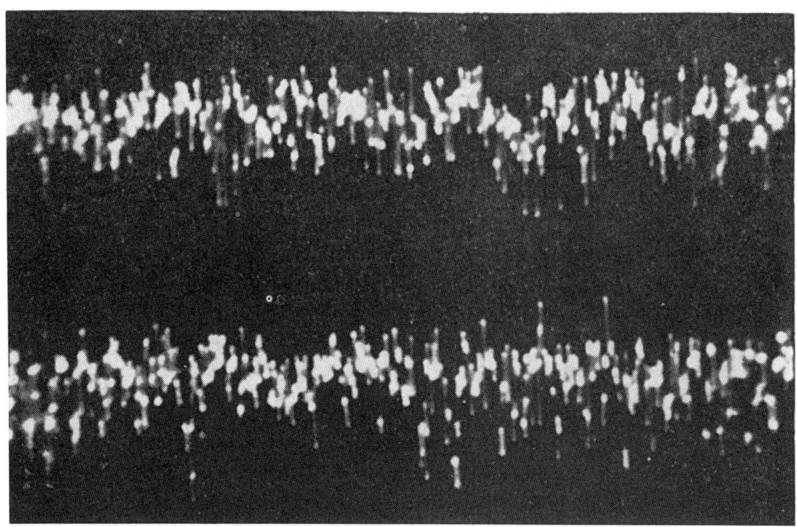

Fig. 17. Wave at top: random noise alone. Wave at bottom: sum of the three waves shown in Fig. 16.

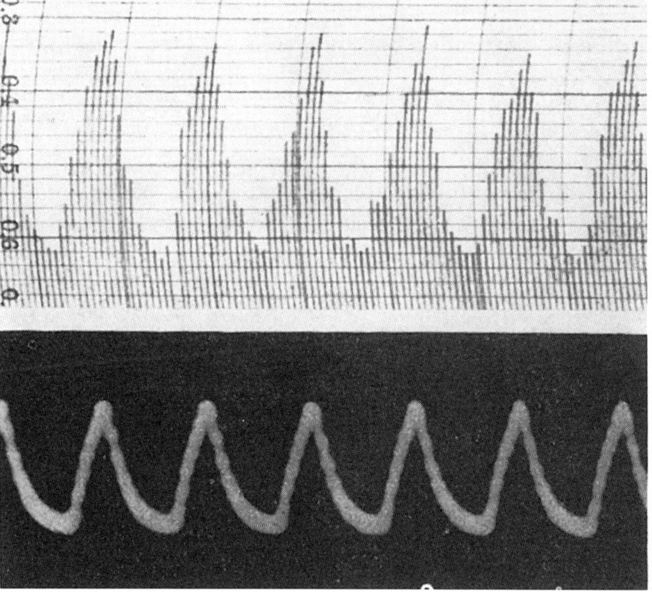

Fig. 18. Sawtooth wave extracted by correlation in comparison with actual wave.

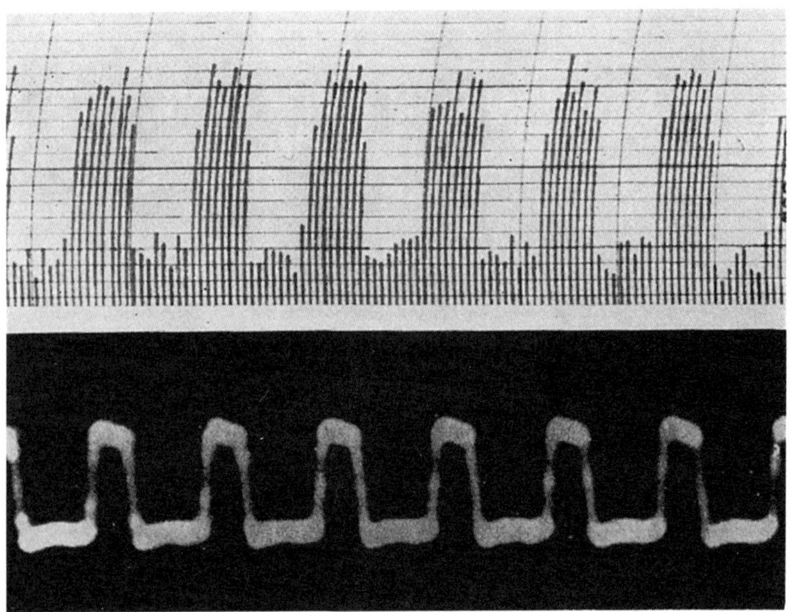

Fig. 19. Rectangular wave extracted by correlation in comparison with actual wave.

6. Detection by Correlation and by Means of Linear Networks

It is interesting to compare the correlation method and the linear filter in the detection of a periodic signal in noise. Since the detection in actual practice involves functions of only finite duration, we shall put the present discussion under this restriction. In other words we shall consider the functions to be aperiodic. For these functions we have seen in Sec. B-5 of Chapter 2 that, if $f_1(t)$ and $f_2(t)$ are aperiodic functions, the crosscorrelation of $f_1(t)$ and $f_2(t)$ is the same as the convolution of $f_1(-t)$ and $f_2(t)$. Repeating (193), Chapter 2, we have

$$\varphi_{12}(\tau) = \int_{-\infty}^{\infty} f_1(t) f_2(t + \tau)\, dt \tag{78}$$

or

$$\varphi_{12}(\tau) = \int_{-\infty}^{\infty} f_1(-t) f_2(\tau - t)\, dt \tag{79}$$

We shall compare the operations involved in (79) with those that a linear system performs. In a linear system, if $f_i(t)$ is its input and $h(t)$ is its unit-impulse response, its output $f_o(t)$ is given by the convolution integral

$$f_o(t) = \int_{-\infty}^{\infty} h(\tau) f_i(t - \tau)\, d\tau \tag{80}$$

By comparing (79) with (80), which are both convolution integrals, we see that the crosscorrelation between $f_1(t)$ and $f_2(t)$ can be accomplished by a linear system if we can make

$$\left.\begin{aligned} h(t) &= f_1(-t) \\ \text{and} \qquad f_i(t) &= f_2(t) \end{aligned}\right\} \tag{81}$$

Under conditions (81), the output of the linear system is equal to the required crosscorrelation function; that is,

$$f_o(t) = \varphi_{12}(t) \tag{82}$$

In the detection of a sinusoidal signal in noise by crosscorrelation with a sinusoid, we have the basic equation

$$\varphi_{fg}(\tau) = \lim_{T \to \infty} \frac{1}{2T} \int_{-T}^{T} f(t) g(t + \tau)\, dt \tag{83}$$

in which the noisy signal is $f(t) = S(t) + N(t)$ and the local sinusoidal signal is $g(t)$. This equation is the same as (11). Since in detection it does not matter whether we consider $\varphi_{fg}(\tau)$ or its reflection $\varphi_{gf}(\tau)$, let us take

$$\varphi_{gf}(\tau) = \lim_{T \to \infty} \frac{1}{2T} \int_{-T}^{T} g(t) f(t + \tau)\, dt \tag{84}$$

for comparison with the linear system. Let the actual interval of integration be $(-T, 0)$ which means that the part of $g(t)$ from $t = -T$ to $t = 0$ is applied to the noisy signal $f(t)$ in the same interval. Let us consider only positive values of τ. We shall assume that $f(t)$ is available from $t = -T$ to $t = \tau$. Under these conditions the approximate expression of (84) is

$$\varphi_{gf}(\tau) = \frac{1}{T} \int_{-T}^{0} g(t) f(t + \tau)\, dt \tag{85}$$

This expression is in the form (78) so that, in accordance with (78) through (82), a linear system will perform the crosscorrelation (85) if we can design the system with the unit-impulse response

$$h(t) = g(-t) \tag{86}$$

and apply to the system, at $t = -T$, the input

$$f_i(t) = f(t) \tag{87}$$

The output of the linear system will then be

$$f_o(t) = T\varphi_{gf}(t) \qquad \text{for } t > 0 \tag{88}$$

The function $g(t)$ is given by (52). To simplify matters, let us note that in the present problem the phase angle of the sinusoid is not an essential quantity so that we may put

$$g(t) = \begin{cases} E_{m2} \cos \omega_1 t & \text{for } -T < t < 0 \\ 0 & \text{elsewhere} \end{cases} \tag{89}$$

Hence the unit-impulse response of the linear system for performing the crosscorrelation (85) is

$$h(t) = g(-t) = \begin{cases} E_{m2} \cos \omega_1 t & \text{for } 0 < t < T \\ 0 & \text{elsewhere} \end{cases} \tag{90}$$

as shown in Fig. 20. The period, $T_1 = 2\pi/\omega_1$, of the sinusoidal section

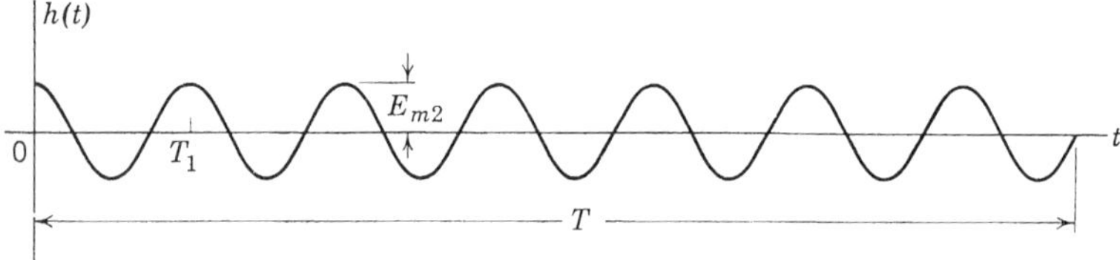

Fig. 20. Unit-impulse response of a linear system for performing the crosscorrelation (85).

is the period of the sinusoid to be detected, and its duration T should be the same as the interval of crosscorrelation on a continuous basis. From the unit-impulse response (90) it is not difficult to show that the system function of the linear system is

$$H(\omega) = \frac{E_{m2}T}{2} \left\{ \frac{\sin\left(\frac{(\omega + \omega_1)T}{2}\right)}{\frac{(\omega + \omega_1)T}{2}} e^{-j(\omega+\omega_1)T/2} + \frac{\sin\left(\frac{(\omega - \omega_1)T}{2}\right)}{\frac{(\omega - \omega_1)T}{2}} e^{-j(\omega-\omega_1)T/2} \right\} \tag{91}$$

Figure 21 shows the amplitude factor of (91) which we denote by $[H(\omega)]_{\text{amp}}$. The longer the interval of crosscorrelation T, the narrower

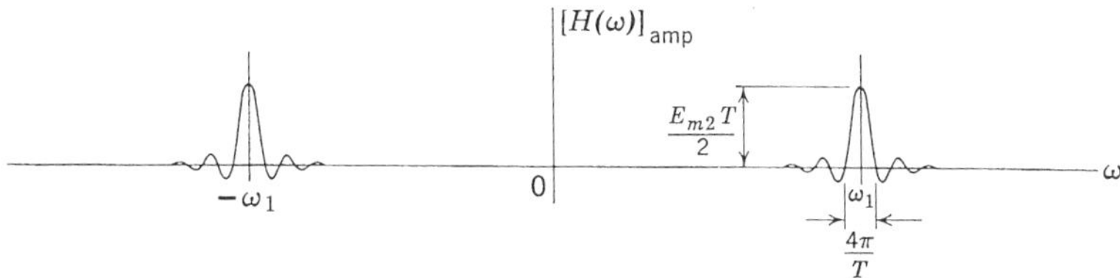

Fig. 21. Amplitude of system function corresponding to unit-impulse response of Fig. 20.

will be the bandwidth of the two peaks since the bandwidth varies as $4\pi/T$. At the same time the height of the peaks increases directly as T. As T tends to infinity, the peaks become lines at the angular frequencies $\pm\omega_1$. This limiting form corresponds to the ideal filter that passes only the sinusoid with angular frequency ω_1.

Whereas electronic circuits are capable of performing the operations of crosscorrelation on either the discrete or the continuous basis with a long averaging interval, linear systems synthesized in the classical manner on the basis of the impulse response have the usual difficulty that precise and long memory is not easy to achieve. It is rather a simple matter to generate a local pure sine wave for crosscorrelation over a long duration, but it is difficult to design and build a classical filter with a unit-impulse response such as that shown in Fig. 20 with a long duration.

The use of periodic unit impulses in crosscorrelation, which is equivalent to periodic sampling, provides an interesting comparison with the corresponding linear system. In a manner similar to the finding of an equivalent linear system for detecting a sinusoid in noise by crosscorrelation with a sinusoid, we find that the linear system that performs the equivalent of periodic sampling for the detection of a periodic wave in noise must have the unit-impulse response

$$h(t) = u(t) + u(t - T_1) + u(t - 2T_1) + \cdots + u[t - (n - 1)T_1] \quad (92)$$

This is simply the sum of n unit impulses separated from each other by time T_1, which is the period of the periodic wave to be detected. A graph of this function is given in Fig. 22. If in crosscorrelation the

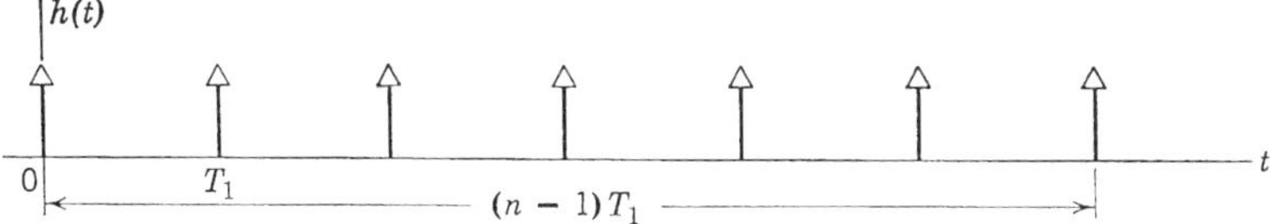

Fig. 22. Unit-impulse response of a linear system for performing the equivalent of periodic sampling in the detection of a periodic waveform in noise.

sample size is n, the interval between the first and last impulses is $(n - 1)T_1$. The system function $H(\omega)$ can be evaluated by noting that

$$H(\omega) = \int_{-\infty}^{\infty} h(t)e^{-j\omega t}\,dt \tag{93}$$

and

$$\int_{-\infty}^{\infty} u(t - kT_1)e^{-j\omega t}\,dt = e^{-jkT_1\omega} \tag{94}$$

Accordingly we find that the linear system whose unit-impulse response is given by (92) has the system function

$$\begin{aligned} H(\omega) &= 1 + e^{-jT_1\omega} + e^{-j2T_1\omega} + \cdots + e^{-j(n-1)T_1\omega} \\ &= \frac{1 - e^{-jnT_1\omega}}{1 - e^{-jT_1\omega}} \end{aligned} \tag{95}$$

The last expression follows from a well-known summation for a power series. It will be helpful in interpretation and in drawing the graph if this expression is put into the form of a product of an amplitude factor and a phase factor. To do this, let us note that the pure phase factor must be $e^{-j(n-1)T_1\omega/2}$. This is explained by the fact that, if $h(t)$ is displaced to the left by the amount $(n - 1)T_1/2$, then $h[t - (n - 1)T_1/2]$ is an even function, and as a consequence its transform will have zero phase for all frequencies. In the frequency domain this displacement is the factor just given. Therefore

$$\begin{aligned} H(\omega) &= \left[\frac{e^{jnT_1\omega/2} - e^{-jnT_1\omega/2}}{e^{jT_1\omega/2} - e^{-jT_1\omega/2}}\right] e^{-j(n-1)T_1\omega/2} \\ &= \frac{\sin\left(\dfrac{nT_1\omega}{2}\right)}{\sin\left(\dfrac{T_1\omega}{2}\right)} e^{-j(n-1)T_1\omega/2} \end{aligned} \tag{96}$$

The graph of the amplitude factor is given in Fig. 23. This graph is drawn for the sample size $n = 20$. If we consider the absolute value of (96)

$$|H(\omega)| = \left|\frac{\sin\left(\dfrac{nT_1\omega}{2}\right)}{\sin\left(\dfrac{T_1\omega}{2}\right)}\right| \tag{97}$$

we find that it is periodic with the fundamental angular frequency $\omega_1 = 2\pi/T_1$. It is observed that $2\pi/T_1$ is the fundamental angular fre-

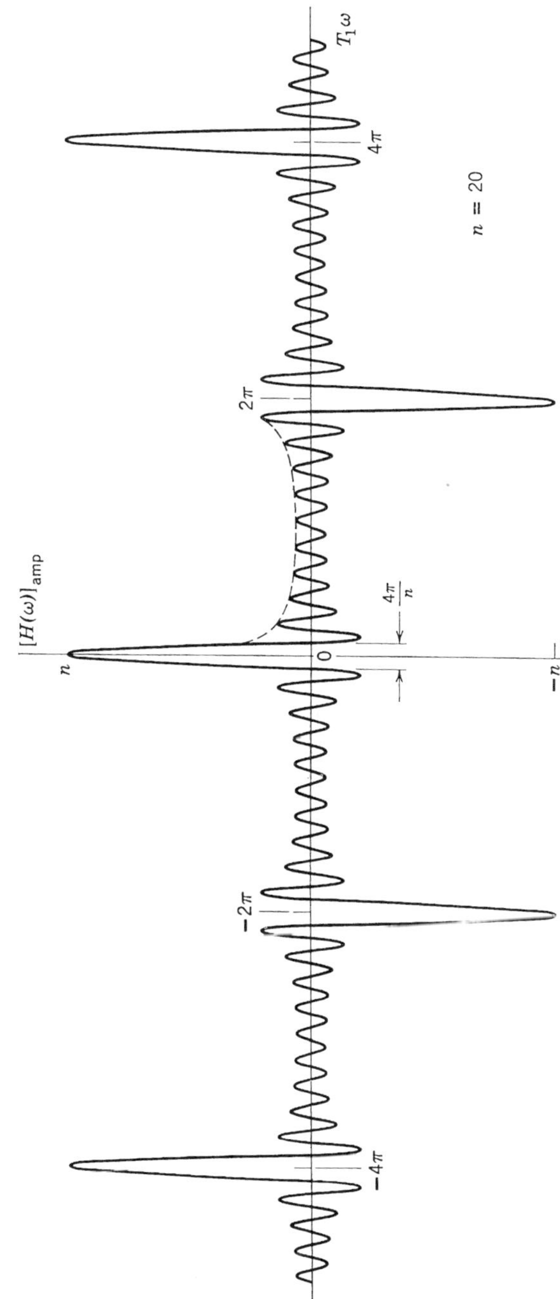

Fig. 23. Amplitude of system function corresponding to unit-impulse response of Fig. 22.

quency of the periodic signal to be detected. The predominant peaks of $|H(\omega)|$ are at $\omega = 0, \pm 2\pi/T_1, \pm 4\pi/T_1, \ldots$. The envelope of the oscillations $\sin (nT_1\omega/_2)$ is $\csc (T_1\omega/2)$. As the sample size n increases, the height of the predominant peaks increases directly as n, but the width of these peaks decreases inversely as n. If the sample size is $n = 1000$ and $T_1 = 1/1000$, we find that the width of the predominant peaks is 2 cycles per second.

From this analysis it is clear that, if a linear filter is to be designed to perform the equivalent of periodic sampling for detecting a periodic wave in noise without distorting its original waveform, the filter should have a system function as shown in Fig. 23. In the limiting form this filter should pass all the harmonics of the periodic wave without attenuation and should attenuate completely all other frequencies in the continuous spectrum. The phase spectrum of the filter should be linear. The practical frequency domain design of such a filter is difficult. However, looking at the same filter in the time domain (Fig. 22), we see that one way to realize the filter is to construct a uniform delay line with taps at the points corresponding to $t = 0, T_1, 2T_1, \ldots, (n - 1)T_1$. The parallel connection of these taps is the output of the required linear system. One point that deserves attention in the comparison of the delay line filter and sampling by electronic means in detection is that the delay line needs a total of delay time $(n - 1)T_1$ for the equivalent of n samples in a sample mean in the sampling method. On the other hand the delay required in the sampling method for any sample size is the displacement τ to cover one period of the periodic signal; that is, $\tau = T_1$.

We see that in general, if one of the two functions being crosscorrelated can be approximated by the unit-impulse response of a linear network, it is possible to perform the crosscorrelation by a linear network. When two random functions are correlated, the irregularity of these functions in general would prevent the practical design of a linear system to perform a crosscorrelation. This statement naturally applies to the autocorrelation of a random function.

In view of the foregoing discussion it is clear that pulses of simple forms can be autocorrelated or crosscorrelated with waves and pulses of various types by a linear system. If a pulse of finite duration is the primary concern of a problem, correlation should be defined on the basis of aperiodic functions. Thus, if $g(t)$ is the pulse in the interval $(-T, 0)$ and zero elsewhere and if $f(t)$ is the pulse plus a random noise $N(t)$ in the interval $(-T, \tau)$, the crosscorrelation of $g(t)$ and $f(t)$ is

$$\varphi_{gf}(\tau) = \int_{-T}^{0} g(t)f(t + \tau)\, dt \tag{98}$$

Since

$$f(t) = g(t) + N(t) \tag{99}$$

(98) is

$$\begin{aligned}\varphi_{gf}(\tau) &= \int_{-T}^{0} g(t)[g(t+\tau) + N(t+\tau)]\,dt \\ &= \varphi_{gg}(\tau) + \varphi_{gN}(\tau)\end{aligned} \tag{100}$$

in which $\varphi_{gg}(\tau)$ is the autocorrelation of the pulse and $\varphi_{gN}(\tau)$ is the cross-correlation of the pulse and the random noise; that is,

$$\varphi_{gg}(\tau) = \int_{-T}^{0} g(t)g(t+\tau)\,dt \tag{101}$$

and

$$\varphi_{gN}(\tau) = \int_{-T}^{0} g(t)N(t+\tau)\,dt \tag{102}$$

Analogous to (85) through (88), we find that, if a linear system is designed with its unit-impulse response

$$h(t) = g(-t) \tag{103}$$

then its output $f_o(t)$ for the input $f(t)$ applied at $t = -T$ is

$$\begin{aligned}f_o(t) &= \varphi_{gf}(t) \\ &= \varphi_{gg}(t) + \varphi_{gN}(t) \qquad \text{for } t > 0\end{aligned} \tag{104}$$

In the absence of the random noise $N(t)$, the output of the system for the input $f(t) = g(t)$ applied at $t = -T$ is

$$f_o(t) = \varphi_{gg}(t) \qquad \text{for } t > 0 \tag{105}$$

PROBLEMS

1. A random rectangular wave with two possible values E and 0 and with a Poisson distribution of jumps is shown in Fig. P1. The autocorrelation function of this wave is to be measured by the electronic correlator which takes essentially independent pairs of samples from the random wave. The measured value of the autocorrelation function is a random variable if the measurement is repeated. At $\tau = 1$ microsecond what is the necessary sample size for a standard deviation that is 1 per cent of the true value of the autocorrelation function? The average number of jumps per second is 100,000, and $E = 10$ volts.

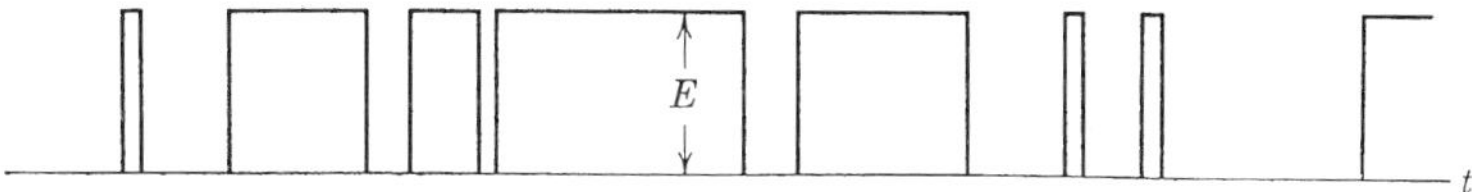

Fig. P1.

2. The autocorrelation function of the random wave of Problem 20, Chapter 8, is to be measured by the electronic correlator which takes pairs of samples of the wave at regular intervals. It is desired to measure the autocorrelation function $\varphi_{11}(\tau)$ at $\tau = 0$ and at $\tau = \tau_o$ (where the function has practically reached its final value) with results of repeated experiments that have the same variance at both points. What should the relative sample sizes for the two points be?

3. An autocorrelation curve for a stationary random wave is to be measured by an electronic correlator which selects samples from the random wave. The pairs of samples are taken at long intervals so that they may be considered to be independent. The random wave is Gaussian with zero mean value and variance σ^2. When repeated measurements are made to determine $\varphi_{11}(\tau)$ for large values of τ where the curve has practically reached its final value, what is the variance of the measured values? When repeated measurements are made at $\tau = 0$, what is the variance of the measured values?

4. One method for obtaining the autocorrelation function $\varphi_{aa}(\tau)$ of $f_a(t)$ is to take the mean square of the sum $f_1(t, \tau) = f_a(t) + f_a(t - \tau)$. Thus, if we let $\psi(\tau)$ be the mean square of $f_1(t, \tau)$, we have

$$\begin{aligned}\psi(\tau) &= \lim_{T\to\infty} \frac{1}{2T} \int_{-T}^{T} f_1{}^2(t, \tau)\, dt \\ &= \lim_{T\to\infty} \frac{1}{2T} \int_{-T}^{T} [f_a(t) + f_a(t - \tau)]^2\, dt \\ &= 2\varphi_{aa}(0) + 2\varphi_{aa}(\tau)\end{aligned}$$

From this relationship $\varphi_{aa}(\tau)$ can be determined from $\psi(\tau)$ and $\varphi_{aa}(0)$. This problem concerns the measurement of $\psi(\tau)$. Let $f_1(t, \tau)$ be fed into the measuring circuit A as shown in Fig. P4. The delay τ is adjusted by circuit D. Circuit A takes a series of independent single samples from $f_1(t, \tau)$, squares them, then sums the squares, and finally records the result automatically. The experiment is carried out for a series of values of τ to obtain the curve of $\psi(\tau)$. Assuming that $f_a(t)$ is Gaussian with the variance σ^2 and the mean m, determine the variance of the results of repeated experiments for $\tau = 0$ and for $\tau = \tau_o$ where $\psi(\tau)$ has practically reached its final value. The sample size is n.

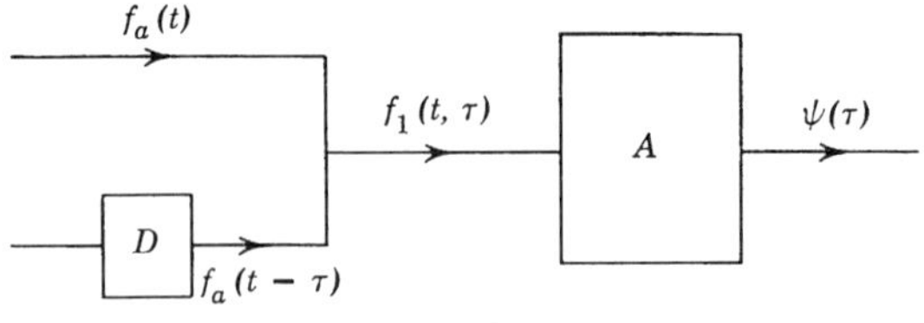

Fig. P4.

5. In Fig. P5, $f_a(t)$ is a Poisson rectangular wave with the two possible values of amplitude $+E$ and $-E$ and the average rate of k zero crossings per second; $f_b(t)$ is a Gaussian noise with the mean value m and the variance σ^2. The network N is a summing circuit so that $f_1(t) = f_a(t) + f_b(t)$.

(a) The autocorrelation function $\varphi_{11}(\tau)$ of $f_1(t)$ is to be measured by the sampling method with the aid of an electronic correlator. If the variance of the measured autocorrelation function at the point $\tau = 0$ is to be A, what is the sample size for the measurement at this point in terms of the given quantities?

(b) Assume that at the point $\tau = \tau_o$ the autocorrelation function has practically reached its final value. If the variance of the measured autocorrelation function at the point $\tau = \tau_o$ is to be A, what is the sample size for the measurement at this point in terms of the given quantities?

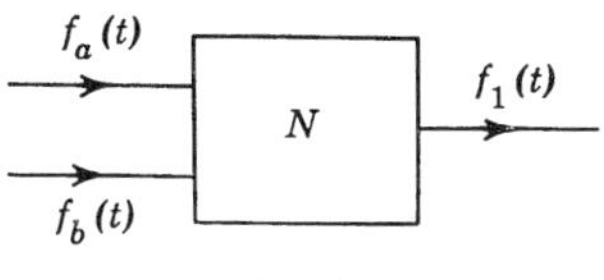

Fig. P5.

6. The second-order autocorrelation function of a random process $f_1(t)$ is defined as

$$\varphi_{111}(\tau_1, \tau_2) = \lim_{T\to\infty} \frac{1}{2T} \int_{-T}^{T} f_1(t)f_1(t + \tau_1)f_1(t + \tau_1 + \tau_2)\, dt$$

(See Fig. P6.) As an ensemble average, the function is

$$\varphi_{111}(\tau_1, \tau_2) = \int_{-\infty}^{\infty} \int_{-\infty}^{\infty} \int_{-\infty}^{\infty} x_1x_2x_3P_{\xi_1\xi_2\xi_3}(x_1, x_2, x_3; \tau_1, \tau_2)\, dx_1\, dx_2\, dx_3$$

When plotted against τ_1 and τ_2, the function is a surface. Measurement of the function by the sampling method is done by taking products of three values instead of two values as in the case of the autocorrelation function $\varphi_{11}(\tau)$. Consider the three points ($\tau_1 = 0$, $\tau_2 = 0$), ($\tau_1 = 0$, $\tau_2 = A$), and ($\tau_1 = A$, $\tau_2 = A$) where measurements are to be made. If the amplitudes of the process are practically independent when they are separated by the time A and if the amplitude probability density is normal (Gaussian) with zero mean, what are the variances of the measured values at the three points in terms of the variance, σ^2, of the process and the sample size, n?

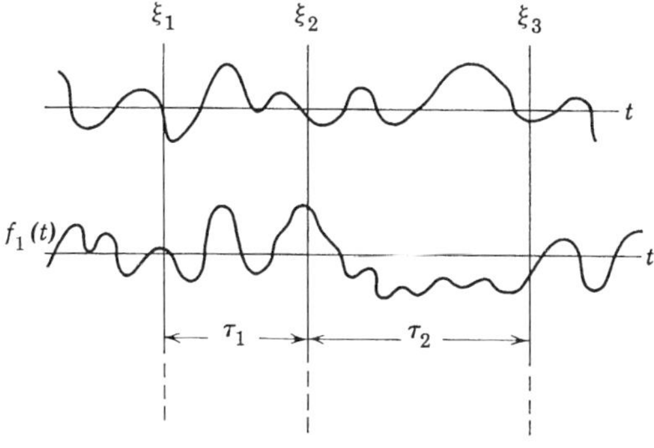

Fig. P6.

7. In the random wave shown in Fig. P7, the probability density $P_\xi(x)$ of the pulse width ξ is uniform from $x = 0$ to $x = \Delta/2$ and linear from $x = \Delta/2$ to $x = \Delta$ as shown. The pulse widths are independent of one another. In measuring the autocorrelation function by an electronic correlator, pairs of samples are selected from the wave at long intervals and at random so that they can be considered as independent samples. It is desired that the variance of the results of repeated measurements of the autocorrelation function $\varphi_{11}(\tau)$ at $\tau = 0$ and the variance of those at $\tau = 2\Delta$ be the same. Determine the ratio of the sample sizes for the two points.

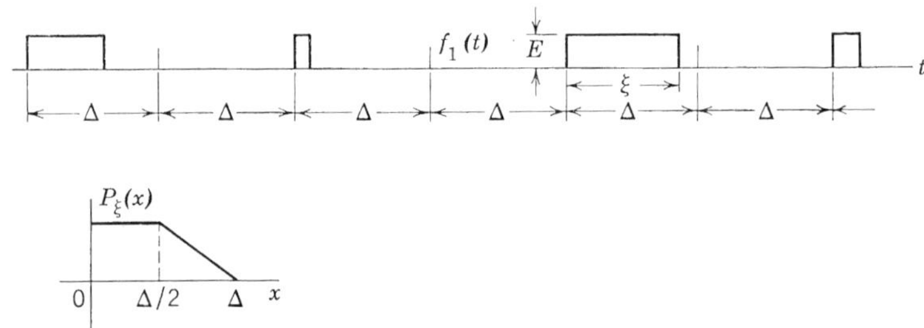

Fig. P7.

chapter 13

Fundamental Relations for a Linear System and Certain Applications

In preparation for the formulation of filtering and prediction problems we develop in this chapter relations that involve a linear system and its input and output. Our particular emphasis is on random inputs and outputs so that they will enter into relations with the linear system not only as instantaneous messages and noise but also as correlation functions and power density spectrums. Interesting and instructive applications are introduced in some of the fundamental relations that we establish in the course of the development of the chapter, but the major problem of filtering and prediction is reserved for study in later chapters.

1. The Instantaneous Input-Output Relation

The convolution integral relating the input, the output, and the characterizing function of a linear system is well known. However, for completeness and in view of the fact that random functions are now included as inputs, the concepts and assumptions on which the integral are based should be briefly reviewed before we proceed to make extensive use of it.

We know that a linear system is characterized by its response to a unit-impulse excitation. When we say that the system is characterized by a certain function, we mean that the output of the linear system for input functions of a general class are expressible in terms of the characteristic function and the input. The inputs and outputs are functions of time. One convenient characterizing function of a linear system is the system's response to a unit-impulse function. We shall consider the input and output to be voltages. A unit-impulse voltage is the limiting form of a rectangular voltage pulse of height A and base b, such that $Ab = 1$ always as A tends to infinity and b becomes dt. The unit impulse $u(t)$ is therefore an impulse of an infinitesimal duration dt, and in this entire duration (not just at a point) the amplitude of the impulse tends to infinity. It is an integrable function, for we have an area of 1 in this extremely long and narrow rectangle. Outside of the interval dt of the impulse, the function is zero. Unless otherwise specified, the unit-impulse function, or the unit impulse, for brevity, has the pulse duration dt located at the origin of the time axis such that $t = 0$ is at the midpoint of dt. In other words, $u(t)$ is an even function. For practical purposes such a function may be approximated satisfactorily by a voltage of extremely short duration.

The response of the linear system to the unit impulse is denoted by $h(t)$. Since $u(t)$ is applied at $t = 0$, it would be physically impossible for it to have any effect on the output for $t < 0$. If the system is initially at rest, which is always the case unless otherwise stated, it is of necessity that $h(t) = 0$ for $t < 0$. The response of the system to $u(t)$ at the output begins at $t = 0$. Figure 1*a* shows the unit-impulse function $u(t)$ which is applied to a linear system A. For the purpose of illustration, suppose that the system unit-impulse response $h(t)$ is as shown.

Now we wish to show that $h(t)$ characterizes the system behavior; that is, we wish to show that for a given input of a general class the system output is expressible in terms of $h(t)$ and the input. In doing this, we must bring out the properties which we assume the system to have. The first important property of the system is that, if the application of the unit impulse is delayed by time $t = \sigma$, the response $h(t)$ will also be delayed by the same length of time; that is, for the input $u(t - \sigma)$, the output is $h(t - \sigma)$ as Fig. 1*b* shows. This is a property of physical laws which enables us to predict the results of future experiments on the basis of past experiments. The invariance in the input-output relation of a system under a translation in time means that if the system were represented by a differential equation involving the input and output, the coefficients of the equation would be constants. We shall confine the discussion to time-invariant systems.

Consider a general class of functions consisting of the periodic, aperiodic, and random functions discussed in previous chapters, and let $f_i(t)$ in Fig. 1*c* be representative of this class acting as the system input. Let the function start at an arbitrary time $t = -a$. We shall divide $f_i(t)$ into elements of infinitesimal widths, dt, as indicated by the fine lines in the figure. At $t = \sigma$, for example, the element has the width $d\sigma$ and the height $f_i(\sigma)$. If this particular element were a unit impulse, we would know that the output due to this element alone would be $h(t - \sigma)$. However, this element is a very small impulse, in the sense that it has the area $f_i(\sigma)\, d\sigma$ which is an infinitesimal. Nevertheless, there is an

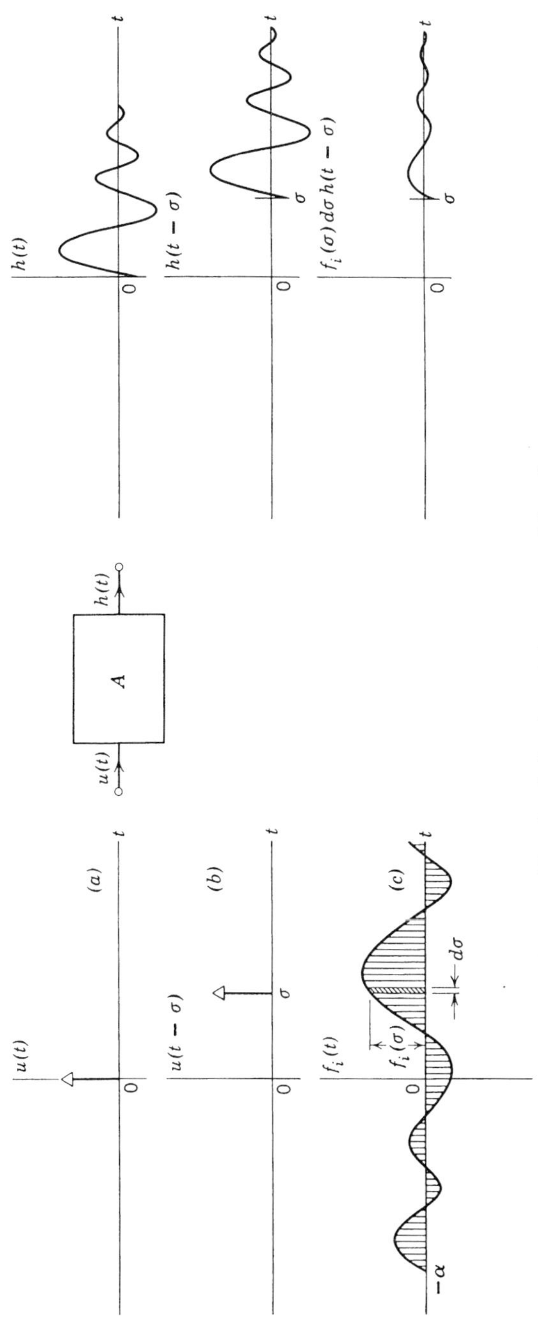

Fig. 1. For the derivation of the superposition theorem.

output due to this infinitesimal impulse, and it is $f_i(\sigma)\,d\sigma\,h(t-\sigma)$. Hence, the differential of the output is

$$df_o(t) = f_i(\sigma)\,d\sigma\,h(t-\sigma) \tag{1}$$

as Fig. 1c shows. In arriving at this output element, we have made the assumption that the system response to an impulse is directly proportional to the area of the impulse. Since the widths of the unit impulse and the infinitesimal impulse at $t = \sigma$ are the same, the assumption amounts to saying that the ratio of the output due to the infinitesimal impulse to the output due to the unit impulse is equal to the ratio of the respective heights of the impulses. Since the height of the infinitesimal impulse is $f_i(\sigma)$ and that of the unit impulse is $1/d\sigma$, the assumption is

$$\frac{df_o(t)}{h(t-\sigma)} = \frac{f_i(\sigma)}{\dfrac{1}{d\sigma}} \tag{2}$$

which reduces to (1). From this assumption it follows that, in general, if for the input $f(t)$ the system output is $g(t)$ and if the input is then multiplied by a real constant k, the output is also multiplied by k. When two quantities are so related, they are said to bear a linear relationship. Graphically the two quantities are represented by a straight line. When this relation holds for a system, it would mean that for various different inputs, the inputs and outputs of the system are expressed by a set of linear equations. Since it is a property of a set of linear equations that superposition holds, the output of a linear system is the superposition of the component outputs produced by the respective component inputs acting individually.

Therefore, by summing all the infinitesimal outputs due to all the elements into which $f_i(t)$ is divided, from the beginning to the time t, we shall have, at time t, the output $f_o(t)$ due to cumulative effects of $f_i(t)$ from the beginning up to time t. The expression is

$$f_o(t) = \int_{-a}^{t} f_i(\sigma)h(t-\sigma)\,d\sigma \tag{3}$$

This is a convolution integral giving the output of a linear system in terms of the system unit-impulse response and the input. Consequently, a system for which this integral holds is a linear system. The integral is also known as the superposition theorem, or superposition integral.

It is important to examine (3) further for physical interpretation. If we put $\nu = t - \sigma$, we shall have

$$f_o(t) = \int_0^{t+a} h(\nu) f_i(t - \nu)\, d\nu \tag{4}$$

In Fig. 2, $f_i(t + \nu)$ is the system input after t seconds have elapsed, counting from the time indicated in Fig. 1c. The portion of $f_i(t + \nu)$ to the left of $\nu = 0$ is the past history of the input up to the present moment. The present moment is indicated by $\nu = 0$. The portion of $f_i(t + \nu)$ to the right of $\nu = 0$ is the future of the function, which has not yet come to the system. In (4) the term $f_i(t - \nu)$ is $f_i(t + \nu)$ folded back with reference to $\nu = 0$ as shown. Now according to (4), $h(\nu)$ is multi-

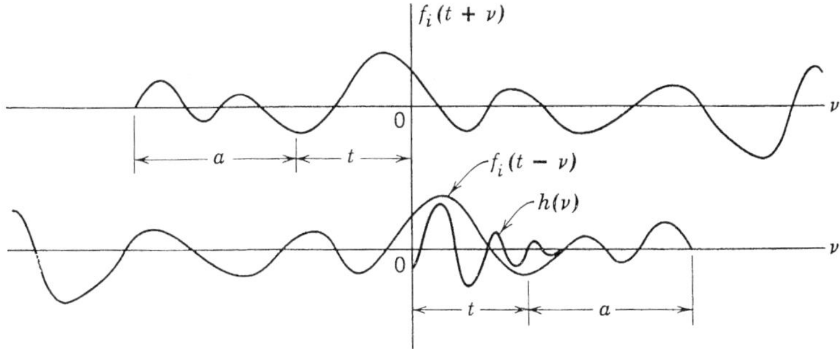

Fig. 2. Physical interpretation of the convolution integral.

plied into the past history of the input up to the present moment, and the product is integrated over the interval of the past history to yield the output for the time t, which is the present moment.

A linear system therefore depends upon what has happened at the input from the beginning to the present to give its output. The output is a summation of the weighted past. The weighting function is the system unit-impulse response. In other words, a linear system scans the history of the input with its unit-impulse response to yield its output. Instead of regarding a linear system as one that is characterized by a pattern of poles and zeros, or by an amplitude spectrum and a phase spectrum, we will find it interesting and helpful in many problems to think of the system in the time domain in the manner described. We shall illustrate this point when we come to prediction and filtering.

Referring to (4), we note that, since $f_i(t)$ is assumed to be zero from $-\infty$ to $-a$ and since $h(\nu) = 0$ for $\nu < 0$, the product $h(\nu)f_i(t - \nu)$ is actually zero outside the limits of integration $(0, t + a)$. Furthermore, if $f_i(t)$ starts from the infinite past, a becomes infinite. Hence, nothing would be added to the integral if we replace these limits by $(-\infty, \infty)$. We shall do this as a matter of convenience. Henceforth (4) will be written

$$f_o(t) = \int_{-\infty}^{\infty} h(\nu) f_i(t - \nu)\, d\nu \tag{5}$$

We recall that in convolution the integral is the same regardless of which one of the functions under the integral sign is displaced and folded back. Hence (5) may be written

$$f_o(t) = \int_{-\infty}^{\infty} f_i(\nu) h(t - \nu)\, d\nu \tag{6}$$

We repeat for emphasis that the integral holds whether $f_i(t)$ and $f_o(t)$ be periodic, aperiodic, or random.

2. The Unit-Impulse Response–System Function Relation

With the aid of physical reasoning, we shall establish the relation between the unit-impulse response of a linear system and its system function. Let Re $[E_i e^{j\omega t}]$ and Re $[E_o e^{j\omega t}]$ be the steady-state sinusoidal input and output respectively of the linear system. In these expressions Re stands for "the real part of," and E_i and E_o are the complex amplitudes of the input and output respectively. The system function $H(\omega)$ of the linear system is defined as

$$H(\omega) = \frac{E_o}{E_i} \tag{7}$$

Hence for the input Re $[E_i e^{j\omega t}]$, the output is

$$\text{Re}\,[E_o e^{j\omega t}] = \text{Re}\,[H(\omega) E_i e^{j\omega t}] \tag{8}$$

To establish the desired relation, first assume that the unit impulse is applied to the system. The unit impulse $u(t)$ is related to its energy density spectrum $U(\omega)$ by

$$u(t) = \int_{-\infty}^{\infty} U(\omega) e^{j\omega t}\, d\omega \tag{9}$$

and

$$U(\omega) = \frac{1}{2\pi} \int_{-\infty}^{\infty} u(t) e^{-j\omega t}\, dt \tag{10}$$

Since we know that

$$U(\omega) = \frac{1}{2\pi} \tag{11}$$

substitution of (11) into (9) gives

$$u(t) = \int_{-\infty}^{\infty} \frac{1}{2\pi} e^{j\omega t}\, d\omega \tag{12}$$

which may be written in the form

$$u(t) = \int_{-\infty}^{\infty} \left(\frac{1}{2\pi}\, d\omega\right) e^{j\omega t} \tag{13}$$

We write (12) in the form (13) to bring out more clearly the fact that according to Fourier theory the unit impulse is a summation of infinitesimal sinusoids of all frequencies. The sinusoids are represented by $e^{j\omega t}$, and in fact $e^{j\omega t}$ may be replaced by $\cos \omega t$ since $u(t)$ is an even function. The amplitudes of the component sinusoids are the same, being $2(1/2\pi)\, d\omega$. These infinitesimal sinusoids of all frequencies, when added up, completely cancel out one another at all values of t except in the infinitesimal interval dt at the origin. This picture seems fantastic, but it is entirely in accordance with the fundamental concept of harmonic analysis. Such a picture is of great help in the formulation of our problem.

Now, the only difference between the sinusoids involved in (7) and those in (13) is the relative amplitude; those in (7) are finite, but those in (13) are infinitesimal. However, a linear system responds to both groups in the same manner with a difference only in amplitude. Applying (8) to infinitesimal sinusoids, we have, for the component input $\text{Re}\left[2\left(\frac{1}{2\pi}\, d\omega\right) e^{j\omega t}\right]$ of $u(t)$, the system output

$$dh(t) = \text{Re}\left[H(\omega) 2 \left(\frac{1}{2\pi}\, d\omega\right) e^{j\omega t}\right] \tag{14}$$

which may be written

$$dh(t) = \frac{1}{2\pi} [\overline{H}(\omega) e^{-j\omega t} + H(\omega) e^{j\omega t}]\, d\omega \tag{15}$$

The term $dh(t)$ is accounted for by the fact that the infinitesimal output in question is the differential of the system's unit-impulse response. By summing all the component outputs due to all the sinusoidal components of $u(t)$, we have

$$h(t) = \frac{1}{2\pi} \int_{-\infty}^{\infty} H(\omega) e^{j\omega t}\, d\omega \tag{16}$$

This is the desired result which relates the unit-impulse response to the system function. The expression is in the form of a Fourier transform. By inverse transformation, we obtain

$$H(\omega) = \int_{-\infty}^{\infty} h(t) e^{-j\omega t}\, dt \tag{17}$$

According to the manner in which the Fourier transform was obtained in Chapter 2, the factor $1/2\pi$ has always been associated with the integral in which the time function is being transformed. Contrary to this general rule, the factor is now associated with the other integral in (16) and (17). This difference is caused by the fact that the time function $h(t)$ has a special significance, being the response to a particular type of input function $u(t)$. As we know, the transform of $u(t)$ is $1/2\pi$. If we had chosen a 2π-impulse function instead of a unit-impulse function as the fundamental excitation function, we would find the factor $1/2\pi$ in (17) instead of (16). In transforming a function which actually represents a system function or a unit-impulse response, we shall follow (16) and (17).

3. The Transform of the Superposition Integral

We have emphasized the fact that, in the superposition integral (5) or (6), $f_o(t)$ and $f_i(t)$ may be periodic, aperiodic, or random. A matter of interest is the transformation of the integral. We shall consider the transformation with respect to the type of function involved. First, if the input function $f_i(t)$ is periodic, $f_o(t)$ is also periodic with the same fundamental period. To transform (5) under the condition of a periodic $f_i(t)$, with the fundamental angular frequency $\omega_1 = 2\pi/T_1$, we follow (13) and (14), Chapter 2. Thus

$$\frac{1}{T_1}\int_{-T_1/2}^{T_1/2} f_o(t)e^{-jn\omega_1 t}\,dt = \frac{1}{T_1}\int_{-T_1/2}^{T_1/2} e^{-jn\omega_1 t}\,dt\int_{-\infty}^{\infty} h(\nu)f_i(t-\nu)\,d\nu \tag{18}$$

Letting $F_o(n)$ and $F_i(n)$ be the complex spectrums of the output and input respectively and making a change of variable $\sigma = t - \nu$, we have

$$F_o(n) = \frac{1}{T_1}\int_{-T_1/2}^{T_1/2} e^{-jn\omega_1(\nu+\sigma)}\,d\sigma\int_{-\infty}^{\infty} h(\nu)f_i(\sigma)\,d\nu \tag{19}$$

In this double integral it is possible to write all terms involving ν under one integral and all terms involving σ under another. Thus

$$F_o(n) = \frac{1}{T_1}\int_{-T_1/2}^{T_1/2} f_i(\sigma)e^{-jn\omega_1\sigma}\,d\sigma\int_{-\infty}^{\infty} h(\nu)e^{-jn\omega_1\nu}\,d\nu \tag{20}$$

As a result of the separation of variables, the double integral becomes a product of two single integrals; that is,

$$F_o(n) = \left[\frac{1}{T_1}\int_{-T_1/2}^{T_1/2} f_i(\sigma)e^{-jn\omega_1\sigma}\,d\sigma\right]\left[\int_{-\infty}^{\infty} h(\nu)e^{-jn\omega_1\nu}\,d\nu\right] \tag{21}$$

The first factor is $F_i(n)$, and the second is, by (17), $H(n)$. Hence the transform of (5) is

$$F_o(n) = H(n)F_i(n) \qquad \text{for } n = 0, \pm 1, \pm 2, \ldots \tag{22}$$

This relation obtained from (5) is seen to be in agreement with the definition of a system function (7), as it should be.

Next, if $f_i(t)$ is a transient function, the transformation of (5) will be performed according to (125) and (126), Chapter 2. We assume that $f_i(t)$, $f_o(t)$, and $h(t)$ have the transforms $F_i(\omega)$, $F_o(\omega)$, and $H(\omega)$. Then

$$\frac{1}{2\pi}\int_{-\infty}^{\infty} f_o(t)e^{-j\omega t}\,dt = \frac{1}{2\pi}\int_{-\infty}^{\infty} e^{-j\omega t}\,dt \int_{-\infty}^{\infty} h(\nu)f_i(t-\nu)\,d\nu \tag{23}$$

Again, with the change of variable, $\sigma = t - \nu$, (23) is

$$\begin{aligned} F_o(\omega) &= \frac{1}{2\pi}\int_{-\infty}^{\infty} e^{-j\omega(\nu+\sigma)}\,d\sigma \int_{-\infty}^{\infty} h(\nu)f_i(\sigma)\,d\nu \\ &= \frac{1}{2\pi}\int_{-\infty}^{\infty} h(\nu)e^{-j\omega\nu}\,d\nu \int_{-\infty}^{\infty} f_i(\sigma)e^{-j\omega\sigma}\,d\sigma \\ &= \left[\int_{-\infty}^{\infty} h(\nu)e^{-j\omega\nu}\,d\nu\right]\left[\frac{1}{2\pi}\int_{-\infty}^{\infty} f_i(\sigma)e^{-j\omega\sigma}\,d\sigma\right] \end{aligned} \tag{24}$$

Since the first factor is the system function $H(\omega)$ and the second is $F_i(\omega)$, the final result is

$$F_o(\omega) = H(\omega)F_i(\omega) \tag{25}$$

Finally, if $f_i(t)$ is a random function, we shall not be able to transform the superposition integral. The reason is that $f_i(t)$ and $f_o(t)$, being random functions, do not possess transforms. When this integral appears in our problems involving random functions and in association with other quantities, we shall always attempt to manipulate the expressions in such a way that the random functions are put into the form of a correlation function before a transformation is made.

4. Input-Output Relation in Autocorrelation

In a linear system with a random input, a relation of great importance is that between the input autocorrelation and the output autocorrelation. To establish this relation, let us write the output autocorrelation function according to definition

$$\varphi_{oo}(\tau) = \lim_{T\to\infty}\frac{1}{2T}\int_{-T}^{T} f_o(t)f_o(t+\tau)\,dt \tag{26}$$

and then attempt to express it in terms of the characteristic of the system and the input. The necessary link between input, output, and the system is the superposition integral (5). Its substitution in (26) yields

$$\varphi_{oo}(\tau) = \lim_{T\to\infty} \frac{1}{2T} \int_{-T}^{T} dt \int_{-\infty}^{\infty} h(\nu) f_i(t-\nu)\, d\nu \int_{-\infty}^{\infty} h(\sigma) f_i(t+\tau-\sigma)\, d\sigma \quad (27)$$

To make this step clearer, let us note that

$$f_o(t) = \int_{-\infty}^{\infty} h(\nu) f_i(t-\nu)\, d\nu \quad (28)$$

and

$$f_o(t+\tau) = \int_{-\infty}^{\infty} h(\nu) f_i(t+\tau-\nu)\, d\nu \quad (29)$$

but, if the two integrals are placed together as in (27), it is necessary to distinguish the variables of integration because they are independent variables. Hence the variable ν in (29) is changed to σ.

By inversion of the order of integration, we shall put (27) into the desired form. Noting that, if $f_i(t-\nu)$ and $f_i(t+\tau-\sigma)$ are grouped together with the limit integral, an autocorrelation would be formed, we write

$$\varphi_{oo}(\tau) = \int_{-\infty}^{\infty} h(\nu)\, d\nu \int_{-\infty}^{\infty} h(\sigma)\, d\sigma \lim_{T\to\infty} \frac{1}{2T} \int_{-T}^{T} f_i(t-\nu) f_i(t+\tau-\sigma)\, dt \quad (30)$$

Since

$$\varphi_{ii}(\tau+\nu-\sigma) = \lim_{T\to\infty} \frac{1}{2T} \int_{-T}^{T} f_i(t-\nu) f_i(t+\tau-\sigma)\, dt \quad (31)$$

(30) is

$$\varphi_{oo}(\tau) = \int_{-\infty}^{\infty} h(\nu)\, d\nu \int_{-\infty}^{\infty} h(\sigma)\, d\sigma\, \varphi_{ii}(\tau+\nu-\sigma) \quad (32)$$

This is the desired relation among the input autocorrelation, the output autocorrelation, and the system unit-impulse response. The expression may be put into a simpler form as we shall show presently. For $\tau = 0$, we have

$$\varphi_{oo}(0) = \int_{-\infty}^{\infty} h(\nu)\, d\nu \int_{-\infty}^{\infty} h(\sigma)\, d\sigma\, \varphi_{ii}(\nu-\sigma)$$

$$= \overline{f_o^2(t)} \quad (33)$$

5. Input-Output Relation in Spectrum

At this point we would want to have the relation (32) expressed in the frequency domain. By the Wiener theorem for autocorrelation, the output power density spectrum $\Phi_{oo}(\omega)$ is

$$\begin{aligned}\Phi_{oo}(\omega) &= \frac{1}{2\pi}\int_{-\infty}^{\infty} \varphi_{oo}(\tau)e^{-j\omega\tau}\,d\tau \\ &= \frac{1}{2\pi}\int_{-\infty}^{\infty} e^{-j\omega\tau}\,d\tau \int_{-\infty}^{\infty} h(\nu)\,d\nu \int_{-\infty}^{\infty} h(\sigma)\,d\sigma\ \varphi_{ii}(\tau+\nu-\sigma)\end{aligned} \tag{34}$$

With the change of variable $\mu = \tau + \nu - \sigma$, followed by a separation of variables, (34) is

$$\begin{aligned}\Phi_{oo}(\omega) &= \frac{1}{2\pi}\int_{-\infty}^{\infty} e^{-j\omega(\mu+\sigma-\nu)}\,d\mu \int_{-\infty}^{\infty} h(\nu)\,d\nu \int_{-\infty}^{\infty} h(\sigma)\,d\sigma\ \varphi_{ii}(\mu) \\ &= \frac{1}{2\pi}\int_{-\infty}^{\infty} h(\nu)e^{j\omega\nu}\,d\nu \int_{-\infty}^{\infty} h(\sigma)e^{-j\omega\sigma}\,d\sigma \int_{-\infty}^{\infty} \varphi_{ii}(\mu)e^{-j\omega\mu}\,d\mu \\ &= \left[\int_{-\infty}^{\infty} h(\nu)e^{j\omega\nu}\,d\nu\right]\left[\int_{-\infty}^{\infty} h(\sigma)e^{-j\omega\sigma}\,d\sigma\right]\left[\frac{1}{2\pi}\int_{-\infty}^{\infty} \varphi_{ii}(\mu)e^{-j\omega\mu}\,d\mu\right]\end{aligned} \tag{35}$$

The first factor is the conjugate of the system function since the conjugate of (17) is

$$\overline{H}(\omega) = \int_{-\infty}^{\infty} h(t)e^{j\omega t}\,dt \tag{36}$$

The bar over H denotes the conjugate. Note that, in the integral of (36), $h(t)$ is always real; the only complex quantity in it is the j in the exponential term. To obtain $\overline{H}(\omega)$, we replace every j in $H(\omega)$ by $-j$. The second factor is $H(\omega)$, and the last factor is the power density spectrum $\Phi_{ii}(\omega)$ of the input. Hence the output power density spectrum of the linear system is

$$\Phi_{oo}(\omega) = \overline{H}(\omega)H(\omega)\Phi_{ii}(\omega) \tag{37}$$

or

$$\Phi_{oo}(\omega) = |H(\omega)|^2\Phi_{ii}(\omega) \tag{38}$$

This is the relation (32) in the frequency domain. It is pointed out in Sec. 3 that the transformation of the superposition integral for random functions does not exist since the transform of a random function does not exist. Consequently, no relation such as (22) for a periodic function or (25) for an aperiodic function is available for a random function. However, we are not stopped by this difficulty; for, by considering the power density spectrum, we are able to establish the relation (38) for problems involving random functions applied to a linear system. This is a very important aspect of the Wiener theorem.

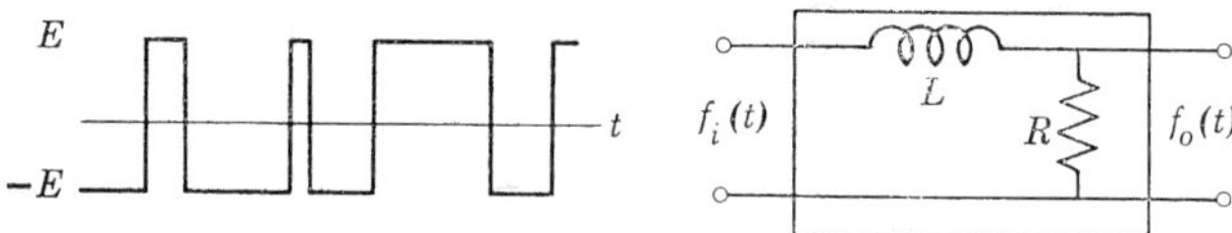

Fig. 3. A Poisson rectangular wave applied to an RL network.

To illustrate (38) by a simple example, consider an RL circuit with the Poisson rectangular wave as its input as shown in Fig. 3. From previous work we have

$$\Phi_{ii}(\omega) = \frac{E^2}{\pi} \frac{2k}{(2k)^2 + \omega^2} \tag{39}$$

where k is the average number of zero crossings per second and E is the amplitude of the wave. The system function of the network is

$$H(\omega) = \frac{R}{R + j\omega L} \tag{40}$$

By (38) the power density spectrum of the network output is

$$\Phi_{oo}(\omega) \doteq \frac{E^2}{\pi} \left(\frac{R^2}{R^2 + \omega^2 L^2}\right) \left(\frac{2k}{4k^2 + \omega^2}\right) \tag{41}$$

6. Another Form of the Input-Output Relation in Autocorrelation

By a change of variable we can show that (32) may be put into a simpler form for certain applications. Thus, if we introduce the change of variable $t = \sigma - \nu$, (32) becomes

$$\varphi_{oo}(\tau) = \int_{-\infty}^{\infty} h(\nu)\, d\nu \int_{-\infty}^{\infty} h(t + \nu)\, dt\, \varphi_{ii}(\tau - t) \tag{42}$$

Inverting the order of integration, we have

$$\varphi_{oo}(\tau) = \int_{-\infty}^{\infty} \varphi_{ii}(\tau - t)\, dt \int_{-\infty}^{\infty} h(\nu) h(\nu + t)\, d\nu \tag{43}$$

In this expression we see that the change of variable and the inversion of the order of integration result in an autocorrelation of the system unit-impulse response as a term in the equation for $\varphi_{oo}(\tau)$. The autocorrelation of the system unit-impulse response $\varphi_{hh}(\tau)$ is defined by

$$\varphi_{hh}(\tau) = \int_{-\infty}^{\infty} h(t) h(t + \tau)\, dt \tag{44}$$

if it exists. In terms of this autocorrelation, (43) is

$$\varphi_{oo}(\tau) = \int_{-\infty}^{\infty} \varphi_{hh}(t)\varphi_{ii}(\tau - t)\, dt \tag{45}$$

The result we have obtained states that the output autocorrelation of a linear system is the convolution of the system unit-impulse response autocorrelation and the input autocorrelation.

A way to express (45) from a slightly different point of view is obtained by changing t in (45) to $-t$. Thus

$$\varphi_{oo}(\tau) = \int_{-\infty}^{\infty} \varphi_{hh}(-t)\varphi_{ii}(\tau + t)\, dt \tag{46}$$

Inasmuch as $\varphi_{hh}(\tau)$ is an even function, (46) is

$$\varphi_{oo}(\tau) = \int_{-\infty}^{\infty} \varphi_{hh}(t)\varphi_{ii}(t + \tau)\, dt \tag{47}$$

Therefore this equivalent to (45) states that the output autocorrelation of a linear system is the crosscorrelation of the system unit-impulse response autocorrelation and the input autocorrelation. This is an interesting result that involves the concepts of the autocorrelation of a random function, the autocorrelation of an aperiodic function, and the crosscorrelation of two aperiodic functions each having been derived from a different type of function. It is interesting to note that, for $\tau = 0$, (47) is

$$\varphi_{oo}(0) = \int_{-\infty}^{\infty} \varphi_{hh}(t)\varphi_{ii}(t)\, dt$$

$$= \overline{f_o{}^2(t)} \tag{48}$$

Before turning to another topic, we return to (44) and consider its transform. We have

$$\frac{1}{2\pi}\int_{-\infty}^{\infty} \varphi_{hh}(\tau)e^{-j\omega\tau}\, d\tau = \frac{1}{2\pi}\int_{-\infty}^{\infty} e^{-j\omega\tau}\, d\tau \int_{-\infty}^{\infty} h(\nu)h(\nu + t)\, d\nu \tag{49}$$

Integrating in the manner illustrated in Sec. 3 and 5 or applying (141), Chapter 2, we find

$$\frac{1}{2\pi}\int_{-\infty}^{\infty} \varphi_{hh}(\tau)e^{-j\omega\tau}\, d\tau = \frac{1}{2\pi}|H(\omega)|^2 \tag{50}$$

so that

$$|H(\omega)|^2 = \int_{-\infty}^{\infty} \varphi_{hh}(\tau)e^{-j\omega\tau}\, d\tau \tag{51}$$

By inverse transformation

$$\varphi_{hh}(\tau) = \frac{1}{2\pi} \int_{-\infty}^{\infty} |H(\omega)|^2 e^{j\omega\tau}\, d\omega \tag{52}$$

Since autocorrelation functions are even functions, the pair of relations (51) and (52) may be replaced by

$$|H(\omega)|^2 = \int_{-\infty}^{\infty} \varphi_{hh}(\tau) \cos \omega\tau\, d\tau \tag{53}$$

and

$$\varphi_{hh}(\tau) = \frac{1}{2\pi} \int_{-\infty}^{\infty} |H(\omega)|^2 \cos \omega\tau\, d\omega \tag{54}$$

By means of these results relating the autocorrelation of the unit-impulse response to the square of the absolute value of the system function, it is simple to show that the transform of (45) or (47) is (38).

7. Application of Input-Output Relation in Autocorrelation to the Determination of Autocorrelation Functions of Poisson Waves *

(a) Poisson Wave of Equally Likely Positive and Negative Pulses. We have seen in Chapter 8 that the determination of the autocorrelation function from a statistical description of a random process is frequently a difficult task. One important random function which is fairly difficult to handle is the wave of Poisson-distributed pulses such as the one shown in Fig. 4*a*. It consists of a series of identical pulses whose initial

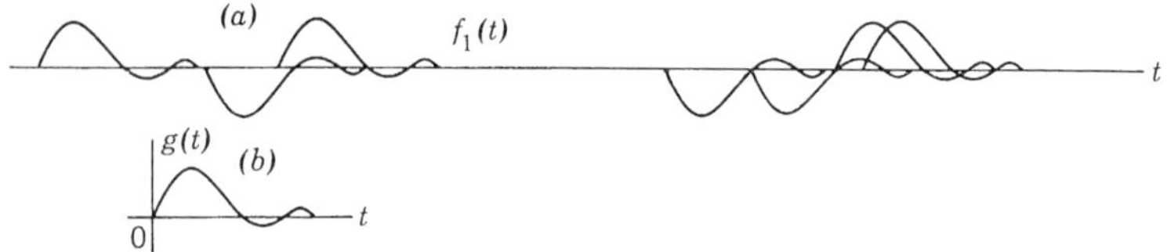

Fig. 4. (*a*) A Poisson wave of equally likely positive and negative pulses; (*b*) form of the pulses in the Poisson wave.

points are Poisson distributed. Positive and negative pulses occur with equal probability. The figure shows the individual pulses, but the actual value of the function at any given time is the algebraic sum of all the component values. We shall call this function a Poisson wave of identical pulses. The autocorrelation function of this wave has been obtained in a rather involved manner.† Here we wish to show that this

* Y. W. Lee, "Autocorrelation Functions of Poisson Waves," *Quarterly Progress Report*, Research Laboratory of Electronics, M.I.T., January 15, 1955, pp. 56–62.

† S. O. Rice, "Mathematical Analysis of Random Noise," *Bell System Tech. Jour.*, **23**, 282–332 (1944).

wave and other similar ones can be handled effectively by an application of the results of Sec. 6.

To obtain the autocorrelation $\varphi_{11}(\tau)$ of the function $f_1(t)$, let a Poisson wave of equally likely positive and negative unit impulses $f_P(t)$ (see Fig. 18*a*, Chapter 8) be applied to a linear system whose unit-impulse response is the same in form as a single pulse in $f_1(t)$ (Fig. 5). Thus, if the form of the pulses in $f_1(t)$ is $g(t)$ as shown in Fig. 4*b*, the unit-impulse response $h(t)$ of the linear system should be

$f_P(t)$ → $h(t)=g(t)$ → $f_o(t) = f_1(t)$

Fig. 5. A Poisson wave of unit-impulses applied to a linear system.

$$h(t) = g(t) \tag{55}$$

We shall refer to the linear system as the pulse-shaping system. Clearly, the output of the system is the given function $f_1(t)$ which is to be correlated. Now, we apply the relation (45) of Sec. 6. Since the quantities corresponding to those in (45) are

$$\left.\begin{aligned} \varphi_{oo}(\tau) &= \varphi_{11}(\tau) \\ \varphi_{hh}(\tau) &= \varphi_{gg}(\tau) \\ \varphi_{ii}(\tau) &= \varphi_{PP}(\tau) \end{aligned}\right\} \tag{56}$$

where $\varphi_{PP}(\tau)$ is the autocorrelation function of the equally likely positive and negative Poisson-distributed unit impulses, equation (45) becomes

$$\varphi_{11}(\tau) = \int_{-\infty}^{\infty} \varphi_{gg}(t)\varphi_{PP}(\tau - t)\,dt \tag{57}$$

From (80), Chapter 8,

$$\varphi_{PP}(\tau) = ku(\tau) \tag{58}$$

so that

$$\varphi_{11}(\tau) = k\int_{-\infty}^{\infty} \varphi_{gg}(t)u(\tau - t)\,dt \tag{59}$$

Inasmuch as the integral is the convolution of $\varphi_{gg}(t)$ and the unit-impulse function, which leaves $\varphi_{gg}(t)$ unchanged, we have the result that

$$\varphi_{11}(\tau) = k\varphi_{gg}(\tau) \tag{60}$$

By definition

$$\varphi_{gg}(\tau) = \int_{-\infty}^{\infty} g(t)g(t + \tau)\,dt \tag{61}$$

Hence the autocorrelation of the Poisson wave of equally likely positive and negative pulses of the form $g(t)$ is

$$\varphi_{11}(\tau) = k\int_{-\infty}^{\infty} g(t)g(t + \tau)\,dt \tag{62}$$

Stated in words, the autocorrelation function is the average number of pulses per second times the autocorrelation of a single pulse.

The power density spectrum of $f_1(t)$ can be obtained from (38) or from the transformation of (62). It is

$$\Phi_{11}(\omega) = \frac{k}{2\pi}|G(\omega)|^2 \tag{63}$$

where

$$G(\omega) = \int_{-\infty}^{\infty} g(t)e^{-j\omega t}\,dt \tag{64}$$

(b) Poisson Wave of Pulses of the Same Sign. If the pulses of the Poisson wave have the same sign, such as is shown in Fig. 6, we take the

Fig. 6. A Poisson wave of pulses of the same sign.

Poisson wave $f_P{}^*(t)$ of unit impulses of the same sign to be the input to the pulse-shaping system. See Fig. 18*b*, Chapter 8, for $f_P{}^*(t)$.

Again, if $g(t)$ is a single pulse in $f_1(t)$, we shall have, similar to (57),

$$\varphi_{11}(\tau) = \int_{-\infty}^{\infty} \varphi_{gg}(t)\overset{*}{\varphi}_{PP}(\tau - t)\,dt \tag{65}$$

in which $\overset{*}{\varphi}_{PP}(\tau)$ is the autocorrelation of $f_P{}^*(t)$ as expressed by (85), Chapter 8. Substitution of this expression into (65) yields

$$\varphi_{11}(\tau) = \int_{-\infty}^{\infty} \varphi_{gg}(t)[ku(\tau - t) + k^2]\,dt \tag{66}$$

Since

$$\begin{aligned}\int_{-\infty}^{\infty} \varphi_{gg}(t)\,dt &= \int_{-\infty}^{\infty} dt \int_{-\infty}^{\infty} g(\sigma)g(\sigma + t)\,d\sigma \\ &= \int_{-\infty}^{\infty} g(\sigma)\,d\sigma \int_{-\infty}^{\infty} g(\sigma + t)\,dt \\ &= \left[\int_{-\infty}^{\infty} g(\sigma)\,d\sigma\right]^2 \end{aligned} \tag{67}$$

we find that (66) is

$$\varphi_{11}(\tau) = k\int_{-\infty}^{\infty} g(t)g(t + \tau)\,dt + \left[k\int_{-\infty}^{\infty} g(\sigma)\,d\sigma\right]^2 \tag{68}$$

In the last term of this expression, k is the average number of pulses per second, and $\int_{-\infty}^{\infty} g(\sigma)\,d\sigma$ is the area of a pulse so that $k\int_{-\infty}^{\infty} g(\sigma)\,d\sigma$ is the average area of $f_1(t)$ per second which is the same as the average value of $f_1(t)$. Hence

$$k \int_{-\infty}^{\infty} g(\sigma)\, d\sigma = \overline{f_1(t)} \tag{69}$$

and (68) becomes

$$\varphi_{11}(\tau) = k \int_{-\infty}^{\infty} g(t) g(t + \tau)\, dt + \overline{f_1(t)}^2 \tag{70}$$

This is the autocorrelation of the Poisson wave $f_1(t)$ of identical all positive or all negative pulses of the form $g(t)$. The result differs from (62) by the constant $\overline{f_1(t)}^2$ which is the square of the d-c component of $f_1(t)$.

The power density spectrum of $f_1(t)$ is

$$\Phi_{11}(\omega) = \frac{k}{2\pi} |G(\omega)|^2 + \overline{f_1(t)}^2 u(\omega) \tag{71}$$

(c) Poisson Wave of Alternately Positive and Negative Pulses. Another interesting application of the method is the determination of the autocorrelation function of a Poisson wave of alternately positive and negative pulses. The input to the pulse-shaping system is the Poisson wave of unit impulses f_P^{**} discussed in Sec. 6*c* and shown in Fig. 18*c*, Chapter 8. Similar to (65)

$$\varphi_{11}(\tau) = \int_{-\infty}^{\infty} \varphi_{gg}(t) \varphi_{PP}^{**}(\tau - t)\, dt \tag{72}$$

where $\varphi_{PP}^{**}(\tau)$ is the autocorrelation of $f_P^{**}(t)$ and is given by (98), Chapter 8. With this expression in (72)

$$\begin{aligned} \varphi_{11}(\tau) &= \int_{-\infty}^{\infty} \varphi_{gg}(t)[k u(\tau - t) - k^2 e^{-2k|\tau - t|}]\, dt \\ &= k \varphi_{gg}(\tau) - k^2 \int_{-\infty}^{\infty} \varphi_{gg}(t) e^{-2k|\tau - t|}\, dt \end{aligned} \tag{73}$$

For a specific example, let

$$g(t) = \begin{cases} E e^{-at} & \text{for } t \geq 0 \\ 0 & \text{for } t < 0 \end{cases} \tag{74}$$

so that $f_1(t)$ is as shown in Fig. 7. Then

$$\begin{aligned} \varphi_{gg}(\tau) &= E^2 \int_0^{\infty} e^{-at} e^{-a(t + |\tau|)}\, dt \\ &= \frac{E^2}{2a} e^{-a|\tau|} \end{aligned} \tag{75}$$

Fig. 7. A Poisson wave of alternately positive and negative pulses.

Substituting (75) into (73) and performing the integration, we find that

$$\varphi_{11}(\tau) = \frac{kE^2}{2(a^2 - 4k^2)} [ae^{-a|\tau|} - 2ke^{-2k|\tau|}] \tag{76}$$

By the Wiener theorem the power density spectrum of the Poisson-distributed exponential pulses is

$$\begin{aligned} \Phi_{11}(\omega) &= \frac{kE^2}{2\pi(a^2 - 4k^2)} \left[\frac{a^2}{a^2 + \omega^2} - \frac{4k^2}{4k^2 + \omega^2}\right] \\ &= \frac{kE^2}{2\pi} \frac{\omega^2}{(a^2 + \omega^2)(4k^2 + \omega^2)} \end{aligned} \tag{77}$$

Let us consider the limiting form of (76) and (77) as $a \to 0$. This would mean that $g(t)$ is a step function of height E and that the pulse-shaping system acts as an integrator. In this circumstance the wave in Fig. 7 becomes the rectangular wave in Fig. 8. As $a \to 0$, (76) becomes

$$\varphi_{11}(\tau) = \frac{E^2}{4} e^{-2k|\tau|} \tag{78}$$

and (77) becomes

$$\Phi_{11}(\omega) = \frac{kE^2}{2\pi} \frac{1}{4k^2 + \omega^2} \tag{79}$$

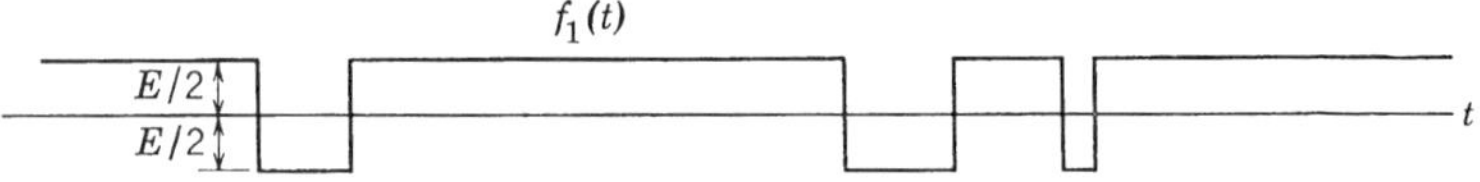

Fig. 8. A Poisson rectangular wave.

These results agree with those we obtained in Chapter 8, Sec. 2.

If we solve the problem in the frequency domain, we shall have, according to (38)

$$\Phi_{11}^{**}(\omega) = |G(\omega)|^2 \Phi_{PP}^{**}(\omega) \tag{80}$$

where $\Phi_{11}(\omega)$ is the power density spectrum of $f_1(t)$, $\Phi_{PP}^{**}(\omega)$ is that of $f_P^{**}(t)$ as given by (99), Chapter 8, and $G(\omega)$ is the system function of the pulse-shaping system. For the particular case (74),

$$G(\omega) = \frac{E}{a + j\omega} \tag{81}$$

so that

$$\Phi_{11}(\omega) = \frac{E^2}{a^2 + \omega^2}\left(\frac{k}{2\pi}\frac{\omega^2}{4k^2 + \omega^2}\right) \tag{82}$$

which is the same as (77).

8. Input-Output Crosscorrelation Theorem

A relation of considerable importance for a linear system, in theory and in practical application, is that involving the input-output crosscorrelation, the input autocorrelation, and the unit-impulse response. To establish this relation, we shall first write the input-output crosscorrelation $\varphi_{io}(\tau)$ of a linear system with the random input $f_i(t)$ and output $f_o(t)$ according to definition

$$\varphi_{io}(\tau) = \lim_{T\to\infty} \frac{1}{2T}\int_{-T}^{T} f_i(t) f_o(t + \tau)\, dt \tag{83}$$

To bring the system unit-impulse response into this expression, we introduce the superposition integral for $f_o(t + \tau)$. Accordingly,

$$\varphi_{io}(\tau) = \lim_{T\to\infty} \frac{1}{2T}\int_{-T}^{T} f_i(t)\, dt \int_{-\infty}^{\infty} h(\nu) f_i(t + \tau - \nu)\, d\nu \tag{84}$$

By inverting the order of integration, we have

$$\varphi_{io}(\tau) = \int_{-\infty}^{\infty} h(\nu)\, d\nu \lim_{T\to\infty} \frac{1}{2T}\int_{-T}^{T} f_i(t) f_i(t + \tau - \nu)\, dt \tag{85}$$

Since

$$\varphi_{ii}(\tau - \nu) = \lim_{T\to\infty} \frac{1}{2T}\int_{-T}^{T} f_i(t) f_i(t + \tau - \nu)\, dt \tag{86}$$

(85) is

$$\varphi_{io}(\tau) = \int_{-\infty}^{\infty} h(\nu) \varphi_{ii}(\tau - \nu)\, d\nu \tag{87}$$

We have therefore shown that *the input-output crosscorrelation of a linear system is the convolution of the unit-impulse response and the input autocorrelation.*

This important relation can be expressed in the frequency domain by a transformation. Thus taking the Fourier transform of both sides of (87), we have

$$\frac{1}{2\pi}\int_{-\infty}^{\infty} \varphi_{io}(\tau) e^{-j\omega\tau}\, d\tau = \frac{1}{2\pi}\int_{-\infty}^{\infty} e^{-j\omega\tau}\, d\tau \int_{-\infty}^{\infty} h(\nu)\varphi_{ii}(\tau - \nu)\, d\nu \tag{88}$$

which yields, in a manner similar to (23) and (24)

$$\Phi_{io}(\omega) = H(\omega)\Phi_{ii}(\omega) \tag{89}$$

This result states that *the input-output cross-power density spectrum of a linear system is the product of the system function and the input power density spectrum.*

The input-output crosscorrelation theorem for a linear system is of considerable importance in prediction and filtering problems which are taken up in later chapters. However, at this point let us consider another interesting application.

9. Determination of the Behavior of a Linear System by Crosscorrelation

The application of the input-output crosscorrelation theorem that we wish to consider is the determination of the behavior of a linear system by exciting the system with random noise.

A random noise whose power density spectrum is flat over a band of frequencies which is considerably wider than the band of the system under investigation is called a white noise. The term "white" is taken from the term "white light" in optics. Of course we assume that the band of the system is within the band of the noise. One type of white noise is the Poisson wave of equally likely positive and negative unit impulses. Physically, the impulses are approximated by pulses of sufficiently short duration. A gas tube is a source of white noise. Since the power density spectrum of a white noise is theoretically the same value K for all frequencies,

$$\Phi_{11}(\omega) = K \tag{90}$$

and it follows that the autocorrelation of the noise is an impulse of area $2\pi K$. Thus

$$\varphi_{11}(\tau) = 2\pi K u(\tau) \tag{91}$$

Now, suppose that the linear system under investigation is excited by this noise. We shall have for its input-output crosscorrelation the relation, in accordance with (87),

$$\begin{aligned}\varphi_{io}(\tau) &= \int_{-\infty}^{\infty} h(\nu) 2\pi K u(\tau - \nu)\, d\nu \\ &= 2\pi K h(\tau)\end{aligned} \tag{92}$$

This is an interesting result, because it states that the input-output crosscorrelation of a linear system under white noise excitation is proportional to the system unit-impulse response. The result suggests that the unit-impulse response may be obtained by a crosscorrelation measurement of the input and output, as shown in Fig. 9, when the input is a white noise.

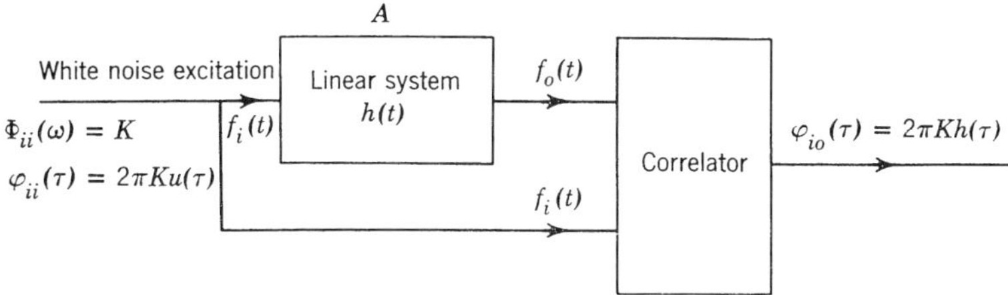

Fig. 9. Determination of the unit-impulse response of a linear system by crosscorrelation.

As we know, the conventional method for measuring the unit-impulse response is to apply an extremely short pulse to the system under test and to record the response on photographic film. An alternative measurement is done by applying a step voltage and recording the response. However, this measurement gives the integral of the unit-impulse response. Instead of the unit-impulse response, we may measure the system function under steady-state sinusoidal excitation. These measurements are equivalent because the unit-impulse response and the system function are Fourier transforms of each other.

The method of crosscorrelation involves different concepts and has some remarkable advantages. First, let us consider a linear system A, as shown in Fig. 10, which has a random input $f_i(t)$, an output $f_o(t)$, and

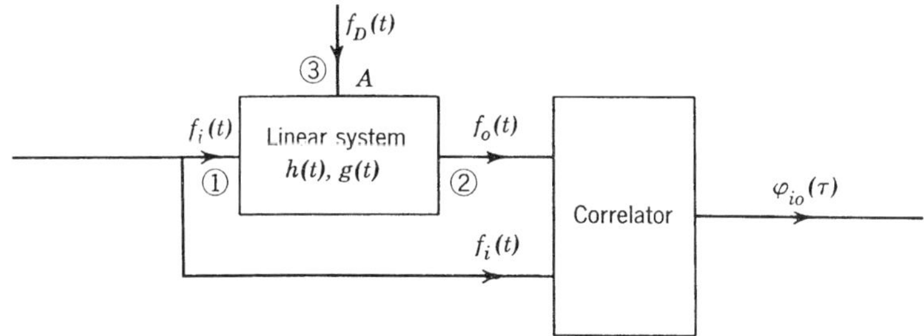

Fig. 10. Analysis of the determination of the unit-impulse response of a linear system by crosscorrelation under a disturbance.

a random disturbance $f_D(t)$. Suppose that the system is connected to a correlator for measuring the unit-impulse response. In this circuit, the output $f_o(t)$ is

$$f_o(t) = \int_{-\infty}^{\infty} h(\nu) f_i(t - \nu)\, d\nu + \int_{-\infty}^{\infty} g(\nu) f_D(t - \nu)\, d\nu \tag{93}$$

where $g(t)$ is the unit-impulse response of system A when a unit impulse is applied at point 3 and the response recorded at point 2. The unit-impulse response $h(t)$ of A has the same meaning as before. The output of the correlator $\varphi_{io}(\tau)$ is

$$\begin{aligned}\varphi_{io}(\tau) &= \lim_{T\to\infty} \frac{1}{2T} \int_{-T}^{T} f_i(t) \left[\int_{-\infty}^{\infty} h(\nu) f_i(t + \tau - \nu)\, d\nu \right. \\ &\qquad \left. + \int_{-\infty}^{\infty} g(\nu) f_D(t + \tau - \nu)\, d\nu \right] dt \\ &= \int_{-\infty}^{\infty} h(\nu)\varphi_{ii}(\tau - \nu)\, d\nu \\ &\qquad + \int_{-\infty}^{\infty} g(\nu)\, d\nu \lim_{T\to\infty} \frac{1}{2T} \int_{-T}^{T} f_i(t) f_D(t + \tau - \nu)\, dt \\ &= \int_{-\infty}^{\infty} h(\nu)\varphi_{ii}(\tau - \nu)\, d\nu + \int_{-\infty}^{\infty} g(\nu)\varphi_{iD}(\tau - \nu)\, d\nu \end{aligned} \tag{94}$$

If $f_i(t)$ and $f_D(t)$ are independent, then

$$\varphi_{iD}(\tau) = \overline{f_i(t)}\;\overline{f_D(t)} \tag{95}$$

and (94) becomes

$$\varphi_{io}(\tau) = \int_{-\infty}^{\infty} h(\nu)\varphi_{ii}(\tau - \nu)\, d\nu + \overline{f_i(t)}\;\overline{f_D(t)} \int_{-\infty}^{\infty} g(\nu)\, d\nu \tag{96}$$

The last term of this expression is a constant so that (96) is

$$\varphi_{io}(\tau) = \int_{-\infty}^{\infty} h(\nu)\varphi_{ii}(\tau - \nu)\, d\nu + \text{const.} \tag{97}$$

Therefore, if an independent disturbance $f_D(t)$ is present as shown in Fig. 10, the equation for $\varphi_{io}(\tau)$ differs from its no-disturbance form (87) by a constant. Furthermore, if any one of the factors $\overline{f_i(t)}$, $\overline{f_D(t)}$, and $\int_{-\infty}^{\infty} g(\nu)\, d\nu$ is zero, the constant becomes zero, and (97) is the same as (87). We conclude that, if the disturbance is independent of the input, the measurement of the system unit-impulse response by the crosscorrelation method, with a white noise excitation, has the theoretical result

$$\varphi_{io}(\tau) = 2\pi K h(\tau) + \text{const.} \tag{98}$$

and the constant can be reduced to zero by the condition just stated. Note that the same conclusion can be drawn if the disturbance is an internal one, provided that it is independent of the input.

The fact that the determination of the unit-impulse response of a linear system by the method of crosscorrelation is independent of external or internal disturbances, as long as they are independent of the excitation, is a remarkable advantage of the method over conventional methods. To test the validity of the method, an *RLC* circuit having the unit-impulse response shown in Fig. 11 * was connected to the correlator

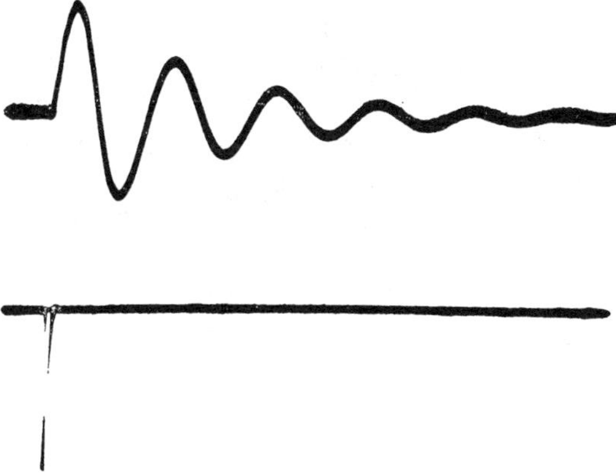

Fig. 11. Photograph of the unit-impulse response of an *RLC* circuit. (Unit impulse shown at bottom of figure.)

as shown in Fig. 10. The system A was the *RLC* circuit. Two independent white noise generators served as the sources for $f_i(t)$ and $f_D(t)$. The first experiment was performed without $f_D(t)$ in the circuit. Single-trace pictures of a portion of $f_i(t)$ and $f_o(t)$ are shown in Fig. 12.* The result of crosscorrelation is shown in Fig. 13.* The sample size for each point of the curve was 30,000. By comparing this crosscorrelation curve with the unit-impulse response produced by the conventional method, we see that the accuracy is satisfactory. This experiment gives strong support to the theoretical result (92).

The second experiment was conducted with $f_D(t)$ introduced. In order to emphasize this disturbance, it was made twice as strong as the excitation in rms values. With the same sample size, the crosscorrelation curve shown in Fig. 14 * was obtained. It is seen that the result does not differ noticeably from that obtained without any disturbance. The experiment was therefore in agreement with (98).

* From *Experimental Determination of System Functions by the Method of Correlation*, J. B. Wiesner and Y. W. Lee, a paper presented at the IRE National Convention, New York, 1950 (unpublished).

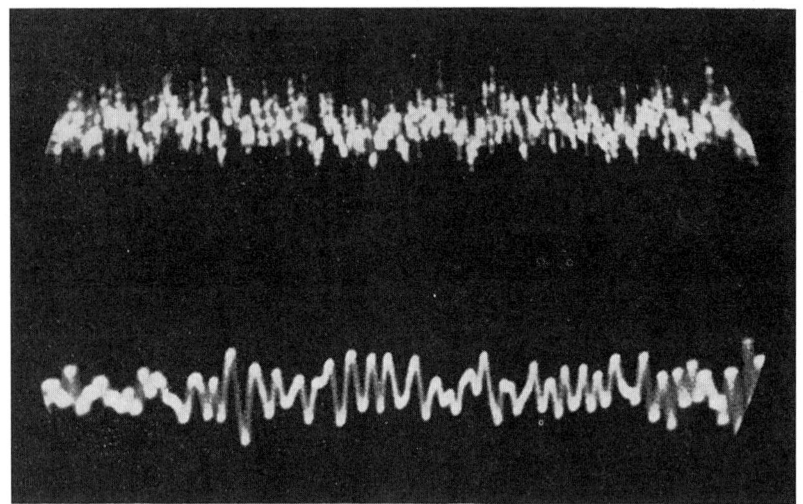

Fig. 12. Single-trace pictures of input (top) and output (bottom) noise of *RLC* circuit in the measurement of unit-impulse response by crosscorrelation.

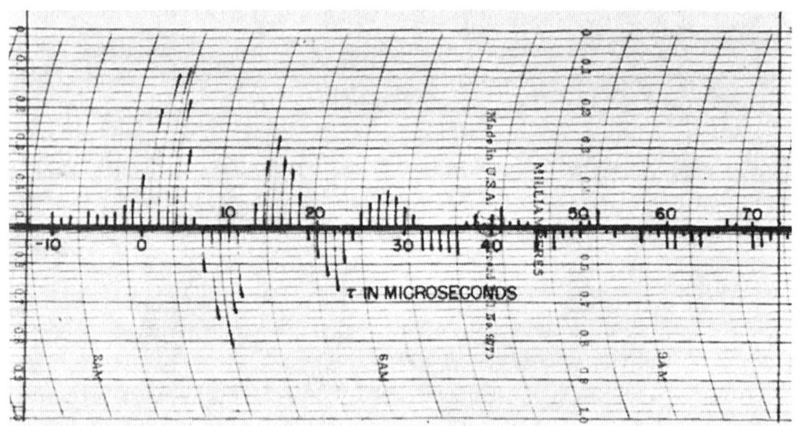

Fig. 13. Unit-impulse response of an *RLC* circuit, referred to in Fig. 11, measured by an electronic correlator by the method of crosscorrelation.

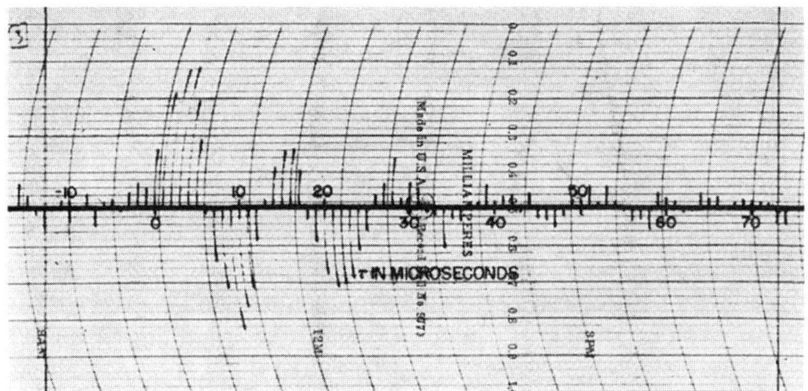

Fig. 14. Unit-impulse response of an RLC circuit, referred to in Fig. 11, measured by an electronic correlator by the method of crosscorrelation in the presence of a strong disturbance.

That a white noise excitation is chosen for the measurement of a unit-impulse response by crosscorrelation is primarily due to the fact that theorem (87) reduces to the simple form (92) by this choice. If a white noise source of sufficient power is available, this method avoids the solution of the convolution integral (87) under general conditions, thereby giving us the advantage of a direct measurement of the desired result.

The problem of the measurement of a system characteristic under general conditions by the application of theorem (87) brings out another advantage of the crosscorrelation method over conventional methods. This advantage is seen from the fact that, if a system, such as a control system, in operation has a stationary random input, this input can be regarded as the excitation for the unit-impulse response measurement. However, the function we seek, $h(t)$, is under the integral sign in (87). But in the frequency domain, we have from (89)

$$H(\omega) = \frac{\Phi_{io}(\omega)}{\Phi_{ii}(\omega)} \tag{99}$$

In this method, where $f_i(t)$ is not necessarily a white noise, the quantities that we measure are the input-output crosscorrelation $\varphi_{io}(\tau)$ and the input autocorrelation $\varphi_{ii}(\tau)$. These measurements are made with the actual input and actual output when the system is in normal operation. Clearly, the measurement of a system characteristic under such circumstances is indeed an advantage. Furthermore, the system under test need not be stopped, and perhaps disconnected from other components, and then excited by a special type of testing force for system characteristic measurement as it is done in accordance with conventional methods. The advantage of being able to measure the system characteristic without interfering with the system's normal and continuous operation is a notable one. We must emphasize that the method is based upon the assumption of a stationary random input. A difficulty of the method is that the solution of the input-output crosscorrelation theorem for $h(t)$ is not a simple one. At least we can say that our electronic techniques for solving a convolution integral have not been developed to the stage where a fairly general class of functions $\varphi_{io}(\tau)$ and $\varphi_{ii}(\tau)$ can be handled readily with satisfactory results. The solution in the frequency domain as indicated by (99) is also difficult. Since the spectrums $\Phi_{io}(\omega)$ and $\Phi_{ii}(\omega)$ involve transformations of the measured $\varphi_{io}(\tau)$ and $\varphi_{ii}(\tau)$, a considerable amount of computation is necessary to produce results of acceptable accuracy.

Because of the special features, it seems that this method of measuring the unit-impulse response or the system function of a linear system has great possibilities of practical and effective application. It is reasonable to expect that difficulties in computational techniques will be overcome as the development of this field of work advances.

In this application of crosscorrelation we have an opportunity to turn to the question of what a cross-power density represents. This question came to our mind when this spectrum was first introduced in an earlier chapter. Here we have a physical system associated with a cross-power density spectrum, and the equation involving them is (89). To obtain a physical interpretation for the cross-power density spectrum, we can think of it in terms of $H(\omega)$ and $\Phi_{ii}(\omega)$, as (89) indicates, as being the product of these quantities. We can point out that the phase spectrum of $\Phi_{io}(\omega)$ is that of $H(\omega)$ since $\Phi_{ii}(\omega)$ has a zero phase spectrum and that the amplitude spectrum of $\Phi_{io}(\omega)$ is the product of the amplitude spectrums of $H(\omega)$ and $\Phi_{ii}(\omega)$. There does not seem to be any necessity in requiring a physical meaning for $\Phi_{io}(\omega)$ in terms of power just because it is called a cross-power density spectrum.

10. Crosscorrelation of Linearly Transformed Random Functions

In Fig. 15 we have two linear systems 1 and 2 whose unit-impulse responses are $h_1(t)$ and $h_2(t)$ respectively. System 1 has the random input $f_1(t)$ and the output $f_a(t)$, and system 2 has the random input $f_2(t)$ and the output $f_b(t)$. We wish to consider the crosscorrelation of the outputs of the two systems in terms of the input and system characteristics. For this problem let us first write in accordance with definition the crosscorrelation $\varphi_{ab}(\tau)$ of the outputs $f_a(t)$ and $f_b(t)$

$$\varphi_{ab}(\tau) = \lim_{T\to\infty} \frac{1}{2T} \int_{-T}^{T} f_a(t) f_b(t+\tau)\, dt \qquad (100)$$

To bring into this function the input and system characteristics, we introduce the convolution integral for a linear system. Thus

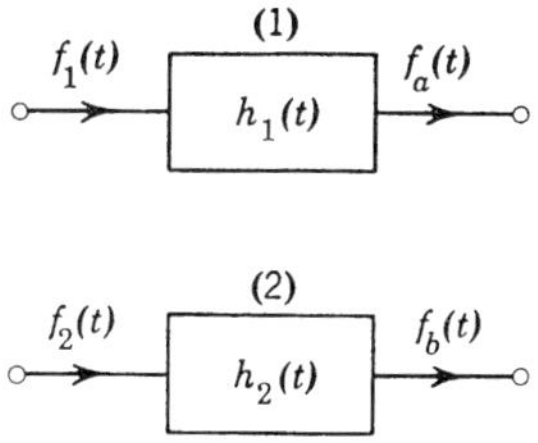

$$\left.\begin{aligned} f_a(t) &= \int_{-\infty}^{\infty} h_1(\nu) f_1(t-\nu)\, d\nu \\ \text{and} \qquad f_b(t) &= \int_{-\infty}^{\infty} h_2(\sigma) f_2(t-\sigma)\, d\sigma \end{aligned}\right\} \qquad (101)$$

Fig. 15. Random functions under linear transformation.

so that substitution of these integrals into (100) results in

$$\varphi_{ab}(\tau) = \lim_{T\to\infty} \frac{1}{2T} \int_{-T}^{T} dt \int_{-\infty}^{\infty} h_1(\nu) f_1(t-\nu)\, d\nu \int_{-\infty}^{\infty} h_2(\sigma) f_2(t+\tau-\sigma)\, d\sigma \qquad (102)$$

By inversion of the order of integration, we get

$$\varphi_{ab}(\tau) = \int_{-\infty}^{\infty} h_1(\nu)\, d\nu \int_{-\infty}^{\infty} h_2(\sigma)\, d\sigma \lim_{T\to\infty} \frac{1}{2T} \int_{-T}^{T} f_1(t-\nu) f_2(t+\tau-\sigma)\, dt \qquad (103)$$

Since

$$\varphi_{12}(\tau+\nu-\sigma) = \lim_{T\to\infty} \frac{1}{2T} \int_{-T}^{T} f_1(t-\nu) f_2(t+\tau-\sigma)\, dt \qquad (104)$$

(103) is

$$\varphi_{ab}(\tau) = \int_{-\infty}^{\infty} h_1(\nu)\, d\nu \int_{-\infty}^{\infty} h_2(\sigma)\, d\sigma\, \varphi_{12}(\tau+\nu-\sigma) \qquad (105)$$

We have thus arrived at an expression which gives the crosscorrelation of the outputs of the linear systems in terms of the crosscorrelation $\varphi_{12}(\tau)$ of the inputs and the unit-impulse responses of the systems. This result is similar, in form, to (32).

To establish the relation (105) in the frequency domain, we take the Fourier transform of both sides of the equation. Thus

$$\frac{1}{2\pi} \int_{-\infty}^{\infty} \varphi_{ab}(\tau) e^{-j\omega\tau}\, d\tau$$
$$= \frac{1}{2\pi} \int_{-\infty}^{\infty} e^{-j\omega\tau}\, d\tau \int_{-\infty}^{\infty} h_1(\nu)\, d\nu \int_{-\infty}^{\infty} h_2(\sigma)\, d\sigma\, \varphi_{12}(\tau+\nu-\sigma) \qquad (106)$$

The left-hand member of the equation is $\Phi_{ab}(\omega)$, the cross-power density spectrum of the outputs. By a change of variable $\mu = \tau + \nu - \sigma$, we treat the right-hand member as we did in (34) and (35). Accordingly

$$\Phi_{ab}(\omega) = \frac{1}{2\pi}\int_{-\infty}^{\infty} e^{-j\omega(\mu+\sigma-\nu)}\,d\mu \int_{-\infty}^{\infty} h_1(\nu)\,d\nu \int_{-\infty}^{\infty} h_2(\sigma)\,d\sigma\,\varphi_{12}(\mu)$$

$$= \left[\int_{-\infty}^{\infty} h_1(\nu)e^{j\omega\nu}\,d\nu\right]\left[\int_{-\infty}^{\infty} h_2(\sigma)e^{-j\omega\sigma}\,d\sigma\right]\left[\frac{1}{2\pi}\int_{-\infty}^{\infty}\varphi_{12}(\mu)e^{-j\omega\mu}\,d\mu\right] \tag{107}$$

Since the system functions $H_1(\omega)$ and $H_2(\omega)$ of systems 1 and 2 are related to the respective unit-impulse responses by

$$H_1(\omega) = \int_{-\infty}^{\infty} h_1(t)e^{-j\omega t}\,dt$$

and

$$H_2(\omega) = \int_{-\infty}^{\infty} h_2(t)e^{-j\omega t}\,dt \tag{108}$$

and since

$$\Phi_{12}(\omega) = \frac{1}{2\pi}\int_{-\infty}^{\infty}\varphi_{12}(\mu)e^{-j\omega\mu}\,d\mu \tag{109}$$

is the cross-power density spectrum of the inputs $f_1(t)$ and $f_2(t)$, we see that (107) is

$$\Phi_{ab}(\omega) = \overline{H}_1(\omega)H_2(\omega)\Phi_{12}(\omega) \tag{110}$$

This is the desired relation in the frequency domain corresponding to (105) in the time domain. We note that (110) is similar, in form, to (37).

The relation (105) can be put into another form. For this purpose let us introduce the change of variable $t = \sigma - \nu$ into (105); we obtain

$$\varphi_{ab}(\tau) = \int_{-\infty}^{\infty} h_1(\nu)\,d\nu \int_{-\infty}^{\infty} h_2(\nu + t)\,dt\,\varphi_{12}(\tau - t) \tag{111}$$

An inversion of the order of integration yields

$$\varphi_{ab}(\tau) = \int_{-\infty}^{\infty}\varphi_{12}(\tau - t)\,dt \int_{-\infty}^{\infty} h_1(\nu)h_2(\nu + t)\,d\nu \tag{112}$$

We note that we have treated the present problem in a manner similar to that in Sec. 6.

In (112) the extreme right-hand integral is the crosscorrelation $\varphi_{h1h2}(\tau)$ of the unit-impulse responses of the two systems by definition; that is,

$$\varphi_{h1h2}(\tau) = \int_{-\infty}^{\infty} h_1(t)h_2(t + \tau)\,dt \tag{113}$$

We assume that this function exists. With the introduction of this crosscorrelation function into (112) we arrive at the following alternative form of (105)

$$\varphi_{ab}(\tau) = \int_{-\infty}^{\infty} \varphi_{h1h2}(t)\varphi_{12}(\tau - t)\, dt \tag{114}$$

This interesting relation states that the output crosscorrelation of the linear systems is the convolution of the input crosscorrelation and the unit-impulse response crosscorrelation. Clearly one conclusion we can draw from this relation is that, if the inputs have a zero crosscorrelation, the outputs must also have a zero crosscorrelation irrespective of the linear transformations.

Finally let us consider the transform of the unit-impulse crosscorrelation of the two linear systems. Taking the transform of both sides of (113), we have

$$\frac{1}{2\pi}\int_{-\infty}^{\infty} \varphi_{h1h2}(\tau)e^{-j\omega\tau}\, d\tau = \frac{1}{2\pi}\int_{-\infty}^{\infty} e^{-j\omega\tau}\, d\tau \int_{-\infty}^{\infty} h_1(t)h_2(t+\tau)\, dt \tag{115}$$

in which the double integral is evaluated by a change of variable as we did before. As a result we obtain

$$\frac{1}{2\pi}\int_{-\infty}^{\infty} \varphi_{h1h2}(\tau)e^{-j\omega\tau}\, d\tau = \frac{1}{2\pi}\overline{H}_1(\omega)H_2(\omega) \tag{116}$$

so that we find

$$\overline{H}_1(\omega)H_2(\omega) = \int_{-\infty}^{\infty} \varphi_{h1h2}(\tau)e^{-j\omega\tau}\, d\tau \tag{117}$$

and by inverse transformation we have

$$\varphi_{h1h2}(\tau) = \frac{1}{2\pi}\int_{-\infty}^{\infty} \overline{H}_1(\omega)H_2(\omega)e^{j\omega\tau}\, d\omega \tag{118}$$

PROBLEMS

1. In the circuit of Fig. P1 the white noise has a power density spectrum of c^2 (volt^2 per radian per second) and the oscillator generates a voltage $e(t) = a \sin t + b \cos 2t$ (volts). Determine and sketch the autocorrelation function and power density spectrum of the output voltage $e_o(t)$.

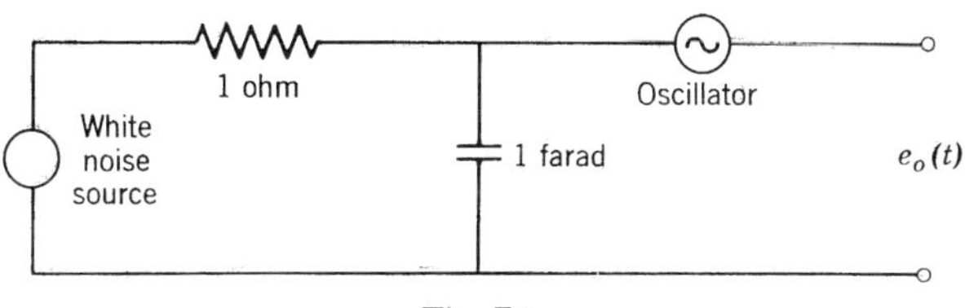

Fig. P1.

2. In Fig. P2, the system input $f_i(t)$ and output $f_o(t)$ are random functions. Derive an expression for the cross-power density spectrum $\Phi_{jk}(\omega)$ in terms of the system functions and the cross-power density spectrum $\Phi_{io}(\omega)$. No other functions should appear in the final expression.

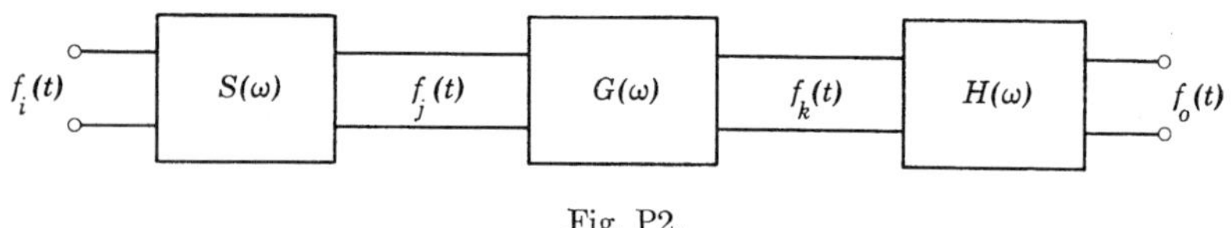

Fig. P2.

3. In Fig. P3, $f_1(t)$ and $f_2(t)$ are random functions whose cross-power density spectrum is $\Phi_{12}(\omega) = 1/(1 + \omega^2)$. By passing $f_1(t)$ through a linear system A with system function $H_a(\omega)$ and $f_2(t)$ through a linear system B with system function $H_b(\omega)$, we wish to produce a cross-power density spectrum of the outputs $f_a(t)$ and $f_b(t)$ in the form $\Phi_{ab}(\omega) = 1/(1 + \omega^2)^3$. Note that $f_1(t) \neq f_2(t)$. Show whether it is possible to do so. If it is possible, what should $H_a(\omega)$ and $H_b(\omega)$ be? If not, why not?

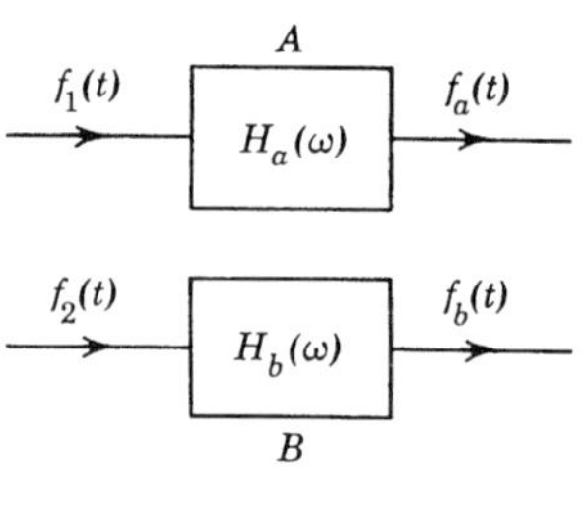

Fig. P3.

4. A random wave is formed by a series of pulses that have the same waveform. The initial points of the pulses have a Poisson distribution, and the positive and negative pulses have the same probability of occurrence. The power density spectrum of the wave is

$$\Phi_{11}(\omega) = \frac{1}{(1 + \omega^2)^4}$$

Determine the waveform of the pulses.

5. A random voltage $f_i(t)$ that alternates between the two possible values $+E$ and $-E$ with zero crossings following the Poisson distribution is applied to the circuit shown in Fig. P5. The autocorrelation curve of the output $f_o(t)$ is to be measured by the electronic correlator which selects essentially independent pairs of samples from $f_o(t)$. Determine the sample size for a standard deviation of 0.01 for large values of τ when the autocorrelation function has practically reached the final value. $E = 10$ volts, $k = 1000$ (average number of zero crossings per second), $1/RC = 1000$.

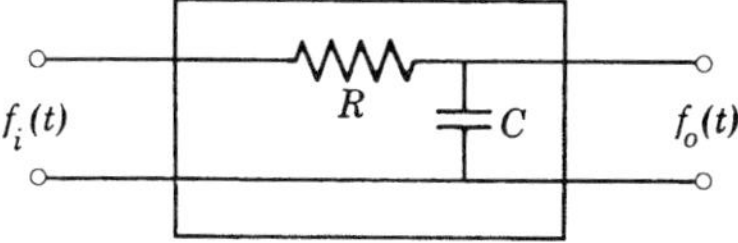

Fig. P5.

6. A rectangular voltage wave $f_i(t)$ that alternates between the two possible values $+E$ and $-E$ with Poisson-distributed zero crossings is applied to the system shown in Fig. P6. The average number of zero crossings per second is k. The input-output crosscorrelation function $\varphi_{io}(\tau)$ of the system is to be measured by the sampling method with the aid of an electronic correlator. If the variance of the measured crosscorrelation function at the point $\tau = 0$ is to be A, what is the necessary sample size for the measurement at this point in terms of E, k, R, C, and A?

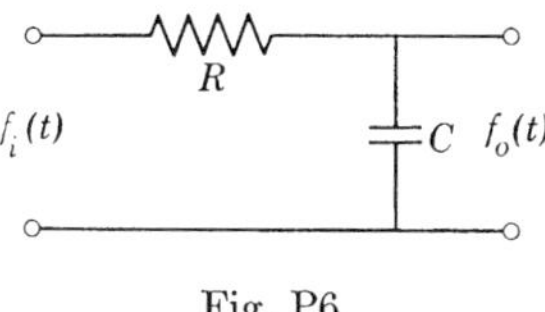

Fig. P6.

7. A noise $N(t)$ consisting of random short pulses, each of the form Ete^{-at}, is fed into a multiplying device as shown in Fig. P7. On the average there are k pulses per second. The occurrence of the pulses follows the Poisson distribution with the possibility of overlapping and equal probability of being positive or negative. A sine wave $A \sin \omega_o t$ is supplied to the device so that is output is $N(t)[A \sin \omega_o t]$. Determine the autocorrelation function of the output.

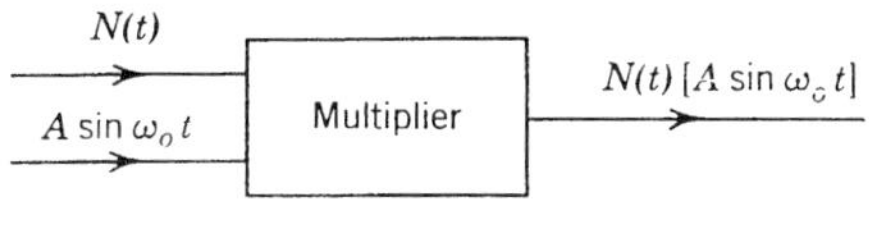

Fig. P7.

8. A random process $f_a(t)$ in the form of a series of Poisson-distributed unit impulses is applied as a current at the point a in the lattice network shown in Fig. P8. The impulses of $f_a(t)$ occur with equal probability of being positive or negative.

(*a*) Determine the waveform of the current $f_b(t)$ at the point b. The derivation of the expression for the system function must be shown. Sketch the waveform to correspond to the impulses at the input as shown in the figure.

(*b*) Determine the power density spectrum of $f_b(t)$.

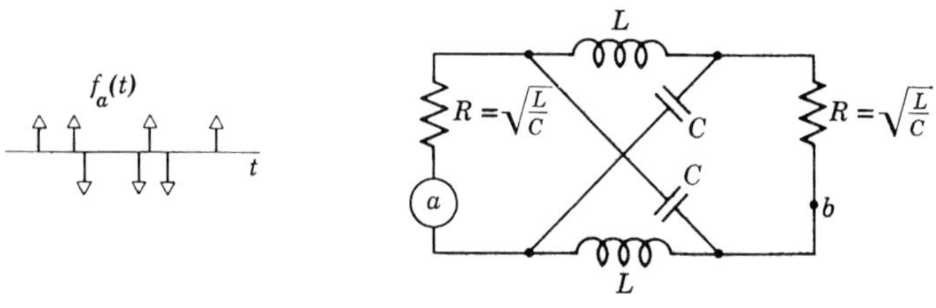

Fig. P8.

9. As shown in Fig. P9 a white Gaussian noise $f_1(t)$ with the power density spectrum $\Phi_{11}(\omega) = K$ is applied to a linear network N whose system function is $H(\omega) = R/(R + j\omega L)$. The output $f_2(t)$ of the network goes to the summing circuit S in two paths, one being direct and one passing through the delay circuit D, so that the output of the summing circuit is $f_o(t) = f_2(t) + f_2(t - \alpha)$ where α is the delay time. Assuming that $f_o(t)$ is Gaussian, determine the amplitude probability density of $f_o(t)$ in terms of the given quantities.

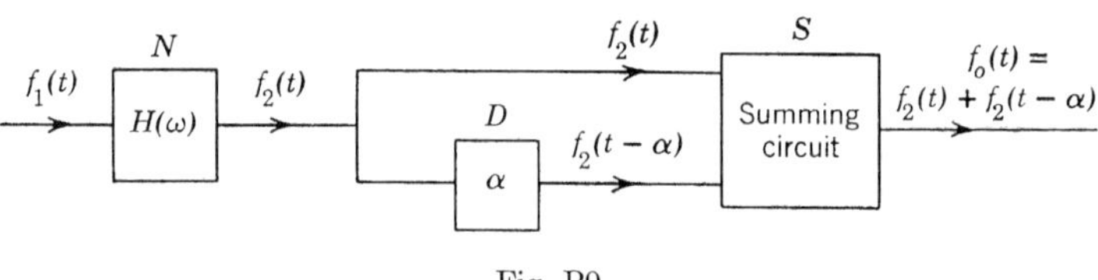

Fig. P9.

10. In Fig. P10, $f_a(t)$ is a Poisson rectangular wave with the average number of zero crossings per second equal to k; $f_b(t)$ is a wave of rectangular pulses each of duration b. At each point where $f_a(t)$ jumps from $-E$ to $+E$, a positive pulse of duration b starts in $f_b(t)$; and, at each point where $f_a(t)$ drops from $+E$ to $-E$, a negative pulse of duration b starts in $f_b(t)$. In cases where a group of rectangular pulses of $f_b(t)$ overlap, the instantaneous values will be the algebraic sum of these pulses. Using the material in this chapter, determine the cross-power density spectrum $\Phi_{ab}(\omega)$ of $f_a(t)$ and $f_b(t)$.

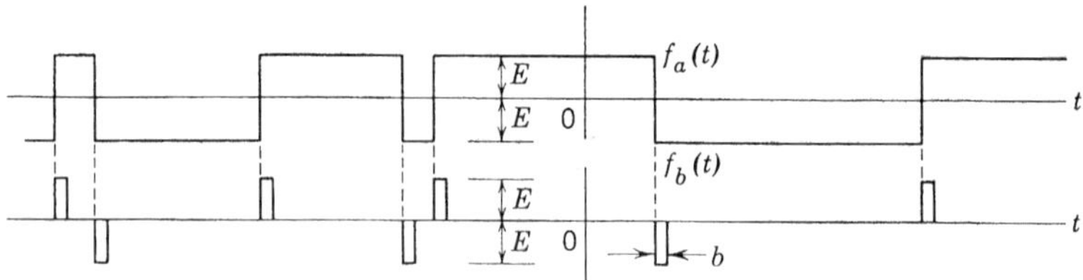

Fig. P10.

chapter 14

Optimum Linear Systems

In classical communication theory an ideal filter has zero attenuation over the band of frequencies of the message and infinite attenuation over other bands occupied by unwanted messages and noise. The practical filter approximates this ideal characteristic. When message and noise spectrums overlap, the design is guided by experience and judgment to achieve the best compromise. Another basic concept in classical theory is that messages and noise are representable by a Fourier series or a Fourier integral. In other words, classical theory regards messages and noise as periodic or aperiodic phenomena. Frequently a "typical" message such as a step is considered as an appropriate representation of the actual message.

Classical filters have played a major role in the development of communication engineering. They will remain, for a long time to come, as an extremely important part of some communication systems. However, a new point of view in the problem of filtering, which incorporates the statistical concept of messages and noise, has led to the development of a statistical theory of linear systems of which filter theory is only a part. The development has given the communication engineer a better insight into a large class of problems as well as their relations to one another.

The statistical theory was originated by Norbert Wiener and published by him in *The Extrapolation, Interpolation and Smoothing of Stationary Time Series.* Beginning with the present chapter, we shall present a detailed account of this interesting and powerful theory.

1. The Filter Problem and Some Generalizations

In the linear filter problem we are given a message that has been corrupted by a noise, and it is desired to extract the message by a linear system with minimum error in accordance with a chosen measure of error. In Fig. 1 the linear system A is the filter we wish to design. By this we mean that either the system function $H(\omega)$ or the unit-impulse

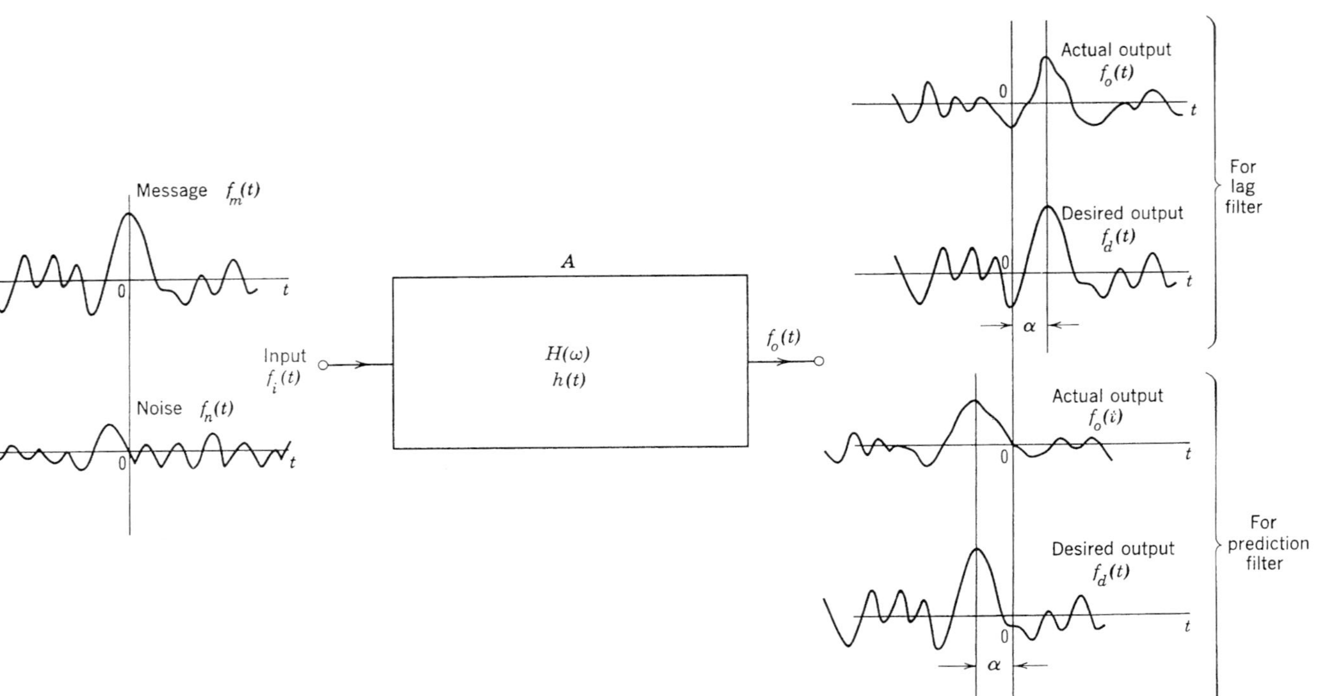

Fig. 1. Illustrating the formulation of the optimum filtering and prediction problem.

response $h(t)$ of the system is to be determined for the purpose stated. With the specification of $H(\omega)$ or $h(t)$ as a result of the solution of the problem, it will be possible to construct the filter by well-known techniques. The specification must be such that, when exact physical realization is impossible, we should be able to approximate it as closely as we please. Our primary problem is the determination of $H(\omega)$ or $h(t)$. The synthesis of the linear system is only a secondary problem.

The input of the system $f_i(t)$ is in general some function of a message $f_m(t)$ and a noise $f_n(t)$, but a common situation is that the input is the sum of the message and the noise. Thus

$$f_i(t) = f_m(t) + f_n(t) \tag{1}$$

as indicated in the figure. Both $f_m(t)$ and $f_n(t)$ are stationary random processes.

Because of the presence of the noise, it is expected that complete recovery of the message by the linear system is impossible, unless the situation is an exceptional one. Nevertheless we can state the ideal output. We shall call this output the *desired output* $f_d(t)$. The *actual output* of the system is denoted by $f_o(t)$.

If we put $f_d(t) = f_m(t)$, it would mean that we desire at the output an instantaneous replica of the message. However, in many applications, such as a telephone circuit, a lag in the reproduction of the message is not a distortion. Allowing for the lag, α, we put $f_d(t) = f_m(t - \alpha)$ as indicated in the figure. A peak in a sample of $f_m(t)$ at the input is arbitrarily chosen as the reference point $t = 0$ to show the relative time at the input and output. The peak in the desired output lags the corresponding peak at the input by the time α. Assuming that it is physically impossible to produce the desired output exactly, we shall determine the system with an actual output $f_o(t)$ that approximates the desired output as closely as possible, in the sense that the error, whose measure has been chosen, is the minimum.

Instead of a lag in the desired output, we may consider a lead. With $f_d(t) = f_m(t + \alpha)$ it would mean that, at the output of the system, we desire the pure input message with an advance in time of α seconds. The situation is indicated in the figure with $f_d(t)$ representing $f_m(t)$ displaced forward in time by α seconds. Taking the particular peak referred to for illustration, we see that, if perfect realization of $f_d(t)$ were possible, the peak would appear at the output α seconds before it actually happens at the input. We are really stating a problem of prediction. Furthermore, since we are formulating the problem with message and noise described statistically, the prediction is a statistical one. The system is to be designed in such a way that the actual output follows the desired output as closely as possible.

In discussing the statement of the filter problem, we have brought out the interesting fact that a filter can be either a lag filter or a prediction filter. Emphasis has been placed on the specification of the desired output in a problem. For the filter we have

$$f_d(t) = f_m(t \pm \alpha) \tag{2}$$

where α is a positive real number representing the lag time or prediction time depending upon whether we desire a lag filter or a prediction filter.

In specifying the desired output, we know that we are usually demanding more than what the system can accomplish. There will be an error. Clearly the instantaneous error is

$$\varepsilon(t) = f_o(t) - f_d(t) \tag{3}$$

An appropriate mean value of $\varepsilon(t)$ shall be taken as a measure of how well the filter performs with respect to the demand. Evidently we must consider some positive function of $\varepsilon(t)$ since a meaningful measure must not allow positive and negative instantaneous errors to cancel each other. The absolute value of $\varepsilon(t)$ and this value raised to an integral power are examples of positive functions of $\varepsilon(t)$. In the original work of Wiener, the mean-square error was chosen. It is defined as

$$\overline{\varepsilon^2(t)} = \lim_{T\to\infty} \frac{1}{2T} \int_{-T}^{T} [f_o(t) - f_d(t)]^2 \, dt \tag{4}$$

The reason for this choice is primarily the subsequent comparative mathematical simplicity in the development of the theory. Although for many practical problems the mean-square error is a meaningful measure of error, it is quite possible that other measures are better for some problems. Unfortunately other measures of error almost always lead to unmanageable difficulties. In the present work we follow (4), but at a later stage we shall consider another measure of error. The linear system that we determine based upon the minimum-mean-square-error criterion will be called the *optimum linear system.*

2. The Mean-Square-Error Functional

To link the filter performance to the mean-square error, we introduce $h(t)$ into (4) through the term $f_o(t)$, in the form of the superposition theorem. In so doing, we have

$$\overline{\varepsilon^2(t)} = \lim_{T\to\infty} \frac{1}{2T} \int_{-T}^{T} \left[\int_{-\infty}^{\infty} h(\tau) f_i(t-\tau) \, d\tau - f_d(t) \right]^2 dt \tag{5}$$

By expansion

$$\overline{\varepsilon^2(t)} = \lim_{T\to\infty} \frac{1}{2T} \int_{-T}^{T} dt \int_{-\infty}^{\infty} h(\tau) f_i(t-\tau)\, d\tau \int_{-\infty}^{\infty} h(\sigma) f_i(t-\sigma)\, d\sigma$$

$$- 2 \lim_{T\to\infty} \frac{1}{2T} \int_{-T}^{T} f_d(t)\, dt \int_{-\infty}^{\infty} h(\tau) f_i(t-\tau)\, d\tau + \lim_{T\to\infty} \frac{1}{2T} \int_{-T}^{T} f_d^2(t)\, dt \quad (6)$$

In this expansion we note that

$$\left[\int_{-\infty}^{\infty} h(\tau) f_i(t-\tau)\, d\tau\right]^2 = \int_{-\infty}^{\infty} h(\tau) f_i(t-\tau)\, d\tau \int_{-\infty}^{\infty} h(\sigma) f_i(t-\sigma)\, d\sigma \quad (7)$$

Inversion of the order of integration in the multiple integrals of (6) yields

$$\overline{\varepsilon^2(t)} = \int_{-\infty}^{\infty} h(\tau)\, d\tau \int_{-\infty}^{\infty} h(\sigma)\, d\sigma \lim_{T\to\infty} \frac{1}{2T} \int_{-T}^{T} f_i(t-\tau) f_i(t-\sigma)\, dt$$

$$- 2 \int_{-\infty}^{\infty} h(\tau)\, d\tau \lim_{T\to\infty} \frac{1}{2T} \int_{-T}^{T} f_d(t) f_i(t-\tau)\, dt + \lim_{T\to\infty} \frac{1}{2T} \int_{-T}^{T} f_d^2(t)\, dt \quad (8)$$

Since

$$\lim_{T\to\infty} \frac{1}{2T} \int_{-T}^{T} f_i(t-\tau) f_i(t-\sigma)\, d\sigma = \varphi_{ii}(\tau - \sigma) \quad (9)$$

$$\lim_{T\to\infty} \frac{1}{2T} \int_{-T}^{T} f_d(t) f_i(t-\tau)\, dt = \varphi_{id}(\tau) \quad (10)$$

$$\lim_{T\to\infty} \frac{1}{2T} \int_{-T}^{T} f_d^2(t)\, dt = \varphi_{dd}(0) \quad (11)$$

(8) becomes

$$\overline{\varepsilon^2(t)} = \int_{-\infty}^{\infty} h(\tau)\, d\tau \int_{-\infty}^{\infty} h(\sigma)\, d\sigma\, \varphi_{ii}(\tau - \sigma) - 2 \int_{-\infty}^{\infty} h(\tau)\, d\tau\, \varphi_{id}(\tau) + \varphi_{dd}(0) \quad (12)$$

We have arrived at an expression of considerable importance and significance. The expression states that the mean-square error between the actual output and the desired output of the filter depends upon the following quantities:

$h(t)$ = the unit-impulse response of the filter
$\varphi_{ii}(\tau)$ = the autocorrelation of the filter input
$\varphi_{id}(\tau)$ = the input–desired-output crosscorrelation of the filter
$\varphi_{dd}(0)$ = the mean square value of the desired filter output

We have not as yet minimized the mean-square error of the filter. In fact, we have taken any linear system with the unit-impulse response $h(t)$ whose input has the autocorrelation $\varphi_{ii}(\tau)$ and have assigned a desired output so that the crosscorrelation $\varphi_{id}(\tau)$ is known, and we have found that the mean-square error between the actual output of the system and the desired output is given by (12). The result shows that aside from the system behavior, the functions that determine the mean-square error are the input autocorrelation and the input–desired-output crosscorrelation. At this point we can see clearly why correlation functions are important in the theory of linear systems.

Since $\varphi_{ii}(\tau)$ and $\varphi_{id}(\tau)$ are completely determined once the input and desired output are specified in a problem, any change in the mean-square error (12) can come only through a change in the system unit-impulse response $h(t)$. Hence (12) has a specific numerical value for each member of the class of linear systems. In other words (12) is a functional of $h(t)$. To emphasize this point, we introduce the subscript h in the left-hand member of (12). Thus

$$\overline{\varepsilon_h{}^2(t)} = \int_{-\infty}^{\infty} h(\tau)\,d\tau \int_{-\infty}^{\infty} h(\sigma)\,d\sigma\,\varphi_{ii}(\tau-\sigma) - 2\int_{-\infty}^{\infty} h(\tau)\,d\tau\,\varphi_{id}(\tau) + \varphi_{dd}(0) \tag{13}$$

The problem now before us is the determination of $h(t)$ in (13) which minimizes the mean-square error. Since the minimization of functionals is the subject of calculus of variations, we now turn to this subject for a brief review.

3. Calculus of Variations

The simplest problem in calculus of variations is the minimization or maximization of the functional

$$I = \int_a^b f(x, y, y')\,dx \tag{14}$$

where $y' = dy/dx$ and f is a given function. The integral is taken along the curve $y(x)$ with

$$\left.\begin{aligned} y(a) &= A \\ \text{and} \qquad y(b) &= B \end{aligned}\right\} \tag{15}$$

Our problem is to find the curve which makes (14) a maximum or a minimum and which satisfies the conditions (15).

Instead of $y(x)$, we consider $y(x)$ plus a change in the whole curve; that is, we consider

$$y(x) + \epsilon\eta(x) \tag{16}$$

which is written simply as

$$y + \epsilon\eta \tag{17}$$

where ϵ is a real parameter, independent of x, and η is any differentiable function of x, satisfying the conditions

$$\left.\begin{aligned} \eta(a) &= 0 \\ \eta(b) &= 0 \end{aligned}\right\} \tag{18}$$

so that at $x = a$ and $x = b$ the conditions (15) are satisfied. The change in the whole curve, $\epsilon\eta$, is called the variation of y and is denoted by δy. In Fig. 2*a*, the curve $y(x)$ is indicated by a heavy solid line. For a particular form of $\eta(x)$ as shown, $y(x) + \epsilon\eta(x)$ with various values of ϵ are indicated by a family of curves in the neighborhood of $y(x)$. Figure 2*b* shows the situation when a different form of $\eta(x)$ is chosen.

As a result of the variation δy in y, the functional I has the variation δI. With these variations, (14) is

$$I + \delta I = \int_a^b f(x, y + \epsilon\eta, y' + \epsilon\eta')\, dx \tag{19}$$

We assume that, for any possible variation η, (19) has a continuous derivative with respect to ϵ. With any $y(x)$ and $\eta(x)$ given, we can find the value of ϵ for which $I + \delta I$ is a maximum or minimum from the condition

$$\frac{d}{d\epsilon}(I + \delta I) = 0 \tag{20}$$

When $y(x)$ is not the curve for maximum or minimum I, ϵ will be nonzero and δI will also be nonzero. However, if the given $y(x)$ were actually the curve for maximum or minimum I, the variation

$$\delta y = \epsilon\eta \tag{21}$$

should be zero so that, no matter what the given η is, ϵ should turn out to be zero when (20) is solved for ϵ. We conclude then that the curve $y(x)$ for which I is maximum or minimum must satisfy the condition

$$\frac{d}{d\epsilon}(I + \delta I)\Big|_{\epsilon=0} = 0 \qquad \text{for all possible } \eta\text{'s} \tag{22}$$

This is a necessary condition. The determination of the sufficiency of the condition is a much more difficult problem. In view of this difficulty

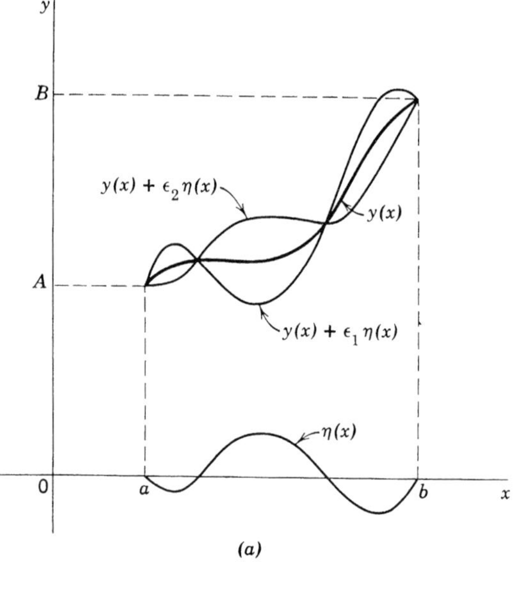

(a)

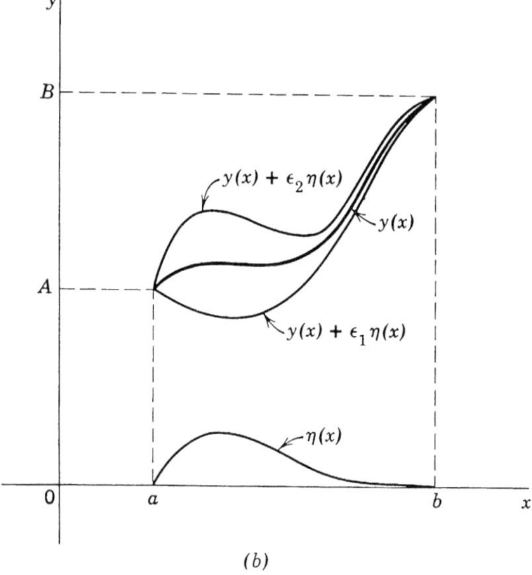

(b)

Fig. 2. Illustrating a problem in calculus of variations.

we shall show later in our optimum system problem that the mathematical proof of sufficiency can be avoided by physical considerations.

In applying the functional (19) to condition (22), let

$$\left.\begin{aligned} f_\epsilon &= f(x, y + \epsilon\eta, y' + \epsilon\eta') \\ y_\epsilon &= y + \epsilon\eta \\ y_\epsilon' &= y' + \epsilon\eta' \end{aligned}\right\} \qquad (23)$$

so that (19) is

$$I + \delta I = \int_a^b f_\epsilon(x, y_\epsilon, y_\epsilon')\, dx \qquad (24)$$

Then differentiating under the integral sign, we have

$$\begin{aligned} \frac{d}{d\epsilon}(I + \delta I) &= \int_a^b \left(\frac{\partial f_\epsilon}{\partial x}\frac{dx}{d\epsilon} + \frac{\partial f_\epsilon}{\partial y_\epsilon}\frac{dy_\epsilon}{d\epsilon} + \frac{\partial f_\epsilon}{\partial y_\epsilon'}\frac{dy_\epsilon'}{d\epsilon}\right) dx \\ &= \int_a^b \left(\frac{\partial f_\epsilon}{\partial y_\epsilon}\eta + \frac{\partial f_\epsilon}{\partial y_\epsilon'}\eta'\right) dx \end{aligned} \qquad (25)$$

When $\epsilon = 0$, (25) is

$$\frac{d}{d\epsilon}(I + \delta I)\Big|_{\epsilon=0} = \int_a^b \left(\frac{\partial f}{\partial y}\eta + \frac{\partial f}{\partial y'}\eta'\right) dx \qquad (26)$$

In accordance with (22), a necessary condition for I to be maximum or minimum is

$$\int_a^b \left(\frac{\partial f}{\partial y}\eta + \frac{\partial f}{\partial y'}\eta'\right) dx = 0 \qquad \text{for all possible } \eta\text{'s} \qquad (27)$$

This condition can be reduced to a simple form by first changing the integrand in the second integral

$$\int_a^b \frac{\partial f}{\partial y'}\eta'\, dx \qquad (28)$$

to one that has η as a factor. This is accomplished by integration by parts. Thus

$$\int_a^b \frac{\partial f}{\partial y'}\eta'\, dx = \eta(x)\frac{\partial f}{\partial y'}\Big|_a^b - \int_a^b \eta\frac{d}{dx}\left(\frac{\partial f}{\partial y'}\right) dx \qquad (29)$$

Since by (18) the first right-hand member is zero, (29) is

$$\int_a^b \frac{\partial f}{\partial y'}\eta'\, dx = -\int_a^b \eta\frac{d}{dx}\left(\frac{\partial f}{\partial y'}\right) dx \qquad (30)$$

Substitution of (30) into (27) yields the condition

$$\int_a^b \left[\frac{\partial f}{\partial y} - \frac{d}{dx}\left(\frac{\partial f}{\partial y'}\right)\right] \eta \, dx = 0 \qquad \text{for all possible } \eta\text{'s} \tag{31}$$

Inasmuch as η is any differentiable function satisfying the end conditions (18), the coefficient of η in (31) must be zero. Clearly, if this coefficient were not zero, an η that has the same sign as the coefficient for all values of x from $x = a$ to $x = b$ can be chosen so that the integral in (31) would be a positive quantity in contradiction to the requirement that it be zero for all possible η's. Therefore we conclude that, if I in (14) has a maximum or a minimum value along a curve $y(x)$, the curve must be a solution of the equation

$$\frac{\partial f}{\partial y} - \frac{d}{dx}\left(\frac{\partial f}{\partial y'}\right) = 0 \tag{32}$$

This is known as the Euler-Lagrange equation.

In applying calculus of variations to our optimum system problem, we do not actually require condition (32), but instead we follow the method just reviewed and obtain a condition for maximum or minimum mean-square error which is analogous to (32).

4. Minimization of the Mean-Square-Error Functional

We return to the mean-square-error functional (13) to determine a necessary condition for its minimum value. We see that $\overline{\varepsilon_h{}^2(t)}$ corresponds to I and $h(t)$ corresponds to $y(x)$, so that the problem is the search for the $h(t)$ that minimizes $\overline{\varepsilon_h{}^2(t)}$. Following the procedure in the review, we let ϵ be a real parameter, independent of t, and $\eta(t)$ be any differentiable function satisfying the condition

$$\eta(t) = 0 \qquad \text{for } t < 0 \tag{33}$$

We let $\delta h(t) = \epsilon\eta(t)$ be the variation of $h(t)$ and $\delta\overline{\varepsilon_h{}^2(t)}$ be the variation of the mean-square-error functional. The condition (33) is necessitated by the physical fact that

$$h(t) = 0 \qquad \text{for } t < 0 \tag{34}$$

Since the system is at rest before the application of the unit impulse at $t = 0$ and since the response of the system cannot precede the excitation, condition (34) must hold. Clearly if a variation of $h(t)$ is a possible one, it must have the property (33); otherwise, we would be introducing changes in the unit-impulse response of a linear system that are physically impossible. Introducing the variations, we find that (13) becomes

$$\overline{\varepsilon_h^2(t)} + \overline{\delta\varepsilon_h^2(t)} = \int_{-\infty}^{\infty} [h(\tau) + \epsilon\eta(\tau)]\, d\tau \int_{-\infty}^{\infty} [h(\sigma) + \epsilon\eta(\sigma)]\, d\sigma\, \varphi_{ii}(\tau - \sigma)$$

$$- 2\int_{-\infty}^{\infty} [h(\tau) + \epsilon\eta(\tau)]\, d\tau\, \varphi_{id}(\tau) + \varphi_{dd}(0) \quad (35)$$

By expansion of the right-hand members, there will be terms that do not contain ϵ. The sum of these terms is $\overline{\varepsilon_h^2(t)}$, as we expect. The remaining terms will have ϵ or ϵ^2 as a factor. These terms appear as a result of a variation in h, and their sum is the variation in mean-square error. Hence

$$\overline{\delta\varepsilon_h^2(t)} = \epsilon\int_{-\infty}^{\infty} \eta(\tau)\, d\tau \int_{-\infty}^{\infty} h(\sigma)\, d\sigma\, \varphi_{ii}(\tau - \sigma)$$

$$+ \epsilon\int_{-\infty}^{\infty} \eta(\sigma)\, d\sigma \int_{-\infty}^{\infty} h(\tau)\, d\tau\, \varphi_{ii}(\tau - \sigma)$$

$$+ \epsilon^2\int_{-\infty}^{\infty} \eta(\tau)\, d\tau \int_{-\infty}^{\infty} \eta(\sigma)\, d\sigma\, \varphi_{ii}(\tau - \sigma)$$

$$- 2\epsilon\int_{-\infty}^{\infty} \eta(\tau)\, d\tau\, \varphi_{id}(\tau) \quad (36)$$

We can show that the first and second right-hand members are equal by simply noting that φ_{ii} is an even function so that the second right-hand member is

$$\epsilon\int_{-\infty}^{\infty} \eta(\sigma)\, d\sigma \int_{-\infty}^{\infty} h(\tau)\, d\tau\, \varphi_{ii}(\tau - \sigma)$$

$$= \epsilon\int_{-\infty}^{\infty} \eta(\sigma)\, d\sigma \int_{-\infty}^{\infty} h(\tau)\, d\tau\, \varphi_{ii}(\sigma - \tau) \quad (37)$$

Since τ and σ in (37) are independent of the τ and σ in the first right-hand member of (36), we see that these double integrals are the same. Combining the first two terms on the right-hand side of (36), we have

$$\overline{\delta\varepsilon_h^2(t)} = 2\epsilon\int_{-\infty}^{\infty} \eta(\tau)\, d\tau \int_{-\infty}^{\infty} h(\sigma)\, d\sigma\, \varphi_{ii}(\tau - \sigma)$$

$$+ \epsilon^2\int_{-\infty}^{\infty} \eta(\tau)\, d\tau \int_{-\infty}^{\infty} \eta(\sigma)\, d\sigma\, \varphi_{ii}(\tau - \sigma) - 2\epsilon\int_{-\infty}^{\infty} \eta(\tau)\, d\tau\, \varphi_{id}(\tau) \quad (38)$$

To simplify the writing of (38) and subsequent manipulations, let

$$\left.\begin{aligned} A &= \int_{-\infty}^{\infty} \eta(\tau)\, d\tau \int_{-\infty}^{\infty} h(\sigma)\, d\sigma\, \varphi_{ii}(\tau - \sigma) \\ B &= \int_{-\infty}^{\infty} \eta(\tau)\, d\tau \int_{-\infty}^{\infty} \eta(\sigma)\, d\sigma\, \varphi_{ii}(\tau - \sigma) \\ C &= \int_{-\infty}^{\infty} \eta(\tau)\, d\tau\, \varphi_{id}(\tau) \end{aligned}\right\} \tag{39}$$

so that (38) reads

$$\overline{\delta \varepsilon_h{}^2(t)} = 2\epsilon A + \epsilon^2 B - 2\epsilon C \tag{40}$$

In applying condition (22) to the mean-square-error functional, we note that, since $\overline{\varepsilon_h{}^2(t)}$ is not a function of ϵ, the condition for minimum value of the functional is

$$\frac{d}{d\epsilon}\, \overline{\delta \varepsilon_h{}^2(t)}\,\big|_{\epsilon=0} = 0 \qquad \text{for all possible } \eta\text{'s} \tag{41}$$

Differentiating (40) with respect to ϵ and then setting $\epsilon = 0$, we have

$$\begin{aligned} \frac{d}{d\epsilon}\, \overline{\delta \varepsilon_h{}^2(t)}\,\big|_{\epsilon=0} &= 2A + 2\epsilon B - 2C\,\big|_{\epsilon=0} \\ &= 2A - 2C \end{aligned} \tag{42}$$

Condition (41) now reads

$$A - C = 0 \qquad \text{for all possible } \eta\text{'s} \tag{43}$$

or, referring back to (39),

$$\int_{-\infty}^{\infty} \eta(\tau)\, d\tau \int_{-\infty}^{\infty} h(\sigma)\, d\sigma\, \varphi_{ii}(\tau - \sigma) - \int_{-\infty}^{\infty} \eta(\tau)\, d\tau\, \varphi_{id}(\tau) = 0$$

$$\text{for all possible } \eta\text{'s} \tag{44}$$

Since $\eta(\tau)$ is common to both left-hand members, their combination reduces (44) to

$$\int_{-\infty}^{\infty} \eta(\tau) \left[\int_{-\infty}^{\infty} h(\sigma) \varphi_{ii}(\tau - \sigma)\, d\sigma - \varphi_{id}(\tau) \right] d\tau = 0$$

$$\text{for all possible } \eta\text{'s} \tag{45}$$

In drawing conclusions from this result, we should realize that the range of the variable of the last integration, τ, is $(-\infty, \infty)$. We shall first consider the interval $(-\infty < \tau < 0)$. In this interval $\eta(\tau) = 0$ by condition

(33). Hence the left-hand member of (45) vanishes, even though the coefficient of $\eta(\tau)$ in brackets may not be zero, thereby satisfying condition (45) for $-\infty < \tau < 0$. We shall see later that the coefficient is, in general, not zero in this range of the variable τ. The important point we have established, therefore, is that in general, for minimum mean-square error,

$$\int_{-\infty}^{\infty} h(\sigma)\varphi_{ii}(\tau - \sigma)\, d\sigma - \varphi_{id}(\tau) \neq 0 \qquad \text{for } \tau < 0 \tag{46}$$

Turning to the interval $(0 \leq \tau < \infty)$, we find that, since $\eta(\tau)$ is any differentiable function in $(-\infty, \infty)$ that satisfies condition (33), the coefficient of $\eta(\tau)$ must be zero in order that (45) be satisfied. As in the case of (31), if the coefficient were not zero, we could choose an η that has the same sign as that of the coefficient for $(0 \leq \tau < \infty)$ with the result that the left-hand member of (45) is a positive number in contradiction to condition (45). We now conclude that, if the mean-square-error functional $\overline{\varepsilon_h^2(t)}$ has a minimum value for a unit-impulse response $h(t)$, this response must satisfy the condition

$$\int_{-\infty}^{\infty} h(\sigma)\varphi_{ii}(\tau - \sigma)\, d\sigma - \varphi_{id}(\tau) = 0 \qquad \text{for } \tau \geq 0 \tag{47}$$

In other words, the unit-impulse response of the optimum linear system is the solution of this integral equation. This equation is of the type known as the Wiener-Hopf equation. An interesting point is that the Wiener-Hopf integral equation in optimum linear system theory corresponds to the Euler-Lagrange differential equation in calculus of variations. It is not superfluous to repeat for emphasis that equation (47) holds only for $\tau \geq 0$. Without this restriction, the equation would not be a Wiener-Hopf equation; and, without understanding thoroughly how the restriction entered into our problem, we would have lost a major point in the analysis. As we know, condition (47) is a necessary one. The question we wish to settle immediately is whether the condition is for minimum or maximum mean-square error. Then finally we shall proceed to solve (47) for the optimum unit-impulse response.

Referring to the expression for mean-square error (4), we can easily see that the error can be made as large as we please simply by increasing the amplitude of the output $f_o(t)$. On the basis of this fact we can intuitively say that, if a solution of (47) exists, it must be for minimum mean-square error. Nevertheless, it is desirable that we answer the question in a more convincing manner. For this purpose we begin with the variation of the mean-square-error functional (38). Suppose that instead of any given h in (38) we specify that it be the one that satisfies condition (47). In so doing we shall see what happens to the mean-square error when a variation is introduced into the h that satisfies (47). To carry out this procedure, we substitute (47) into (38) by writing the last integral of (38) as

$$\int_{-\infty}^{\infty} \eta(\tau)\, d\tau\, \varphi_{id}(\tau) = \int_{-\infty}^{\infty} \eta(\tau)\, d\tau \int_{-\infty}^{\infty} h(\sigma)\varphi_{ii}(\tau - \sigma)\, d\sigma \tag{48}$$

This substitution leads to the cancellation of the first and last right-hand members of (38), leaving

$$\overline{\delta \varepsilon_h^{\,2}(t)} = \epsilon^2 \int_{-\infty}^{\infty} \eta(\tau)\, d\tau \int_{-\infty}^{\infty} \eta(\sigma)\, d\sigma\, \varphi_{ii}(\tau - \sigma) \tag{49}$$

Since (47) holds only for $\tau \geq 0$, we might inquire as to what happens to the substitution for $\tau < 0$. Because of condition (33) we see immediately that the integrals in (38) with $\eta(\tau)$ as a factor in the integrands are zero for $\tau < 0$. In other words, there is no contribution of mean-square error from the integrals for $\tau < 0$ under any circumstances. Hence (49) is the variation of the mean-square error when the variation $\epsilon\eta$ is made on the unit-impulse response that satisfies (47). Now, if we can show that (49) is positive for all possible $\epsilon\eta$, we can conclude that (47) is a condition for minimum mean-square error only.

To show that (49) is positive for all possible $\epsilon\eta$, we refer to Chapter 13, equation (33). If we consider a linear system whose input is $f_i(t)$ and whose unit-impulse response is $\eta(t)$, the output autocorrelation function for $\tau = 0$ is

$$\varphi_{oo}(0) = \int_{-\infty}^{\infty} \eta(\nu)\, d\nu \int_{-\infty}^{\infty} \eta(\sigma)\, d\sigma\, \varphi_{ii}(\nu - \sigma) \tag{50}$$

which is the mean-square value of the system output. Since this expression is the same as the double integral in (49), we write

$$\overline{\delta \varepsilon_h^{\,2}(t)} = \epsilon^2 \varphi_{oo}(0) \tag{51}$$

Inasmuch as ϵ is real and $\varphi_{oo}(0) > 0$, we have shown that, when h satisfies the Wiener-Hopf equation (47), then

$$\overline{\delta \varepsilon_h^{\,2}(t)} > 0 \qquad \text{for all possible } \eta\text{'s} \tag{52}$$

If we obtain a solution for h from (47), we would be assured by (52) that it is for minimum mean-square error and that it is the only solution because, if another solution were possible, it would be a variation of the first solution, a situation which (52) does not permit. Therefore the solution of (47) yields the optimum linear system. To emphasize this fact, we write (47) with h_{opt} for the unit-impulse response of the optimum system. Thus the Wiener-Hopf equation is

$$\varphi_{id}(\tau) = \int_{-\infty}^{\infty} h_{\text{opt}}(\sigma)\varphi_{ii}(\tau - \sigma)\, d\sigma \qquad \text{for } \tau \geq 0 \tag{53}$$

5. Some Aspects of the Wiener-Hopf Equation

Before going into the solution of the Wiener-Hopf integral equation, it seems desirable to point out some general aspects of the equation. First, let us observe that the integral in (53) is a convolution integral. We might jump to the conclusion that the equation could be solved by taking Fourier transforms of both sides of the equation. This conclusion is erroneous because the equation holds only for $\tau \geq 0$. This restriction makes the solution difficult but extremely interesting.

In the expression for mean-square error we pointed out the important fact that the characteristics concerning the input and output that affect the error are the input autocorrelation and the input–desired-output crosscorrelation. It is interesting that these two characteristics completely control the design of the optimum system as expressed by the Wiener-Hopf equation. Since a filter problem and a prediction problem differ merely in the desired output, we see that both problems have been combined in the equation. Furthermore, other desired outputs such as the derivative of the message or the integral of the message, with a lag or a lead, are included in the input–desired-output crosscorrelation so that the Wiener-Hopf equation actually covers a large class of problems. As long as the desired output and the input have a crosscorrelation that is not a constant or zero, (53) determines the optimum system for the problem.

Another aspect of the Wiener-Hopf equation (53) is its similarity to the input-output crosscorrelation theorem [Chapter 13, Sec. 8, equation (87)] in certain respects. Let us examine (53) with the theorem in mind. The right-hand member of (53) is the convolution of the optimum unit-impulse response and the input autocorrelation as shown in Figs. 3*a* and 3*b*. According to the crosscorrelation theorem this convolution is equal to the input-output crosscorrelation of the optimum system; that is,

$$\int_{-\infty}^{\infty} h_{\text{opt}}(\sigma)\varphi_{ii}(\tau - \sigma)\, d\sigma = [\varphi_{io}(\tau)]_{h(\text{opt})} \tag{54}$$

Hence the Wiener-Hopf equation is

$$\varphi_{id}(\tau) = [\varphi_{io}(\tau)]_{h(\text{opt})} \qquad \text{for } \tau \geq 0 \tag{55}$$

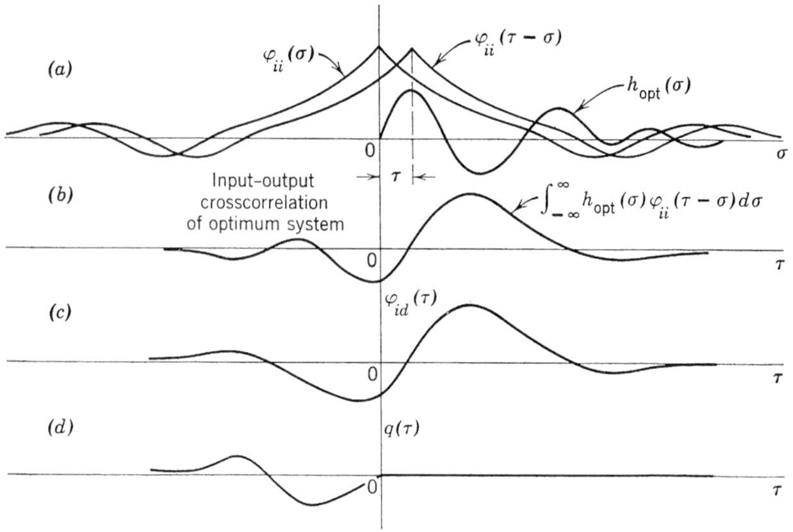

Fig. 3. Interpretation of the Wiener-Hopf equation.

Thus with $\varphi_{id}(\tau)$ shown in Fig. 3*c*, the two functions in Figs. 3*b* and 3*c* must be equal for $(0 \leq \tau < \infty)$. Note that for $(-\infty < \tau < 0)$ nothing is stated about their relations.

When the desired output is specified in a problem, it is generally a goal which we cannot achieve completely by the linear system. We try to do the best we can by minimizing the mean-square error. If the desired output could be obtained without any error, the input-output crosscorrelation of the optimum linear system would indeed be the input–desired-output crosscorrelation itself so that (55) would hold not only for $\tau \geq 0$ but also for $\tau < 0$. The problem becomes a trivial one, because now we can take the Fourier transforms of both sides of (53), in the manner shown in Chapter 13, Sec. 3, to obtain

$$\Phi_{id}(\omega) = H_{\text{opt}}(\omega)\Phi_{ii}(\omega) \tag{56}$$

where $H_{\text{opt}}(\omega)$ is the system function of the optimum linear system. We then have

$$H_{\text{opt}}(\omega) = \frac{\Phi_{id}(\omega)}{\Phi_{ii}(\omega)} \tag{57}$$

as the solution of our problem. Because we do not know exactly what desired output, appropriate to a given problem, can be obtained without error through a linear system, we demand more than the system can do so that minimization of mean-square error will give us the best approximation to the demand. In some ways this method of trying to get the best we can in a problem is similar to bargaining in a business transaction.

When we demand more than a linear system can achieve, the input-output crosscorrelation function of the optimum linear system cannot be equal to the input–desired-output crosscorrelation function for all values of the argument. According to (55), if the system is to be designed on the basis of minimum mean-square error, the system unit-impulse response should be such that these functions become equal for $\tau \geq 0$, allowing that portion of $[\varphi_{io}(\tau)]_{h(\text{opt})}$ for $\tau < 0$ to fall however it may. The freedom in the interval $(-\infty, 0)$ is necessitated by the restriction in the interval $(0, \infty)$. Thus, if $q(\tau)$ is the difference between the input–desired-output crosscorrelation and the input-output crosscorrelation of the optimum linear system, we have

$$q(\tau) = \varphi_{id}(\tau) - \int_{-\infty}^{\infty} h_{\text{opt}}(\sigma)\varphi_{ii}(\tau - \sigma)\, d\sigma \tag{58}$$

which, according to (53), is a function that is in general nonzero for $\tau < 0$ but vanishes for $\tau \geq 0$. This function is shown in Fig. 3*d*. We shall see that the vanishing of $q(\tau)$ for $\tau \geq 0$ is a key to the solution of (53) for the unit-impulse response of the optimum system.

6. Location of Singularities of the Complex Fourier Transform of a Function Vanishing over a Half-Line

Since the solution of the Wiener-Hopf equation involves $q(\tau)$ and other similar functions which vanish over half-lines, we shall consider the transforms of these functions and the location of singularities when the functions are represented in the complex domain.

If an aperiodic function $f(t)$, as shown in Fig. 4*a*, is a function that vanishes to the left of the origin, that is,

$$f(t) = 0 \qquad \text{for } t < 0 \tag{59}$$

and satisfies the condition

$$\int_{0}^{\infty} |f(t)|\, dt < \infty \tag{60}$$

then its Fourier transform is

$$F(\omega) = \frac{1}{2\pi}\int_{0}^{\infty} f(t)e^{-j\omega t}\, dt \tag{61}$$

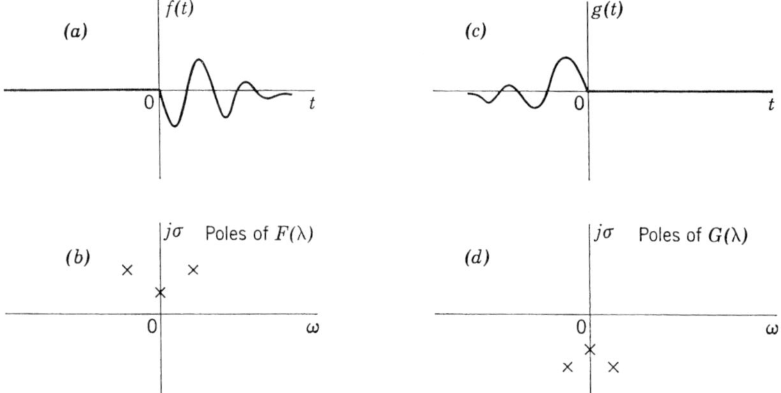

Fig. 4. Functions that vanish over half-lines and their poles.

The integral converges for all values of ω in the range $(-\infty, \infty)$. The inverse of (61) is

$$f(t) = \int_{-\infty}^{\infty} F(\omega) e^{j\omega t}\, d\omega \tag{62}$$

Letting a complex variable λ be

$$\lambda = \omega + j\sigma \tag{63}$$

where ω and σ are real and independent variables, we write the complex Fourier transform of $f(t)$ as

$$F(\lambda) = \frac{1}{2\pi} \int_{0}^{\infty} f(t) e^{-j\lambda t}\, dt \tag{64}$$

Rewriting (64) in terms of ω and σ, we have

$$\begin{aligned} F(\lambda) &= \frac{1}{2\pi} \int_{0}^{\infty} f(t) e^{-j(\omega + j\sigma)t}\, dt \\ &= \frac{1}{2\pi} \int_{0}^{\infty} [f(t)e^{\sigma t}] e^{-j\omega t}\, dt \end{aligned} \tag{65}$$

Since $f(t)$ satisfies (60), clearly

$$\int_{0}^{\infty} |f(t)e^{\sigma t}|\, dt < \infty \qquad \text{for } \sigma \leq 0 \tag{66}$$

By this fact and the fact that (61) converges for $-\infty < \omega < \infty$, we conclude that the integral in (65) converges for $\sigma \leq 0$, and $-\infty < \omega < \infty$. In other words $F(\lambda)$ has no singularities over the entire lower half-plane including the ω-axis. Figure 4*b* shows the λ-plane with ω represented by the real axis and σ by the imaginary axis in accordance with (63). Since a function of a complex variable must have at least one singularity, unless it is a constant or zero, the singularities of $F(\lambda)$ must be located in the upper half-plane as indicated in Fig. 4*b*. The inverse of (64) is (62) with ω replaced by $\omega + j\sigma$ so that

$$f(t) = \int_{-\infty+j\sigma_1}^{\infty+j\sigma_1} F(\omega + j\sigma)e^{j(\omega+j\sigma)t}\, d(\omega + j\sigma) \qquad \text{for } \sigma_1 \leq 0 \tag{67}$$

or

$$f(t) = \int_{-\infty+j\sigma_1}^{\infty+j\sigma_1} F(\lambda)e^{j\lambda t}\, d\lambda \qquad \text{for } \sigma_1 \leq 0 \tag{68}$$

Suppose that $f(t) = 0$ for $t < 0$ and that

$$\int_0^\infty |f(t)|\, dt \tag{69}$$

diverges but that

$$\int_0^\infty |f(t)|e^{\sigma t}\, dt < \infty \qquad \text{for } \sigma < 0 \tag{70}$$

For instance, $f(t)$ may grow as t^n. Under condition (70)

$$F(\lambda) = \frac{1}{2\pi}\int_0^\infty [f(t)e^{\sigma t}]e^{-j\omega t}\, dt \tag{71}$$

converges for $\sigma < 0$ and $-\infty < \omega < \infty$, so that $F(\lambda)$ is free of singularities in the lower half-plane excluding the ω-axis. Therefore the singularities of $F(\lambda)$ are on the ω-axis, and those (if any) that do not lie on the ω-axis must be in the upper half-plane. The inverse of (71) is

$$f(t) = \int_{-\infty+j\sigma_1}^{\infty+j\sigma_1} F(\lambda)e^{j\lambda t}\, d\lambda \qquad \text{for } \sigma_1 < 0 \tag{72}$$

In a similar manner we consider the function $g(t)$ (Fig. 4*c*) which vanishes to the right of the origin. We have

$$g(t) = 0 \qquad \text{for } t > 0 \tag{73}$$

and assume that

$$\int_{-\infty}^0 |g(t)|\, dt < \infty \tag{74}$$

The Fourier transform of $g(t)$ is

$$G(\omega) = \frac{1}{2\pi}\int_{-\infty}^0 g(t)e^{-j\omega t}\, dt \tag{75}$$

which converges for all values of ω in the range $(-\infty, \infty)$. The inverse of (75) is

$$g(t) = \int_{-\infty}^{\infty} G(\omega) e^{j\omega t}\, d\omega \tag{76}$$

In terms of the complex variable (63), (75) is

$$\begin{aligned} G(\lambda) &= \frac{1}{2\pi} \int_{-\infty}^{0} g(t) e^{-j\lambda t}\, dt \\ &= \frac{1}{2\pi} \int_{-\infty}^{0} [g(t) e^{\sigma t}] e^{-j\omega t}\, dt \end{aligned} \tag{77}$$

Because of the assumption (74), the expression (77) for $G(\lambda)$ converges for $\sigma \geq 0$ and $-\infty < \omega < \infty$. Hence $G(\lambda)$ has no singularities over the entire upper half-plane, including the ω-axis. Unless $G(\lambda)$ is a constant or zero, its poles must therefore be located in the lower half-plane as indicated in Fig. 4*d*. Similar to (68), the inverse of (77) is

$$g(t) = \int_{-\infty+j\sigma_1}^{\infty+j\sigma_1} G(\lambda) e^{j\lambda t}\, d\lambda \qquad \text{for } \sigma_1 \geq 0 \tag{78}$$

Now, consider the case $g(t) = 0$ for $t > 0$, and

$$\int_{-\infty}^{0} |g(t)|\, dt \tag{79}$$

diverges, but

$$\int_{-\infty}^{0} |g(t)|\, e^{\sigma t}\, dt < \infty \qquad \text{for } \sigma > 0 \tag{80}$$

With condition (80) applied to equation (77) for $G(\lambda)$, we find that $G(\lambda)$ converges for $\sigma > 0$ and $-\infty < \omega < \infty$. This convergence means that $G(\lambda)$ is free of singularities in the upper half-plane excluding the ω-axis. We conclude that the singularities of $G(\lambda)$, under conditions (79) and (80), are on the ω-axis and that those singularities (if any) that do not fall on the ω-axis must lie on the lower half-plane. Analogous to (72), the inverse of (77) is

$$g(t) = \int_{-\infty+j\sigma_1}^{\infty+j\sigma_1} G(\lambda) e^{j\lambda t}\, d\lambda \qquad \text{for } \sigma_1 > 0 \tag{81}$$

7. Complex Fourier Transform of the Difference between the Input–Desired-Output Crosscorrelation and the Input-Output Crosscorrelation of the Optimum System

Returning to (58) from which we wish to determine the unit-impulse response of the optimum system, we recall that

$$q(\tau) = 0 \qquad \text{for } \tau \geq 0 \tag{82}$$

This is the type of function that we referred to as $g(t)$ in (73). With the assumption (80) for (82), that is,

$$\int_{-\infty}^{0} |q(\tau)| e^{\sigma\tau} \, d\tau < \infty \qquad \text{for } \sigma > 0 \tag{83}$$

the complex Fourier transform of $q(\tau)$ is, as we have shown, free from singularities in the upper half-plane excluding the ω-axis. Let the complex Fourier transform of $q(\tau)$ be

$$Q(\lambda) = \frac{1}{2\pi} \int_{-\infty}^{\infty} q(\tau) e^{-j\lambda\tau} \, d\tau \tag{84}$$

so that

$$q(\tau) = \int_{-\infty+j\sigma_1}^{\infty+j\sigma_1} Q(\lambda) e^{j\lambda\tau} \, d\lambda \qquad \text{for } \sigma_1 > 0 \tag{85}$$

Taking the complex Fourier transform of both sides of (58), we have

$$\begin{aligned} Q(\lambda) &= \frac{1}{2\pi} \int_{-\infty}^{\infty} e^{-j\lambda\tau} \, d\tau \, [\varphi_{id}(\tau) - \int_{-\infty}^{\infty} h_{\text{opt}}(\sigma) \varphi_{ii}(\tau - \sigma) \, d\sigma] \\ &= \frac{1}{2\pi} \int_{-\infty}^{\infty} \varphi_{id}(\tau) e^{-j\lambda\tau} \, d\tau \\ &\quad - \frac{1}{2\pi} \int_{-\infty}^{\infty} e^{-j\lambda\tau} \, d\tau \int_{-\infty}^{\infty} h_{\text{opt}}(\sigma) \varphi_{ii}(\tau - \sigma) \, d\sigma \end{aligned} \tag{86}$$

By definition the first right-hand member is the input–desired-output cross-power density spectrum as a function of λ

$$\Phi_{id}(\lambda) = \frac{1}{2\pi} \int_{-\infty}^{\infty} \varphi_{id}(\tau) e^{-j\lambda\tau} \, d\tau \tag{87}$$

The second right-hand member of (86) can be shown, as in Chapter 13, Sec. 3, to be

$$\frac{1}{2\pi} \int_{-\infty}^{\infty} e^{-j\lambda\tau} \, d\tau \int_{-\infty}^{\infty} h_{\text{opt}}(\sigma) \varphi_{ii}(\tau - \sigma) \, d\sigma = H_{\text{opt}}(\lambda) \Phi_{ii}(\lambda) \tag{88}$$

where $H_{\text{opt}}(\lambda)$ is the system function of the optimum linear system which is related to $h_{\text{opt}}(t)$ by the expression

$$H_{\text{opt}}(\lambda) = \int_{-\infty}^{\infty} h_{\text{opt}}(t) e^{-j\lambda t}\, dt \tag{89}$$

and $\Phi_{ii}(\lambda)$ is the input power density spectrum as a function of λ

$$\Phi_{ii}(\lambda) = \frac{1}{2\pi} \int_{-\infty}^{\infty} \varphi_{ii}(\tau) e^{-j\lambda\tau}\, d\tau \tag{90}$$

Therefore, (86) becomes

$$Q(\lambda) = \Phi_{id}(\lambda) - H_{\text{opt}}(\lambda)\Phi_{ii}(\lambda) \tag{91}$$

To proceed from here, we need the method of spectrum factorization.

8. Spectrum Factorization

Before we discuss the method of spectrum factorization, it is necessary that we consider the singularities and zeros of the input power density spectrum of the optimum linear system as a function of the complex variable λ. This is the spectrum to be factorized. We take as an example the spectrum

$$\Phi_{ii}(\omega) = \frac{1}{b^2\omega^2 + 1} \tag{92}$$

where b is a positive real constant. We have seen that among other random waves the Poisson rectangular wave has this power density spectrum. Writing (92) as the product of two factors in terms of the roots of the denominator, we have

$$\Phi_{ii}(\omega) = \left(\frac{1}{b\omega - j}\right)\left(\frac{1}{b\omega + j}\right) \tag{93}$$

As a function of the complex variable λ, (93) is

$$\Phi_{ii}(\lambda) = \left(\frac{1}{b\lambda - j}\right)\left(\frac{1}{b\lambda + j}\right) \tag{94}$$

The first factor has a first-order pole when

$$\left.\begin{aligned} b\lambda - j &= 0 \\ \text{or} \qquad b(\omega + j\sigma) &= j \end{aligned}\right\} \tag{95}$$

so that the pole is at $\omega = 0$, $\sigma = 1/b$. This pole is indicated in Fig. 5.

Similarly, the second factor has a pole when

$$\left.\begin{aligned} b\lambda + j &= 0 \\ b(\omega + j\sigma) &= -j \end{aligned}\right\} \tag{96}$$

or

so that the pole is at $\omega = 0$, $\sigma = -1/b$ as shown in Fig. 5. The zero of $\Phi_{ii}(\lambda)$ is at infinity.

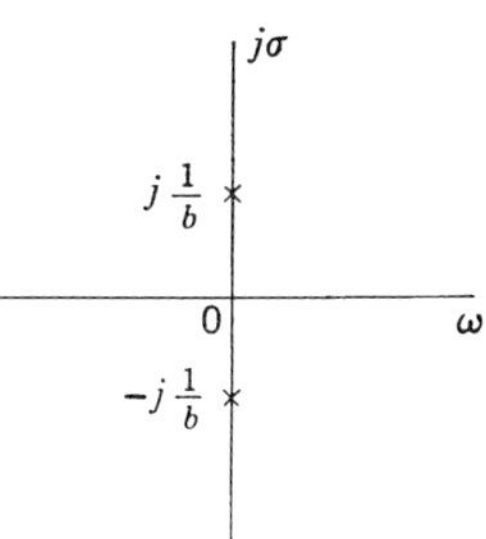

Fig. 5. Poles of a power density spectrum.

As a second example we consider an input consisting of a rectangular Poisson wave plus a white noise with zero crosscorrelation with the rectangular wave. If the Poisson wave has the power density spectrum

$$\Phi_{mm}(\omega) = \frac{a^2}{b^2\omega^2 + 1} \tag{97}$$

where a and b are positive real constants and the noise has the spectrum

$$\Phi_{nn}(\omega) = 1 \tag{98}$$

then the input power density spectrum is

$$\begin{aligned} \Phi_{ii}(\omega) &= \frac{a^2}{b^2\omega^2 + 1} + 1 \\ &= \frac{b^2\omega^2 + a^2 + 1}{b^2\omega^2 + 1} \end{aligned} \tag{99}$$

and as a function of λ it is

$$\Phi_{ii}(\lambda) = \frac{b^2\lambda^2 + a^2 + 1}{b^2\lambda^2 + 1} \tag{100}$$

Factoring both the numerator and denominator of this expression in the manner of (94), we have

$$\begin{aligned} \Phi_{ii}(\lambda) &= \left(\frac{b\lambda - j\sqrt{a^2 + 1}}{b\lambda - j}\right)\left(\frac{b\lambda + j\sqrt{a^2 + 1}}{b\lambda + j}\right) \\ &= \left(\frac{\lambda - j\alpha}{\lambda - j\beta}\right)\left(\frac{\lambda + j\alpha}{\lambda + j\beta}\right) \end{aligned} \tag{101}$$

where $\alpha = \sqrt{a^2 + 1}/b$, and $\beta = 1/b$. With $a^2 = 2$, $b^2 = 100$, the first factor has a pole at $\omega = 0$, $\sigma = \alpha = 0.173$, and a zero at $\omega = 0$, $\sigma = \beta = 0.100$ as shown in Fig. 6. The second factor has a pole at $\omega = 0$, $\sigma = -\alpha = -0.173$, and a zero at $\omega = 0$, $\sigma = -\beta = -0.100$. In the two

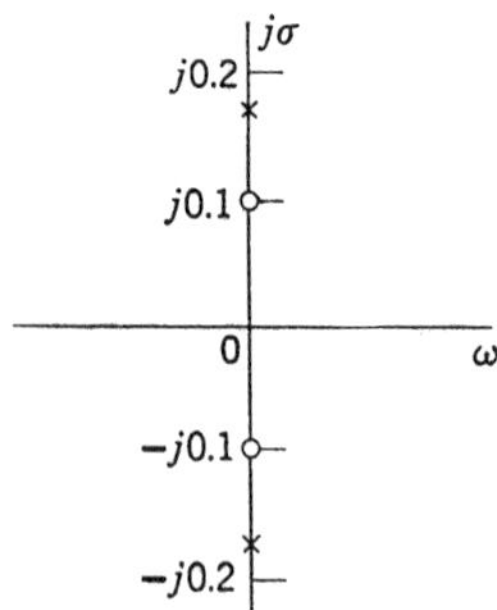

Fig. 6. Poles and zeros of an input power density spectrum.

examples the poles and zeros are all located on the σ-axis. Both zeros and poles appear as symmetrical pairs with respect to the origin.

We can generalize these examples by considering the expression for $\Phi_{ii}(\lambda)$ as

$$\Phi_{ii}(\lambda) = \frac{P(\lambda)}{Q(\lambda)} \tag{102}$$

in which $P(\lambda)$ and $Q(\lambda)$ are polynomials so that $\Phi_{ii}(\lambda)$ is represented by a rational function. For the present, the poles and zeros are confined to the imaginary axis. Assuming that $\Phi_{ii}(\lambda)$ can be so represented under this restriction, we see that (102) takes the form

$$\Phi_{ii}(\lambda) = A^2 \frac{\prod_{k=1}^{m} (\lambda^2 + d_k^2)}{\prod_{i=1}^{n} (\lambda^2 + b_i^2)} \tag{103}$$

in which A, d_k, and b_i are real, positive constants. Since both the numerator and denominator are real, positive, and even functions of ω when λ is replaced by ω, the expression (103) is an appropriate representation of the input power density spectrum which has such characteristics. In the factored form, (103) is

$$\Phi_{ii}(\lambda) = A^2 \frac{\prod_{k=1}^{m} (\lambda - jd_k)(\lambda + jd_k)}{\prod_{i=1}^{n} (\lambda - jb_i)(\lambda + jb_i)} \tag{104}$$

Since the poles and zeros are on the imaginary axis, the roots $\pm jd_k$ and $\pm jb_i$ of $P(\lambda)$ and $Q(\lambda)$ respectively are pure imaginary. For each pole in the upper half-plane there is a reflection of it in the lower half-plane, since jd_k and $-jd_k$ are conjugates. Similarly for each zero in the upper half-plane there is a reflection of it in the lower half-plane.

If the poles and zeros are not restricted to lie on the imaginary axis but are allowed to be located anywhere on the plane, (104) is generalized to the form

$$\Phi_{ii}(\lambda) = A^2 \frac{\prod_{k=1}^{m} [\lambda - (c_k + jd_k)]}{\prod_{i=1}^{n} [\lambda - (a_i + jb_i)]}$$

$$\times \frac{[\lambda - (c_k - jd_k)][\lambda - (-c_k + jd_k)][\lambda - (-c_k - jd_k)]}{[\lambda - (a_i - jb_i)][\lambda - (-a_i + jb_i)][\lambda - (-a_i - jb_i)]} \quad (105)$$

in which $c_k + jd_k$, $c_k - jd_k$, $-c_k + jd_k$, $-c_k - jd_k$ are the roots of $P(\lambda)$ with c_k and d_k being real and positive, and $a_i + jb_i$, $a_i - jb_i$, $-a_i + jb_i$, $-a_i - jb_i$ are the roots of $Q(\lambda)$ with a_i and b_i being real and positive. In each set of the roots, the first two roots are conjugates, and the last two roots are also conjugates. Furthermore, the first and third roots in each set are mirror images of each other with respect to the imaginary axis. The second and fourth roots are mirror images in the same manner. As we know, the roots of $P(\lambda)$ are the zeros of $\Phi_{ii}(\lambda)$, and the roots of $Q(\lambda)$ are the poles. These zeros and poles for one value of i are shown in Fig. 7. The zeros are at the corners of a rectangle with its center at the origin, and the poles are at the corners of another rectangle.

To see that (105) is a real, positive, and even function of ω in agreement with the properties of a power density spectrum, we reduce (105) to the

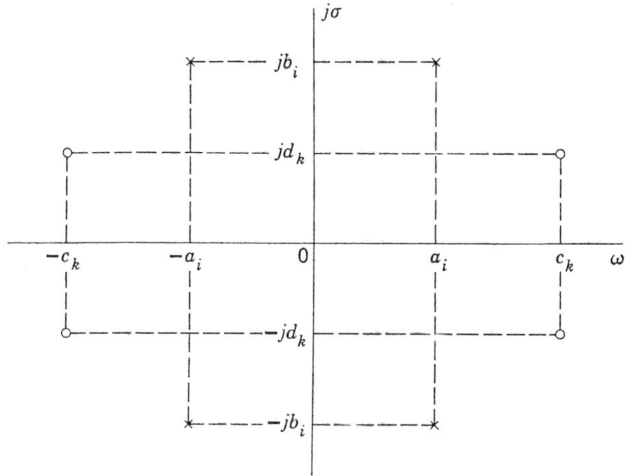

Fig. 7. Illustrating the poles and zeros of an input power density spectrum.

form

$$\Phi_{ii}(\lambda) = A^2 \frac{\prod_{k=1}^{m} [(\lambda - c_k)^2 + d_k{}^2][(\lambda + c_k)^2 + d_k{}^2]}{\prod_{i=1}^{n} [(\lambda - a_i)^2 + b_i{}^2][(\lambda + a_i)^2 + b_i{}^2]} \tag{106}$$

Clearly, when λ is replaced by ω, both numerator and denominator of the rational function are real, positive, and even functions of the real angular frequency ω.

Now, the first step in the method of spectrum factorization for the solution of the Wiener-Hopf equation consists in writing (105) as the product of two factors as follows

$$\Phi_{ii}(\lambda) = \Phi_{ii}^{+}(\lambda)\Phi_{ii}^{-}(\lambda) \tag{107}$$

with

$$\Phi_{ii}^{+}(\lambda) = A \frac{\prod_{k=1}^{m} [\lambda - (c_k + jd_k)][\lambda - (-c_k + jd_k)]}{\prod_{i=1}^{n} [\lambda - (a_i + jb_i)][\lambda - (-a_i + jb_i)]} \tag{108}$$

and

$$\Phi_{ii}^{-}(\lambda) = A \frac{\prod_{k=1}^{m} [\lambda - (c_k - jd_k)][\lambda - (-c_k - jd_k)]}{\prod_{i=1}^{n} [\lambda - (a_i - jb_i)][\lambda - (-a_i - jb_i)]} \tag{109}$$

When $\Phi_{ii}(\lambda)$ is expressed by (104), then (108) and (109) become

$$\Phi_{ii}^{+}(\lambda) = A \frac{\prod_{k=1}^{m} (\lambda - jd_k)}{\prod_{i=1}^{n} (\lambda - jb_i)} \tag{110}$$

and

$$\Phi_{ii}^{-}(\lambda) = A \frac{\prod_{k=1}^{m} (\lambda + jd_k)}{\prod_{i=1}^{n} (\lambda + jb_i)} \tag{111}$$

We have grouped the factors in such a manner that

$$\left.\begin{array}{l}\Phi_{ii}^{+}(\lambda) \text{ contains all the zeros and poles of } \Phi_{ii}(\lambda) \\ \qquad \text{that are in the upper half-plane} \\ \text{and} \\ \Phi_{ii}^{-}(\lambda) \text{ contains all the zeros and poles of } \Phi_{ii}(\lambda) \\ \qquad \text{that are in the lower half-plane}\end{array}\right\} \tag{112}$$

and that they are conjugates; that is,

$$\left.\begin{array}{l}\Phi_{ii}^{-}(\omega) = \overline{\Phi}_{ii}^{+}(\omega) \\ \text{or} \\ \Phi_{ii}^{+}(\omega) = \overline{\Phi}_{ii}^{-}(\omega)\end{array}\right\} \tag{113}$$

so that (107) is

$$\begin{aligned}\Phi_{ii}(\omega) &= \Phi_{ii}^{+}(\omega)\Phi_{ii}^{-}(\omega) \\ &= |\Phi_{ii}^{\pm}(\omega)|^2\end{aligned} \tag{114}$$

We have considered only first-order poles and zeros, but it is sufficiently clear that, if they are of higher orders, then all the factors in (103) through (106) and (108) through (111), for each value of i and k, are raised to the power equal to the order of the corresponding pole or zero.

As an example of factorization in a combination of forms (104) and (105), consider $f_i(t) = f_1(t) + f_2(t)$. Let the power density spectrum of $f_1(t)$ be

$$\Phi_{11}(\omega) = \frac{1}{a^2 + \omega^2} \tag{115}$$

and that of $f_2(t)$ be

$$\Phi_{22}(\omega) = \frac{1}{2}\left[\frac{1}{a^2 + (\omega - \omega_o)^2} + \frac{1}{a^2 + (\omega + \omega_o)^2}\right] \tag{116}$$

Let $\Phi_{12}(\omega) = 0$. The spectrum (116) is that of a rectangular Poisson wave multiplied by a sinusoid of angular frequency ω_o. We have

$$\begin{aligned}\Phi_{ii}(\omega) &= \frac{1}{a^2 + \omega^2} + \frac{1}{2}\left[\frac{1}{a^2 + (\omega - \omega_o)^2} + \frac{1}{a^2 + (\omega + \omega_o)^2}\right] \\ &= \frac{(a^2 + \omega^2 + \omega_o{}^2)(2a^2 + 2\omega^2 + \omega_o{}^2) - 4\omega^2\omega_o{}^2}{(a^2 + \omega^2)[a^2 + (\omega - \omega_o)^2][a^2 + (\omega + \omega_o)^2]}\end{aligned} \tag{117}$$

To simplify the algebraic work in this example, we let $a = 100$ and $\omega_o = 600$. With these values, (117) is factorized to

$$\Phi_{ii}(\lambda) = 2\frac{\left\{\begin{array}{l}[\lambda - (415 + j304)][\lambda - (415 - j304)] \\ \quad \times [\lambda - (-415 + j304)][\lambda - (-415 - j304)]\end{array}\right\}}{\left\{\begin{array}{l}(\lambda - j100)(\lambda + j100)[\lambda - (600 + j100)][\lambda - (600 - j100)] \\ \quad \times [\lambda - (-600 + j100)][\lambda - (-600 - j100)]\end{array}\right\}} \tag{118}$$

so that

$$\Phi_{ii}^{+}(\lambda) = \sqrt{2}\frac{[\lambda - (415 + j304)][\lambda - (-415 + j304)]}{(\lambda - j100)[\lambda - (600 + j100)][\lambda - (-600 + j100)]} \tag{119}$$

and

$$\Phi_{ii}^{-}(\lambda) = \sqrt{2}\frac{[\lambda - (415 - j304)][\lambda - (-415 - j304)]}{(\lambda + j100)[\lambda - (600 - j100)][\lambda - (-600 - j100)]} \tag{120}$$

The poles and zeros of $\Phi_{ii}(\lambda)$, $\Phi_{ii}^{+}(\lambda)$, and $\Phi_{ii}^{-}(\lambda)$ are shown in Fig. 8.

One reason for factoring the power density spectrum in the manner described is that by so doing we shall obtain one factor whose transform vanishes over the left half-line and another factor whose transform vanishes over the right half-line. To consider this matter, let $\varphi_{ii}^{+}(\tau)$ and $\varphi_{ii}^{-}(\tau)$ be the transforms of $\Phi_{ii}^{+}(\lambda)$ and $\Phi_{ii}^{-}(\lambda)$ respectively; that is,

$$\varphi_{ii}^{+}(\tau) = \int_{-\infty+j\sigma_1}^{\infty+j\sigma_1} \Phi_{ii}^{+}(\lambda)e^{j\lambda\tau}\,d\lambda \qquad \text{for } \sigma_1 < 0 \tag{121}$$

so that

$$\Phi_{ii}^{+}(\lambda) = \frac{1}{2\pi}\int_{-\infty}^{\infty} \varphi_{ii}^{+}(\tau)e^{-j\lambda\tau}\,d\tau \tag{122}$$

and

$$\varphi_{ii}^{-}(\tau) = \int_{-\infty+j\sigma_1}^{\infty+j\sigma_1} \Phi_{ii}^{-}(\lambda)e^{j\lambda\tau}\,d\lambda \qquad \text{for } \sigma_1 > 0 \tag{123}$$

so that

$$\Phi_{ii}^{-}(\lambda) = \frac{1}{2\pi}\int_{-\infty}^{\infty} \varphi_{ii}^{-}(\tau)e^{-j\lambda\tau}\,d\tau \tag{124}$$

Since all the poles of $\Phi_{ii}^{+}(\lambda)$ are in the upper half-plane,

$$\varphi_{ii}^{+}(\tau) = 0 \qquad \text{for } \tau < 0 \tag{125}$$

and, since all the poles of $\Phi_{ii}^{-}(\lambda)$ are in the lower half-plane,

$$\varphi_{ii}^{-}(\tau) = 0 \qquad \text{for } \tau > 0 \tag{126}$$

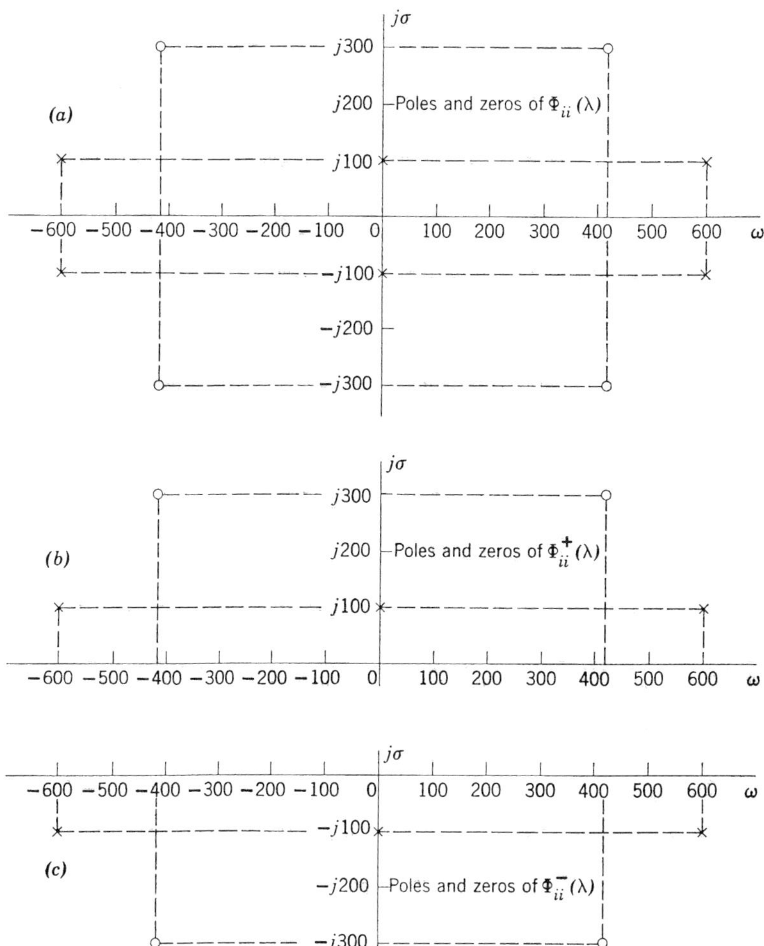

Fig. 8. Poles and zeros of $\Phi_{ii}(\lambda)$, $\Phi_{ii}^{+}(\lambda)$, and $\Phi_{ii}^{-}(\lambda)$ in an example.

These results follow from the discussion in Sec. 6. Furthermore, the functions $\varphi_{ii}^{+}(\tau)$ and $\varphi_{ii}^{-}(\tau)$ are real. This fact can be established from the particular manner in which the poles and zeros of $\Phi_{ii}^{+}(\lambda)$ and $\Phi_{ii}^{-}(\lambda)$ are located.

A situation that merits discussion is that in which the poles of $\Phi_{ii}(\lambda)$ are on the ω-axis. For example,

$$\Phi_{ii}(\lambda) = \frac{1}{\lambda^{2n}} \qquad n = 1, 2, 3, \ldots \tag{127}$$

The factorization of this spectrum is

$$\Phi_{ii}(\lambda) = \left(\frac{1}{j\lambda}\right)^n \left(\frac{1}{-j\lambda}\right)^n \tag{128}$$

in which

$$\Phi_{ii}^{+}(\lambda) = \left(\frac{1}{j\lambda}\right)^n \tag{129}$$

and

$$\Phi_{ii}^{-}(\lambda) = \left(\frac{1}{-j\lambda}\right)^n \tag{130}$$

To see that this factorization is in agreement with our objective of expressing $\Phi_{ii}(\lambda)$ as the product of two factors whose transforms vanish over opposite half-lines, we note that (129) is the transform of

$$\varphi_{ii}^{+}(\tau) = \begin{cases} \dfrac{2\pi}{(n-1)!}\tau^{n-1} & \text{for } \tau \geq 0 \\ 0 & \text{for } \tau < 0 \end{cases} \tag{131}$$

that is,

$$\begin{aligned} \Phi_{ii}^{+}(\lambda) &= \frac{1}{2\pi}\frac{2\pi}{(n-1)!}\int_0^\infty \tau^{n-1}e^{-j\lambda\tau}\,d\tau \\ &= \left(\frac{1}{j\lambda}\right)^n \end{aligned} \tag{132}$$

Inasmuch as

$$\int_0^\infty |\tau^{n-1}|e^{\sigma\tau}\,d\tau < \infty \qquad \text{for } \sigma < 0 \tag{133}$$

the transform of (132) is, in accordance with (121),

$$\varphi_{ii}^{+}(\tau) = \int_{-\infty+j\sigma_1}^{\infty+j\sigma_1}\left(\frac{1}{j\lambda}\right)^n e^{j\lambda\tau}\,d\lambda \qquad \text{for } \sigma_1 < 0 \tag{134}$$

which has the expression (131). Therefore the transform of (129) is a function that vanishes for $\tau < 0$. Similarly, we see that (130) is the transform of

$$\varphi_{ii}^{-}(\tau) = \begin{cases} \dfrac{2\pi}{(n-1)!}|\tau|^{n-1} & \text{for } \tau \leq 0 \\ 0 & \text{for } \tau > 0 \end{cases} \tag{135}$$

that is,

$$\begin{aligned} \Phi_{ii}^{-}(\lambda) &= \frac{1}{2\pi}\frac{2\pi}{(n-1)!}\int_{-\infty}^{0} |\tau|^{n-1}e^{-j\lambda\tau}\,d\tau \\ &= \left(\frac{1}{-j\lambda}\right)^n \end{aligned} \tag{136}$$

Since

$$\int_{-\infty}^{0} |\tau|^{n-1} e^{\sigma\tau}\, d\tau < \infty \qquad \text{for } \sigma > 0 \tag{137}$$

the transform of (136) is, in accordance with (123),

$$\varphi_{ii}^{-}(\tau) = \int_{-\infty+j\sigma_1}^{\infty+j\sigma_1} \left(\frac{1}{-j\lambda}\right)^n e^{j\lambda\tau}\, d\lambda \qquad \text{for } \sigma_1 > 0 \tag{138}$$

This integral has the expression (135) and shows that the transform of (130) is a function that vanishes for $\tau > 0$.

This discussion shows that (121), (123), (125), and (126) hold even when the poles lie on the ω-axis if the spectrum is factorized in the manner described. We may consider the pole of (129) to be in the upper half-plane at the origin and the pole of (130) to be in the lower half-plane at the origin. The pole in the upper half-plane at the origin is approached by considering $1/(a + j\lambda)^n$, which has a pole of order n in the upper half-plane at $\omega = 0$, $j\sigma = ja$, and then by letting $a \to 0$. Similarly, the pole in the lower half-plane at the origin is approached by considering $1/(a - j\lambda)^n$, which has a pole of order n in the lower half-plane at $\omega = 0$, $j\sigma = -ja$, and then by letting $a \to 0$. Note that (121) and (123) have been chosen to include the situation where poles lie on the ω-axis by making $\sigma_1 < 0$ in (121) and $\sigma_1 > 0$ in (123).

At this point we wish to introduce a theorem that is helpful in our discussion. It states that, if the Fourier transform of a real function $f(t)$ is $F(\omega)$, the Fourier transform of $f(-t)$ is $\overline{F}(\omega)$. The function $f(t)$ need not vanish over a half-line. To see this, we start with

$$f(t) = \int_{\infty}^{\infty} F(\omega) e^{j\omega t}\, d\omega \tag{139}$$

and

$$F(\omega) = \frac{1}{2\pi} \int_{-\infty}^{\infty} f(t) e^{-j\omega t}\, dt \tag{140}$$

If we take the conjugate of both sides of (139), then, since $f(t)$ is real, it remains unchanged, but the conjugate of the integral is

$$\int_{-\infty}^{\infty} \overline{F}(\omega) e^{-j\omega t}\, d\omega \tag{141}$$

Hence, writing (139) for $-t$ after the conjugation, we have

$$f(-t) = \int_{-\infty}^{\infty} \overline{F}(\omega) e^{j\omega t}\, d\omega \tag{142}$$

and the inverse of this expression is

$$\bar{F}(\omega) = \frac{1}{2\pi}\int_{-\infty}^{\infty} f(-t)e^{-j\omega t}\,dt \tag{143}$$

In accordance with this theorem, since $\Phi_{ii}^{+}(\omega)$ and $\Phi_{ii}^{-}(\omega)$ are conjugates, their transforms $\varphi_{ii}^{+}(\tau)$ and $\varphi_{ii}^{-}(\tau)$ are related by the expression

$$\varphi_{ii}^{\pm}(\tau) = \varphi_{ii}^{\mp}(-\tau) \tag{144}$$

In other words $\varphi_{ii}^{+}(\tau)$ and $\varphi_{ii}^{-}(\tau)$ are reflections of each other with respect to the vertical axis.

Another situation that merits discussion is one in which the pole is at infinity. Consider the theoretical spectrum

$$\Phi_{ii}(\lambda) = \lambda^{2n} \qquad n = 1, 2, 3, \ldots \tag{145}$$

The factorization of this spectrum is

$$\Phi_{ii}(\lambda) = (j\lambda)^n(-j\lambda)^n \tag{146}$$

so that

$$\Phi_{ii}^{+}(\lambda) = (j\lambda)^n \tag{147}$$

and

$$\Phi_{ii}^{-}(\lambda) = (-j\lambda)^n \tag{148}$$

It is known that (147), except for a constant factor, is the transform of the nth derivative of the unit impulse. This fact can be shown readily by referring to Chapter 2, Sec. B-5, equation (199), for the unit impulse expressed as the transform

$$u(\tau) = \frac{1}{2\pi}\int_{-\infty}^{\infty} e^{j\omega\tau}\,d\omega \tag{149}$$

Although strictly speaking this integral is improper, it is helpful in the interpretation of certain singular functions in terms of their spectrums that have been obtained by limiting processes. Taking the nth derivative of both sides of (149) with respect to τ, we have *

$$u^{(n)}(\tau) = \frac{1}{2\pi}\int_{-\infty}^{\infty} (j\omega)^n e^{j\omega\tau}\,d\omega \tag{150}$$

which gives us the result that $(1/2\pi)(j\omega)^n$ is the transform of $u^{(n)}(\tau)$. The function $u^{(n)}(\tau)$ is the nth derivative of the unit-impulse function. The first derivative of the unit-impulse function $u'(\tau)$ is called the unit doublet impulse, and the second derivative $u''(\tau)$ is called the unit triplet

* The formal operations here are more fully discussed in E. A. Guillemin, *The Mathematics of Circuit Analysis*, John Wiley and Sons, New York, 1949, pp. 531–547.

impulse. By (142) we have

$$u^{(n)}(-\tau) = \frac{1}{2\pi}\int_{-\infty}^{\infty} (-j\omega)^n e^{j\omega\tau}\,d\omega \tag{151}$$

which shows that $(1/2\pi)(-j\omega)^n$ is the transform of $u^{(n)}(-\tau)$. Therefore, when (145) is factorized into (147) and (148), the interpretation of (121) and (123) will be based upon (150) and (151) respectively resulting in

$$\varphi_{ii}^{+}(\tau) = 2\pi u^{(n)}(\tau) \tag{152}$$

and

$$\varphi_{ii}^{-}(\tau) = 2\pi u^{(n)}(-\tau) \tag{153}$$

As we know, the unit impulse and its derivatives are functions that vanish everywhere except in an infinitesimal interval at the origin. Clearly, because of this characteristic of impulse functions, (152) and (153) satisfy the important requirements (125) and (126) in spectrum factorization. A further point to be noted is that since the unit-impulse is an even function, we have

$$u^{(n)}(-\tau) = u^{(n)}(\tau) \qquad \text{for } n = 2, 4, 6, \ldots \tag{154}$$

$$u^{(n)}(-\tau) = -u^{(n)}(\tau) \qquad \text{for } n = 1, 3, 5, \ldots \tag{155}$$

We shall see later that, when the pole is at infinity, it is not necessary to assign it to either half-plane.

Another important property of $\Phi_{ii}^{+}(\lambda)$ and $\Phi_{ii}^{-}(\lambda)$ upon which we depend in the solution of the Wiener-Hopf equation is that not only their transforms vanish over opposite half-lines but also the transforms of their reciprocals vanish over opposite half-lines. We note that all the poles and zeros of $\Phi_{ii}^{+}(\lambda)$ can lie only on the upper half-plane, including the ω-axis and the point at infinity. Since, when we take the reciprocal of $\Phi_{ii}^{+}(\lambda)$, a pole becomes a zero and a zero becomes a pole, the poles and zeros of $1/\Phi_{ii}^{+}(\lambda)$ can only lie on the upper half-plane, including the ω-axis and the point at infinity. Similarly we find that all the poles and zeros of $1/\Phi_{ii}^{-}(\lambda)$ can only lie on the lower half-plane, including the ω-axis and the point at infinity. To discuss this matter further, let

$$\Gamma_{ii}(\lambda) = \frac{1}{\Phi_{ii}(\lambda)} \tag{156}$$

$$\Gamma_{ii}^{+}(\lambda) = \frac{1}{\Phi_{ii}^{+}(\lambda)} \tag{157}$$

$$\Gamma_{ii}^{-}(\lambda) = \frac{1}{\Phi_{ii}^{-}(\lambda)} \tag{158}$$

so that the reciprocal of (107) is

$$\Gamma_{ii}(\lambda) = \Gamma_{ii}^{+}(\lambda)\Gamma_{ii}^{-}(\lambda) \tag{159}$$

the reciprocal of (113) is

$$\Gamma_{ii}^{\mp}(\omega) = \overline{\Gamma}_{ii}^{\pm}(\omega) \tag{160}$$

and the reciprocal of (114) is

$$\begin{aligned} \Gamma_{ii}(\omega) &= \Gamma_{ii}^{+}(\omega)\Gamma_{ii}^{-}(\omega) \\ &= |\Gamma_{ii}^{\pm}(\omega)|^2 \end{aligned} \tag{161}$$

Let $\gamma_{ii}^{+}(\tau)$ and $\gamma_{ii}^{-}(\tau)$ be the transforms of $\Gamma_{ii}^{+}(\lambda)$ and $\Gamma_{ii}^{-}(\lambda)$ respectively; that is,

$$\gamma_{ii}^{+}(\tau) = \int_{-\infty+j\sigma_1}^{\infty+j\sigma_1} \Gamma_{ii}^{+}(\lambda)e^{j\lambda\tau}\,d\lambda \qquad \text{for } \sigma_1 < 0 \tag{162}$$

so that

$$\Gamma_{ii}^{+}(\lambda) = \frac{1}{2\pi}\int_{-\infty}^{\infty} \gamma_{ii}^{+}(\tau)e^{-j\lambda\tau}\,d\tau \tag{163}$$

and

$$\gamma_{ii}^{-}(\tau) = \int_{-\infty+j\sigma_1}^{\infty+j\sigma_1} \Gamma_{ii}^{-}(\lambda)e^{j\lambda\tau}\,d\lambda \qquad \text{for } \sigma_1 > 0 \tag{164}$$

so that

$$\Gamma_{ii}^{-}(\lambda) = \frac{1}{2\pi}\int_{-\infty}^{\infty} \gamma_{ii}^{-}(\tau)e^{-j\lambda\tau}\,d\tau \tag{165}$$

Because of the manner in which the poles and zeros of $\Gamma_{ii}^{\pm}(\lambda)$ are located, as described before, the functions $\gamma_{ii}^{\pm}(\tau)$ are real, just as $\varphi_{ii}^{\pm}(\tau)$ are real. Since $\Gamma_{ii}^{+}(\lambda)$ and $\Gamma_{ii}^{-}(\lambda)$ are conjugates, it follows from (142) and (143) that

$$\gamma_{ii}^{\pm}(\tau) = \gamma_{ii}^{\mp}(-\tau) \tag{166}$$

In view of our discussion concerning $\Phi_{ii}^{+}(\lambda)$ and $\Phi_{ii}^{-}(\lambda)$ with poles not only on the finite plane but also on the axis and at infinity, it is sufficiently clear that

$$\gamma_{ii}^{+}(\tau) = 0 \qquad \text{for } \tau < 0 \tag{167}$$

and

$$\gamma_{ii}^{-}(\tau) = 0 \qquad \text{for } \tau > 0 \tag{168}$$

These properties are analogous to (125) and (126).

9. Solution of the Wiener-Hopf Equation by the Method of Spectrum Factorization

We now return to Sec. 7 to continue with the solution of (91). Into this expression we introduce the factored form of $\Phi_{ii}(\lambda)$. Thus

$$Q(\lambda) = [\Phi_{id}(\lambda) - H_{\text{opt}}(\lambda)\Phi_{ii}^{+}(\lambda)\Phi_{ii}^{-}(\lambda)] \tag{169}$$

The next step in the solution is to divide both sides of the equation by $\Phi_{ii}^{-}(\lambda)$ so that, in terms of $\Gamma_{ii}^{+}(\lambda)$ and $\Gamma_{ii}^{-}(\lambda)$, (169) becomes

$$Q(\lambda)\Gamma_{ii}^{-}(\lambda) = \Phi_{id}(\lambda)\Gamma_{ii}^{-}(\lambda) - H_{\text{opt}}(\lambda)\Phi_{ii}^{+}(\lambda) \tag{170}$$

We now consider the transform of this equation because certain important conclusions can be drawn from it. The transform of (170) is

$$\int_{-\infty+j\sigma_1}^{\infty+j\sigma_1} Q(\lambda)\Gamma_{ii}^{-}(\lambda)e^{j\lambda\tau}\,d\lambda = \int_{-\infty+j\sigma_2}^{\infty+j\sigma_2} \Phi_{id}(\lambda)\Gamma_{ii}^{-}(\lambda)e^{j\lambda\tau}\,d\lambda - \int_{-\infty+j\sigma_3}^{\infty+j\sigma_3} H_{\text{opt}}(\lambda)\Phi_{ii}^{+}(\lambda)e^{j\lambda\tau}\,d\lambda \tag{171}$$

The left-hand member of this equation is

$$\begin{aligned}\int_{-\infty+j\sigma_1}^{\infty+j\sigma_1} Q(\lambda)\Gamma_{ii}^{-}(\lambda)e^{j\lambda\tau}\,d\lambda &= \int_{-\infty+j\sigma_1}^{\infty+j\sigma_1} \Gamma_{ii}^{-}(\lambda)e^{j\lambda\tau}\,d\lambda\,\frac{1}{2\pi}\int_{-\infty}^{\infty} q(t)e^{-j\lambda t}\,dt \\ &= \frac{1}{2\pi}\int_{-\infty}^{\infty} q(t)\,dt \int_{-\infty+j\sigma_1}^{\infty+j\sigma_1} \Gamma_{ii}^{-}(\lambda)e^{j(\tau-t)\lambda}\,d\lambda \qquad \text{for } \sigma_1 > 0 \\ &= \frac{1}{2\pi}\int_{-\infty}^{\infty} q(t)\gamma_{ii}^{-}(\tau - t)\,dt \end{aligned} \tag{172}$$

Similarly, the first right-hand member of (171) is

$$\begin{aligned}\int_{-\infty+j\sigma_2}^{\infty+j\sigma_2} \Phi_{id}(\lambda)\Gamma_{ii}^{-}(\lambda)e^{j\lambda\tau}\,d\lambda &= \int_{-\infty+j\sigma_2}^{\infty+j\sigma_2} \Gamma_{ii}^{-}(\lambda)e^{j\lambda\tau}\,d\lambda\,\frac{1}{2\pi}\int_{-\infty}^{\infty} \varphi_{id}(t)e^{-j\lambda t}\,dt \\ &= \frac{1}{2\pi}\int_{-\infty}^{\infty} \varphi_{id}(t)\,dt \int_{-\infty+j\sigma_2}^{\infty+j\sigma_2} \Gamma_{ii}^{-}(\lambda)e^{j(\tau-t)\lambda}\,d\lambda \qquad \text{for } \sigma_2 > 0 \\ &= \frac{1}{2\pi}\int_{-\infty}^{\infty} \varphi_{id}(t)\gamma_{ii}^{-}(\tau - t)\,dt \end{aligned} \tag{173}$$

and the last term of (171) is

$$\int_{-\infty+j\sigma_3}^{\infty+j\sigma_3} H_{\text{opt}}(\lambda)\Phi_{ii}^{+}(\lambda)e^{j\lambda\tau}\,d\lambda$$
$$= \int_{-\infty+j\sigma_3}^{\infty+j\sigma_3} \Phi_{ii}^{+}(\lambda)e^{j\lambda\tau}\,d\lambda \int_{-\infty}^{\infty} h_{\text{opt}}(t)e^{-j\lambda t}\,dt$$
$$= \int_{-\infty}^{\infty} h_{\text{opt}}(t)\,dt \int_{-\infty+j\sigma_3}^{\infty+j\sigma_3} \Phi_{ii}^{+}(\lambda)e^{j(\tau-t)\lambda}\,d\lambda \qquad \text{for } \sigma_3 < 0$$
$$= \int_{-\infty}^{\infty} h_{\text{opt}}(t)\varphi_{ii}^{+}(\tau - t)\,dt \tag{174}$$

Thus equation (171) becomes

$$\frac{1}{2\pi}\int_{-\infty}^{\infty} q(t)\gamma_{ii}^{-}(\tau - t)\,dt$$
$$= \frac{1}{2\pi}\int_{-\infty}^{\infty} \varphi_{id}(t)\gamma_{ii}^{-}(\tau - t)\,dt - \int_{-\infty}^{\infty} h_{\text{opt}}(t)\varphi_{ii}^{+}(\tau - t)\,dt \tag{175}$$

We shall consider the properties of the three integrals in this equation with the object of finding an equation for h_{opt} in terms of φ_{id}, γ_{ii}^{+}, and γ_{ii}^{-} which are obtained from the specification of the optimum system problem.

First, let the left-hand integral of (175) be

$$g(\tau) = \frac{1}{2\pi}\int_{-\infty}^{\infty} q(t)\gamma_{ii}^{-}(\tau - t)\,dt \tag{176}$$

Since $q(\tau) = 0$ for $\tau \geq 0$ in accordance with (82) and since $\gamma_{ii}^{-}(\tau) = 0$ for $\tau > 0$ as given by (168), the convolution integral in (176) has the property that

$$g(\tau) = 0 \qquad \text{for } \tau \geq 0 \tag{177}$$

Next, let the last integral in (175) be

$$f(\tau) = \int_{-\infty}^{\infty} h_{\text{opt}}(t)\varphi_{ii}^{+}(\tau - t)\,dt \tag{178}$$

Since $h_{\text{opt}}(t)$ is the response of the optimum linear system to a unit impulse applied at $t = 0$, it is necessary that $h_{\text{opt}}(t) = 0$ for $t < 0$. As given by (125), $\varphi_{ii}^{+}(\tau) = 0$ for $\tau < 0$. Hence the convolution integral in (178) has the property that

$$f(\tau) = 0 \qquad \text{for } \tau < 0 \tag{179}$$

It should be remarked here that we permit $h_{\text{opt}}(t)$ to have singularity functions in the interval $(0 \leq t < \infty)$.

Finally, let the first right-hand integral in (175) be

$$\psi(\tau) = \frac{1}{2\pi} \int_{-\infty}^{\infty} \varphi_{id}(t) \gamma_{ii}^{-}(\tau - t)\, dt \tag{180}$$

Inasmuch as $\varphi_{id}(\tau)$, in general, does not vanish over a half-line although $\gamma_{ii}^{-}(\tau) = 0$ for $\tau > 0$, the integral in (180) does not vanish over a half-line. Let us put

$$\psi^{+}(\tau) = \begin{cases} \psi(\tau) & \text{for } \tau \geq 0 \\ 0 & \text{for } \tau < 0 \end{cases} \tag{181}$$

$$\psi^{-}(\tau) = \begin{cases} 0 & \text{for } \tau \geq 0 \\ \psi(\tau) & \text{for } \tau < 0 \end{cases} \tag{182}$$

From these considerations it is clear that (175) involves four components each vanishing over a half-line as shown in Fig. 9, and the equation is

$$g(\tau) = \psi^{+}(\tau) + \psi^{-}(\tau) - f(\tau) \tag{183}$$

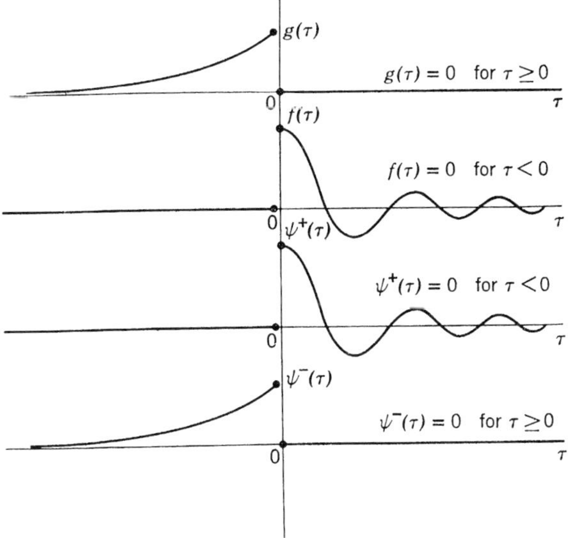

Fig. 9. Illustrating a step in the solution of the Wiener-Hopf equation.

Because in the interval $(-\infty < \tau < 0)$ both $f(\tau)$ and $\psi^{+}(\tau)$ are zero and in the interval $(0 \leq \tau < \infty)$ both $g(\tau)$ and $\psi^{-}(\tau)$ are zero, equation (183) yields

$$g(\tau) - \psi^{-}(\tau) \tag{184}$$

By similar reasoning we obtain from (183) the relation

$$f(\tau) = \psi^{+}(\tau) \tag{185}$$

The result (185) will lead to the final solution of our problem.

Taking the transform of both sides of (185), we have

$$\frac{1}{2\pi}\int_{-\infty}^{\infty} f(\tau)e^{-j\lambda\tau}\,d\tau = \frac{1}{2\pi}\int_{-\infty}^{\infty} \psi^{+}(\tau)e^{-j\lambda\tau}\,d\tau \tag{186}$$

The left-hand member of equation (186) is the transform of (178)

$$\begin{aligned}\frac{1}{2\pi}\int_{-\infty}^{\infty} f(\tau)e^{-j\lambda\tau}\,d\tau &= \frac{1}{2\pi}\int_{-\infty}^{\infty} e^{-j\lambda\tau}\,d\tau\int_{-\infty}^{\infty} h_{\text{opt}}(t)\varphi_{ii}^{+}(\tau - t)\,dt \\ &= H_{\text{opt}}(\lambda)\Phi_{ii}^{+}(\lambda)\end{aligned} \tag{187}$$

The right-hand member of equation (186) is, in accordance with (181),

$$\frac{1}{2\pi}\int_{-\infty}^{\infty} \psi^{+}(\tau)e^{-j\lambda\tau}\,d\tau = \frac{1}{2\pi}\int_{0}^{\infty} \psi(\tau)e^{-j\lambda\tau}\,d\tau \tag{188}$$

Thus (186) is

$$H_{\text{opt}}(\lambda)\Phi_{ii}^{+}(\lambda) = \frac{1}{2\pi}\int_{0}^{\infty} \psi(\tau)e^{-j\lambda\tau}\,d\tau \tag{189}$$

which gives us the optimum linear system function $H_{\text{opt}}(\lambda)$ as

$$H_{\text{opt}}(\lambda) = \frac{1}{2\pi\Phi_{ii}^{+}(\lambda)}\int_{0}^{\infty} \psi(\tau)e^{-j\lambda\tau}\,d\tau \tag{190}$$

in which $\psi(\tau)$ is given by (180) or the equivalent expression

$$\psi(\tau) = \int_{-\infty+j\sigma_1}^{\infty+j\sigma_1} \frac{\Phi_{id}(\lambda)}{\Phi_{ii}^{-}(\lambda)}\,e^{j\lambda\tau}\,d\lambda \tag{191}$$

This expression for $\psi(\tau)$ is actually the first right-hand member of (171). Equations (190) and (191) can be combined into a double integral so that the optimum system function is

$$H_{\text{opt}}(\lambda) = \frac{1}{2\pi\Phi_{ii}^{+}(\lambda)}\int_{0}^{\infty} e^{-j\lambda\tau}\,d\tau\int_{-\infty+jv_1}^{\infty+jv_1} \frac{\Phi_{id}(w)}{\Phi_{ii}^{-}(w)}\,e^{jw\tau}\,dw \tag{192}$$

in which the complex variable of integration w is $w = u + jv$ with u and v being independent real variables.

We have now arrived at an expression which gives the optimum linear system function explicitly in terms of $\Phi_{id}(\lambda)$ and the factorized components of $\Phi_{ii}(\lambda)$. This is the solution of the Wiener-Hopf equation (53).

10. Verification of the Solution of the Wiener-Hopf Equation

At this point it is instructive to see how (53) reduces to an identity when (192) is substituted into it. Since

$$h_{\text{opt}}(\sigma) = \frac{1}{2\pi} \int_{-\infty+j\sigma_1}^{\infty+j\sigma_1} H_{\text{opt}}(\lambda) e^{j\lambda\sigma}\, d\lambda \tag{193}$$

the right-hand side of the Wiener-Hopf equation (53) is

$$\int_{-\infty}^{\infty} h_{\text{opt}}(\sigma)\varphi_{ii}(\tau - \sigma)\, d\sigma = \int_{-\infty}^{\infty} \varphi_{ii}(\tau - \sigma)\, d\sigma \, \frac{1}{2\pi} \int_{-\infty+j\sigma_1}^{\infty+j\sigma_1} H_{\text{opt}}(\lambda) e^{j\lambda\sigma}\, d\lambda \tag{194}$$

Substituting the solution (192) of the Wiener-Hopf equation into (194), we have [with τ in (192) changed to t in order to avoid confusion with the τ in (53)]

$$\int_{-\infty}^{\infty} h_{\text{opt}}(\sigma)\varphi_{ii}(\tau - \sigma)\, d\sigma = \int_{-\infty}^{\infty} \varphi_{ii}(\tau - \sigma)\, d\sigma \, \frac{1}{2\pi} \int_{-\infty+j\sigma_1}^{\infty+j\sigma_1} e^{j\lambda\sigma}\, d\lambda$$
$$\times \frac{1}{2\pi\Phi_{ii}^{+}(\lambda)} \int_{0}^{\infty} e^{-j\lambda t}\, dt \int_{-\infty+jv_1}^{\infty+jv_1} \frac{\Phi_{id}(w)}{\Phi_{ii}^{-}(w)} e^{jwt}\, dw \tag{195}$$

To simplify this expression, we invert the order of integration as follows:

$$\int_{-\infty}^{\infty} h_{\text{opt}}(\sigma)\varphi_{ii}(\tau - \sigma)\, d\sigma = \frac{1}{2\pi} \int_{-\infty+jv_1}^{\infty+jv_1} \frac{\Phi_{id}(w)}{\Phi_{ii}^{-}(w)}\, dw \int_{0}^{\infty} e^{jwt}\, dt$$
$$\times \int_{-\infty+j\sigma_1}^{\infty+j\sigma_1} \frac{1}{\Phi_{ii}^{+}(\lambda)} e^{-j\lambda t}\, d\lambda \, \frac{1}{2\pi} \int_{-\infty}^{\infty} \varphi_{ii}(\tau - \sigma) e^{j\lambda\sigma}\, d\sigma \tag{196}$$

By the change of variable $\nu = \tau - \sigma$, (196) becomes

$$\int_{-\infty}^{\infty} h_{\text{opt}}(\sigma)\varphi_{ii}(\tau - \sigma)\, d\sigma = \frac{1}{2\pi} \int_{-\infty+jv_1}^{\infty+jv_1} \frac{\Phi_{id}(w)}{\Phi_{ii}^{-}(w)}\, dw \int_{0}^{\infty} e^{jwt}\, dt$$
$$\times \int_{-\infty+j\sigma_1}^{\infty+j\sigma_1} \frac{1}{\Phi_{ii}^{+}(\lambda)} e^{-j(t-\tau)\lambda}\, d\lambda \, \frac{1}{2\pi} \int_{-\infty}^{\infty} \varphi_{ii}(\nu) e^{-j\lambda\nu}\, d\nu \tag{197}$$

Since the extreme right-hand integral is $\Phi_{ii}(\lambda)$ and $\Phi_{ii}(\lambda)/\Phi_{ii}^{+}(\lambda) = \Phi_{ii}^{-}(\lambda)$, (197) becomes

$$\int_{-\infty}^{\infty} h_{\text{opt}}(\sigma)\varphi_{ii}(\tau - \sigma)\, d\sigma$$
$$= \frac{1}{2\pi} \int_{-\infty+jv_1}^{\infty+jv_1} \frac{\Phi_{id}(w)}{\Phi_{ii}^{-}(w)}\, dw \int_{0}^{\infty} e^{jwt}\, dt \int_{-\infty+j\sigma_1}^{\infty+j\sigma_1} \Phi_{ii}(\lambda) e^{-j(t-\tau)\lambda}\, d\lambda \tag{198}$$

By identifying the extreme right-hand integral of (198) with (123), we write

$$\int_{-\infty}^{\infty} h_{\text{opt}}(\sigma)\varphi_{ii}(\tau - \sigma)\,d\sigma = \frac{1}{2\pi}\int_{-\infty+jv_1}^{\infty+jv_1} \frac{\Phi_{id}(w)}{\Phi_{ii}^{-}(w)}\,dw \int_{0}^{\infty} \varphi_{ii}^{-}(\tau - t)e^{jwt}\,dt \tag{199}$$

Introducing the change of variable $x = \tau - t$ in (199), we obtain

$$\int_{-\infty}^{\infty} h_{\text{opt}}(\sigma)\varphi_{ii}(\tau - \sigma)\,d\sigma = \frac{1}{2\pi}\int_{-\infty+jv_1}^{\infty+jv_1} \frac{\Phi_{id}(w)}{\Phi_{ii}^{-}(w)}\,e^{jw\tau}\,dw \int_{-\infty}^{\tau} \varphi_{ii}^{-}(x)e^{-jwx}\,dx \tag{200}$$

Inasmuch as $\varphi_{ii}^{-}(x)$ is a function that vanishes for $x > 0$ as given by (126), the extreme right-hand integral of (200) is the transform (124) if the upper limit τ of the integral in question is greater than or equal to zero. In other words

$$\frac{1}{2\pi}\int_{-\infty}^{\tau} \varphi_{ii}^{-}(x)e^{-jwx}\,dx = \Phi_{ii}^{-}(w) \qquad \text{for } \tau \geq 0 \tag{201}$$

This restriction on the range of τ is of great interest to us because such a restriction is an exceedingly important feature of the Wiener-Hopf equation as we have pointed out before. With equation (201) substituted into (200), we have

$$\int_{-\infty}^{\infty} h_{\text{opt}}(\sigma)\varphi_{ii}(\tau - \sigma)\,d\sigma = \int_{-\infty+jv_1}^{\infty+jv_1} \frac{\Phi_{id}(w)}{\Phi_{ii}^{-}(w)}\Phi_{ii}^{-}(w)e^{jw\tau}\,dw \qquad \text{for } \tau \geq 0 \tag{202}$$

It must be emphasized that this equation does not hold for $\tau < 0$. Finally, we arrive at the result that

$$\begin{aligned}\int_{-\infty}^{\infty} h_{\text{opt}}(\sigma)\varphi_{ii}(\tau - \sigma)\,d\sigma &= \int_{-\infty+jv_1}^{\infty+jv_1} \Phi_{id}(w)e^{jw\tau}\,dw \qquad \text{for } \tau \geq 0 \\ &= \varphi_{id}(\tau) \qquad \text{for } \tau \geq 0\end{aligned} \tag{203}$$

This is the identity that we sought in verifying the solution (192) of the Wiener-Hopf equation.

11. Expression of the Solution of the Wiener-Hopf Equation in the Time Domain

To express the solution of the Wiener-Hopf equation (192), which is the optimum system function, as the time response of the optimum system to a unit-impulse excitation, we begin with (190). This equation may be expressed in the form

$$H_{\text{opt}}(\lambda) = \frac{1}{2\pi} \Gamma_{ii}^{+}(\lambda) \int_{-\infty}^{\infty} \psi^{+}(\tau) e^{-j\lambda\tau} \, d\tau \tag{204}$$

By transforming both sides of this equation, we have

$$\frac{1}{2\pi} \int_{-\infty+j\sigma_1}^{\infty+j\sigma_1} H_{\text{opt}}(\lambda) e^{j\lambda t} \, d\lambda = \frac{1}{4\pi^2} \int_{-\infty+j\sigma_1}^{\infty+j\sigma_1} e^{j\lambda t} \, d\lambda \, \Gamma_{ii}^{+}(\lambda) \int_{-\infty}^{\infty} \psi^{+}(\tau) e^{-j\lambda\tau} \, d\tau \tag{205}$$

The left-hand member of this equation is the impulse response of the optimum system. Hence (205) is

$$h_{\text{opt}}(t) = \frac{1}{4\pi^2} \int_{-\infty}^{\infty} \psi^{+}(\tau) \, d\tau \int_{-\infty+j\sigma_1}^{\infty+j\sigma_1} \Gamma_{ii}^{+}(\lambda) e^{j\lambda(t-\tau)} \, d\lambda \tag{206}$$

which yields

$$h_{\text{opt}}(t) = \frac{1}{4\pi^2} \int_{-\infty}^{\infty} \psi^{+}(\tau) \gamma_{ii}^{+}(t - \tau) \, d\tau \tag{207}$$

In this expression of the optimum system impulse response, $\psi^{+}(\tau)$ is defined by (181) where $\psi(\tau)$ is given by (191), and $\gamma_{ii}^{+}(\tau)$ is given by (162).

PROBLEM

1. An optimum linear filter A has the input $f_i(t) = f_m(t) + f_n(t)$ where $f_m(t)$ is the message and $f_n(t)$ is the noise; the desired output is $f_d(t) = f_m(t - \alpha)$ where α is the lag time, and the actual output is $f_o(t)$. Assume that the crosscorrelation between message and noise is zero. The power density spectrum of $f_i(t)$ is $\Phi_{ii}(\omega)$, and the input–desired-output cross-power density spectrum is $\Phi_{id}(\omega)$. As usual, we let

$$\psi(\tau) = \int_{-\infty+j\sigma_1}^{\infty+j\sigma_1} \frac{\Phi_{id}(\lambda)}{\Phi_{ii}^{-}(\lambda)} e^{j\lambda\tau} \, d\lambda$$

(*a*) Consider the optimum filter A under the condition that only the message $f_m(t)$ is applied to the input. Let the crosscorrelation between the input and the output of the filter, under this condition, be $\varphi_{mm'}(\tau)$. Consider also the crosscorrelation $\varphi_{do}(\tau)$ between $f_d(t)$ and $f_o(t)$ when the input is $f_i(t) = f_m(t) + f_n(t)$. Determine the relation between $\varphi_{mm'}(\tau)$ and $\varphi_{do}(\tau)$.

(*b*) Determine the expression for $\varphi_{mm'}(\tau)$ in terms of $\psi(\tau)$ only. No other functions should appear in the final expression.

chapter 15

Optimum Filtering and Prediction

The solution of the Wiener-Hopf equation presented in the last chapter is applicable to a large class of problems. Among the most important ones are filtering and prediction. In this chapter we shall study filtering with lag, filtering with prediction, and pure prediction.

1. The Optimum Filter

To apply the optimum system function (192), Chapter 14, to a specific problem, let us consider filtering. We have already stated that a common form of the input $f_i(t)$ is the sum of a message $f_m(t)$ and a noise $f_n(t)$. These functions are stationary random processes, of course. In filtering, the desired output is the message delayed by α seconds or advanced by α seconds depending upon whether a lag filter or a prediction filter is wanted. Hence as in (1) and (2), Chapter 14,

$$f_i(t) = f_m(t) + f_n(t) \tag{1}$$

and

$$f_d(t) = f_m(t \pm \alpha) \tag{2}$$

These statements are necessary in specifying a filter problem. Since the optimum system function is expressed in terms of $\Phi_{ii}(\lambda)$ and $\Phi_{id}(\lambda)$, we proceed with the determination of these functions from (1) and (2). The input autocorrelation $\varphi_{ii}(\tau)$ is

$$\begin{aligned}
\varphi_{ii}(\tau) &= \overline{f_i(t)f_i(t+\tau)} \\
&= \overline{[f_m(t) + f_n(t)][f_m(t+\tau) + f_n(t+\tau)]} \\
&= \overline{f_m(t)f_m(t+\tau)} + \overline{f_n(t)f_n(t+\tau)} + \overline{f_m(t)f_n(t+\tau)} + \overline{f_n(t)f_m(t+\tau)} \\
&= \varphi_{mm}(\tau) + \varphi_{nn}(\tau) + \varphi_{mn}(\tau) + \varphi_{nm}(\tau)
\end{aligned} \tag{3}$$

It follows immediately that

$$\Phi_{ii}(\omega) = \frac{1}{2\pi}\int_{-\infty}^{\infty} \varphi_{ii}(\tau)\cos\omega\tau\, d\tau$$

$$= \Phi_{mm}(\omega) + \Phi_{nn}(\omega) + \Phi_{mn}(\omega) + \Phi_{nm}(\omega) \tag{4}$$

For $\Phi_{id}(\lambda)$ we shall first determine $\varphi_{id}(\tau)$ which is

$$\varphi_{id}(\tau) = \overline{f_i(t)f_d(t+\tau)}$$

$$= \overline{f_i(t)f_m(t+\tau\pm\alpha)}$$

$$= \varphi_{im}(\tau\pm\alpha) \tag{5}$$

Although φ_{im} can be further resolved into

$$\varphi_{im}(\tau\pm\alpha) = \varphi_{mm}(\tau\pm\alpha) + \varphi_{nm}(\tau\pm\alpha) \tag{6}$$

we shall use the shorter expression (5). Transforming (5), we have

$$\Phi_{id}(\omega) = \frac{1}{2\pi}\int_{-\infty}^{\infty} \varphi_{im}(\tau\pm\alpha)e^{-j\omega\tau}\, d\tau$$

$$= e^{\pm j\omega\alpha}\Phi_{im}(\omega) \tag{7}$$

where

$$\Phi_{im}(\omega) = \Phi_{mm}(\omega) + \Phi_{nm}(\omega) \tag{8}$$

Writing the optimum filter system function in the forms (190) and (191), Chapter 14, we have

$$H_{\text{opt}}(\lambda) = \frac{1}{2\pi\Phi_{ii}^{+}(\lambda)}\int_{0}^{\infty} \psi(\tau\pm\alpha)e^{-j\lambda\tau}\, d\tau \tag{9}$$

where $\Phi_{ii}(\lambda)$ is obtained from (4) and

$$\psi(\tau\pm\alpha) = \int_{-\infty+j\sigma_1}^{\infty+j\sigma_1} \frac{\Phi_{im}(\lambda)}{\Phi_{ii}^{-}(\lambda)}\, e^{j(\tau\pm\alpha)\lambda}\, d\lambda \tag{10}$$

2. Illustrative Example in Optimum Filtering

To gain a better insight of the theory, we shall consider an illustrative example in statistical filtering. The example will be based upon (9) and (10). Let us assume that the message has the power density spectrum

$$\Phi_{mm}(\omega) = \frac{a^2}{b^2\omega^2 + 1} \tag{11}$$

that the noise is a white noise with the power density spectrum

$$\Phi_{nn}(\omega) = c^2 \tag{12}$$

and that the message and noise crosscorrelation is zero; that is,

$$\Phi_{mn}(\omega) = 0 \tag{13}$$

These assumptions are similar to those in an example in factorization. [See (97) and (98), Chapter 14, Sec. 8.] From these quantities the input power density spectrum can be shown to have the following form when factorized

$$\Phi_{ii}(\lambda) = \left(\frac{bc\lambda - j\sqrt{a^2 + c^2}}{b\lambda - j}\right)\left(\frac{bc\lambda + j\sqrt{a^2 + c^2}}{b\lambda + j}\right) \tag{14}$$

so that

$$\Phi_{ii}^{+}(\lambda) = \frac{bc\lambda - j\sqrt{a^2 + c^2}}{b\lambda - j} \tag{15}$$

and

$$\Phi_{ii}^{-}(\lambda) = \frac{bc\lambda + j\sqrt{a^2 + c^2}}{b\lambda + j} \tag{16}$$

Clearly

$$\begin{aligned} \Phi_{im}(\lambda) &= \Phi_{mm}(\lambda) \\ &= \frac{a^2}{b^2\lambda^2 + 1} \end{aligned} \tag{17}$$

The first integral to be evaluated is (10). We have

$$\begin{aligned} \psi(\tau \pm \alpha) \\ &= \int_{-\infty + j\sigma_1}^{\infty + j\sigma_1} \frac{a^2}{(b^2\lambda^2 + 1)} \frac{b\lambda + j}{(bc\lambda + j\sqrt{a^2 + c^2})} e^{j(\tau \pm \alpha)\lambda}\, d\lambda \\ &= a^2 \int_{-\infty + j\sigma_1}^{\infty + j\sigma_1} \frac{1}{(1 + jb\lambda)(\sqrt{a^2 + c^2} - jbc\lambda)} e^{j(\tau \pm \alpha)\lambda}\, d\lambda \\ &= \frac{a^2}{b(c + \sqrt{a^2 + c^2})} \int_{-\infty + j\sigma_1}^{\infty + j\sigma_1} \left(\frac{1}{\frac{1}{b} + j\lambda} + \frac{1}{\frac{\sqrt{a^2 + c^2}}{bc} - j\lambda}\right) e^{j(\tau \pm \alpha)\lambda}\, d\lambda \end{aligned} \tag{18}$$

Concerning the evaluation of this integral, we note that, if a function $f(t)$ is

$$f(t) = \begin{cases} e^{-kt} & \text{for } t \geq 0 \\ 0 & \text{for } t < 0 \end{cases} \tag{19}$$

then its transform is

$$F(\lambda) = \frac{1}{2\pi}\int_0^\infty e^{-kt}e^{-j\lambda t}\,dt$$

$$= \frac{1}{2\pi}\frac{1}{k+j\lambda} \tag{20}$$

so that

$$f(t) = \int_{-\infty+j\sigma_1}^{\infty+j\sigma_1} \frac{1}{2\pi}\frac{1}{k+j\lambda} e^{j\lambda t}\,d\lambda$$

$$= \begin{cases} e^{-kt} & \text{for } t \geq 0 \\ 0 & \text{for } t < 0 \end{cases} \tag{21}$$

and furthermore

$$f(t \pm \alpha) = \int_{-\infty+j\sigma_1}^{\infty+j\sigma_1} \frac{1}{2\pi}\frac{1}{k+j\lambda} e^{j(t\pm\alpha)\lambda}\,d\lambda$$

$$= \begin{cases} e^{-k(t\pm\alpha)} & \text{for } t \geq \mp\alpha \\ 0 & \text{for } t < \mp\alpha \end{cases} \tag{22}$$

It follows immediately that the first integral in (18) is

$$\int_{-\infty+j\sigma_1}^{\infty+j\sigma_1} \frac{1}{\dfrac{1}{b}+j\lambda} e^{j(\tau\pm\alpha)\lambda}\,d\lambda = \begin{cases} 2\pi e^{-(\tau\pm\alpha)/b} & \text{for } \tau \geq \mp\alpha \\ 0 & \text{for } \tau < \mp\alpha \end{cases} \tag{23}$$

In a manner similar to the evaluation of the first integral and with the aid of (142) and (143), Chapter 14, we find that the second integral in (18) is

$$\int_{-\infty+j\sigma_1}^{\infty+j\sigma_1} \frac{1}{\dfrac{\sqrt{a^2+c^2}}{bc} - j\lambda} e^{j(\tau\pm\alpha)\lambda}\,d\lambda$$

$$= \begin{cases} 0 & \text{for } \tau > \mp\alpha \\ 2\pi \exp\left[\dfrac{\sqrt{a^2+c^2}}{bc}\right](\tau\pm\alpha) & \text{for } \tau \leq \mp\alpha \end{cases} \tag{24}$$

Hence (18) becomes

$$\psi(\tau \pm \alpha) = \begin{cases} Ae^{-k_1(\tau\pm\alpha)} & \text{for } \tau \geq \mp\alpha \\ Ae^{k_2(\tau\pm\alpha)} & \text{for } \tau \leq \mp\alpha \end{cases} \tag{25}$$

where $A = 2\pi a^2/b(c + \sqrt{a^2+c^2}\,)$, $k_1 = 1/b$, and $k_2 = \sqrt{a^2+c^2}/bc$.

Figure 1 shows graphs of $\psi(\tau \pm \alpha)$ for particular values of the constants in the message and noise power density spectrums.

The next integral to be evaluated is the integral

$$\frac{1}{2\pi} \int_0^\infty \psi(\tau \pm \alpha) e^{-j\lambda\tau} \, d\tau \tag{26}$$

in the expression (9). Since (9) and (10) apply to both the lag filter and the prediction filter depending upon the choice of $-\alpha$ and $+\alpha$ re-

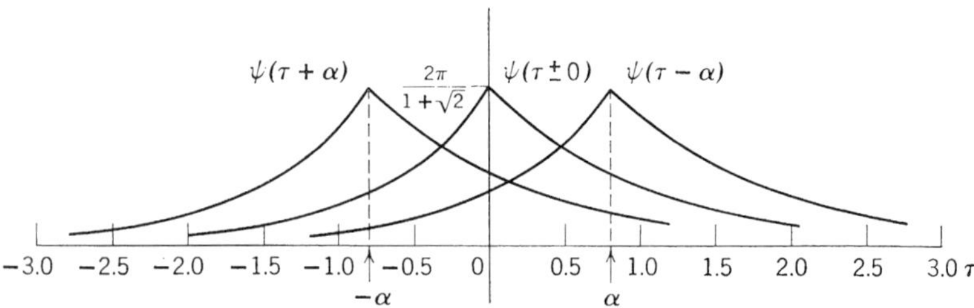

Fig. 1. Graphs of equation (25). Drawn for $a = 1$, $b = 1$, $c = 1$.

spectively, we shall consider first the lag filter and then the prediction filter. For the lag filter we evaluate the integral (26) with $-\alpha$ in the argument of ψ. The integral is

$$\begin{aligned} \frac{1}{2\pi} \int_0^\infty \psi(\tau - \alpha) e^{-j\lambda\tau} \, d\tau \\ &= \frac{A}{2\pi} \left[\int_0^\alpha e^{k_2(\tau-\alpha)} e^{-j\lambda\tau} \, d\tau + \int_\alpha^\infty e^{-k_1(\tau-\alpha)} e^{-j\lambda\tau} \, d\tau \right] \\ &= \frac{A}{2\pi} \frac{(k_1 + k_2) e^{-j\lambda\alpha} - (k_1 + j\lambda) e^{-k_2\alpha}}{(k_1 + j\lambda)(k_2 - j\lambda)} \end{aligned} \tag{27}$$

The final step in determining the system function for the optimum lag filter is to multiply (27) by $1/\Phi_{ii}^{+}(\lambda)$ in accordance with (9). Therefore, with

$$\begin{aligned} \Phi_{ii}^{+}(\lambda) &= \frac{bc\lambda - j\sqrt{a^2 + c^2}}{b\lambda - j} \\ &= \frac{c(k_2 + j\lambda)}{k_1 + j\lambda} \end{aligned} \tag{28}$$

the optimum lag filter system function is

$$H_{\text{opt}}(\lambda) = \frac{k_1 + j\lambda}{c(k_2 + j\lambda)} \frac{A[(k_1 + k_2)e^{-j\lambda\alpha} - (k_1 + j\lambda)e^{-k_2\alpha}]}{2\pi(k_1 + j\lambda)(k_2 - j\lambda)}$$

$$= k_1{}^2(k_2 - k_1) \frac{(k_1 + k_2)e^{-j\lambda\alpha} - (k_1 + j\lambda)e^{-k_2\alpha}}{k_2{}^2 + \lambda^2} \tag{29}$$

Although it is difficult to visualize the actual network that this expression represents, it is the exact system function of the optimum lag filter. There are various methods for reducing an expression of this type to an approximate expression in the form of a rational function that corresponds to an actual network. This is a problem of network synthesis. To gain a better idea of the behavior of the optimum lag filter without going into the approximation of (29) by a rational function, we shall, in Sec. 9, express (29) in the time domain by transformation. But before doing this, let us consider a simple situation for which (29) reduces to a form that is readily recognized as the system function of an actual network.

3. Optimum Zero-Lag Filter

The simple situation in question is that of an optimum zero-lag filter. If we put $\alpha = 0$, the desired output is the exact replica of the input message without any lag or lead; that is,

$$f_d(t) = f_m(t) \tag{30}$$

For this specification (29) reduces to

$$H_{\text{opt}}(\lambda) = k_1{}^2(k_2 - k_1) \frac{(k_1 + k_2) - (k_1 + j\lambda)}{k_2{}^2 + \lambda^2}$$

$$= k_1{}^2(k_2 - k_1) \frac{1}{k_2 + j\lambda} \tag{31}$$

This is the system function of the optimum zero-lag filter. Clearly this corresponds to an RL circuit or an RC circuit such as the one shown in Fig. 2. The resistor, the capacitor, and the attenuator can be made to have such values that the system function of the circuit has the expression (31). This extremely simple circuit is of course the

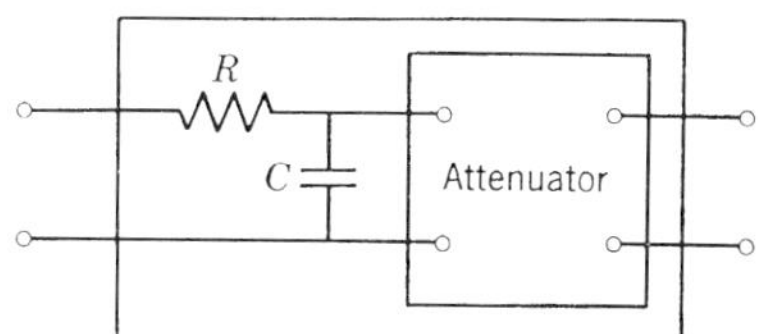

Fig. 2. An optimum zero-lag filter.

result of the simple forms of the message and noise spectrums and the absence of a lag in the desired output. We should emphasize at this point that, for the given desired output, (31) is the only system function that yields the minimum mean-square error. Any linear system whose system function differs from (31) will surely yield a greater mean-square error because we have shown that any variation of the optimum system results in an additional mean-square error. Even though (31) has the advantage of simplicity, it is not the most desirable in a filter problem. The reason is that, if a lag is permissible, we should take advantage of this lag to reduce the mean-square error. The proof that the mean-square error decreases with the increase of lag will be given when we consider the matter of errors in optimum linear systems.

Instead of obtaining the optimum-zero-lag-filter characteristic from the general expression (29), we can go back to (9) and (10) and find a simpler procedure. For $\alpha = 0$, (9) and (10) are

$$H_{\text{opt}}(\lambda) = \frac{1}{2\pi\Phi_{ii}^{+}(\lambda)} \int_0^\infty \psi(\tau \pm 0) e^{-j\lambda\tau}\, d\tau \tag{32}$$

and

$$\psi(\tau \pm 0) = \int_{-\infty+j\sigma_1}^{\infty+j\sigma_1} \frac{\Phi_{im}(\lambda)}{\Phi_{ii}^{-}(\lambda)} e^{j\tau\lambda}\, d\lambda \tag{33}$$

Let us put

$$\Psi(\lambda) = \frac{\Phi_{im}(\lambda)}{\Phi_{ii}^{-}(\lambda)} \tag{34}$$

and suppose that by partial fractions we can express (34) as

$$\Psi(\lambda) = \Psi^{+}(\lambda) + \Psi^{-}(\lambda) \tag{35}$$

in which $\Psi^{+}(\lambda)$ and $\Psi^{-}(\lambda)$ have the respective transforms

$$\psi^{+}(\tau \pm 0) = \int_{-\infty+j\sigma_1}^{\infty+j\sigma_1} \Psi^{+}(\lambda) e^{j\lambda\tau}\, d\lambda \tag{36}$$

and

$$\psi^{-}(\tau \pm 0) = \int_{-\infty+j\sigma_1}^{\infty+j\sigma_1} \Psi^{-}(\lambda) e^{j\lambda\tau}\, d\lambda \tag{37}$$

and these transforms have the properties that

$$\psi^{+}(\tau \pm 0) = 0 \qquad \text{for } \tau < 0 \tag{38}$$

and

$$\psi^{-}(\tau \pm 0) = 0 \qquad \text{for } \tau > 0 \tag{39}$$

Then, writing (32) and (33) in the form

$$H_{\text{opt}}(\lambda) = \frac{1}{2\pi\Phi_{ii}^{+}(\lambda)} \int_0^\infty e^{-j\lambda\tau}\, d\tau \int_{-\infty+j\sigma_1}^{\infty+j\sigma_1} [\Psi^{+}(w) + \Psi^{-}(w)] e^{jw\tau}\, dw \tag{40}$$

we see that $\Psi^-(w)$ does not enter into the final result at all since the integral with respect to τ is taken over $(0, \infty)$ and the transform of $\Psi^-(w)$ is zero in this interval. Therefore the system function of the optimum zero-lag filter is

$$H_{\text{opt}}(\lambda) = \frac{1}{\Phi_{ii}^+(\lambda)} \Psi^+(\lambda) \tag{41}$$

It is shown in (18) that for the example under consideration

$$\Psi(\lambda) = \frac{a^2}{b(c + \sqrt{a^2 + c^2})} \left[\frac{1}{\frac{1}{b} + j\lambda} + \frac{1}{\frac{\sqrt{a^2 + c^2}}{bc} - j\lambda} \right]$$

$$= \frac{A}{2\pi} \left[\frac{1}{k_1 + j\lambda} + \frac{1}{k_2 - j\lambda} \right] \tag{42}$$

Clearly

$$\Psi^+(\lambda) = \frac{A}{2\pi} \frac{1}{k_1 + j\lambda} \tag{43}$$

and

$$\Psi^-(\lambda) = \frac{A}{2\pi} \frac{1}{k_2 - j\lambda} \tag{44}$$

Therefore in accordance with (41) the optimum-zero-lag-filter system function is

$$H_{\text{opt}}(\lambda) = \frac{k_1 + j\lambda}{c(k_2 + j\lambda)} \frac{A}{2\pi} \frac{1}{k_1 + j\lambda}$$

$$= k_1{}^2(k_2 - k_1) \frac{1}{k_2 + j\lambda} \tag{45}$$

which is the result we obtained before from the general optimum lag filter.

4. Optimum Prediction Filter

If the filter is a prediction filter, the optimum system function is

$$H_{\text{opt}}(\lambda) = \frac{1}{2\pi\Phi_{ii}^+(\lambda)} \int_0^\infty \psi(\tau + \alpha) e^{-j\lambda\tau} \, d\tau \tag{46}$$

where

$$\psi(\tau + \alpha) = \int_{-\infty + j\sigma_1}^{\infty + j\sigma_1} \frac{\Phi_{im}(\lambda)}{\Phi_{ii}^-(\lambda)} e^{j(\tau+\alpha)\lambda} \, d\lambda \tag{47}$$

As in the analysis of the zero-lag filter, we let

$$\frac{\Phi_{im}(\lambda)}{\Phi_{ii}^{-}(\lambda)} = \Psi(\lambda)$$

$$= \Psi^{+}(\lambda) + \Psi^{-}(\lambda) \tag{48}$$

whose transforms are (36) and (37). These transforms have the properties (38) and (39).

Since the transformation of $\psi(\tau + \alpha)$ in (46) is taken in the interval $(0 \leq \tau < \infty)$, that part of $\psi(\tau + \alpha)$ for $\tau < 0$ does not enter into the solution of the problem. This situation is similar to that of the zero-lag filter. Therefore (47) can be written, for $\tau \geq 0$

$$\psi(\tau + \alpha) = \int_{-\infty + j\sigma_1}^{\infty + j\sigma_1} \Psi^{+}(\lambda) e^{j(\tau+\alpha)\lambda}\, d\lambda$$

$$= \psi^{+}(\tau + \alpha) \tag{49}$$

and the optimum-prediction-filter system function is

$$H_{\text{opt}}(\lambda) = \frac{1}{2\pi \Phi_{ii}^{+}(\lambda)} \int_0^\infty \psi^{+}(\tau + \alpha) e^{-j\lambda\tau}\, d\tau \tag{50}$$

where

$$\psi^{+}(\tau + \alpha) = \int_{-\infty + j\sigma_1}^{\infty + j\sigma_1} \Psi^{+}(\lambda) e^{j(\tau+\alpha)\lambda}\, d\lambda \tag{51}$$

In the example of Sec. 2, we have from (25)

$$\psi^{+}(\tau + \alpha) = \begin{cases} Ae^{-k_1(\tau+\alpha)} & \text{for } \tau \geq -\alpha \\ 0 & \text{for } \tau < -\alpha \end{cases} \tag{52}$$

If the filter of Sec. 2 is to be designed with a prediction time of α seconds, its optimum system function is, in accordance with (50),

$$H_{\text{opt}}(\lambda) = \frac{1}{2\pi \Phi_{ii}^{+}(\lambda)} \int_0^\infty Ae^{-k_1(\tau+\alpha)} e^{-j\lambda\tau}\, d\tau$$

$$= \frac{k_1 + j\lambda}{c(k_2 + j\lambda)} \frac{Ae^{-k_1\alpha}}{2\pi(k_1 + j\lambda)}$$

$$= k_1{}^2(k_2 - k_1)e^{-k_1\alpha} \frac{1}{k_2 + j\lambda} \tag{53}$$

As $\alpha \to 0$, this expression takes the form (45) for the zero-lag filter, as we expected. However, it is interesting to note that the prediction time

α enters the expression (53) only in the factor $e^{-k_1\alpha}$. This means that the circuit elements of the optimum predictor remain unchanged for all values of prediction time ($0 < \alpha < \infty$); the only variable in the system to be adjusted in accordance with the desired prediction time is the attenuation factor. Thus in the example shown in Fig. 2, the values of R and C are the same for all prediction time; the only necessary adjustment to be made for any given prediction time is the attenuation constant. This result seems to have failed to do what we wish because simply decreasing the output amplitude of the system for increasing prediction time does not agree with our idea that prediction means displacing the message forward in time. The situation is puzzling at present, but we shall consider this matter again for a better understanding when we come to the subject of pure prediction. For the present we just state that the result (53) is actually in agreement with the method of formulation of the problem and the minimum-mean-square-error criterion. Moreover, we must point out that the situation under discussion is an unusual one. Nevertheless, the situation merits attention.

There is another interesting matter that should be brought out at this time. We note that the expression (29) for the lag filter is considerably more involved than that for the prediction filter. As we have seen, this difference is due to the fact that the lag filter is dependent upon both $\psi^+(\tau \pm 0)$ and $\psi^-(\tau \pm 0)$ in the integral (27). Expressed in terms of (36) and (37), the equation that we have is

$$\begin{aligned}\frac{1}{2\pi}\int_0^\infty \psi(\tau - \alpha)e^{-j\lambda\tau}\,d\tau \\ &= \frac{1}{2\pi}\int_{-\alpha}^\infty [\psi^+(\tau \pm 0) + \psi^-(\tau \pm 0)]e^{-j\lambda\tau}\,d\tau \\ &= \frac{1}{2\pi}\int_{-\alpha}^0 \psi^-(\tau \pm 0)e^{-j\lambda\tau}\,d\tau + \frac{1}{2\pi}\int_0^\infty \psi^+(\tau \pm 0)e^{-j\lambda\tau}\,d\tau \quad (54)\end{aligned}$$

On the other hand, for a prediction filter, the corresponding integral is

$$\begin{aligned}\frac{1}{2\pi}\int_0^\infty \psi(\tau + \alpha)e^{-j\lambda\tau}\,d\tau &= \frac{1}{2\pi}\int_{\alpha}^\infty [\psi^+(\tau \pm 0) + \psi^-(\tau \pm 0)]e^{-j\lambda\tau}\,d\tau \\ &= \frac{1}{2\pi}\int_{\alpha}^\infty \psi^+(\tau \pm 0)e^{-j\lambda\tau}\,d\tau \quad (55)\end{aligned}$$

Because of the forward displacement of $\psi(\tau)$ by α in a prediction filter, the integral (55) involves only $\psi^+(\tau \pm 0)$. We have thus shown that, in general, the system function of a prediction filter is simpler than that of a lag filter. It is hardly necessary to say that we assume the same conditions for both filters with the only difference being $f_d(t) = f_m(t - \alpha)$ for the lag filter and $f_d(t) = f_m(t + \alpha)$ for the prediction filter.

5. The Optimum Pure Predictor

In a problem of pure prediction we assume that noise is absent so that the input to the predictor is simply the message itself; that is,

$$f_i(t) = f_m(t) \tag{56}$$

The desired output is the message advanced by the prediction time α seconds. Thus

$$f_d(t) = f_m(t + \alpha) \tag{57}$$

To determine the system function for this special problem, we first find the expressions for $\Phi_{ii}(\lambda)$ and $\Phi_{id}(\lambda)$ from the specifications (56) and (57) and then substitute them into (190) and (191), Chapter 14, in the same manner that we treated the filter problem. The input autocorrelation in the present problem is simply

$$\begin{aligned} \varphi_{ii}(\tau) &= \overline{f_m(t)f_m(t+\tau)} \\ &= \varphi_{mm}(\tau) \end{aligned} \tag{58}$$

so that

$$\Phi_{ii}(\lambda) = \Phi_{mm}(\lambda) \tag{59}$$

The input–desired-output crosscorrelation for pure prediction is

$$\begin{aligned} \varphi_{id}(\tau) &= \overline{f_m(t)f_m(t+\tau+\alpha)} \\ &= \varphi_{mm}(\tau+\alpha) \end{aligned} \tag{60}$$

which leads to

$$\begin{aligned} \Phi_{id}(\omega) &= \frac{1}{2\pi}\int_{-\infty}^{\infty} \varphi_{mm}(\tau+\alpha)e^{-j\omega\tau}\,d\tau \\ &= e^{j\omega\alpha}\Phi_{mm}(\omega) \end{aligned} \tag{61}$$

Therefore the pure-predictor system function is

$$H_{\text{opt}}(\lambda) = \frac{1}{2\pi\Phi_{mm}^{+}(\lambda)}\int_0^{\infty} \psi(\tau+\alpha)e^{-j\lambda\tau}\,d\tau \tag{62}$$

where

$$\psi(\tau+\alpha) = \int_{-\infty+j\sigma_1}^{\infty+j\sigma_1} \Phi_{mm}^{+}(\lambda)e^{j(\tau+\alpha)\lambda}\,d\lambda \tag{63}$$

Note that in pure prediction the only function required is $\Phi_{mm}^{+}(\lambda)$.

As an example, consider a Poisson wave of equally likely positive and negative pulses each of the form

$$g(t) = \begin{cases} Ete^{-at} & \text{for } t \geq 0 \\ 0 & \text{for } t < 0 \end{cases} \tag{64}$$

According to (63), Chapter 13, Sec. 7a, this wave has the power density spectrum

$$\Phi_{mm}(\omega) = \frac{kE^2}{2\pi} \frac{1}{(a^2 + \omega^2)^2} \tag{65}$$

where k is the average number of pulses per second. The factorization of this spectrum is

$$\Phi_{mm}(\lambda) = E\sqrt{\frac{k}{2\pi}} \frac{1}{(a + j\lambda)^2} E\sqrt{\frac{k}{2\pi}} \frac{1}{(a - j\lambda)^2} \tag{66}$$

so that

$$\Phi_{mm}^{+}(\lambda) = E\sqrt{\frac{k}{2\pi}} \frac{1}{(a + j\lambda)^2} \tag{67}$$

Let us note that, if $\Phi_{mm}^{+}(\lambda)$ of (63) has a constant factor, this factor will eventually be canceled by the same factor in $\Phi_{mm}^{+}(\lambda)$ of (62). In other words the final expression of the optimum predictor is independent of the constant factor of $\Phi_{mm}^{+}(\lambda)$. In view of this fact and for simplicity in algebraic work, let us put $kE^2/2\pi = 1$.

Since we know that

$$\frac{1}{2\pi} \int_0^\infty te^{-at} e^{-j\lambda t}\, dt = \frac{1}{2\pi(a + j\lambda)^2} \tag{68}$$

and its inverse is

$$\int_{-\infty + j\sigma_1}^{\infty + j\sigma_1} \frac{1}{2\pi(a + j\lambda)^2} e^{j\lambda t}\, d\lambda = \begin{cases} te^{-at} & \text{for } t \geq 0 \\ 0 & \text{for } t < 0 \end{cases} \tag{69}$$

we have for (63)

$$\psi(\tau + \alpha) = \int_{-\infty + j\sigma_1}^{\infty + j\sigma_1} \frac{1}{(a + j\lambda)^2} e^{j(\tau + \alpha)\lambda}\, d\lambda$$

$$= \begin{cases} 2\pi(\tau + \alpha)e^{-a(\tau + \alpha)} & \text{for } \tau \geq -\alpha \\ 0 & \text{for } \tau < -\alpha \end{cases} \tag{70}$$

and the integral in (62) is

$$\frac{1}{2\pi}\int_0^\infty \psi(\tau+\alpha)e^{-j\lambda\tau}\,d\tau = \frac{1}{2\pi}\int_0^\infty 2\pi(\tau+\alpha)e^{-a(\tau+\alpha)}e^{-j\lambda\tau}\,d\tau$$

$$= e^{-a\alpha}\left(\frac{1}{(a+j\lambda)^2}+\frac{\alpha}{a+j\lambda}\right) \tag{71}$$

Finally in accordance with (62)

$$H_{\text{opt}}(\lambda) = (a+j\lambda)^2 e^{-a\alpha}\left(\frac{1}{(a+j\lambda)^2}+\frac{\alpha}{a+j\lambda}\right)$$

$$= e^{-a\alpha}[(1+a\alpha)+j\alpha\lambda] \tag{72}$$

This is the system function of the optimum predictor for the Poisson wave. A simple way to represent this expression as a circuit is given in

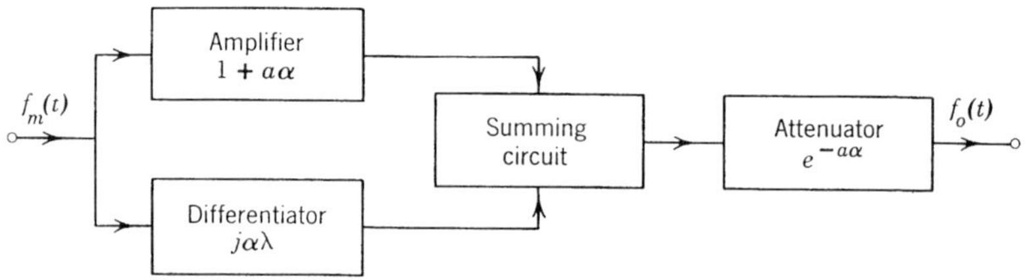

Fig. 3. An optimum predictor.

Fig. 3. In this example all three components of the circuit are dependent upon the prediction time α. It is interesting to observe that the amplitude of the predictor output should tend to zero exponentially as the prediction time tends to infinity. When we discussed the mean-square error in the formulation of the optimum system problem, we pointed out that, even if two waves are of the same waveform, the mean-square error between the two waves is zero only when they have the same amplitude. Hence the optimum system function based on the minimum-mean-square-error criterion must include the exact amplitude factor.

Another helpful example in the understanding of the optimum predictor is the prediction of a Poisson wave of equally likely positive and negative pulses each of the form

$$g(t) = \begin{cases} Ee^{-at} & \text{for } t \geq 0 \\ 0 & \text{for } t < 0 \end{cases} \tag{73}$$

Referring to (63), Chapter 13, Sec. 7*a*, we have

$$\Phi_{mm}(\lambda) = \frac{kE^2}{2\pi}\frac{1}{a^2+\lambda^2} \tag{74}$$

Factoring this spectrum, we have

$$\Phi_{mm}(\lambda) = \left(E\sqrt{\frac{k}{2\pi}}\frac{1}{a+j\lambda}\right)\left(E\sqrt{\frac{k}{2\pi}}\frac{1}{a-j\lambda}\right) \tag{75}$$

in which

$$\Phi^+_{mm}(\lambda) = E\sqrt{\frac{k}{2\pi}}\frac{1}{a+j\lambda} \tag{76}$$

As explained in the preceding example, we let

$$\Phi^+_{mm}(\lambda) = \frac{1}{a+j\lambda} \tag{77}$$

Then (63) is

$$\begin{aligned}\psi(\tau+\alpha) &= \int_{-\infty+j\sigma_1}^{\infty+j\sigma_1} \frac{1}{a+j\lambda}e^{j(\tau+\alpha)\lambda}\,d\lambda \\ &= \begin{cases}2\pi e^{-a(\tau+\alpha)} & \text{for } \tau > -\alpha \\ 0 & \text{for } \tau < -\alpha\end{cases}\end{aligned} \tag{78}$$

and the transform of this function over $(0, \infty)$ is

$$\frac{1}{2\pi}\int_0^\infty 2\pi e^{-a(\tau+\alpha)}e^{-j\lambda\tau}\,d\tau = e^{-a\alpha}\frac{1}{a+j\lambda} \tag{79}$$

Finally we have

$$\begin{aligned}H_{\text{opt}}(\lambda) &= (a+j\lambda)e^{-a\alpha}\frac{1}{a+j\lambda} \\ &= e^{-a\alpha}\end{aligned} \tag{80}$$

This result states that the optimum predictor for the Poisson wave of equally likely positive and negative exponential pulses is simply an attenuator that reduces the amplitude of the wave exponentially as $e^{-a\alpha}$. In the study of the optimum prediction filter we came to a similar situation. Although this result does not seem to serve our purpose, on the basis of the formulation of the optimum system problem and the minimum-mean-square-error criterion, the result actually satisfies the demands of the problem. We shall discuss this matter in the next section.

6. An Interpretation of the Pure Predictor

It is interesting to examine the pure-predictor system function (62) and (63) for an insight into the operations involved in the problem. We shall consider the prediction of messages in the form of Poisson waves. These waves may be generated in the manner shown in the block indicated as message source in Fig. 4. A generator N_a generates a Poisson wave $f_P(t)$ of equally likely positive and negative unit impulses. This unit-impulse wave is applied to the pulse-shape network N_b whose unit-

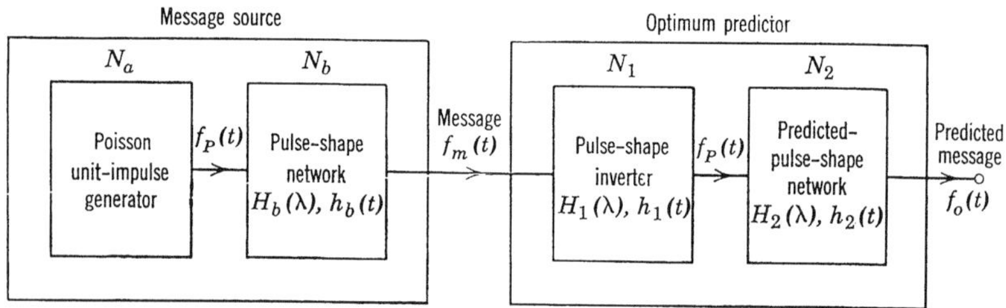

Fig. 4. An interpretation of pure prediction.

impulse response $h_b(t)$ is designed to have the given form $g(t)$; that is,

$$h_b(t) = g(t) \tag{81}$$

The output of N_b is clearly a Poisson wave of equally likely positive and negative pulses each of the form $g(t)$. This Poisson wave with pulses of any given form is the message $f_m(t)$ that we wish to predict. Denoting the system function of N_b by $H_b(\lambda)$ and the Fourier transform of $g(t)$ by $G(\lambda)$, we have

$$H_b(\lambda) = G(\lambda) \tag{82}$$

For the power density spectrum of $f_m(t)$ we refer to equation (63), Chapter 13, Sec. 7*a*. We find that the message power density spectrum is

$$\Phi_{mm}(\omega) = \frac{k}{2\pi} |G(\omega)|^2 \tag{83}$$

where k is the average number of pulses per second. The spectrum factorization required in the problem is

$$\Phi_{mm}(\lambda) = \sqrt{\frac{k}{2\pi}}\, G^+(\lambda) \sqrt{\frac{k}{2\pi}}\, G^-(\lambda) \tag{84}$$

so that

$$\Phi_{mm}^{+}(\lambda) = \sqrt{\frac{k}{2\pi}}\, G^{+}(\lambda) \tag{85}$$

and

$$\Phi_{mm}^{-}(\lambda) = \sqrt{\frac{k}{2\pi}}\, G^{-}(\lambda) \tag{86}$$

Here $G^{+}(\lambda)$ is the factor in $|G(\lambda)|^2$ that contains all the poles and zeros in the upper half-plane as in our previous work, and $G^{-}(\lambda)$ is the factor in $|G(\lambda)|^2$ that contains all the poles and zeros in the lower half-plane. Furthermore

$$G^{+}(\lambda) = \overline{G}^{-}(\lambda) \tag{87}$$

We should be careful in distinguishing among $G(\lambda)$, $\overline{G}(\lambda)$, $G^{+}(\lambda)$, and $G^{-}(\lambda)$. Although

$$\Phi_{mm}(\lambda) = \frac{k}{2\pi} G^{+}(\lambda)G^{-}(\lambda)$$

$$= \frac{k}{2\pi} G(\lambda)\overline{G}(\lambda) \tag{88}$$

we must not conclude that $G^{+}(\lambda)$ and $G(\lambda)$ are the same and that $G^{-}(\lambda)$ and $\overline{G}(\lambda)$ are the same, in general. It is a fact that, if $G(\lambda)$ has no zeros in the lower half-plane, then $G(\lambda) = G^{+}(\lambda)$ and $\overline{G}(\lambda) = G^{-}(\lambda)$. However, zeros of $G(\lambda)$ may appear in the lower half-plane. For example,

$$G(\lambda) = \frac{2 - j\lambda}{(1 + j\lambda)(4 + j\lambda)} \tag{89}$$

has a zero at $\omega = 0$, $j\sigma = -j2$. For this example we have

$$\overline{G}(\lambda) = \frac{2 + j\lambda}{(1 - j\lambda)(4 - j\lambda)} \tag{90}$$

$$G^{+}(\lambda) = \frac{2 + j\lambda}{(1 + j\lambda)(4 + j\lambda)} \tag{91}$$

and

$$G^{-}(\lambda) = \frac{2 - j\lambda}{(1 - j\lambda)(4 - j\lambda)} \tag{92}$$

In the present discussion let us assume that $G(\lambda)$ has no zeros in the lower half-plane so that $G^{+}(\lambda) = G(\lambda)$ and $G^{-}(\lambda) = \overline{G}(\lambda)$. With this assumption we have

$$\Phi_{mm}^{+}(\lambda) = \sqrt{\frac{k}{2\pi}}\, G(\lambda) \tag{93}$$

Inasmuch as the constant factor $\sqrt{k/2\pi}$ does not enter into the final expression for $H_{\text{opt}}(\lambda)$ as we pointed out before, let us put, for simplicity,

$$\Phi_{mm}^{+}(\lambda) = G(\lambda) \tag{94}$$

With this result, (62) and (63) become

$$H_{\text{opt}}(\lambda) = \frac{1}{2\pi G(\lambda)} \int_0^\infty \psi(\tau + \alpha) e^{-j\lambda\tau}\, d\tau \tag{95}$$

$$\psi(\tau + \alpha) = \int_{-\infty + j\sigma_1}^{\infty + j\sigma_1} G(\lambda) e^{j(\tau+\alpha)\lambda}\, d\lambda \tag{96}$$

We shall represent the optimum predictor as a cascade connection of two networks N_1 and N_2 as shown in Fig. 4. To the network N_1 we assign the system function

$$H_1(\lambda) = \frac{1}{G(\lambda)} \tag{97}$$

Since $G(\lambda)$ has no zeros in the lower half-plane, its reciprocal is a possible system function. To the network N_2 we assign the system function

$$H_2(\lambda) = \frac{1}{2\pi} \int_0^\infty e^{-j\lambda\tau}\, d\tau \int_{-\infty + jv_1}^{\infty + jv_1} G(w) e^{j(\tau+\alpha)w}\, dw \tag{98}$$

The optimum predictor system function is therefore

$$H_{\text{opt}}(\lambda) = H_1(\lambda) H_2(\lambda) \tag{99}$$

One important aspect of the analysis is that the output of the network N_1 is the unit-impulse Poisson wave $f_P(t)$. To see this, we need only consider the system function of N_b and N_1 in cascade. This system function is

$$H_b(\lambda) H_1(\lambda) = G(\lambda) \frac{1}{G(\lambda)} = 1 \tag{100}$$

Since the input of N_b is $f_P(t)$, the output of N_1 must be $f_P(t)$. We call network N_1 the pulse-shape inverter.

Concerning N_2 we find that its system function as given in (98) is

$$H_2(\lambda) = \int_0^\infty g(\tau + \alpha) e^{-j\lambda\tau}\, d\tau \tag{101}$$

which means that

$$h_2(t) = \begin{cases} g(t + \alpha) & \text{for } t \geq 0 \\ 0 & \text{for } t < 0 \end{cases} \tag{102}$$

This is the unit-impulse response of the network N_2. Since the input to N_2 has been shown to be the unit-impulse Poisson wave, $h_2(t)$ given by (102) is actually the shape of the individual pulses in the predicted message. Hence we call N_2 the predicted-pulse-shape network.

This analysis shows that, if Poisson waves of equally likely positive and negative pulses of the same shape are to be predicted, the optimum

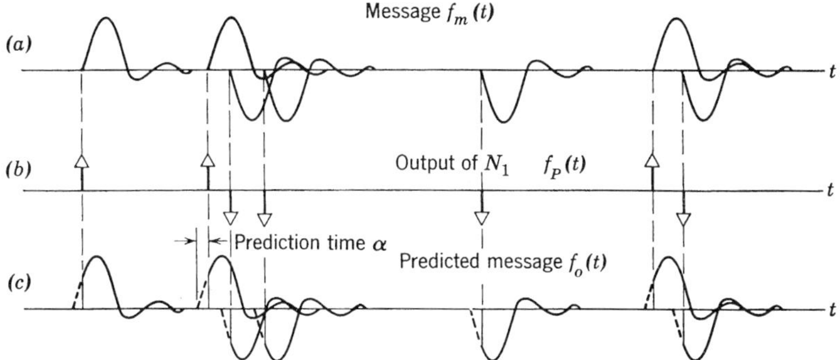

Fig. 5. Waveforms involved in pure prediction. Instantaneous value of wave (*a*) or wave (*c*) is the algebraic sum of the pulses.

predictor first converts the pulses into impulses and then these impulses are shaped into the predicted form. According to the theory that we have presented, the predicted shape of the pulses is simply the shape of the pulses in the given message displaced forward by the prediction time α but with that part of the displaced pulse for $t < 0$ reduced to zero. This elimination is of course a physical necessity since the output of the network N_2 cannot precede the cause. In accordance with this analysis, the waveforms involved in the problem are illustrated in Fig. 5. The first wave (*a*) is the message to be predicted; then the pulses are transformed into impulses as in (*b*), and then these impulses are transformed into pulses of the predicted form as in (*c*).

On the basis of this interpretation of optimum prediction the result (72) for the first example can be obtained in a simple manner and at the same time with an understanding of the basic operations involved. In this example, the system function of the pulse-shape inverter is

$$H_1(\lambda) = \frac{1}{E}(a + j\lambda)^2 \tag{103}$$

and the impulse response of the predicted-pulse-shape network in accordance with (102) is

$$h_2(t) = \begin{cases} E(t+\alpha)e^{-a(t+\alpha)} & \text{for } t \geq 0 \\ 0 & \text{for } t < 0 \end{cases} \tag{104}$$

Similarly for the second example

$$H_1(\lambda) = \frac{1}{E}(a + j\lambda) \tag{105}$$

and

$$h_2(t) = \begin{cases} Ee^{-a(t+\alpha)} = e^{-a\alpha}Ee^{-at} & \text{for } t \geq 0 \\ 0 & \text{for } t < 0 \end{cases} \tag{106}$$

Since it is a property of the exponential function that a displacement merely changes the function by an amplitude factor, the predicted message is simply the same as the message decreased by the factor $e^{-a\alpha}$. This is the explanation for the result (80). The waveforms of the input and output of the predictor in the second example are given in Fig. 6.

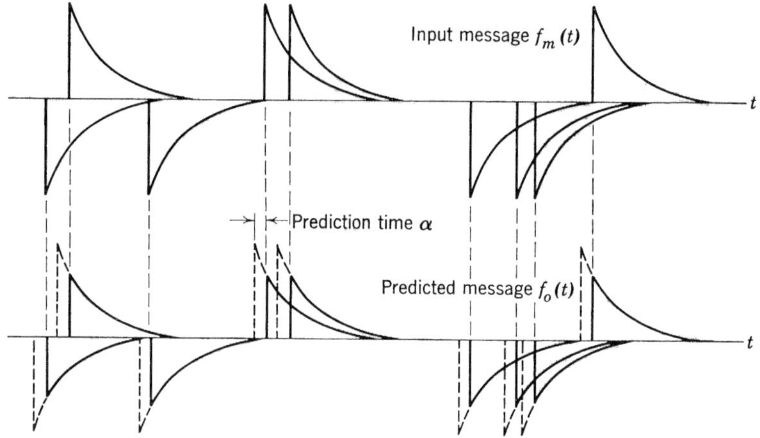

Fig. 6. Waveforms of the input and output of a predictor. Instantaneous value of each wave is the algebraic sum of the pulses.

The analysis of the optimum prediction of Poisson waves of equally likely positive and negative pulses of the same shape shows that, if the transform of the pulse shape has no zeros in the lower half-plane so that (94) holds, the prediction of each pulse from its initial point to the time α (the prediction time) is always zero but that the prediction from that point on is always without error. Under the given conditions, the error in prediction lies only in the beginning of each pulse. We shall discuss prediction error in another section, but at this time it is interesting to note this feature of the theory. To put the observation in another way, we say that the predictor does not predict the initial points of the Poisson wave in any way; the prediction of each pulse begins at time α after the occurrence of the initial point of the pulse. The fact that the initial points of the pulses are not predicted is to be expected, because in a Poisson wave the occurrence of an initial point in any given dt interval is independent of the number of initial points in any interval before the given dt interval.

We continue the analysis by considering the situation where the transform of the pulse shape has one or more zeros in the lower half-plane. This situation requires that (62) and (63) take the forms

$$H_{\text{opt}}(\lambda) = \frac{1}{2\pi G^{+}(\lambda)} \int_0^{\infty} \psi(\tau + \alpha) e^{-j\lambda\tau}\, d\tau \tag{107}$$

$$\psi(\tau + \alpha) = \int_{-\infty+j\sigma_1}^{\infty+j\sigma_1} G^{+}(\lambda) e^{j(\tau+\alpha)\lambda}\, d\lambda \tag{108}$$

The optimum predictor is represented by two networks N_1 and N_2 in cascade as shown in Fig. 4. The message source is the same as before. In a manner similar to (97) we write the system function of N_1 as

$$H_1(\lambda) = \frac{1}{G^{+}(\lambda)} \tag{109}$$

This is a possible system function because $G^{+}(\lambda)$ has no zeros in the lower half-plane. Note that under present conditions $1/G(\lambda)$ is not a possible system function. The system function of the network N_2 will be

$$H_2(\lambda) = \frac{1}{2\pi} \int_0^{\infty} e^{-j\lambda\tau}\, d\tau \int_{-\infty+jv_1}^{\infty+jv_1} G^{+}(w) e^{j(\tau+\alpha)w}\, dw \tag{110}$$

Again, we have

$$H_{\text{opt}}(\lambda) = H_1(\lambda) H_2(\lambda) \tag{111}$$

When we consider networks N_b and N_1 in cascade, we find that

$$H_b(\lambda) H_1(\lambda) = \frac{G(\lambda)}{G^{+}(\lambda)} \tag{112}$$

To interpret this expression, let us note that $G^{+}(\lambda)$ is obtained by replacing those factors in $G(\lambda)$ for zeros in the lower half-plane by factors for zeros in the upper half-plane that are reflected from the lower half-plane. For instance, in the example in Sec. 6, if

$$G(\lambda) = \frac{2 - j\lambda}{(1 + j\lambda)(4 + j\lambda)} \tag{113}$$

with a zero in the lower half-plane at $\omega = 0$, $j\sigma = -j2$, then

$$G^+(\lambda) = \frac{2 + j\lambda}{(1 + j\lambda)(4 + j\lambda)} \tag{114}$$

with a zero in the upper half-plane at $\omega = 0$, $j\sigma = j2$ which is the reflection of the zero in $G(\lambda)$. The ratio of (113) and (114) is

$$\frac{G(\lambda)}{G^+(\lambda)} = \frac{2 - j\lambda}{2 + j\lambda} \tag{115}$$

We can generalize this result and state that, if $G(\lambda)$ has zeros at $-ja_1$, $-ja_2$, . . ., $-ja_n$ on the imaginary axis, then

$$\frac{G(\lambda)}{G^+(\lambda)} = \frac{(a_1 - j\lambda)(a_2 - j\lambda) \cdots (a_n - j\lambda)}{(a_1 + j\lambda)(a_2 + j\lambda) \cdots (a_n + j\lambda)} \tag{116}$$

This expression can be extended to include zeros in the lower half-plane that are not on the imaginary axis, but we need not do this in the present discussion. We know that (116) represents a pure phase-shift system with its absolute value of amplitude equal to unity. When $G(\lambda)$ has no zeros in the lower half-plane, the output of N_1 is simply the unit-impulse Poisson wave. However, when $G(\lambda)$ has zeros in the lower half-plane, the output of N_1 does not enjoy this simplicity because of the fact that N_b and N_1 in cascade result in a pure phase-shift system. It can be shown that this output is a combination of impulses and exponentials.

Since the input to N_2 is no longer the unit-impulse Poisson wave and since $H_2(\lambda)$ is given in terms of $G^+(\lambda)$ instead of $G(\lambda)$, it becomes difficult to see the predicted pulse shape without some analysis. A further discussion on this matter is given in Chapter 16, Sec. 3.

Although we have discussed the prediction of only Poisson waves, we may obtain valuable information in a problem concerning a wave of another type by considering a Poisson wave with the same power density spectrum. We know that the optimum pure predictor depends upon only the power density spectrum of the wave irrespective of its waveform. If the power density spectrum of a wave is $\Phi_{mm}(\omega)$, clearly we can construct a Poisson wave that has this power density spectrum by finding the pulse shape from the expression

$$\Phi_{mm}(\lambda) = \frac{k}{2\pi} G(\lambda)\overline{G}(\lambda) \tag{117}$$

If $\Phi_{mm}(\lambda)$ is expressed as a rational function, then by factorization we can obtain $G(\lambda)$ from which we determine the pulse shape $g(t)$.

7. A Laboratory Demonstration of Prediction

A laboratory demonstration of pure prediction of random noise has been carried out.* The block diagram for the demonstration is given in Fig. 7. Random noise from a Type 6D4 gas tube was to be predicted.

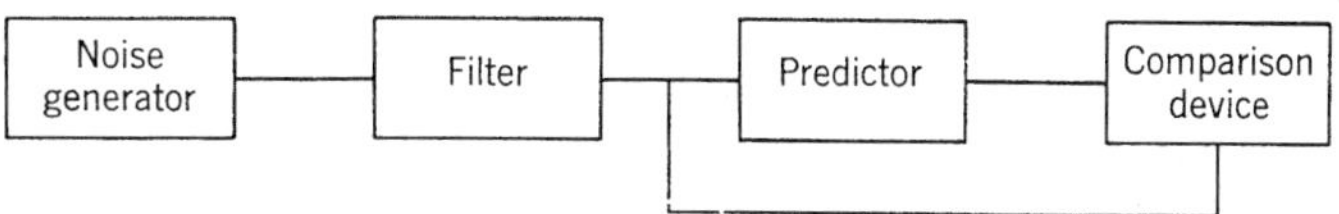

Fig. 7. Block diagram for demonstration of prediction.

The filter shown was a circuit for shaping the spectrum of the noise to a simple form for representation. The predictor circuit designed in accordance with the theory presented was a simple one, and the basic elements are shown in Fig. 8. The comparison device was a high-speed

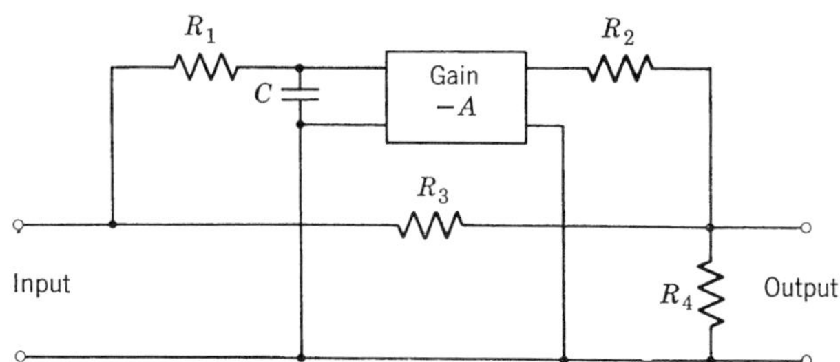

Fig. 8. Basic circuit diagram of predictor for demonstration.

camera which recorded continuously the input and output random waves with the aid of a two-beam oscilloscope. This instantaneous recording

* Y. W. Lee and C. A. Stutt, *Statistical Prediction of Noise*, Technical Report No. 129, Research Laboratory of Electronics, M.I.T., 1949; *Proc. NEC*, **5**, 342–365 (1949).

permitted the comparison of actual and predicted values of the random noise. Sample records are shown in Fig. 9 for three prediction times.

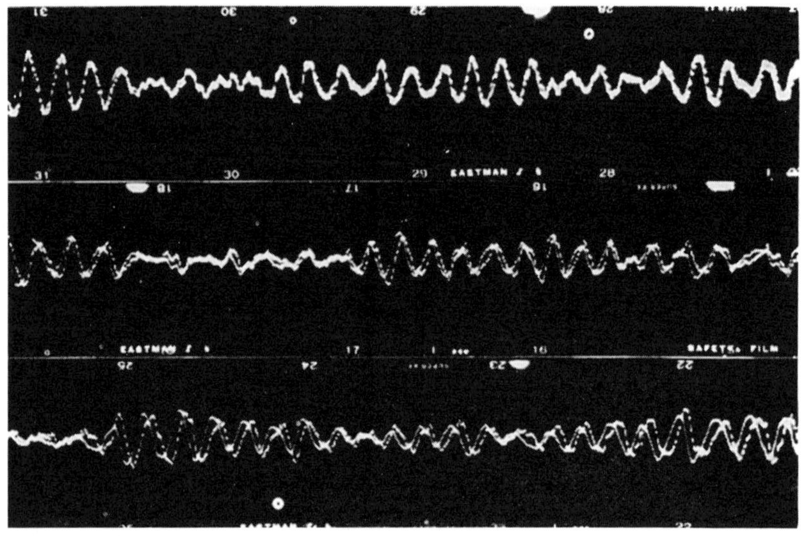

Fig. 9. Demonstration of prediction. Predictor input is dotted; predictor output is solid. Prediction times (reading down): 0.06 millisecond, 0.12 millisecond, 0.18 millisecond. Past is to the left, future to the right. Flash marks on edge of film are 1/60 second apart.

8. Contour Integration

In optimum filter theory the evaluation of certain integrals may often be facilitated by the method of contour integration. We shall not go into details but simply state some of the well-known results for reference.

The Cauchy residue theorem states that the integral of an analytic function $f(\lambda)$ around a closed path C enclosing poles at $\lambda_1, \lambda_2, \ldots, \lambda_n$ is equal to $2\pi j$ times the sum of the residues of $f(\lambda)$ at $\lambda_1, \lambda_2, \ldots, \lambda_n$; that is,

$$\oint_C f(\lambda)\, d\lambda = 2\pi j(R_1 + R_2 + \cdots + R_n) \tag{118}$$

The path C is in the counterclockwise direction as shown in Fig. 10. If the integral is taken in the clockwise direction, the value of the integral is $-2\pi j(R_1 + R_2 + \cdots + R_n)$.

Among the rules for evaluating the residues of a function, a convenient one states that, if $f(\lambda)$ has a pole of order m at λ_1, then letting

$$\phi(\lambda) = (\lambda - \lambda_1)^m f(\lambda) \tag{119}$$

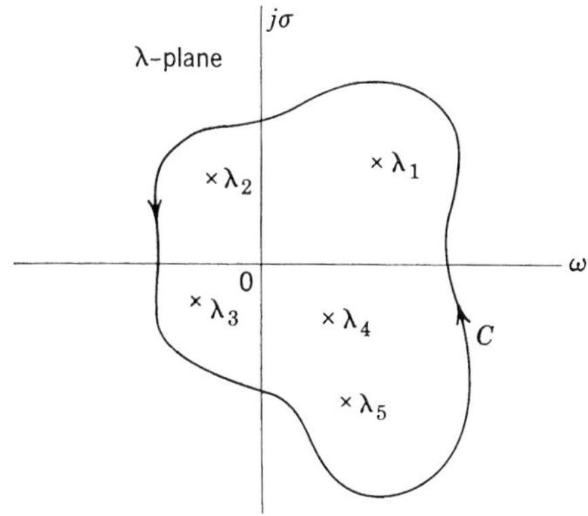

Fig. 10. Illustrating contour integration.

we find that the residue of $f(\lambda)$ at λ_1 is

$$R_1 = \frac{1}{(m-1)!}\frac{d^{m-1}}{d\lambda^{m-1}}\phi(\lambda)\,|_{\lambda=\lambda_1} \tag{120}$$

For the first-order pole ($m = 1$) the residue of $f(\lambda)$ at λ_1 is

$$R_1 = (\lambda - \lambda_1)f(\lambda)\,|_{\lambda=\lambda_1} \tag{121}$$

An integral that comes into our work frequently is the integral

$$\psi(\tau \pm \alpha) = \int_{-\infty+j\sigma_1}^{\infty+j\sigma_1} \Psi(\lambda)e^{j(\tau\pm\alpha)\lambda}\,d\lambda \tag{122}$$

If $|\psi(\lambda)| \to 0$ at least as fast as $1/|\lambda|$ as $|\lambda| \to \infty$ and if $\Psi(\lambda)$ is a rational function, the integral can be replaced by a contour integral taken along two closed paths, one for $\tau \pm \alpha > 0$ and the other for $\tau \pm \alpha < 0$. We have

$$\int_{-\infty+j\sigma_1}^{\infty+j\sigma_1} \Psi(\lambda)e^{j(\tau\pm\alpha)\lambda}\,d\lambda = \begin{cases} \oint_{C^+} \Psi(\lambda)e^{j(\tau\pm\alpha)\lambda}\,d\lambda & \text{for } \tau \pm \alpha > 0 \\ \oint_{C^-} \Psi(\lambda)e^{j(\tau\pm\alpha)\lambda}\,d\lambda & \text{for } \tau \pm \alpha < 0 \end{cases} \tag{123}$$

As shown in Fig. 11, the integral for $\tau \pm \alpha > 0$ is taken along a counterclockwise path C^+ that is formed by the ω-axis and a semi-circle in the upper half-plane enclosing all the poles of $\Psi(\lambda)$ in the upper half-plane. On the other hand, the integral for $\tau \pm \alpha < 0$ is taken along a clockwise path C^- that is formed by the ω-axis and a semi-circle in the lower half-

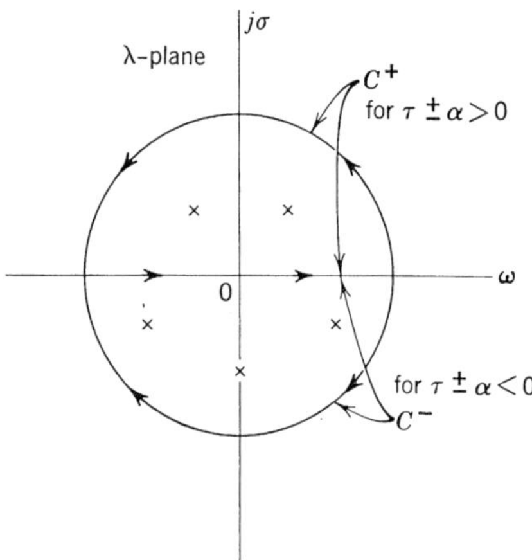

Fig. 11. Illustrating the evaluation of equation (123) by contour integration.

plane enclosing all the poles of $\Psi(\lambda)$ in the lower half-plane. In terms of residues, we have

$$\oint_{C^+} \Psi(\lambda) e^{j(\tau \pm \alpha)\lambda}\, d\lambda = 2\pi j \text{ [sum of residues of } \Psi(\lambda) e^{j(\tau\pm\alpha)\lambda} \text{ in the upper half-plane]} \quad \text{for } \tau \pm \alpha > 0 \tag{124}$$

$$\oint_{C^-} \Psi(\lambda) e^{j(\tau \pm \alpha)\lambda}\, d\lambda = -2\pi j \text{ [sum of residues of } \Psi(\lambda) e^{j(\tau\pm\alpha)\lambda} \text{ in the lower half-plane]} \quad \text{for } \tau \pm \alpha < 0 \tag{125}$$

9. Example of Optimum System Expressed in the Time Domain

For the illustrative example of optimum filtering in Sec. 2, we found that the optimum filter system function has the expression (29). This expression does not readily give us an idea as to the behavior of the system. At the first glance at this function we might even doubt its physical realization, because the denominator is $k_2{}^2 + \lambda^2$ which becomes $(k_2 + j\lambda)(k_2 - j\lambda)$ when factorized. It might seem that the optimum system requires a pole in the upper half-plane and a pole in the lower half-plane thus making the system physically impossible. Actually this is not the case, and the reason for it is that, when $k_2 - j\lambda = 0$, the expression (29) becomes indeterminate and should be evaluated by differentiation.

Looking at the optimum system in the time domain is interesting and clears up many questions. Furthermore, certain features of the system are better displayed in the time domain. From the standpoint of network synthesis, a time domain specification of the optimum system may be advantageous. To express (29) in the time domain, we may simply transform the expression or follow the formula (207), Chapter 14, Sec. 11.

In accordance with the method of transformation, we have

$$h_{\text{opt}}(t) = \frac{1}{2\pi} k_1{}^2(k_2 - k_1) \int_{-\infty+j\sigma_1}^{\infty+j\sigma_1} \frac{(k_1 + k_2)e^{-j\lambda\alpha} - (k_1 + j\lambda)e^{-k_2\alpha}}{k_2{}^2 + \lambda^2} e^{j\lambda t}\, d\lambda$$

$$= k_1{}^2(k_2 - k_1) \left[\frac{1}{2\pi} \int_{-\infty+j\sigma_1}^{\infty+j\sigma_1} \frac{k_1 + k_2}{k_2{}^2 + \lambda^2} e^{j(t-\alpha)\lambda}\, d\lambda \right.$$

$$\left. - \frac{1}{2\pi} \int_{-\infty+j\sigma_1}^{\infty+j\sigma_1} e^{-k_2\alpha} \frac{(k_1 + j\lambda)}{k_2{}^2 + \lambda^2} e^{j\lambda t}\, d\lambda \right]$$

$$= k_1{}^2(k_2 - k_1)[A(t) - B(t)] \tag{126}$$

in which $A(t)$ is the first integral in the brackets and $B(t)$ is the second integral. Since both $A(t)$ and $B(t)$ satisfy the conditions for contour integration which we stated in connection with (122), we shall determine $A(t)$ and $B(t)$ by this method. For $A(t)$, the integrand has a pole in the upper half-plane when $\lambda - jk_2 = 0$, and it is the only pole in the upper half-plane. Hence

$$R_1 = (\lambda - jk_2) \frac{k_1 + k_2}{(\lambda - jk_2)(\lambda + jk_2)} e^{j(t-\alpha)\lambda} |_{\lambda = jk_2}$$

$$= \frac{k_1 + k_2}{2jk_2} e^{-k_2(t-\alpha)} \qquad \text{for } t - \alpha > 0 \tag{127}$$

Evaluating the residue of the integrand in this manner for $t - \alpha < 0$, we have

$$A(t) = \begin{cases} \dfrac{k_1 + k_2}{2k_2} e^{-k_2(t-\alpha)} & \text{for } t - \alpha > 0 \\[2ex] \dfrac{k_1 + k_2}{2k_2} e^{k_2(t-\alpha)} & \text{for } t - \alpha < 0 \end{cases} \tag{128}$$

Similarly we find that

$$B(t) = \begin{cases} \dfrac{e^{-k_2\alpha}(k_1 - k_2)}{2k_2} e^{-k_2 t} & \text{for } t > 0 \\[2ex] \dfrac{e^{-k_2\alpha}(k_1 + k_2)}{2k_2} e^{k_2 t} & \text{for } t < 0 \end{cases} \tag{129}$$

We shall determine $h_{\text{opt}}(t)$ first for the interval $(-\infty, 0)$, then for the interval $(0, \alpha)$, and finally for the interval (α, ∞). For the first interval we have in accordance with (126)

$$h_{\text{opt}}(t) = k_1^2(k_2 - k_1)\left[\frac{k_1 + k_2}{2k_2} e^{k_2(t-\alpha)} - \frac{e^{-k_2\alpha}(k_1 + k_2)}{2k_2} e^{k_2 t}\right]$$

$$= 0 \qquad \text{for } t < 0 \tag{130}$$

This result is what it should be, otherwise we would be in grave trouble. The expression for the second interval is

$$h_{\text{opt}}(t) = k_1^2(k_2 - k_1)\left[\frac{k_1 + k_2}{2k_2} e^{k_2(t-\alpha)} - \frac{e^{-k_2\alpha}(k_1 - k_2)}{2k_2} e^{-k_2 t}\right]$$

$$= \frac{k_1^2(k_2 - k_1)}{2k_2} e^{-k_2\alpha}[k_2(e^{k_2 t} + e^{-k_2 t}) + k_1(e^{k_2 t} - e^{-k_2 t})]$$

$$= \frac{k_1^2(k_2 - k_1)}{k_2} e^{-k_2\alpha}[k_2 \cosh k_2 t + k_1 \sinh k_2 t] \qquad \text{for } 0 \le t \le \alpha \tag{131}$$

Finally the expression for the third interval is

$$h_{\text{opt}}(t) = k_1^2(k_2 - k_1)\left[\frac{k_1 + k_2}{2k_2} e^{-k_2(t-\alpha)} - \frac{e^{-k_2\alpha}(k_1 - k_2)}{2k_2} e^{-k_2 t}\right]$$

$$= \frac{k_1^2(k_2 - k_1)}{k_2} e^{-k_2 t}[k_2 \cosh k_2\alpha + k_1 \sinh k_2\alpha] \qquad \text{for } \alpha \le t < \infty \tag{132}$$

This expression is simply an exponential function; and it is interesting that (131) and (132) are symmetrical since, if α is replaced by t and t by α, (131) becomes (132). Graphs of $h_{\text{opt}}(t)$ for a set of constants that are chosen solely for convenience in numerical computation and for various values of lag time α are given in Fig. 12. These curves show the behavior of $h_{\text{opt}}(t)$ as a function of α. The envelope of the peak at α is obtained by setting $t = \alpha$ in (131) or in (132), and it has the expression

$$s(\alpha) = \frac{k_1^2(k_2 - k_1)}{2k_2}[(k_1 + k_2) + (k_2 - k_1)e^{-2k_2\alpha}] \tag{133}$$

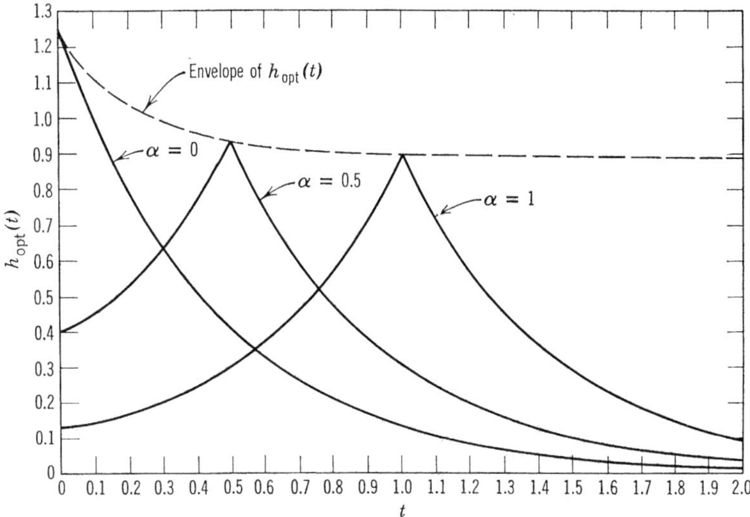

Fig. 12. Graphs of equations (131), (132), and (133). Drawn for $a = 1$, $b = 1$, $c = \frac{1}{2}$.

As $\alpha \to \infty$, the envelope approaches the value

$$s(\infty) = \frac{k_1{}^2(k_2{}^2 - k_1{}^2)}{2k_2} \tag{134}$$

To find the initial value of $h_{\text{opt}}(t)$ as a function of α, we put $t = 0$ in (131) and obtain

$$h_{\text{opt}}(0, \alpha) = k_1{}^2(k_2 - k_1)e^{-k_2\alpha} \tag{135}$$

It is instructive to apply the expression for the optimum system in the time domain to the example that we have been considering. The expression is a convolution integral and is given by (207), Chapter 14, Sec. 11. We repeat the integral here for convenient reference. When it is applied to a lag filter, $\psi^+(\tau)$ becomes $\psi^+(\tau - \alpha)$ so that

$$h_{\text{opt}}(t) = \frac{1}{4\pi^2}\int_{-\infty}^{\infty} \psi^+(\tau - \alpha)\gamma_{ii}^+(t - \tau)\,d\tau \tag{136}$$

From (25) we have

$$\psi^+(\tau - \alpha) = \begin{cases} Ae^{-k_1(\tau-\alpha)} & \text{for } \alpha \le \tau < \infty \\ Ae^{k_2(\tau-\alpha)} & \text{for } 0 \le \tau \le \alpha \\ 0 & \text{for } -\infty < \tau < 0 \end{cases} \tag{137}$$

and from (28) we have

$$\Gamma_{ii}^+(\lambda) = \frac{k_1 + j\lambda}{c(k_2 + j\lambda)} \tag{138}$$

so that

$$\gamma_{ii}^{+}(\tau) = \int_{-\infty+j\sigma_1}^{\infty+j\sigma_1} \frac{k_1 + j\lambda}{c(k_2 + j\lambda)} e^{j\lambda\tau}\, d\lambda$$

$$= \frac{1}{c}\int_{-\infty+j\sigma_1}^{\infty+j\sigma_1} \left[1 + \frac{k_1 - k_2}{k_2 + j\lambda}\right] e^{j\lambda\tau}\, d\lambda \tag{139}$$

The first integral of (139) is to be interpreted as in (149), Chapter 14. It follows immediately that

$$\gamma_{ii}^{+}(\tau) = \begin{cases} \dfrac{2\pi}{c}[u(\tau) + (k_1 - k_2)e^{-k_2\tau}] & \text{for } \tau \geq 0 \\ 0 & \text{for } \tau < 0 \end{cases} \tag{140}$$

For the interval $(0 \leq t \leq \alpha)$ the convolution is illustrated in Fig. 13. Concerning the graph of $\gamma_{ii}^{+}(\tau)$, we note that $k_1 - k_2 =$

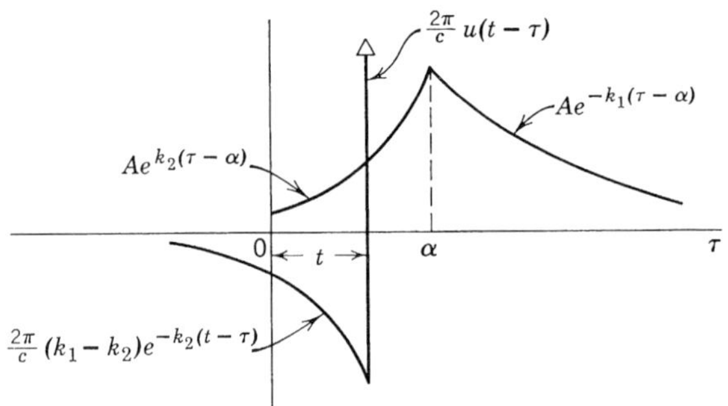

Fig. 13. Illustrating the evaluation of equation (136) for $(0 \leq t \leq \alpha)$.

$(c - \sqrt{a^2 + c^2}\,)/bc < 0$. In accordance with (136)

$$h_{\text{opt}}(t) = \frac{1}{4\pi^2}\int_0^t Ae^{k_2(\tau-\alpha)}\,\frac{2\pi}{c}\,[u(t-\tau) + (k_1 - k_2)e^{-k_2(t-\tau)}]\, d\tau$$

$$= k_1{}^2(k_2 - k_1)\left[e^{k_2(t-\alpha)} + \frac{k_1 - k_2}{2k_2}\, e^{-k_2\alpha}(e^{k_2t} - e^{-k_2t})\right]$$

$$= \frac{k_1{}^2(k_2 - k_1)}{2k_2}\, e^{-k_2\alpha}[k_2(e^{k_2t} + e^{-k_2t}) + k_1(e^{k_2t} - e^{-k_2t})]$$

$$= \frac{k_1{}^2(k_2 - k_1)}{k_2}\, e^{-k_2\alpha}[k_2 \cosh k_2t + k_1 \sinh k_2t] \qquad \text{for } 0 \leq t \leq \alpha \tag{141}$$

which is the same as (131).

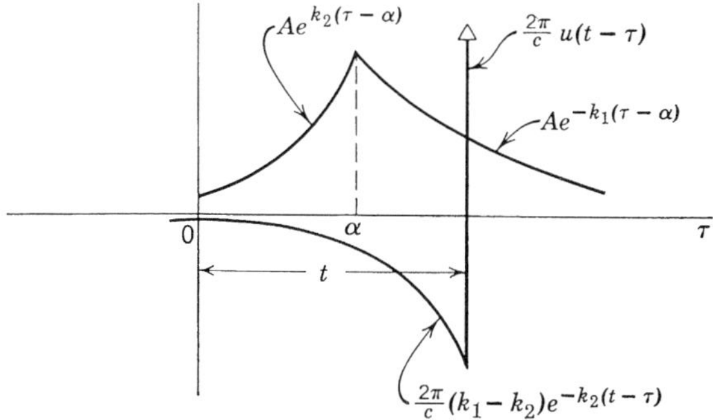

Fig. 14. Illustrating the evaluation of equation (136) for ($\alpha \le t < \infty$).

The determination of $h_{\text{opt}}(t)$ for the interval ($\alpha \le t < \infty$) is carried out in a similar manner with the aid of Fig. 14. Thus

$$\begin{aligned}
h_{\text{opt}}(t) &= \frac{1}{4\pi^2}\int_\alpha^\infty Ae^{-k_1(\tau-\alpha)}\,\frac{2\pi}{c}\,u(t-\tau)\,d\tau \\
&\quad + \frac{1}{4\pi^2}\int_0^\alpha Ae^{k_2(\tau-\alpha)}\,\frac{2\pi}{c}\,(k_1-k_2)e^{-k_2(t-\tau)}\,d\tau \\
&\quad + \frac{1}{4\pi^2}\int_\alpha^t Ae^{-k_1(\tau-\alpha)}\,\frac{2\pi}{c}\,(k_1-k_2)e^{-k_2(t-\tau)}\,d\tau \\
&= k_1{}^2(k_2-k_1)\left[e^{-k_1(t-\alpha)} + \frac{(k_1-k_2)}{2k_2}\right. \\
&\qquad \left. \times\,(e^{k_2\alpha}-e^{-k_2\alpha})e^{-k_2t} + e^{-k_2(t-\alpha)} - e^{-k_1(t-\alpha)}\right] \\
&= \frac{k_1{}^2(k_2-k_1)}{2k_2}\,e^{-k_2t}[k_2(e^{k_2\alpha}+e^{-k_2\alpha}) + k_1(e^{k_2\alpha}-e^{-k_2\alpha})] \\
&= \frac{k_1{}^2(k_2-k_1)}{k_2}\,e^{-k_2t}[k_2\cosh k_2\alpha + k_1\sinh k_2\alpha] \\
&\qquad\qquad \text{for } \alpha \le t < \infty \qquad (142)
\end{aligned}$$

This result agrees with (132) which was obtained by transformation. This work has been done not merely to check our results but also to point out that in handling a convolution integral we must know exactly what is happening and how to represent the operations involved. This point is particularly important when the functions with which we deal are singularity functions and functions that are described piecemeal.

PROBLEMS

1. An optimum filter A as shown in Fig. P1 is to be designed. The filter input is a message $f_m(t)$. A disturbance $f_n(t)$ is introduced into the system at an output terminal of the filter as indicated. The message and disturbance are stationary random processes and are correlated. The desired output $f_d(t)$ is the message delayed by α seconds. Derive a necessary condition for minimum mean-square error, in the form of an integral equation of the Wiener-Hopf type. For practice in handling integrals, give the derivation in detail. Then check the result by applying the general relations pertaining to the optimum system given in the text.

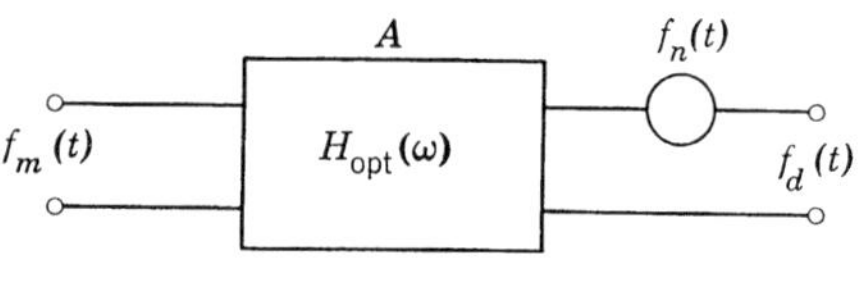

Fig. P1.

2. Given a message power density spectrum $\Phi_{mm}(\omega) = \omega^2/(1+\omega^2)^2$, a white noise $\Phi_{nn}(\omega) = 1/16$, and $\Phi_{mn}(\omega) = 0$. Determine the optimum filter system function for zero delay ($\alpha = 0$).

3. In the circuit shown in Fig. P3, the input to the given linear network A, whose system function is $H_1(\omega)$, is $f_1(t) = f_{m1}(t) + f_{n1}(t)$ where $f_{m1}(t)$ is a message and $f_{n1}(t)$ is a noise. Similarly, the input to network B is $f_2(t) = f_{m2}(t) + f_{n2}(t)$ where $f_{m2}(t)$ and $f_{n2}(t)$ represent a message and a noise respectively. The crosscorrelation between a message and a noise is zero. But $f_{m1}(t)$ and $f_{m2}(t)$ are correlated, and so are $f_{n1}(t)$ and $f_{n2}(t)$. The instantaneous difference between the outputs of A and B is $f_\varepsilon(t)$. Determine the system function $H(\omega)$ such that the mean-square value of $f_\varepsilon(t)$ is the minimum.

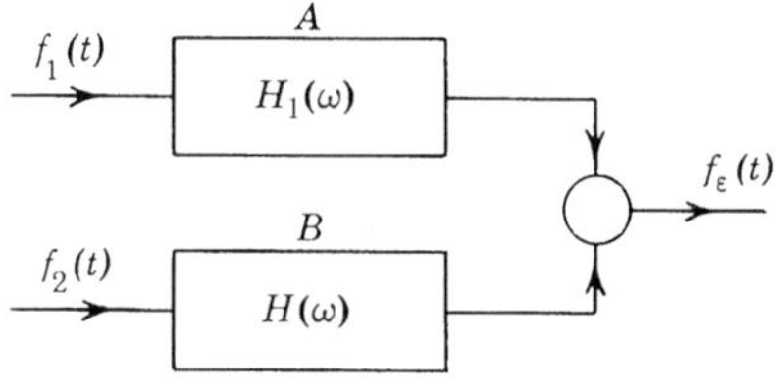

Fig. P3.

4. In the design of an optimum linear system, if the input is $f_i(t) = f_m(t)f_n(t)$ (a product) where $f_m(t)$ and $f_n(t)$ represent a message and a noise (not necessarily white) respectively and are independent stationary random processes with zero means, what is the power density spectrum of the input in terms of the spectrums $\Phi_{mm}(\omega)$ and $\Phi_{nn}(\omega)$ for $f_m(t)$ and $f_n(t)$? If the desired output is $f_d(t) = f_m(t) + f_n(t)$, what is the crosscorrelation between the input and the desired output, and what is the optimum system? Discuss the result.

chapter 16

Errors in Optimum Systems

One of the most important problems in the theory of optimum systems is the analysis of the minimum mean-square error. It is shown in this chapter that this error not only has a simple expression but also can be evaluated without the actual determination of the optimum system. We shall pay particular attention to the effect of the lag in the desired output on the error and shall show that the performance of a filter improves with an increase in the lag. One interesting fact is that the error tends to an irremovable error as the lag tends to infinity. The analysis of the error in prediction will be carried out in conjunction with the interpretation of prediction discussed in Chapter 15.

1. The Minimum Mean-Square Error

To analyze the minimum mean-square error, we shall begin with the general system without specifying the form of the desired output. After having obtained the general expression, we shall study its application to the filter and the predictor.

The starting point is the expression given in (13), Chapter 14. We repeat it here for convenience.

$$\overline{\varepsilon_h{}^2(t)} = \int_{-\infty}^{\infty} h(\tau)\, d\tau \int_{-\infty}^{\infty} h(\sigma)\, d\sigma\, \varphi_{ii}(\tau - \sigma) - 2\int_{-\infty}^{\infty} h(\tau)\, d\tau\, \varphi_{id}(\tau) + \varphi_{dd}(0) \tag{1}$$

This is the mean-square error for any given system whose unit-impulse response is $h(t)$, whose input has the autocorrelation function $\varphi_{ii}(\tau)$ and whose input–desired-output crosscorrelation is $\varphi_{id}(\tau)$. Clearly the minimum mean-square error will be given by (1) when $h(t)$ in the expression is $h_{\text{opt}}(t)$. In other words, for minimum mean-square error $h(t)$ in (1) must be $h_{\text{opt}}(t)$ in the Wiener-Hopf equation

$$\int_{-\infty}^{\infty} h_{\text{opt}}(\sigma)\varphi_{ii}(\tau - \sigma)\, d\sigma - \varphi_{id}(\tau) = 0 \qquad \text{for } \tau \geq 0 \tag{2}$$

Placing this condition in (1), we have the mean-square error of the optimum system which is the same as the minimum mean-square error

$$\begin{aligned}\overline{\varepsilon^2_{h(\text{opt})}(t)} &= \int_{-\infty}^{\infty} h_{\text{opt}}(\tau)\, d\tau\, \varphi_{id}(\tau) - 2\int_{-\infty}^{\infty} h_{\text{opt}}(\tau)\, d\tau\, \varphi_{id}(\tau) + \varphi_{dd}(0) \\ &= \varphi_{dd}(0) - \int_{-\infty}^{\infty} h_{\text{opt}}(\tau)\, d\tau\, \varphi_{id}(\tau)\end{aligned} \tag{3}$$

To reduce this expression to a simple form, it is necessary to write $h_{\text{opt}}(\tau)$ in the integral in terms of the optimum system function. That is, we need the expressions

$$H_{\text{opt}}(\lambda) = \frac{1}{2\pi\Phi_{ii}^{+}(\lambda)} \int_{0}^{\infty} \psi(t) e^{-j\lambda t}\, dt \tag{4}$$

$$\psi(t) = \int_{-\infty+j\sigma_1}^{\infty+j\sigma_1} \frac{\Phi_{id}(\lambda)}{\Phi_{ii}^{-}(\lambda)} e^{j\lambda t}\, d\lambda \tag{5}$$

and

$$h_{\text{opt}}(\tau) = \frac{1}{2\pi} \int_{-\infty+j\sigma_1}^{\infty+j\sigma_1} H_{\text{opt}}(\lambda) e^{j\lambda\tau}\, d\lambda \tag{6}$$

Substitution of these expressions into (3) yields

$$\overline{\varepsilon^2_{h(\text{opt})}(t)} = \varphi_{dd}(0) - \int_{-\infty}^{\infty} \varphi_{id}(\tau)\, d\tau \frac{1}{2\pi} \int_{-\infty+j\sigma_1}^{\infty+j\sigma_1} e^{j\lambda\tau}\, d\lambda \frac{1}{2\pi\Phi_{ii}^{+}(\lambda)} \int_{0}^{\infty} \psi(t) e^{-j\lambda t}\, dt \tag{7}$$

By inverting the order of integration, we have

$$\overline{\varepsilon^2_{h(\text{opt})}(t)} = \varphi_{dd}(0) - \frac{1}{2\pi} \int_{0}^{\infty} \psi(t)\, dt \int_{-\infty+j\sigma_1}^{\infty+j\sigma_1} \frac{1}{\Phi_{ii}^{+}(\lambda)} e^{-j\lambda t}\, d\lambda \times \frac{1}{2\pi} \int_{-\infty}^{\infty} \varphi_{id}(\tau) e^{j\lambda\tau}\, d\tau \tag{8}$$

Since the extreme right-hand integral is the conjugate of the input–desired-output cross-power density spectrum, we have

$$\overline{\varepsilon^2_{h(\text{opt})}(t)} = \varphi_{dd}(0) - \frac{1}{2\pi} \int_{0}^{\infty} \psi(t)\, dt \int_{-\infty+j\sigma_1}^{\infty+j\sigma_1} \frac{\overline{\Phi}_{id}(\lambda)}{\Phi_{ii}^{+}(\lambda)} e^{-j\lambda t}\, d\lambda \tag{9}$$

The extreme right-hand integral has the general appearance of (5), and in fact every factor in the integrand of one is the conjugate of the corresponding factor in the other. Thus, if we conjugate every factor in the integrand of (5), we get

$$\bar{\psi}(t) = \int_{-\infty+j\sigma_1}^{\infty+j\sigma_1} \frac{\bar{\Phi}_{id}(\lambda)}{\bar{\Phi}_{ii}^{-}(\lambda)} e^{-j\lambda t}\, d\lambda$$

$$= \int_{-\infty+j\sigma_1}^{\infty+j\sigma_1} \frac{\bar{\Phi}_{id}(\lambda)}{\Phi_{ii}^{+}(\lambda)} e^{-j\lambda t}\, d\lambda \tag{10}$$

which is the extreme right-hand integral in (9). We note further that $\bar{\psi}(t) = \psi(t)$ because it is a real function. Therefore (9) reduces to [with t in the integral replaced by τ so as to agree with the notations in (191), Chapter 14]

$$\overline{\varepsilon_{h(\text{opt})}^2(t)} = \varphi_{dd}(0) - \frac{1}{2\pi} \int_0^\infty \psi^2(\tau)\, d\tau \tag{11}$$

This is the expression that we wanted to establish for the minimum mean-square error of the optimum system. In this form we can determine the minimum mean-square error without any knowledge of the optimum system itself. We see that, even if we actually do not design systems in accordance with the present theory, the theory is helpful in providing an exact measure of performance.

2. Minimum Mean-Square Error and Irremovable Error in Optimum Filtering

To apply the minimum mean-square error expression to optimum filtering, we have $f_d(t) = f_m(t \pm \alpha)$ for the lag or prediction filter and $\psi(\tau)$ is given by (10), Chapter 15. Accordingly the minimum mean-square error of the optimum filter is

$$\overline{\varepsilon_{h(\text{opt})}^2(t)} = \varphi_{mm}(0) - \frac{1}{2\pi} \int_0^\infty \psi^2(\tau \pm \alpha)\, d\tau \tag{12}$$

or

$$\overline{\varepsilon_{h(\text{opt})}^2(t)} = \varphi_{mm}(0) - \frac{1}{2\pi} \int_{\pm\alpha}^\infty \psi^2(\tau \pm 0)\, d\tau \tag{13}$$

In this expression the lower limit $-\alpha$ is for the lag filter, and $+\alpha$ is for the prediction filter. The integral in this formula is represented in Fig. 1 which is for the illustrative example that has accompanied the presenta-

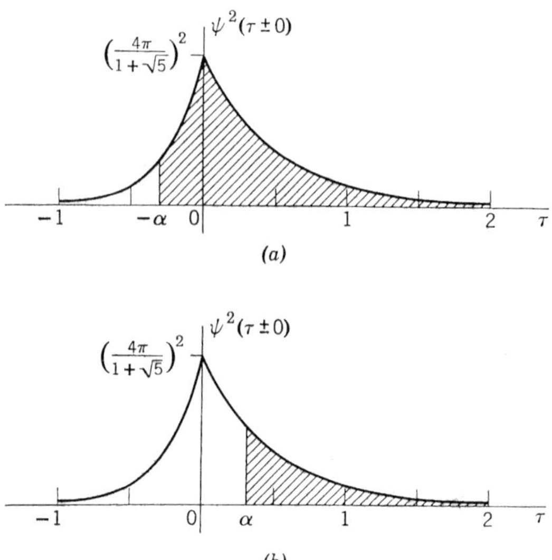

Fig. 1. Illustrating the evaluation of minimum mean-square error for (*a*) a lag filter and (*b*) a prediction filter. Drawn for $a = 1$, $b = 1$, $c = \frac{1}{2}$.

tion of the theory. Clearly $\psi^2(\tau \pm 0)$ is a non-negative function. For the lag filter the minimum mean-square error is the message power minus $1/2\pi$ times the shaded area in Fig. 1*a* between $-\alpha$ and $+\infty$, and for the prediction filter it is the message power minus $1/2\pi$ times the shaded area in Fig. 1*b* between $+\alpha$ and $+\infty$. Therefore we see that the minimum mean-square error of a lag filter is always less than that of a prediction filter.

Since the mean-square error is a non-negative quantity, we have from (13) that

$$\varphi_{mm}(0) \geq \frac{1}{2\pi} \int_{-\infty}^{\infty} \psi^2(\tau \pm 0)\, d\tau \tag{14}$$

The minimum mean-square error for a given desired output is a numerical value. However, if the desired output is the message with a time displacement whose effect on the minimum mean-square error is of interest to us, we shall consider the error as a function of the displacement. Therefore, when we speak of the maximum and minimum values of the minimum mean-square error, we mean the largest and smallest values of the error when considered as a function of the displacement $\pm\alpha$.

An important point in the analysis of errors is that the maximum value of the minimum mean-square error in the optimum filter is equal to the message power. We can see this by noting that the minimum value of the integral in (13) is zero when $+\alpha \to +\infty$, that is, when prediction time tends to infinity. Under this condition the minimum mean-square error is equal to the message power. Hence we write

$$\begin{aligned} \overline{\varepsilon^2_{h(\text{opt})}(t)}\,\Big|_{\max} &= \overline{\varepsilon^2_{h(\text{opt})}(t)}\,\Big|_{+\alpha\to+\infty} \\ &= \varphi_{mm}(0) \end{aligned} \tag{15}$$

Another situation that has the same result is that in which the input and desired output have a zero crosscorrelation. Obviously this situation reduces the integrand of the integral in (10), Chapter 15, to zero for any value of $\pm\alpha$. However, this is a trivial case. On the basis of the minimum-mean-square-error criterion, infinite-time prediction is the worst theoretical condition for a filter.

It can be shown that as the prediction time tends to infinity the output of the optimum system tends to zero. In order to show this fact, let us write the optimum system function of the filter as follows:

$$H_{\text{opt}}(\lambda) = \frac{1}{2\pi\Phi_{ii}^{+}(\lambda)} \int_0^\infty e^{-j\lambda t}\, dt \int_{-\infty+jv_1}^{\infty+jv_1} \frac{\Phi_{im}(w)}{\Phi_{ii}^{-}(w)}\, e^{j(t\pm\alpha)w}\, dw \tag{16}$$

With the change of variable $x = t \pm \alpha$, this equation becomes

$$H_{\text{opt}}(\lambda) = \frac{e^{\pm j\alpha\lambda}}{2\pi\Phi_{ii}^{+}(\lambda)} \int_{\pm\alpha}^\infty e^{-j\lambda x}\, dx \int_{-\infty+jv_1}^{\infty+jv_1} \frac{\Phi_{im}(w)}{\Phi_{ii}^{-}(w)}\, e^{jxw}\, dw \tag{17}$$

If the filter is an infinite-time prediction filter, the lower limit of the second integral should be $+\alpha \to +\infty$. Under this condition the upper and lower limits of the second integral are identical so that the value of the integral is zero. Consequently $H_{\text{opt}}(\lambda) = 0$, which means that the output of the optimum system should be zero. This characteristic of the optimum prediction filter is seen in the examples we have given.

Because of the fact that the greatest value of the minimum mean-square error is the message power, it is often convenient to normalize the expression (13) to this value. Denoting the normalized value by an asterisk, we have

$$\overline{\varepsilon^2_{h(\text{opt})}(t)}^{\,*} = 1 - \frac{1}{2\pi\varphi_{mm}(0)} \int_{\pm\alpha}^\infty \psi^2(\tau \pm 0)\, d\tau \tag{18}$$

Let us now consider the minimum mean-square error as a function of the lag. Since $\psi^2(\tau \pm 0)$ is non-negative, the value of the integral in (13) increases as the lag, $-\alpha$, increases, and the minimum mean-square error decreases accordingly. This clearly shows that the performance of an optimum filter improves with increasing lag. As the lag tends to infinity, the integral in (13) tends to its maximum and the minimum mean-square error tends to its minimum. This is the ultimate value, and we term it the *irremovable error*. Hence the minimum of the minimum mean-square error is the error of the optimum infinite lag filter. We express our result as follows

$$\overline{\varepsilon^2_{h(\text{opt})}(t)}\,\Big|_{\min} = \overline{\varepsilon^2_{h(\text{opt})}(t)}\,\Big|_{-\alpha\to-\infty} = \varphi_{mm}(0) - \frac{1}{2\pi}\int_{-\infty}^{\infty} \psi^2(\tau \pm 0)\,d\tau \tag{19}$$

or

$$\overline{\varepsilon^2_{h(\text{opt})}(t)}^{\,*}\,\Big|_{\min} = \overline{\varepsilon^2_{h(\text{opt})}(t)}^{\,*}\,\Big|_{-\alpha\to-\infty} = 1 - \frac{1}{2\pi\varphi_{mm}(0)}\int_{-\infty}^{\infty} \psi^2(\tau \pm 0)\,d\tau \tag{20}$$

A plot of the minimum mean-square error as a function of $\pm\alpha$, in accordance with (18) and (20) for the optimum filter of Sec. 2, Chapter 15, is given in Fig. 2.

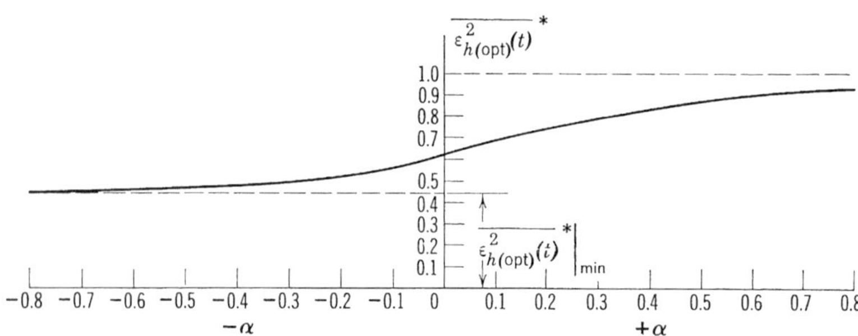

Fig. 2. Normalized minimum mean-square error in an optimum filter as a function of lag and lead. Drawn for $a = 1$, $b = 1$, $c = \frac{1}{2}$.

Since the irremovable error is an important quantity, we shall derive an alternative expression for it. We start with (19). In this expression the function $\psi(\tau \pm 0)$ is

$$\psi(\tau \pm 0) = \int_{-\infty+j\sigma_1}^{\infty+j\sigma_1} \frac{\Phi_{im}(\lambda)}{\Phi_{ii}^{-}(\lambda)}\, e^{j\lambda\tau}\,d\lambda \tag{21}$$

whose inverse is

$$\frac{\Phi_{im}(\lambda)}{\Phi_{ii}^{-}(\lambda)} = \frac{1}{2\pi}\int_{-\infty}^{\infty} \psi(\tau \pm 0)e^{-j\lambda\tau}\,d\tau \tag{22}$$

We shall also need the equation

$$\varphi_{mm}(0) = \int_{-\infty}^{\infty} \Phi_{mm}(\omega)\, d\omega \tag{23}$$

Applying the Parseval theorem (142), Chapter 2, to $\psi(\tau \pm 0)$ and its transform $\Phi_{im}(\omega)/\Phi_{ii}^{-}(\omega)$, we have

$$\frac{1}{2\pi}\int_{-\infty}^{\infty} \psi^2(\tau \pm 0)\, d\tau = \int_{-\infty}^{\infty} \frac{\Phi_{im}(\omega)}{\Phi_{ii}^{-}(\omega)}\, \frac{\overline{\Phi}_{im}(\omega)}{\Phi_{ii}^{+}(\omega)}\, d\omega \tag{24}$$

Substitution of this expression and (23) into (19) yields the irremovable error

$$\begin{aligned} \overline{\varepsilon_{h(\text{opt})}^2(t)}\,\big|_{-\alpha\to-\infty} &= \int_{-\infty}^{\infty} \left[\Phi_{mm}(\omega) - \frac{|\Phi_{im}(\omega)|^2}{\Phi_{ii}(\omega)}\right] d\omega \\ &= \int_{-\infty}^{\infty} \frac{\Phi_{ii}(\omega)\Phi_{mm}(\omega) - |\Phi_{im}(\omega)|^2}{\Phi_{ii}(\omega)}\, d\omega \end{aligned} \tag{25}$$

If $f_i(t) = f_m(t) + f_n(t)$, then $\Phi_{ii}(\omega) = \Phi_{mm}(\omega) + \Phi_{nn}(\omega) + \Phi_{mn}(\omega) + \Phi_{nm}(\omega)$, and $\Phi_{im}(\omega) = \Phi_{mm}(\omega) + \Phi_{nm}(\omega)$. When these expressions are substituted into (25), we have the irremovable error

$$\overline{\varepsilon_{h(\text{opt})}^2(t)}\,\big|_{-\alpha\to-\infty} = \int_{-\infty}^{\infty} \frac{\Phi_{mm}(\omega)\Phi_{nn}(\omega) - |\Phi_{nm}(\omega)|^2}{\Phi_{mm}(\omega) + \Phi_{nn}(\omega) + \Phi_{mn}(\omega) + \Phi_{nm}(\omega)}\, d\omega \tag{26}$$

Furthermore, if $\Phi_{mn}(\omega) = 0$

$$\overline{\varepsilon_{h(\text{opt})}^2(t)}\,\big|_{-\alpha\to-\infty} = \int_{-\infty}^{\infty} \frac{\Phi_{mm}(\omega)\Phi_{nn}(\omega)}{\Phi_{mm}(\omega) + \Phi_{nn}(\omega)}\, d\omega \tag{27}$$

3. Minimum Mean-Square Error in Pure Prediction

Referring to the general expression (11) for minimum mean-square error, we find that for pure prediction we should put

$$\varphi_{dd}(0) = \varphi_{mm}(0) \tag{28}$$

and

$$\begin{aligned} \psi(\tau) &= \psi(\tau + \alpha) \\ &= \int_{-\infty+j\sigma_1}^{\infty+j\sigma_1} \Phi_{mm}^{+}(\lambda) e^{j(\tau+\alpha)\lambda}\, d\lambda \end{aligned} \tag{29}$$

as given by (10), Chapter 15.

Before making the substitutions into the general expression, let us observe that

$$\psi(\tau + 0) = \int_{-\infty + j\sigma_1}^{\infty + j\sigma_1} \Phi_{mm}^{+}(\lambda) e^{j\tau\lambda} \, d\lambda \tag{30}$$

and its inverse is

$$\Phi_{mm}^{+}(\lambda) = \frac{1}{2\pi} \int_{-\infty}^{\infty} \psi(\tau + 0) e^{-j\lambda\tau} \, d\tau \tag{31}$$

Writing (11) for pure prediction, we have

$$\begin{aligned} \overline{\varepsilon_{h(\text{opt})}^2(t)} &= \varphi_{mm}(0) - \frac{1}{2\pi} \int_0^{\infty} \psi^2(\tau + \alpha) \, d\tau \\ &= \varphi_{mm}(0) - \frac{1}{2\pi} \int_{\alpha}^{\infty} \psi^2(\tau + 0) \, d\tau \end{aligned} \tag{32}$$

We can reduce this equation to a simple form by expressing $\varphi_{mm}(0)$ in terms of $\psi(\tau + 0)$. To do this, we first relate $\varphi_{mm}(0)$ to its spectrum and then factorize the spectrum. Thus

$$\begin{aligned} \varphi_{mm}(0) &= \int_{-\infty}^{\infty} \Phi_{mm}(\omega) \, d\omega \\ &= \int_{-\infty}^{\infty} \Phi_{mm}^{+}(\omega) \Phi_{mm}^{-}(\omega) \, d\omega \end{aligned} \tag{33}$$

By an application of the Parseval theorem to $\psi(\tau + 0)$ and its transform $\Phi_{mm}^{+}(\lambda)$ as given in (30) and (31), we have

$$\frac{1}{2\pi} \int_0^{\infty} \psi^2(\tau + 0) \, d\tau = \int_{-\infty}^{\infty} \Phi_{mm}^{+}(\omega) \Phi_{mm}^{-}(\omega) \, d\omega \tag{34}$$

The lower limit 0 of the left-hand integral is accounted for by the fact that $\psi(\tau + 0) = 0$ for $\tau < 0$. We see in (30) that $\psi(\tau + 0)$ is the transform of a function whose poles are in only the upper half-plane. Now we have

$$\varphi_{mm}(0) = \frac{1}{2\pi} \int_0^{\infty} \psi^2(\tau + 0) \, d\tau \tag{35}$$

which leads to

$$\overline{\varepsilon^2_{h(\text{opt})}(t)} = \frac{1}{2\pi}\int_0^\infty \psi^2(\tau + 0)\, d\tau - \frac{1}{2\pi}\int_\alpha^\infty \psi^2(\tau + 0)\, d\tau$$

$$= \frac{1}{2\pi}\int_0^\alpha \psi^2(\tau + 0)\, d\tau \tag{36}$$

This is the minimum mean-square error in pure prediction.

In connection with (36) it is interesting to consider the prediction of a Poisson wave $f_m(t)$ of equally likely positive and negative pulses of the same shape. The pulses are assumed to be generated by applying a unit-impulse Poisson wave to a network whose unit-impulse response has the form $g(t)$ so that its spectrum is

$$G(\lambda) = \int_{-\infty}^{\infty} g(t)e^{-j\lambda t}\, dt \tag{37}$$

and

$$g(t) = \frac{1}{2\pi}\int_{-\infty+j\sigma_1}^{\infty+j\sigma_1} G(\lambda)e^{j\lambda t}\, d\lambda \tag{38}$$

This Poisson wave has the power density spectrum (83), Chapter 15; and, if we assume that $G(\lambda)$ has no zeros in the lower half-plane, then according to (93), Chapter 15,

$$\Phi^+_{mm}(\lambda) = \sqrt{\frac{k}{2\pi}}\, G(\lambda) \tag{39}$$

When $G(\lambda)$ has no zeros in the lower half-plane, we shall say that it has minimum phase. Substituting (39) into (30), we have

$$\psi(\tau + 0) = \sqrt{\frac{k}{2\pi}}\int_{-\infty+j\sigma_1}^{\infty+j\sigma_1} G(\lambda)e^{j\tau\lambda}\, d\lambda$$

$$= \sqrt{2\pi k}\, g(\tau) \tag{40}$$

so that

$$\overline{\varepsilon^2_{h(\text{opt})}(t)} = k\int_0^\alpha g^2(\tau)\, d\tau \tag{41}$$

This result states that the minimum mean-square prediction error is k times the integral of the square of the pulse from zero to the prediction time. It agrees with our understanding of the manner in which the pulses are predicted. We have seen that each predicted pulse has zero value from zero to α, and it is perfectly predicted from α to infinity.

Hence the error is entirely in the interval $(0, \alpha)$ for each pulse. We further note that

$$k \int_0^\infty g^2(\tau)\, d\tau \tag{42}$$

is the mean square value of the Poisson wave in accordance with equation (62), Chapter 13.

As an example, consider the minimum-phase spectrum

$$G_1(\lambda) = \frac{2 + j\lambda}{(1 + j\lambda)(4 + j\lambda)} \tag{43}$$

whose transform gives the pulse shape

$$g_1(t) = \begin{cases} \frac{1}{3}(e^{-t} + 2e^{-4t}) & \text{for } t \geq 0 \\ 0 & \text{for } t < 0 \end{cases} \tag{44}$$

Since (43) has minimum phase, we are able to state immediately that each predicted pulse has the form

$$g_{1p}(t) = \begin{cases} g_1(t + \alpha) = \frac{1}{3}[e^{-(t+\alpha)} + 2e^{-4(t+\alpha)}] & \text{for } t \geq 0 \\ 0 & \text{for } t < 0 \end{cases} \tag{45}$$

The minimum mean-square error is

$$\begin{aligned} \overline{\varepsilon^2_{h(\text{opt})}(t)} &\doteq \frac{k}{9} \int_0^\alpha (e^{-t} + 2e^{-4t})^2\, dt \\ &= k[\tfrac{1}{5} - \tfrac{1}{18}(e^{-2\alpha} + \tfrac{8}{5}e^{-5\alpha} + e^{-8\alpha})] \end{aligned} \tag{46}$$

Next we consider the nonminimum-phase spectrum

$$\begin{aligned} G_2(\lambda) &= G_1(\lambda) \frac{2 - j\lambda}{2 + j\lambda} \\ &= \frac{2 - j\lambda}{(1 + j\lambda)(4 + j\lambda)} \end{aligned} \tag{47}$$

Here $G_2(\lambda)$ has a zero in the lower half-plane at the point of reflection of the zero in $G_1(\lambda)$. Transforming (47), we have the pulse shape

$$g_2(t) = \begin{cases} e^{-t} - 2e^{-4t} & \text{for } t \geq 0 \\ 0 & \text{for } t < 0 \end{cases} \tag{48}$$

An interesting point is that the minimum mean-square prediction error for the Poisson wave with the pulse shape (48) is the same as that for the Poisson wave with the pulse shape (44). The reason is that the power density spectrum of $f_m(t)$ in the nonminimum-phase case is

$$\begin{aligned}\Phi_{mm}(\lambda) &= \frac{k}{2\pi} G_2(\lambda)\overline{G}_2(\lambda) \\ &= \frac{k}{2\pi} G_1(\lambda)\overline{G}_1(\lambda) \frac{2 - j\lambda}{2 + j\lambda}\frac{2 + j\lambda}{2 - j\lambda} \\ &= \frac{k}{2\pi} G_1(\lambda)\overline{G}_1(\lambda) \qquad (49)\end{aligned}$$

so that

$$\Phi^{+}_{mm}(\lambda) = \sqrt{\frac{k}{2\pi}}\, G_1(\lambda) \qquad (50)$$

which leads to the same minimum mean-square prediction error.

Unlike the minimum-phase case, the prediction error in the nonminimum-phase case is only partly due to the predicted pulse being zero from time zero to time α. The other part lies in the imperfect prediction of the pulse from α to infinity. The error due to the zero value of the predicted pulse from 0 to α is

$$\begin{aligned}A &= k\int_0^{\alpha} (e^{-t} - 2e^{-4t})^2\, dt \\ &= k[\tfrac{1}{5} - \tfrac{1}{2}(e^{-2\alpha} - \tfrac{8}{5}e^{-5\alpha} + e^{-8\alpha})] \qquad (51)\end{aligned}$$

This error is less than the corresponding error in the minimum-phase case in the interval $(0, \alpha)$. We can show that the difference between (46) and (51) is

$$\overline{\varepsilon^2_{h(\text{opt})}(t)} - A = \tfrac{4}{9}k(e^{-\alpha} - e^{-4\alpha})^2 \qquad (52)$$

To find the remainder of the error, we shall first determine the shape of the predicted pulse. Applying (62) and (63), Chapter 15, we can show that the optimum predictor system function is

$$H_{\text{opt}}(\lambda) = \frac{1}{3}\left[e^{-\alpha}\frac{4 + j\lambda}{2 + j\lambda} + 2e^{-4\alpha}\frac{1 + j\lambda}{2 + j\lambda}\right] \qquad (53)$$

Since the transform of the input pulse is (47), the transform of the predicted pulse is

$$G_{2p}(\lambda) = G_2(\lambda)H_{\text{opt}}(\lambda)$$

$$= \frac{1}{3}\left[e^{-\alpha}\left(\frac{3}{1+j\lambda} - \frac{4}{2+j\lambda}\right) + 2e^{-4\alpha}\left(\frac{2}{2+j\lambda} - \frac{3}{4+j\lambda}\right)\right] \tag{54}$$

from which we obtain the predicted pulse shape

$$g_{2p}(t) = \begin{cases} \frac{1}{3}[e^{-\alpha}(3e^{-t} - 4e^{-4t}) + 2e^{-4\alpha}(2e^{-2t} - 3e^{-4t})] & \text{for } t \geq 0 \\ 0 & \text{for } t < 0 \end{cases} \tag{55}$$

Clearly the error due to the difference between the predicted pulse and the desired-output pulse, in the interval from α to infinity of the desired-output pulse, is

$$B = k\int_0^\infty [g_{2p}(t) - g_2(t+\alpha)]^2\,dt$$

$$= k\int_0^\infty [g_{2p}(t) - e^{-(t+\alpha)} + 2e^{-4(t+\alpha)}]^2\,dt$$

$$= \tfrac{4}{9}k(e^{-\alpha} - e^{-4\alpha})^2 \tag{56}$$

This is exactly the difference (52).

PROBLEMS

1. With an input consisting of the sum of a message and a noise, both of which are stationary random processes, whose power density spectrums are $\Phi_{mm}(\omega) = 1/(1+\omega^2)$ and $\Phi_{nn}(\omega) = \frac{1}{4}$ respectively, and $\Phi_{mn}(\omega) = 0$, an optimum linear system is found to be

$$H_{\text{opt}}(\omega) = \sqrt{\frac{12}{7}}\,\frac{1}{2+j\omega}$$

for a certain desired output. The nature of the desired output is not disclosed. Determine the normalized minimum mean-square error of the optimum system. The normalization is to be based upon the power of the desired output which has the given value π.

2. A linear system with the unit-impulse response

$$h(t) = \begin{cases} Ae^{-t} & \text{for } t \geq 0 \\ 0 & \text{for } t < 0 \end{cases}$$

has the input $f_i(t) = f_m(t) + f_n(t)$ where $f_m(t)$ is a message and $f_n(t)$ is a noise. The power density spectrums of $f_m(t)$ and $f_n(t)$ are $\Phi_{mm}(\omega) = 1/(1+\omega^2)$ and $\Phi_{nn}(\omega) = \frac{1}{4}$ respectively. The message and the noise are uncorrelated. If the desired output of the system is the message with lag α, for what values of A and α is the mean-square error the minimum?

3. In Fig. P3 a message $f_m(t)$ passes through a linear system A whose system function is

$$G(\omega) = \frac{3}{1 + j\omega}$$

The difference between the output of A and the message, that is, $f_g(t) - f_m(t)$, is fed to the linear system B. Show, in accordance with the Wiener theory, whether or not it is possible to find an optimum linear system B which restores the message with a mean-square error as small as we desire. If it is possible to do so, what is the optimum system unit-impulse response $h_{\text{opt}}(t)$? If not, why not?

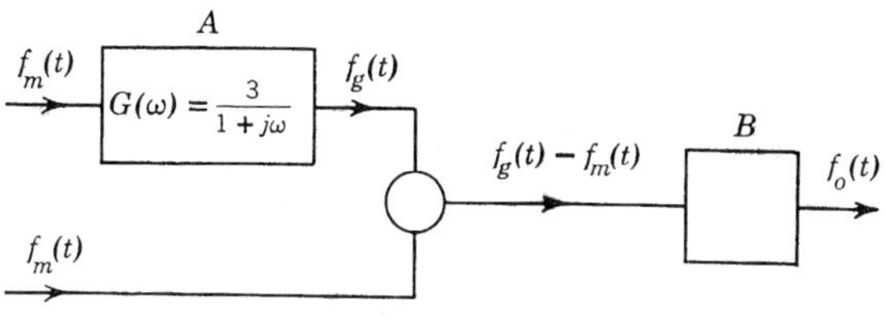

Fig. P3.

4. The input of an optimum linear filter is $f_i(t) = f_m(t) + f_n(t)$ where $f_m(t)$ is the message and $f_n(t)$ the noise. The message and the noise are stationary random functions and are uncorrelated. The desired output is the message with lag α seconds. If both the message and the noise have the same power density spectrum:

(*a*) What is the optimum system function? State in words what the optimum system does to the input.

(*b*) How much (numerical value) is the irremovable mean-square error when normalized to the message power?

chapter 17

Some Problems in the Theory of Optimum Systems

This chapter is concerned with four problems that are of considerable importance, namely, optimum compensation, a generalization of the input and desired output, a new criterion for the optimum system, and optimum systems with transient inputs. The optimum compensation problem deals with the improvement of an existing system and should be of practical interest. In generalizing the input and the desired output, we find that a large number of problems can be expressed in a simple manner by the general solution of the Wiener-Hopf equation. In the problem of finding a new criterion that leads to a simple solution, we have chosen the minimum-integral-square-crosscorrelation-error criterion. Finally, by an extension of the Wiener theory, optimum systems with transient inputs are obtained.

1. Optimum Compensation *

The problem of optimum compensation concerns the minimization of mean-square error for a system to operate in cascade with a given fixed system. In Fig. 1 the system N_1 is a given fixed system whose system function is $H_1(\lambda)$ and whose unit-impulse response is $h_1(t)$. This system is in cascade with system N_2 which is to be designed. With the cascade connection denoted by N, it is clear that the problem is to determine the optimum system N whose input and desired output are $f_i(t)$ and $f_d(t)$ respectively. For the analysis of this problem, let the output of N_1, which is the input of N_2, be $f_j(t)$.

* Y. W. Lee, "The Statistical Theory of Message Transmission," *Proceedings of the Second Symposium in Applied Mathematics of the American Mathematical Society*, held at M.I.T., July 1948, Vol. II, p. 90.

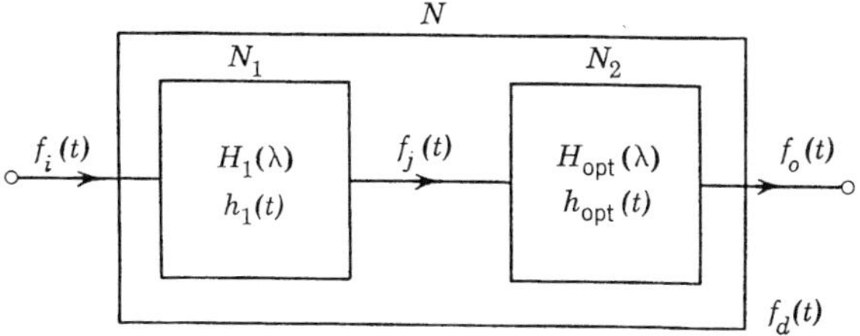

Fig. 1. Block diagram for optimum compensation problem.

With the input $f_j(t)$ and the desired output $f_d(t)$, we can write the optimum system function for N_2 immediately as

$$H_{\text{opt}}(\lambda) = \frac{1}{2\pi\Phi_{jj}^{+}(\lambda)} \int_0^\infty \psi(\tau) e^{-j\lambda\tau} \, d\tau \tag{1}$$

where

$$\psi(\tau) = \int_{-\infty+j\sigma_1}^{\infty+j\sigma_1} \frac{\Phi_{jd}(\lambda)}{\Phi_{jj}^{-}(\lambda)} e^{j\lambda\tau} \, d\lambda \tag{2}$$

The remainder of our problem is the determination of $\Phi_{jd}(\lambda)$, $\Phi_{jj}^{+}(\lambda)$, and $\Phi_{jj}^{-}(\lambda)$ in terms of quantities which specify the input to N and the given system N_1.

By definition we have

$$\varphi_{jd}(\tau) = \lim_{T\to\infty} \frac{1}{2T} \int_{-T}^{T} f_j(t) f_d(t+\tau) \, dt \tag{3}$$

Since

$$f_j(t) = \int_{-\infty}^{\infty} h_1(\sigma) f_i(t-\sigma) \, d\sigma \tag{4}$$

its substitution into (3) gives

$$\begin{aligned} \varphi_{jd}(\tau) &= \lim_{T\to\infty} \frac{1}{2T} \int_{-T}^{T} f_d(t+\tau) \, dt \int_{-\infty}^{\infty} h_1(\sigma) f_i(t-\sigma) \, d\sigma \\ &= \int_{-\infty}^{\infty} h_1(\sigma) \, d\sigma \lim_{T\to\infty} \frac{1}{2T} \int_{-T}^{T} f_i(t-\sigma) f_d(t+\tau) \, dt \\ &= \int_{-\infty}^{\infty} h_1(\sigma) \varphi_{id}(\tau+\sigma) \, d\sigma \end{aligned} \tag{5}$$

By transformation we obtain

$$\Phi_{jd}(\lambda) = \overline{H}_1(\lambda) \Phi_{id}(\lambda) \tag{6}$$

We now have $\Phi_{jd}(\lambda)$ expressed in terms of the system function of N_1 and the input–desired-output cross-power density spectrum.

To express $\Phi_{jj}^{+}(\lambda)$ and $\Phi_{jj}^{-}(\lambda)$ in terms of the given quantities, we write the relation between the input and output power density spectrums of N_1. By (38), Chapter 13, Sec. 5,

$$\Phi_{jj}(\lambda) = |H_1(\lambda)|^2\Phi_{ii}(\lambda) \tag{7}$$

The factorization of $\Phi_{ii}(\lambda)$ is

$$\Phi_{ii}(\lambda) = \Phi_{ii}^{+}(\lambda)\Phi_{ii}^{-}(\lambda) \tag{8}$$

and that of $|H_1(\lambda)|^2$ is

$$|H_1(\lambda)|^2 = H_1{}^{+}(\lambda)H_1{}^{-}(\lambda) \tag{9}$$

in which $H_1{}^{+}(\lambda)$ contains all the poles and zeros of $|H_1(\lambda)|^2$ in the upper half-plane and $H_1{}^{-}(\lambda)$ contains all the poles and zeros of $|H_1(\lambda)|^2$ in the lower half-plane. Furthermore

$$H_1{}^{+}(\lambda) = \overline{H}_1{}^{-}(\lambda) \tag{10}$$

This factorization has been explained in Chapter 15, Sec. 6. Equation (7) then becomes

$$\begin{aligned}\Phi_{jj}(\lambda) &= H_1{}^{+}(\lambda)\Phi_{ii}^{+}(\lambda)H_1{}^{-}(\lambda)\Phi_{ii}^{-}(\lambda)\\ &= \Phi_{jj}^{+}(\lambda)\Phi_{jj}^{-}(\lambda)\end{aligned} \tag{11}$$

where

$$\Phi_{jj}^{+}(\lambda) = H_1{}^{+}(\lambda)\Phi_{ii}^{+}(\lambda) \tag{12}$$

and

$$\Phi_{jj}^{-}(\lambda) = H_1{}^{-}(\lambda)\Phi_{ii}^{-}(\lambda) \tag{13}$$

Obviously

$$\Phi_{jj}^{+}(\lambda) = \overline{\Phi}_{jj}^{-}(\lambda) \tag{14}$$

Substitution of (6), (12), and (13) into (1) and (2) yields the optimum system function of N_2 as

$$H_{\text{opt}}(\lambda) = \frac{1}{2\pi H_1{}^{+}(\lambda)\Phi_{ii}^{+}(\lambda)}\int_0^\infty \psi(\tau)e^{-j\lambda\tau}\,d\tau \tag{15}$$

where

$$\psi(\tau) = \int_{-\infty+j\sigma_1}^{\infty+j\sigma_1} \frac{\overline{H}_1(\lambda)\Phi_{id}(\lambda)}{H_1{}^{-}(\lambda)\Phi_{ii}^{-}(\lambda)}\, e^{j\lambda\tau}\,d\lambda \tag{16}$$

This is the system function of the optimum compensator.

The expression that we have just obtained should be considered with respect as to whether the fixed system is a minimum-phase system (no zeros in the lower half-plane). As pointed out in Chapter 15, Sec. 6, if $H_1(\lambda)$ has minimum phase, then

$$H_1^{+}(\lambda) = H_1(\lambda) \tag{17}$$

so that

$$H_1^{-}(\lambda) = \bar{H}_1(\lambda) \tag{18}$$

Therefore, if $H_1(\lambda)$ has minimum phase, the optimum compensator system function becomes

$$H_{\text{opt}}(\lambda) = \frac{1}{2\pi H_1(\lambda)\Phi_{ii}^{+}(\lambda)} \int_0^\infty \psi(\tau)e^{-j\lambda\tau}\,d\tau \tag{19}$$

$$\psi(\tau) = \int_{-\infty}^{\infty} \frac{\Phi_{id}(\lambda)}{\Phi_{ii}^{-}(\lambda)} e^{j\lambda\tau}\,d\lambda \tag{20}$$

In effect this result states that, under the minimum-phase condition, optimum compensation requires the removal of N_1 and that the cascade system N be consisted of the optimum system N_2 only. This conclusion clearly comes from the fact that $1/H_1(\lambda)$ in (19) cancels $H_1(\lambda)$ of N_1 when N_1 and N_2 are in cascade and the fact that (20) is independent of $H_1(\lambda)$.

If the fixed system is a nonminimum-phase system, then as pointed out in (116), Chapter 15, Sec. 6, the general expressions for $H_1(\lambda)/H_1^{+}(\lambda)$ and $\bar{H}_1(\lambda)/H_1^{-}(\lambda)$ are

$$\frac{H_1(\lambda)}{H_1^{+}(\lambda)} = \frac{(a_1 - j\lambda)(a_2 - j\lambda)\cdots(a_n - j\lambda)}{(a_1 + j\lambda)(a_2 + j\lambda)\cdots(a_n + j\lambda)} \tag{21}$$

and

$$\frac{\bar{H}_1(\lambda)}{H_1^{-}(\lambda)} = \frac{(a_1 + j\lambda)(a_2 + j\lambda)\cdots(a_n + j\lambda)}{(a_1 - j\lambda)(a_2 - j\lambda)\cdots(a_n - j\lambda)} \tag{22}$$

if $H_1(\lambda)$ has zeros at $-ja_1, -ja_2, \ldots, -ja_n$ on the imaginary axis. For example, if

$$H_1(\lambda) = \frac{2 - j\lambda}{(1 + j\lambda)(4 + j\lambda)} \tag{23}$$

then

$$\frac{H_1(\lambda)}{H_1^{+}(\lambda)} = \frac{2 - j\lambda}{2 + j\lambda} \tag{24}$$

and

$$\frac{\bar{H}_1(\lambda)}{H_1^{-}(\lambda)} = \frac{2 + j\lambda}{2 - j\lambda} \tag{25}$$

It is interesting that (21), which is the result of cascading N_1 and that part of the optimum compensator which has the system function $1/H_1^{+}(\lambda)$, is a pure phase-shift network.

2. A Generalization of the Input and Desired Output *

So far, we have considered only one specific form of the input to an optimum system, and that is the additive combination of a message and a noise. We have pointed out that the theory is not restricted to this form. It only requires that the spectrum of the input can be factorized and that the input–desired-output cross-power density spectrum be nontrivial. One general form of the input is the additive mixture of a noise $f_n(t)$ and a message $f_m(t)$ that has gone through a given linear operation. For instance the input may be a noise plus the derivative of a message. If we let $G_1(\lambda)$ be a linear operator and its transform be $g_1(t)$ and consider $G_1(\lambda)$ to be the system function of a linear system and $g_1(t)$ its unit-impulse response, then

$$G_1(\lambda) = \int_{-\infty}^{\infty} g_1(t)e^{-j\lambda t}\,dt \tag{26}$$

and

$$g_1(t) = \frac{1}{2\pi}\int_{-\infty+j\sigma_1}^{\infty+j\sigma_1} G_1(\lambda)e^{j\lambda t}\,d\lambda \tag{27}$$

so that our more general input is

$$f_i(t) = \int_{-\infty}^{\infty} g_1(\tau)f_m(t-\tau)\,d\tau + f_n(t) \tag{28}$$

In a similar manner, instead of specifying that the desired output be the message with a lag or a lead, we state our desired output in a more general form by saying that it is the message that has gone through a linear operation. For example the desired output could be the derivative of the message with a lag or lead. We take a linear system with the system function $G_2(\lambda)$ and unit-impulse response $g_2(t)$ to represent the linear operation so that equations similar to (26) and (27) can be written for $G_2(\lambda)$ and $g_2(t)$. The more general form of the desired output is then

$$f_d(t) = \int_{-\infty}^{\infty} g_2(\tau)f_m(t-\tau)\,d\tau \tag{29}$$

To apply (28) and (29) to the optimum system formulas (190) and (191), Chapter 14, we first determine $\Phi_{ii}(\lambda)$ and then $\Phi_{id}(\lambda)$. Substituting (28) into the definition for $\varphi_{ii}(\tau)$, we have

* Y. W. Lee, "On Wiener Filters and Predictors," *Proceedings of the Symposium on Information Networks*, Polytechnic Institute of Brooklyn, April 1954, pp. 19–29.

$$\varphi_{ii}(\tau) = \lim_{T\to\infty} \frac{1}{2T} \int_{-T}^{T} dt \left[\int_{-\infty}^{\infty} g_1(v) f_m(t - v)\, dv + f_n(t) \right]$$
$$\times \left[\int_{-\infty}^{\infty} g_1(u) f_m(t + \tau - u)\, du + f_n(t + \tau) \right]$$
$$= \int_{-\infty}^{\infty} g_1(v)\, dv \int_{-\infty}^{\infty} g_1(u)\, du \lim_{T\to\infty} \frac{1}{2T} \int_{-T}^{T} f_m(t - v) f_m(t + \tau - u)\, dt$$
$$+ \int_{-\infty}^{\infty} g_1(v)\, dv \lim_{T\to\infty} \frac{1}{2T} \int_{-T}^{T} f_m(t - v) f_n(t + \tau)\, dt$$
$$+ \int_{-\infty}^{\infty} g_1(u)\, du \lim_{T\to\infty} \frac{1}{2T} \int_{-T}^{T} f_n(t) f_m(t + \tau - u)\, dt$$
$$+ \lim_{T\to\infty} \frac{1}{2T} \int_{-T}^{T} f_n(t) f_n(t + \tau)\, dt$$
$$= \int_{-\infty}^{\infty} g_1(v)\, dv \int_{-\infty}^{\infty} g_1(u)\, du\, \varphi_{mm}(\tau + v - u)$$
$$+ \int_{-\infty}^{\infty} g_1(v)\, dv\, \varphi_{mn}(\tau + v)$$
$$+ \int_{-\infty}^{\infty} g_1(u)\, du\, \varphi_{nm}(\tau - u) + \varphi_{nn}(\tau) \tag{30}$$

The corresponding equation in the frequency domain as a result of transformation is

$$\Phi_{ii}(\lambda) = |G_1(\lambda)|^2 \Phi_{mm}(\lambda) + \bar{G}_1(\lambda)\Phi_{mn}(\lambda) + G_1(\lambda)\bar{\Phi}_{mn}(\lambda) + \Phi_{nn}(\lambda) \tag{31}$$

For $\Phi_{id}(\lambda)$ we begin with

$$\varphi_{id}(\tau) = \lim_{T\to\infty} \frac{1}{2T} \int_{-T}^{T} dt \left[\int_{-\infty}^{\infty} g_1(v) f_m(t - v)\, dv + f_n(t) \right]$$
$$\times \left[\int_{-\infty}^{\infty} g_2(u) f_m(t + \tau - u)\, du \right] \tag{32}$$

which is written in accordance with the definition of $\varphi_{id}(\tau)$ when $f_i(t)$ and $f_d(t)$ are given by (28) and (29). By inversion of the order of integration, (32) becomes

$$\begin{aligned}\varphi_{id}(\tau) &= \int_{-\infty}^{\infty} g_1(v)\,dv \int_{-\infty}^{\infty} g_2(u)\,du \lim_{T\to\infty} \frac{1}{2T}\int_{-T}^{T} f_m(t-v)f_m(t+\tau-u)\,dt \\ &\quad + \int_{-\infty}^{\infty} g_2(u)\,du \lim_{T\to\infty}\frac{1}{2T}\int_{-T}^{T} f_n(t)f_m(t+\tau-u)\,dt \\ &= \int_{-\infty}^{\infty} g_1(v)\,dv \int_{-\infty}^{\infty} g_2(u)\,du\, \varphi_{mm}(\tau+v-u) \\ &\quad + \int_{-\infty}^{\infty} g_2(u)\,du\,\varphi_{nm}(\tau-u) \qquad (33)\end{aligned}$$

The transformation of this equation gives us

$$\Phi_{id}(\lambda) = \overline{G}_1(\lambda)G_2(\lambda)\Phi_{mm}(\lambda) + G_2(\lambda)\Phi_{nm}(\lambda) \qquad (34)$$

With (31) and (34) for $\Phi_{ii}(\lambda)$ and $\Phi_{id}(\lambda)$ in the optimum expression (190) and (191), Chapter 14, we include a large class of problems. For example, the desired output may take the form of the nth derivative or the nth integral of the message with a lag or a lead, provided that these derivatives or integrals exist. For the nth derivative of the message with lag or lead, $G_2(\lambda)$ is

$$G_2(\lambda) = (j\lambda)^n e^{\pm j\alpha\lambda} \qquad (35)$$

and for the nth integral of the message with lag or lead, $G_2(\lambda)$ is

$$G_2(\lambda) = \frac{1}{(j\lambda)^n}\, e^{\pm j\alpha\lambda} \qquad (36)$$

Obviously $G_1(\lambda)$ may also take these forms if it involves the nth derivative and the nth integral of the input message.

As an illustrative example, let the input be a noise $f_n(t)$ plus a message $f_m(t)$ that has gone through a linear circuit whose unit-impulse response is

$$h_1(t) = \begin{cases} Ee^{-at} & \text{for } t \geq 0 \\ 0 & \text{for } t < 0 \end{cases} \qquad (37)$$

Let us suppose that the desired output is the first derivative of the message with a lag or a lead. For simplicity let $\Phi_{mn}(\lambda) = 0$. Under these conditions we have

$$g_1(t) = h_1(t) \qquad (38)$$

so that

$$G_1(\lambda) = \frac{E}{a + j\lambda} \qquad (39)$$

and, according to (31),

$$\Phi_{ii}(\lambda) = \frac{E^2}{a^2 + \lambda^2}\Phi_{mm}(\lambda) + \Phi_{nn}(\lambda) \tag{40}$$

Moreover, since

$$G_2(\lambda) = j\lambda e^{\pm j\alpha\lambda} \tag{41}$$

the cross-power density spectrum (34) becomes

$$\Phi_{id}(\lambda) = \left(\frac{E}{a - j\lambda}\right)(j\lambda e^{\pm j\alpha\lambda})\Phi_{mm}(\lambda) \tag{42}$$

The optimum system will then be specified by (190) and (191), Chapter 14, with $\Phi_{ii}(\lambda)$ and $\Phi_{id}(\lambda)$ expressed by (40) and (42). Clearly this example involves not only filtering and prediction but also differentiation and compensation for the distortion in the input message caused by the input circuit. All these operations are performed by the system with the least mean-square error. Examples of this type certainly draw our attention to the power and versatility of the theory.

We should like to describe briefly another interesting example. Suppose that a linear system with the system function $H_1(\lambda)$ is simulated by a computer for study under an input $f_m(t)$. Because of practical limitations, such as the necessity of rounding off numbers, the input is applied with an error $f_n(t)$ so that the output of the computer is corrupted by a corresponding error. To remedy this situation, one method is to determine an optimum system function and replace $H_1(\lambda)$ by it. In this problem the desired output is the output of $H_1(\lambda)$ with $f_m(t)$ as the input. Hence for the optimum system function, which is to replace the actual system function $H_1(\lambda)$, we have $G_1(\lambda) = 1$ and $G_2(\lambda) = H_1(\lambda)$ so that, in accordance with (31),

$$\Phi_{ii}(\lambda) = \Phi_{mm}(\lambda) + \Phi_{mn}(\lambda) + \overline{\Phi}_{mn}(\lambda) + \Phi_{nn}(\lambda) \tag{43}$$

and, in accordance with (34),

$$\Phi_{id}(\lambda) = H_1(\lambda)[\Phi_{mm}(\lambda) + \Phi_{nm}(\lambda)] \tag{44}$$

Therefore we see that, in view of the unavoidable error in feeding the input data to the simulated system under study, the output error will be reduced to the minimum if instead of the actual system we simulate the optimum system determined according to (190) and (191), Chapter 14, with $\Phi_{ii}(\lambda)$ given by (43) and $\Phi_{id}(\lambda)$ given by (44). In other words the mean-square error in our problem can be reduced to the minimum if we distort the simulation of the actual system in accordance with the theory of optimum systems.

We have given here some extensions of the theory to show that the solution of the Wiener-Hopf equation is a basic step in the attack of a large class of problems. We have by no means exhausted the possibilities of its application.

3. Optimum Systems on the Minimum-Integral-Square-Crosscorrelation-Error Criterion *

In the study of optimum linear systems we often wonder if other measures of error besides the mean-square measure may be introduced. One difficulty we can see is that, if higher powers of the instantaneous error are considered, correlation functions of higher orders will be required. These higher order functions will complicate our problems considerably, both in theory and from the standpoint of physical measurements. However, when the criterion of an optimum system is changed, it does not necessarily follow that higher order correlation functions are involved. We shall consider here a new criterion which leads to a solution that depends upon the same autocorrelation and crosscorrelation as those in the minimum-mean-square-error system.

The criterion that we wish to consider is the minimum-integral-square-crosscorrelation-error criterion. Having specified an input $f_i(t)$ and a desired output $f_d(t)$ from which we obtain the input–desired-output crosscorrelation $\varphi_{id}(\tau)$, we consider the integral-square-crosscorrelation error which is defined as

$$I = \int_{-\infty}^{\infty} [\varphi_{io}(\tau) - \varphi_{id}(\tau)]^2 \, d\tau \tag{45}$$

where $\varphi_{io}(\tau)$ is the input–actual-output crosscorrelation function. We shall determine the linear system which minimizes this error.

It is well for us to consider this error in conjunction with the mean-square error before we proceed with the development of the theory. We recall that the minimum-mean-square-error criterion requires the optimum system to have such a system function that its input-output crosscorrelation is exactly equal to the input–desired-output crosscorrelation in the interval $(0 \leq \tau < \infty)$. This requirement is expressed by the Wiener-Hopf equation. In the new criterion we are not attempting to minimize the mean-square error between the desired output and the actual output but are attempting to minimize the integral square of the errors between the two crosscorrelation functions in the interval $(-\infty < \tau < \infty)$. Therefore, although both criteria involve the "match-

* Y. W. Lee, "On Wiener Filters and Predictors," *Proceedings of the Symposium on Information Networks*, Polytechnic Institute of Brooklyn, April 1954, pp. 19–29.

ing" of crosscorrelation functions, the requirements in the matching and the objectives of the criteria are different.

To reduce (45) to a suitable form before minimization, we need the input-output crosscorrelation theorem given in (87), Chapter 13, Sec. 8. When this theorem is applied to $\varphi_{io}(\tau)$ in (45) the expression becomes

$$I = \int_{-\infty}^{\infty} \left[\int_{-\infty}^{\infty} h(u)\varphi_{ii}(\tau - u)\, du - \varphi_{id}(\tau) \right]^2 d\tau \tag{46}$$

The expansion of this equation is

$$I = \int_{-\infty}^{\infty} d\tau \left[\int_{-\infty}^{\infty} h(u)\varphi_{ii}(\tau - u)\, du \int_{-\infty}^{\infty} h(v)\varphi_{ii}(\tau - v)\, dv \right.$$
$$\left. - 2\varphi_{id}(\tau) \int_{-\infty}^{\infty} h(u)\varphi_{ii}(\tau - u)\, du + \varphi_{id}^2(\tau) \right] \tag{47}$$

Inversion of the order of integration yields

$$I = \int_{-\infty}^{\infty} h(u)\, du \int_{-\infty}^{\infty} h(v)\, dv \int_{-\infty}^{\infty} \varphi_{ii}(\tau - u)\varphi_{ii}(\tau - v)\, d\tau$$
$$- 2\int_{-\infty}^{\infty} h(u)\, du \int_{-\infty}^{\infty} \varphi_{ii}(\tau - u)\varphi_{id}(\tau)\, d\tau + \int_{-\infty}^{\infty} \varphi_{id}^2(\tau)\, d\tau \tag{48}$$

Let us define the autocorrelation of the autocorrelation function $\varphi_{ii}(\tau)$ as

$$\varphi_{ii\text{-}ii}(u) = \int_{-\infty}^{\infty} \varphi_{ii}(\tau)\varphi_{ii}(\tau + u)\, d\tau \tag{49}$$

and the crosscorrelation of $\varphi_{ii}(\tau)$ and $\varphi_{id}(\tau)$ as

$$\varphi_{ii\ id}(u) - \int_{-\infty}^{\infty} \varphi_{ii}(\tau)\varphi_{id}(\tau + u)\, d\tau \tag{50}$$

In terms of these functions (48) is

$$I = \int_{-\infty}^{\infty} h(u)\, du \int_{-\infty}^{\infty} h(v)\, dv\ \varphi_{ii\text{-}ii}(u - v)$$
$$- 2\int_{-\infty}^{\infty} h(u)\, du\ \varphi_{ii\text{-}id}(u) + \varphi_{id\text{-}id}(0) \tag{51}$$

The last term of this equation comes from the definition

$$\varphi_{id\text{-}id}(u) = \int_{-\infty}^{\infty} \varphi_{id}(\tau)\varphi_{id}(\tau + u)\, d\tau \tag{52}$$

when $u = 0$ so that

$$\varphi_{id\text{-}id}(0) = \int_{-\infty}^{\infty} \varphi_{id}^2(\tau)\, d\tau \tag{53}$$

We observe that (51) for the integral-square-crosscorrelation error is similar in all respects to (13), Chapter 14, for the mean-square error; the functions $\varphi_{ii\text{-}ii}(u)$, $\varphi_{ii\text{-}id}(u)$ and the constant $\varphi_{id\text{-}id}(0)$ in (51) correspond, respectively, to $\varphi_{ii}(\tau)$, $\varphi_{id}(\tau)$, and $\varphi_{dd}(0)$ in (13), Chapter 14. Because of the similarity and the fact that minimization of (13), Chapter 14, leads to condition (53), Chapter 14, it follows immediately that the condition for the minimization of the integral-square-crosscorrelation error is

$$\int_{-\infty}^{\infty} h_{\text{opt}}^*(v)\varphi_{ii\text{-}ii}(u - v)\, dv - \varphi_{ii\text{-}id}(u) = 0 \qquad \text{for } u \geq 0 \tag{54}$$

Here $h_{\text{opt}}^*(t)$ is the unit-impulse response of the optimum system on the basis of the new criterion.

Letting $H_{\text{opt}}^*(\lambda)$ be the optimum system function under the minimum-integral-square-crosscorrelation-error criterion so that

$$H_{\text{opt}}^*(\lambda) = \int_{-\infty}^{\infty} h_{\text{opt}}^*(v) e^{-j\lambda v}\, dv \tag{55}$$

and letting

$$\Phi_{ii\text{-}ii}(\lambda) = \frac{1}{2\pi} \int_{-\infty}^{\infty} \varphi_{ii\text{-}ii}(u) e^{-j\lambda u}\, du \tag{56}$$

and

$$\Phi_{ii\text{-}id}(\lambda) = \frac{1}{2\pi} \int_{-\infty}^{\infty} \varphi_{ii\text{-}id}(u) e^{-j\lambda u}\, du \tag{57}$$

we shall, without repeating the details, in a manner analogous to (190) and (191), Chapter 14, write the solution of the integral equation (54) as

$$H_{\text{opt}}^*(\lambda) = \frac{1}{2\pi\Phi_{ii\text{-}ii}^+(\lambda)} \int_0^{\infty} \psi^*(\tau) e^{-j\lambda\tau}\, d\tau \tag{58}$$

in which

$$\psi^*(\tau) = \int_{-\infty + j\sigma_1}^{\infty + j\sigma_1} \frac{\Phi_{ii\text{-}id}(\lambda)}{\Phi_{ii\text{-}ii}^-(\lambda)} e^{j\lambda\tau}\, d\lambda \tag{59}$$

In these expressions it is clear that

$$\Phi_{ii\text{-}ii}(\lambda) = \Phi_{ii\text{-}ii}^+(\lambda)\Phi_{ii\text{-}ii}^-(\lambda) \tag{60}$$

with $\Phi_{ii\text{-}ii}^+(\lambda)$ containing all the poles and zeros of $\Phi_{ii\text{-}ii}(\lambda)$ that are in the upper half-plane and $\Phi_{ii\text{-}ii}^-(\lambda)$ containing all the poles and zeros of

$\Phi_{ii\text{-}ii}(\lambda)$ in the lower half-plane and with

$$\Phi^+_{ii\text{-}ii}(\lambda) = \overline{\Phi}^-_{ii\text{-}ii}(\lambda) \tag{61}$$

It is interesting that $\Phi^+_{ii\text{-}ii}(\lambda)$, $\Phi^-_{ii\text{-}ii}(\lambda)$, and $\Phi_{ii\text{-}id}(\lambda)$ may be expressed in terms of $\Phi^+_{ii}(\lambda)$, $\Phi^-_{ii}(\lambda)$, $\Phi_{ii}(\lambda)$, and $\Phi_{id}(\lambda)$. With the definition (49) in (56) we can readily show that

$$\begin{aligned} \Phi_{ii\text{-}ii}(\lambda) &= \frac{1}{2\pi}\int_{-\infty}^{\infty} e^{-j\lambda u}\, du \int_{-\infty}^{\infty} \varphi_{ii}(\tau)\varphi_{ii}(\tau+u)\, d\tau \\ &= 2\pi\Phi^2_{ii}(\lambda) \end{aligned} \tag{62}$$

Similarly, with (50) in (57) we find

$$\begin{aligned} \Phi_{ii\text{-}id}(\lambda) &= \frac{1}{2\pi}\int_{-\infty}^{\infty} e^{-j\lambda u}\, du \int_{-\infty}^{\infty} \varphi_{ii}(\tau)\varphi_{id}(\tau+u)\, d\tau \\ &= 2\pi\Phi_{ii}(\lambda)\Phi_{id}(\lambda) \end{aligned} \tag{63}$$

The factorization of $\Phi_{ii\text{-}ii}(\lambda)$ as given by (60) is clearly

$$\Phi_{ii\text{-}ii}(\lambda) = 2\pi[\Phi^+_{ii}(\lambda)]^2[\Phi^-_{ii}(\lambda)]^2 \tag{64}$$

from which we obtain

$$\Phi^+_{ii\text{-}ii}(\lambda) = \sqrt{2\pi}\,[\Phi^+_{ii}(\lambda)]^2 \tag{65}$$

and

$$\Phi^-_{ii\text{-}ii}(\lambda) = \sqrt{2\pi}\,[\Phi^-_{ii}(\lambda)]^2 \tag{66}$$

By substituting (63), (65), and (66) into (58) and (59), we obtain the following alternative expressions for the optimum system on the new criterion:

$$H^*_{\text{opt}}(\lambda) = \frac{1}{2\pi\sqrt{2\pi}\,[\Phi^+_{ii}(\lambda)]^2}\int_0^{\infty} \psi^*(\tau)e^{-j\lambda\tau}\, d\tau \tag{67}$$

$$\psi^*(\tau) = \sqrt{2\pi}\int_{-\infty+j\sigma_1}^{\infty+j\sigma_1} \frac{\Phi^+_{ii}(\lambda)}{\Phi^-_{ii}(\lambda)}\Phi_{id}(\lambda)e^{j\lambda\tau}\, d\lambda \tag{68}$$

This result shows the interesting fact that the new criterion and the minimum-mean-square-error criterion require the same characteristics of the input and desired output, namely, the autocorrelation of the input and the input–desired-output crosscorrelation. Another point of interest is the fact that the factor $\Phi^+_{ii}(\lambda)/\Phi^-_{ii}(\lambda)$ in the integrand of $\psi^*(\tau)$ is a pure phase-shift factor.

4. The Minimum Integral-Square-Crosscorrelation Error

Clearly, when the condition (54) for minimum integral-square-crosscorrelation error is satisfied in the general expression for integral-square-crosscorrelation error (51), the error reduces to its minimum. Hence replacing

$$\int_{-\infty}^{\infty} h(v)\,dv\,\varphi_{ii}(u-v) \tag{69}$$

in the first integral of (51) by $\varphi_{ii\text{-}id}(u)$ in accordance with (54) and replacing h by h^*_{opt}, we have the minimum integral-square-crosscorrelation error

$$I_{h^*(\text{opt})} = \varphi_{id\text{-}id}(0) - \int_{-\infty}^{\infty} h^*_{\text{opt}}(u)\varphi_{ii\text{-}id}(u)\,du \tag{70}$$

To put this expression into another form, we apply the relation

$$h^*_{\text{opt}}(u) = \frac{1}{2\pi}\int_{-\infty+j\sigma_1}^{\infty+j\sigma_1} H^*_{\text{opt}}(\lambda)e^{j\lambda u}\,d\lambda \tag{71}$$

and formula (58) in (70) thereby obtaining

$$I_{h^*(\text{opt})} = \varphi_{id\text{-}id}(0) - \int_{-\infty}^{\infty} \varphi_{ii\text{-}id}(u)\,du\,\frac{1}{2\pi}\int_{-\infty+j\sigma_1}^{\infty+j\sigma_1} e^{j\lambda u}\,d\lambda \times \frac{1}{2\pi\Phi^+_{ii\text{-}ii}(\lambda)}\int_0^{\infty} \psi^*(\tau)e^{-j\lambda\tau}\,d\tau \tag{72}$$

Inversion of the order of integration yields

$$I_{h^*(\text{opt})} = \varphi_{id\text{-}id}(0) - \frac{1}{2\pi}\int_0^{\infty} \psi^*(\tau)\,d\tau\int_{-\infty+j\sigma_1}^{\infty+j\sigma_1} \frac{1}{\Phi^+_{ii\text{-}ii}(\lambda)}\,e^{-j\lambda\tau}\,d\lambda \times \frac{1}{2\pi}\int_{-\infty}^{\infty} \varphi_{ii\text{-}id}(u)e^{j\lambda u}\,du \tag{73}$$

By (57) the extreme right-hand integral is $\overline{\Phi}_{ii\text{-}id}(\lambda)$ so that

$$I_{h^*(\text{opt})} = \varphi_{id\text{-}id}(0) - \frac{1}{2\pi}\int_0^{\infty} \psi^*(\tau)\,d\tau\int_{-\infty+j\sigma_1}^{\infty+j\sigma_1} \frac{\overline{\Phi}_{ii\text{-}id}(\lambda)}{\Phi^+_{ii\text{-}ii}(\lambda)}\,e^{-j\lambda\tau}\,d\lambda \tag{74}$$

The extreme right-hand integral in (74) is the integral in (59) if every term of the integrand is conjugated. Since in (59) $\psi^*(\tau)$ is real, conjugation does not change the value of the integral. Therefore we have finally the minimum integral-square-crosscorrelation error

$$I_{h^*(\text{opt})} = \varphi_{id\text{-}id}(0) - \frac{1}{2\pi}\int_0^\infty [\psi^*(\tau)]^2 \, d\tau \tag{75}$$

It is interesting to consider the behavior of this minimum error as a function of the lag in the desired output. Suppose that the desired output is the message with a lag, that is, $f_d(t) = f_m(t - \alpha)$, so that $\Phi_{id}(\lambda) = \Phi_{im}(\lambda)e^{-j\alpha\lambda}$. With this condition, (68) becomes

$$\psi^*(\tau - \alpha) = \sqrt{2\pi}\int_{-\infty+j\sigma_1}^{\infty+j\sigma_1} \frac{\Phi_{ii}^+(\lambda)}{\Phi_{ii}^-(\lambda)} \Phi_{im}(\lambda)e^{j\lambda(\tau-\alpha)} \, d\lambda \tag{76}$$

and (75) becomes

$$I_{h^*(\text{opt})} = \varphi_{id\text{-}id}(0) - \frac{1}{2\pi}\int_{-\alpha}^\infty [\psi^*(\tau - 0)]^2 \, d\tau \tag{77}$$

Inspection of this result shows that an increase in the lag of the desired output causes an increase in the value of the integral in (77), thereby reducing $I_{h^*(\text{opt})}$. In other words the performance of the optimum system improves with the increase of lag in the desired output. We have seen that the optimum filter on the minimum-mean-square-error criterion has the same property. However, whereas the minimum mean-square error tends to the irremovable error as the lag tends to infinity, the minimum integral-square-crosscorrelation error tends to zero.

To show that (77) tends to zero as the lag tends to infinity, we first note that the minimum error as the lag tends to infinity is

$$I_{h^*(\text{opt})}\big|_{-\alpha\to-\infty} = \varphi_{id\text{-}id}(0) - \frac{1}{2\pi}\int_{-\infty}^\infty [\psi^*(\tau - 0)]^2 \, d\tau \tag{78}$$

By the Parseval theorem we obtain from (76)

$$\begin{aligned} \frac{1}{2\pi}\int_{-\infty}^\infty [\psi^*(\tau - 0)]^2 \, d\tau &= \int_{-\infty}^\infty 2\pi \left| \frac{\Phi_{ii}^+(\omega)}{\Phi_{ii}^-(\omega)} \Phi_{im}(\omega) \right|^2 d\omega \\ &= 2\pi\int_{-\infty}^\infty |\Phi_{im}(\omega)|^2 \, d\omega \end{aligned} \tag{79}$$

and by the same theorem we obtain from (53)

$$\begin{aligned} \varphi_{id\text{-}id}(0) &= 2\pi\int_{-\infty}^\infty |\Phi_{im}(\omega)e^{-j\omega\alpha}|^2 \, d\omega \\ &= 2\pi\int_{-\infty}^\infty |\Phi_{im}(\omega)|^2 \, d\omega \end{aligned} \tag{80}$$

Substitution of (79) and (80) into (78) gives us the result that

$$I_{h^*(\text{opt})}\Big|_{-\alpha\to-\infty} = 0 \tag{81}$$

5. Optimum Systems with Transient Inputs on the Minimum-Integral-Square-Error Criterion *

Although the Wiener theory was originally developed for application to stationary random processes, it is not difficult to extend the theory to include transient messages and noise.

If the input of a linear system is a transient function $f_i(t)$ [$f_i(t) = 0$ for $t < 0$] whose Fourier transform is $F_i(\lambda)$, and we desire a transient output $f_d(t)$ [$f_d(t) = 0$ for $t < 0$] whose Fourier transform is $F_d(\lambda)$, then

$$f_d(t) = \int_{-\infty}^{\infty} h(\sigma) f_i(t - \sigma)\, d\sigma \tag{82}$$

in which $h(t)$ is the unit-impulse response of the system whose system function is $H(\lambda)$. Let us first assume that, with the given input, there is a physically realizable system which can produce the desired output exactly. In the frequency domain the relation (82) is

$$F_d(\lambda) = H(\lambda)F_i(\lambda) \tag{83}$$

so that

$$H(\lambda) = \frac{F_d(\lambda)}{F_i(\lambda)} \tag{84}$$

If $F_d(\lambda)/F_i(\lambda)$ has no poles in the lower half-plane, the system is realizable.

However, if $F_i(\lambda)$ has zeros in the lower half-plane, and $F_d(\lambda)$ has no zeros that cancel the lower half-plane poles of $1/F_i(\lambda)$, then the system is unrealizable. Under this condition $H(\lambda)$ in (84) has poles in the lower half-plane, and we have the problem of finding a realizable system to produce an output that approximates the desired output in an optimum manner.

We shall choose the integral-square error defined by the integral

$$J = \int_{-\infty}^{\infty} [f_0(t) - f_d(t)]^2\, dt \tag{85}$$

as the measure of error in this problem. In the integral (85), $f_0(t)$ is the

* Y. W. Lee, "Application of Wiener's Techniques in Transient Synthesis," *Quarterly Progress Report*, Research Laboratory of Electronics, M.I.T., January 15, 1950, pp. 62–64, and April 15, 1950, p. 52.

actual transient output. With the relation

$$f_0(t) = \int_{-\infty}^{\infty} h(\tau) f_i(t - \tau)\, d\tau \tag{86}$$

it is not difficult to show, by techniques similar to those shown in Sec. 3, that

$$J = \int_{-\infty}^{\infty} h(\tau)\, d\tau \int_{-\infty}^{\infty} h(\sigma)\, d\sigma\, \varphi_{ii}(\tau - \sigma) - 2\int_{-\infty}^{\infty} h(\tau)\, d\tau\, \varphi_{id}(\tau) + \varphi_{dd}(0) \tag{87}$$

in which

$$\varphi_{ii}(\tau) = \int_{-\infty}^{\infty} f_i(t) f_i(t + \tau)\, dt \tag{88}$$

$$\varphi_{id}(\tau) = \int_{-\infty}^{\infty} f_i(t) f_d(t + \tau)\, dt \tag{89}$$

and

$$\varphi_{dd}(0) = \int_{-\infty}^{\infty} f_d^{\,2}(t)\, dt \tag{90}$$

By arguments similar to those in Sec. 3 in connection with (54), we can show that the minimization of J results in the condition

$$\int_{-\infty}^{\infty} h_{\text{opt}}^{**}(\sigma) \varphi_{ii}(\tau - \sigma)\, d\sigma - \varphi_{id}(\tau) = 0 \qquad \text{for } \tau \geq 0 \tag{91}$$

Here $h_{\text{opt}}^{**}(t)$ is the unit-impulse response of the optimum system (with a transient input) on the basis of minimum integral-square error. Letting the system function of the optimum system be $H_{\text{opt}}^{**}(\lambda)$ so that

$$H_{\text{opt}}^{**}(\lambda) = \int_{-\infty}^{\infty} h_{\text{opt}}^{**}(\tau) e^{-j\lambda\tau}\, d\tau \tag{92}$$

and letting

$$\Phi_{ii}(\lambda) = \frac{1}{2\pi} \int_{-\infty}^{\infty} \varphi_{ii}(\tau) e^{-j\lambda\tau}\, d\tau \tag{93}$$

and

$$\Phi_{id}(\lambda) = \frac{1}{2\pi} \int_{-\infty}^{\infty} \varphi_{id}(\tau) e^{-j\lambda\tau}\, d\tau \tag{94}$$

we can express the solution of (91) in the familiar form

$$H_{\text{opt}}^{**}(\lambda) = \frac{1}{2\pi \Phi_{ii}^{+}(\lambda)} \int_{0}^{\infty} \psi^{**}(\tau) e^{-j\lambda\tau}\, d\tau \tag{95}$$

where

$$\psi^{**}(\tau) = \int_{-\infty+j\sigma_1}^{\infty+j\sigma_1} \frac{\Phi_{id}(\lambda)}{\Phi_{ii}^{-}(\lambda)} e^{j\lambda\tau}\, d\lambda \tag{96}$$

To reduce this solution to a simpler form, we obtain from (93) and (88) the expression

$$\begin{aligned}\Phi_{ii}(\lambda) &= 2\pi \overline{F}_i(\lambda) F_i(\lambda) = 2\pi |F_i(\lambda)|^2 \\ &= \Phi_{ii}^{+}(\lambda)\Phi_{ii}^{-}(\lambda)\end{aligned} \tag{97}$$

and from (94) and (89), the expression

$$\Phi_{id}(\lambda) = 2\pi \overline{F}_i(\lambda) F_d(\lambda) \tag{98}$$

For the factorization of (97) we let

$$|F_i(\lambda)|^2 = F_i^{+}(\lambda) F_i^{-}(\lambda) \tag{99}$$

in which $F_i^{+}(\lambda)$ contains all the poles and zeros of $|F_i(\lambda)|^2$ in the upper half-plane and $F_i^{-}(\lambda)$ contains all the poles and zeros of $|F_i(\lambda)|^2$ in the lower half-plane, and

$$F_i^{+}(\lambda) = \overline{F}_i^{-}(\lambda) \tag{100}$$

Hence the factors of (97) are

$$\Phi_{ii}^{+}(\lambda) = \sqrt{2\pi}\, F_i^{+}(\lambda) \tag{101}$$

and

$$\Phi_{ii}^{-}(\lambda) = \sqrt{2\pi}\, F_i^{-}(\lambda) \tag{102}$$

Now, expressing (95) and (96) in terms of these results, we have, respectively,

$$H_{\text{opt}}^{**}(\lambda) = \frac{1}{2\pi\sqrt{2\pi}\, F_i^{+}(\lambda)} \int_0^{\infty} \psi^{**}(\tau) e^{-j\lambda\tau}\, d\tau \tag{103}$$

and

$$\psi^{**}(\tau) = \sqrt{2\pi} \int_{-\infty+j\sigma_1}^{\infty+j\sigma_1} \frac{\overline{F}_i(\lambda)}{F_i^{-}(\lambda)} F_d(\lambda) e^{j\lambda\tau}\, d\lambda \tag{104}$$

This completes the specification of the optimum linear system, whose input and desired output are transient functions, on the minimum-integral-square-error criterion. We note that if $F_i(\lambda)$ has no zeros in the lower half-plane, then $\overline{F}_i(\lambda) = F_i^{-}(\lambda)$ and $F_i^{+}(\lambda) = F(\lambda)$, so that (103) reduces to the expected result

$$H_{\text{opt}}^{**}(\lambda) = \frac{F_d(\lambda)}{F_i(\lambda)} \tag{105}$$

PROBLEMS

1. In Fig. P1 the network A is a given linear system with unit-impulse response $g(t)$ and system function $G(\omega)$; the network B is the optimum linear system with unit-impulse response $h_{\text{opt}}(t)$ and system function $H_{\text{opt}}(\omega)$ for the desired output $f_d(t)$. The system function $G(\lambda)$ has zeros in both half-planes.

(*a*) Determine the minimum mean-square error between the actual output $f_o(t)$ and the desired output $f_d(t)$ in terms of $g(t)$, $h_{\text{opt}}(t)$, and the input–desired-output crosscorrelation $\varphi_{id}(\tau)$.

(*b*) If $f_i(t) = f_m(t) + f_n(t)$, show in detail whether the irremovable error in optimum filtering with A in cascade with B as shown is greater than, equal to, or less than the irremovable error of the optimum linear filter designed to operate alone with A completely removed.

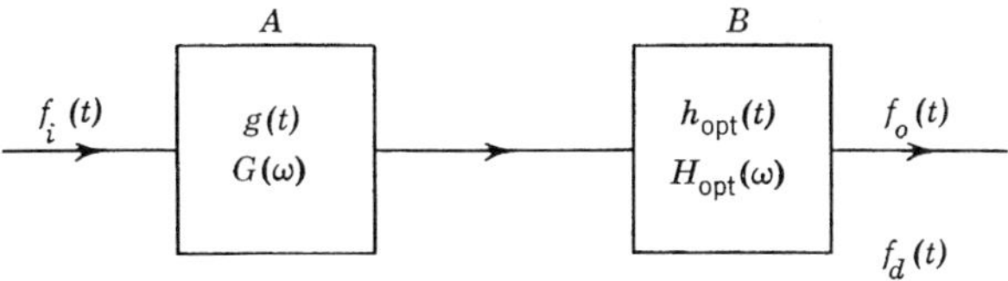

Fig. P1.

2. In the design of an optimum system one general way for expressing the desired output is that it is the result of the input message after a given linear operation. If the input is the sum of a message and a noise, derive the expression for the irremovable error as a single integral involving the given linear operator, the message and noise power density spectrums, and the cross-power density spectrums of the message and noise. Does the crosscorrelation between message and noise increase or decrease the irremovable error? Why?

3. Two independent messages $f_a(t)$ and $f_b(t)$ are fed into the system A as shown in Fig. P3, and the output is $f_a(t) + f_b^{(-1)}(t)$ where $f_b^{(-1)}(t)$ is the integral of $f_b(t)$. It is desired to convert this output to the sum $f_a^{(-1)}(t + \alpha_1) + f_b(t - \alpha_2)$ by an optimum linear system. Here $f_a^{(-1)}(t + \alpha_1)$ is the integral of $f_a(t)$ advanced by α_1 seconds and $f_b(t - \alpha_2)$ is $f_b(t)$ delayed by α_2 seconds. Determine the optimum system function in terms of the power density spectrums of $f_a(t)$ and $f_b(t)$.

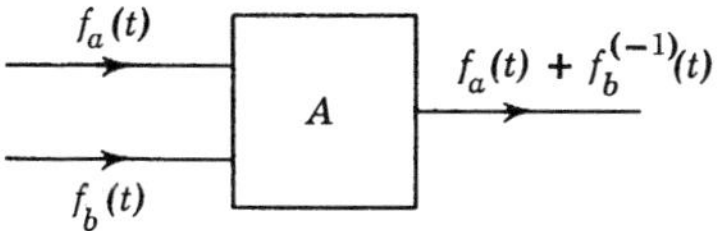

Fig. P3.

4. An optimum linear system has the input $f_i(t) = f_m(t) - f_m'(t)$ where $f_m(t)$ is the message whose power density spectrum is $1/(1 + \omega^2)$ and where $f_m'(t)$ is the derivative of the message. If the desired output is: (*a*) $f_d(t) = f_m(t - \alpha)$ where

α is the lag time, (b) $f_d(t) = f_m'(t - \alpha)$, and (c) $f_d(t) = f_m^{(-1)}(t - \alpha)$ which is the integral of $f_m(t)$ with lag α, what is the irremovable error in each of the three cases?

5. The input of an optimum system is $f_i(t) = f_m(t) + f_n(t)$ where $f_m(t)$ is the message and $f_n(t)$ is the noise. Both the message and the noise are random processes. A general method for expressing the desired output is to let it be the message after a linear operation. This linear operation may be represented by a linear system whose unit-impulse response is $g(t)$. We assume that

$$\int_{-\infty}^{\infty} g^2(t)\, dt = A \quad \text{and} \quad A \neq 0$$

Derive a necessary condition which $g(t)$ must satisfy for the minimization of the irremovable error. The condition must involve quantities that are obtainable from $f_m(t)$ and $f_n(t)$.

6. It is desired to investigate whether or not it is possible to reduce the irremovable error in filtering by giving the message an appropriate linear operation before it enters the optimum system. Thus we have a message that has gone through a linear operation plus a noise (instead of the usual message plus noise) at the input terminals of the optimum system. The desired output is the message with a lag. We assume that the message and the noise have a zero crosscorrelation. Show whether or not the irremovable error can be reduced (from that with message plus noise at the input) by the method suggested. If it is possible to do so, derive the expression by which the reduction can be calculated. Give the conditions if there are any. If it is impossible to do so, show clearly why. The answer to this problem must include a definite and explicit statement as to whether it is possible or impossible.

7. In the design of a linear system for the transmission of messages in the presence of noise, a general way of expressing the objective is to say that the physically realizable system should yield, on the average, an instantaneous output which is as nearly as possible a replica of the instantaneous desired output. Messages and noise are assumed to be stationary random processes. *With this objective in view,* discuss each of the following three specific measures of error:

(1) $$\varepsilon_1 = \int_{-\infty}^{\infty} [\Phi_{ii}(\omega)\,|H(\omega)|^2 - \Phi_{dd}(\omega)]^2\, d\omega$$

Here $H(\omega)$ is the system function, and $\Phi_{ii}(\omega)$ and $\Phi_{dd}(\omega)$ are the input power density spectrum and the desired-output power density spectrum respectively.

(2) $$\varepsilon_2 = \int_{-\infty}^{\infty} [\varphi_{oo}(\tau) - \varphi_{dd}(\tau)]^2\, d\tau$$

where $\varphi_{oo}(\tau)$ is the autocorrelation function of the actual output and $\varphi_{dd}(\tau)$ that of the desired output.

(3) $$\varepsilon_3 = \int_{-\infty}^{\infty} [\varphi_{od}(\tau) - \varphi_{dd}(\tau)]^2\, d\tau$$

in which $\varphi_{dd}(\tau)$ has the same meaning as that given in (2) and $\varphi_{od}(\tau)$ is the crosscorrelation function of the actual output and the desired output.

Of the three measures, which ones are equivalent? Give a proof.

chapter 18

Representation of Correlation Functions and Power Density Spectrums by Orthonormal Functions

When autocorrelation functions are experimentally obtained or when they are represented by expressions whose transforms cannot be readily factorized, it is desirable, in optimum system determination, to expand these functions in terms of functions whose transforms can be easily factorized. To facilitate the actual determination of the optimum system function, it is also desirable to represent a crosscorrelation function by an expansion that is particularly suitable for Fourier transformation.

To meet this need, we shall consider the representation of correlation functions by orthonormal functions that transform into rational functions. Before we go into the details of this work, let us note that, since an autocorrelation function is an even function, it is only necessary to consider the right-hand half of the function in actual computation. On the other hand, the crosscorrelation function is generally not an even function. For convenience we shall consider this function as being the sum of its right-hand half and its left-hand half. On this basis we see that in theoretical discussions we can confine ourselves to the representation of a real function which vanishes over a half-line.

The material in this chapter is based upon the joint work * of N. Wiener and the author, which was originally developed for network synthesis.

* N. Wiener and Y. W. Lee, U. S. Patents No. 2,024,900 (1935), No. 2,128,257 (1938), No. 2,124,599 (1938).

Y. W. Lee, "Synthesis of Electric Networks by Means of the Fourier Transforms of Laguerre's Functions," *Jour. Math. and Physics, M.I.T.*, **11**, 83–113 (1932).

1. Orthonormal Functions

A set of real and continuous functions $w_1(\tau)$, $w_2(\tau)$, $\ldots$ is said to be *orthonormal* in the range (a, b) if

$$\int_a^b w_m(\tau)w_n(\tau)\,d\tau = \begin{cases} 1 & \text{if } m = n \\ 0 & \text{if } m \neq n \end{cases} \tag{1}$$

When the functions satisfy the first condition, they are said to be *normal;* and, when they satisfy the second condition, they are said to be *orthogonal.*

If a real and continuous function $f(\tau)$ is represented in the range (a, b) by the series

$$f(\tau) = \sum_{n=1}^{\infty} c_n w_n(\tau) \tag{2}$$

then

$$c_n = \int_a^b f(\tau)w_n(\tau)\,d\tau \tag{3}$$

That the coefficients c_n have the expression (3) comes from the fact that, if (2) is multiplied by $w_m(\tau)$ and integrated over (a, b), we have

$$\int_a^b f(\tau)w_m(\tau)\,d\tau = \int_a^b \sum_{n=1}^{\infty} c_n w_n(\tau)w_m(\tau)\,d\tau \tag{4}$$

which yields in accordance with (1)

$$\int_a^b f(\tau)w_m(\tau)\,d\tau = \begin{cases} c_m & \text{for } m = n \\ 0 & \text{for } m \neq n \end{cases} \tag{5}$$

The orthonormal set $\{w_n(\tau)\}$, with

$$\int_a^b w_n^2(\tau)\,d\tau < \infty \qquad \text{for } n = 1, 2, \ldots \tag{6}$$

is called complete or closed if either of the following statements is true:

(1) There exists no function $x(\tau)$ with

$$\int_a^b x^2(\tau)\,d\tau < \infty \tag{7}$$

such that

$$\int_a^b x(\tau)w_n(\tau)\,d\tau = 0 \qquad \text{for } n = 1, 2, \ldots \tag{8}$$

(2) For any piecewise continuous function $f(\tau)$ with

$$\int_a^b f^2(\tau)\, d\tau < \infty \tag{9}$$

and an $\epsilon > 0$, however small, there exists an integer N and a polynomial

$$\sum_{n=1}^{N} c_n w_n(\tau) \tag{10}$$

such that

$$\int_a^b \left| f(\tau) - \sum_{n=1}^{N} c_n w_n(\tau) \right|^2 d\tau < \epsilon \tag{11}$$

When the set is complete the orthonormal series representation of $f(\tau)$ converges in the mean to $f(\tau)$. All the sets discussed in this chapter are complete.

2. Orthonormalization in the Time Domain

It is known that, given n real, continuous, and linearly independent functions $g_1(\tau), g_2(\tau), \ldots, g_n(\tau)$ satisfying the condition

$$\int_a^b g_k{}^2(\tau)\, d\tau < \infty \qquad \text{for } k = 1, 2, \ldots, n \tag{12}$$

a set of n linear combinations of these functions, $w_1(\tau), w_2(\tau), \ldots, w_n(\tau)$, can be found with the orthonormal properties (1).

An interesting set of orthonormal functions may be obtained from the sequence

$$\left.\begin{aligned} g_1(\tau) &= e^{-p\tau} \\ g_2(\tau) &= e^{-2p\tau} \\ &\vdots \\ g_n(\tau) &= e^{-np\tau} \\ &\vdots \end{aligned}\right\} \tag{13}$$

where p is a real positive constant. Let the orthonormal set to be derived from (13) be $\{u_n(\tau)\}$ and the range of the set be $(0 \leq \tau < \infty)$. Thus the properties of the set are

$$\int_0^\infty u_m(\tau) u_n(\tau)\, d\tau = \begin{cases} 1 & \text{if } m = n \\ 0 & \text{if } m \neq n \end{cases} \tag{14}$$

For the first term of the set we put

$$u_1(\tau) = A_1 e^{-p\tau} \tag{15}$$

In accordance with the normal property we must have

$$\int_0^\infty u_1^2(\tau)\,d\tau = 1 \tag{16}$$

so that

$$\int_0^\infty A_1^2 e^{-2p\tau}\,d\tau = 1$$

$$= \frac{A_1^2}{2p} \tag{17}$$

from which we obtain

$$A_1 = \sqrt{2p} \tag{18}$$

Hence (15) becomes

$$u_1(\tau) = \sqrt{2p}\,e^{-p\tau} \tag{19}$$

We need not be concerned with the orthogonal property at this moment because all other terms of the set, yet to be determined, will be made orthogonal to the first term.

For the second term of the set we put

$$u_2(\tau) = B_1 e^{-p\tau} + B_2 e^{-2p\tau} \tag{20}$$

To determine B_1 and B_2, we need two equations, one from the normal property requiring that

$$\int_0^\infty u_2^2(\tau)\,d\tau = 1 \tag{21}$$

and one from the orthogonal property requiring that

$$\int_0^\infty u_1(\tau)u_2(\tau)\,d\tau = 0 \tag{22}$$

Therefore we have the equations

$$\int_0^\infty (B_1 e^{-p\tau} + B_2 e^{-2p\tau})^2\,d\tau = 1 \tag{23}$$

and

$$\int_0^\infty e^{-p\tau}(B_1 e^{-p\tau} + B_2 e^{-2p\tau})\,d\tau = 0 \tag{24}$$

from which we find $B_1 = 4\sqrt{p}$ and $B_2 = -6\sqrt{p}$ so that the second term of the set is

$$u_2(\tau) = 2\sqrt{p}\,(2e^{-p\tau} - 3e^{-2p\tau}) \tag{25}$$

By putting the third term of the set in the form

$$u_3(\tau) = C_1e^{-p\tau} + C_2e^{-2p\tau} + C_3e^{-3p\tau} \tag{26}$$

we see that in accordance with (1), the three equations that determine the three coefficients are

$$\int_0^\infty u_3{}^2(\tau)\,d\tau = 1 \tag{27}$$

$$\int_0^\infty u_1(\tau)u_3(\tau)\,d\tau = 0 \tag{28}$$

$$\int_0^\infty u_2(\tau)u_3(\tau)\,d\tau = 0 \tag{29}$$

The solution of these equations yields $C_1 = 3\sqrt{6p}$, $C_2 = -12\sqrt{6p}$, $C_3 = 10\sqrt{6p}$ so that

$$u_3(\tau) = \sqrt{6p}\,(3e^{-p\tau} - 12e^{-2p\tau} + 10e^{-3p\tau}) \tag{30}$$

The following are a few of the succeeding terms of the set:

$$\begin{aligned}
u_4(\tau) &= 2\sqrt{2p}\,(4e^{-p\tau} - 30e^{-2p\tau} + 60e^{-3p\tau} - 35e^{-4p\tau}) \\
u_5(\tau) &= \tfrac{1}{2}\sqrt{10p}\,(10e^{-p\tau} - 120e^{-2p\tau} + 420e^{-3p\tau} - 560e^{-4p\tau} + 252e^{-5p\tau}) \\
u_6(\tau) &= 2\sqrt{3p}\,(6e^{-p\tau} - 105e^{-2p\tau} + 560e^{-3p\tau} - 1260e^{-4p\tau} + 1260e^{-5p\tau} \\
&\qquad - 462e^{-6p\tau}) \\
u_7(\tau) &= \sqrt{14p}\,(7e^{-p\tau} - 168e^{-2p\tau} + 1260e^{-3p\tau} - 4200e^{-4p\tau} \\
&\qquad + 6930e^{-5p\tau} - 5544e^{-6p\tau} + 1716e^{-7p\tau}) \\
u_8(\tau) &= \tfrac{4}{5}\sqrt{p}\,(40e^{-p\tau} - 1260e^{-2p\tau} + 12{,}600e^{-3p\tau} - 57{,}750e^{-4p\tau} \\
&\qquad + 138{,}600e^{-5p\tau} - 180{,}180e^{-6p\tau} + 120{,}120e^{-7p\tau} \\
&\qquad + 32{,}175e^{-8p\tau}) \\
u_9(\tau) &= 3\sqrt{2p}\,(9e^{-p\tau} - 360e^{-2p\tau} + 4620e^{-3p\tau} - 27{,}720e^{-4p\tau} \\
&\qquad + 90{,}090e^{-5p\tau} - 168{,}168e^{-6p\tau} + 180{,}180e^{-7p\tau} \\
&\qquad - 102{,}960e^{-8p\tau} + 24{,}310e^{-9p\tau})
\end{aligned} \tag{31}$$

In accordance with (2) and (3), if a function $f(\tau)$ which vanishes for $\tau < 0$ is represented as

$$f(\tau) = \begin{cases} \sum_{n=1}^{\infty} c_n u_n(\tau) & \text{for } 0 \leq \tau < \infty \\ 0 & \text{for } -\infty < \tau < 0 \end{cases} \tag{32}$$

then

$$c_n = \int_0^\infty f(\tau) u_n(\tau)\, d\tau \tag{33}$$

3. Representation of Correlation Functions by Orthonormal Functions

To expand the autocorrelation function $\varphi_{11}(\tau)$ in terms of the orthonormal set $\{u_n(\tau)\}$, let the right-hand half of $\varphi_{11}(\tau)$ be $[\varphi_{11}(\tau)]^+$; that is, let

$$[\varphi_{11}(\tau)]^+ = \begin{cases} \varphi_{11}(\tau) & \text{for } 0 \leq \tau < \infty \\ 0 & \text{for } -\infty < \tau < 0 \end{cases} \tag{34}$$

If we put

$$[\varphi_{11}(\tau)]^+ = \begin{cases} \sum_{n=1}^{\infty} c_n u_n(\tau) & \text{for } 0 \leq \tau < \infty \\ 0 & \text{for } -\infty < \tau < 0 \end{cases} \tag{35}$$

then the coefficients c_n are

$$c_n = \int_0^\infty [\varphi_{11}(\tau)]^+ u_n(\tau)\, d\tau \tag{36}$$

Since the autocorrelation function is an even function and since its right-hand half has the expression (35), the whole function should have the expression

$$\varphi_{11}(\tau) = \sum_{n=1}^{\infty} c_n u_n(|\tau|) \tag{37}$$

where

$$c_n = \int_0^\infty \varphi_{11}(\tau) u_n(\tau)\, d\tau \tag{38}$$

Inasmuch as the terms of (37) can be regrouped according to the components $e^{-p\tau}$, $e^{-2p\tau}$, ..., we can write (37) in the form

$$\varphi_{11}(\tau) = \sum_{n=1}^{\infty} C_n e^{-np|\tau|} \tag{39}$$

where

$$C_1 = \sqrt{p}\,(\sqrt{2}\,c_1 + 4c_2 + 3\sqrt{6}\,c_3 + 8\sqrt{2}\,c_4 + 5\sqrt{10}\,c_5$$

$$+ 12\sqrt{3}\,c_6 + 7\sqrt{14}\,c_7 + 32c_8 + 27\sqrt{2}\,c_9 + \cdots)$$

$$C_2 = -\sqrt{p}\,(6c_2 + 12\sqrt{6}\,c_3 + 60\sqrt{2}\,c_4 + 60\sqrt{10}\,c_5 + 210\sqrt{3}\,c_6$$

$$+ 168\sqrt{14}\,c_7 + 1008c_8 + 1080\sqrt{2}\,c_9 + \cdots)$$

$$C_3 = \sqrt{p}\,(10\sqrt{6}\,c_3 + 120\sqrt{2}\,c_4 + 210\sqrt{10}\,c_5 + 1120\sqrt{3}\,c_6$$

$$+ 1260\sqrt{14}\,c_7 + 10{,}080c_8 + 13{,}860\sqrt{2}\,c_9 + \cdots)$$

$$C_4 = -\sqrt{p}\,(70\sqrt{2}\,c_4 + 280\sqrt{10}\,c_5 + 2520\sqrt{3}\,c_6 + 4200\sqrt{14}\,c_7$$

$$+ 46{,}200c_8 + 83{,}160\sqrt{2}\,c_9 + \cdots)$$

$$C_5 = \sqrt{p}\,(126\sqrt{10}\,c_5 + 2520\sqrt{3}\,c_6 + 6930\sqrt{14}\,c_7 + 110{,}880c_8$$

$$+ 270{,}270\sqrt{2}\,c_9 + \cdots)$$

$$C_6 = -\sqrt{p}\,(924\sqrt{3}\,c_6 + 5544\sqrt{14}\,c_7 + 144{,}144c_8$$

$$+ 504{,}504\sqrt{2}\,c_9 + \cdots)$$

$$C_7 = \sqrt{p}\,(1716\sqrt{14}\,c_7 + 96{,}096c_8 + 540{,}540\sqrt{2}\,c_9 + \cdots)$$

$$C_8 = -\sqrt{p}\,(25{,}740c_8 + 308{,}880\sqrt{2}\,c_9 + \cdots)$$

$$C_9 = \sqrt{p}\,(72{,}930\sqrt{2}\,c_9 + \cdots) \tag{40}$$

It should be noted that in a practical problem the scale factor p should be chosen in accordance with the general behavior of the autocorrelation function with the objective of obtaining the best approximation with the smallest number of terms. Although to do this precisely is a difficult problem, a satisfactory expansion can be obtained by a judicious choice of the value of the scale factor.

If a crosscorrelation function is not an even function, it is quite clear that, by expressing a crosscorrelation function $\varphi_{12}(\tau)$ as

$$\varphi_{12}(\tau) = [\varphi_{12}(\tau)]^+ + [\varphi_{12}(\tau)]^- \tag{41}$$

where

$$[\varphi_{12}(\tau)]^+ = \begin{cases} \varphi_{12}(\tau) & \text{for } \tau \geq 0 \\ 0 & \text{for } \tau < 0 \end{cases} \tag{42}$$

and

$$[\varphi_{12}(\tau)]^- = \begin{cases} 0 & \text{for } \tau \geq 0 \\ \varphi_{12}(\tau) & \text{for } \tau < 0 \end{cases} \tag{43}$$

and then representing (42) and (43) separately by the orthonormal set $\{u_n(\tau)\}$, we have

$$\varphi_{12}(\tau) = \begin{cases} \sum_{n=1}^{\infty} a_n u_n(\tau) & \text{for } 0 \leq \tau < \infty \\ \sum_{n=1}^{\infty} b_n u_n(-\tau) & \text{for } -\infty < \tau < 0 \end{cases} \tag{44}$$

where

$$a_n = \int_0^{\infty} \varphi_{12}(\tau) u_n(\tau)\, d\tau \tag{45}$$

and

$$b_n = \int_{-\infty}^{0} \varphi_{12}(\tau) u_n(-\tau)\, d\tau \tag{46}$$

By collecting terms according to $e^{-p\tau}$, $e^{-2p\tau}$, . . . as in (39), we can write (44) in the form

$$\varphi_{12}(\tau) = \begin{cases} \sum_{n=1}^{\infty} A_n e^{-np\tau} & \text{for } 0 \leq \tau < \infty \\ \sum_{n=1}^{\infty} B_n e^{np\tau} & \text{for } -\infty < \tau < 0 \end{cases} \tag{47}$$

In this expression A_n are the coefficients given by (40) with C_n replaced by A_n and c_n replaced by a_n; and B_n are the coefficients given by (40) with C_n replaced by B_n and c_n replaced by b_n.

4. Representation of Power Density Spectrums by Orthonormal Functions

One important advantage in the representation of correlation functions by an orthonormal set whose components are exponential functions is that the exponential functions transform into rational functions. Clearly these rational functions facilitate the factorization of the power density spectrum in the determination of an optimum system. Since the Fourier transform of $e^{-np|\tau|}$ is

$$\frac{1}{2\pi}\int_{-\infty}^{\infty} e^{-np|\tau|}e^{-j\omega\tau}\,d\tau = \frac{1}{\pi}\frac{np}{(np)^2+\omega^2} \tag{48}$$

the transform of (39) gives us the power density spectrum in the form

$$\Phi_{11}(\omega) = \frac{1}{\pi}\sum_{n=1}^{\infty} C_n \frac{np}{(np)^2+\omega^2} \tag{49}$$

In a similar manner we can express a cross-power density spectrum $\Phi_{12}(\omega)$ in the following form by transforming (47):

$$\Phi_{12}(\omega) = \frac{1}{2\pi}\sum_{n=1}^{\infty} A_n \frac{1}{np+j\omega} + \frac{1}{2\pi}\sum_{n=1}^{\infty} B_n \frac{1}{np-j\omega} \tag{50}$$

Now, let us consider the representation of power density spectrums by orthonormal functions in the frequency domain on the basis of the given spectrums without referring back to the corresponding correlation functions. For this discussion we refer back to the set $\{u_n(\tau)\}$ and obtain the Fourier transforms of the set. If we let $U_n(\omega)$ be the Fourier transform of $u_n(\tau)$ for $(0 \le \tau < \infty)$, we find that the set $\{U_n(\omega)\}$ obtained from (19), (25), (30), and (31) is

$$\begin{aligned}
U_1(\omega) &= \frac{\sqrt{2p}}{2\pi}\frac{1}{p+j\omega} \\
U_2(\omega) &= \frac{\sqrt{p}}{\pi}\left(\frac{2}{p+j\omega} - \frac{3}{2p+j\omega}\right) \\
U_3(\omega) &= \frac{\sqrt{6p}}{2\pi}\left(\frac{3}{p+j\omega} - \frac{12}{2p+j\omega} + \frac{10}{3p+j\omega}\right) \\
U_4(\omega) &= \frac{\sqrt{2p}}{\pi}\left(\frac{4}{p+j\omega} - \frac{30}{2p+j\omega} + \frac{60}{3p+j\omega} - \frac{35}{4p+j\omega}\right)
\end{aligned} \tag{51}$$

and so on.

Letting $F(\omega)$ be the Fourier transform of $f(\tau)$ in (32) and transforming both sides of the equation, we have

$$F(\omega) = \sum_{n-1}^{\infty} c_n U_n(\omega) \tag{52}$$

Note that the coefficients c_n are the same in both (32) and (52). To obtain an expression for c_n in terms of $F(\omega)$ and $U_n(\omega)$, we apply Parseval's theorem to the functions $u_n(\tau)$ and their transforms $U_n(\omega)$. According to this theorem,

$$\int_0^\infty u_m(\tau)u_n(\tau)\,d\tau = 2\pi\int_{-\infty}^\infty \overline{U}_m(\omega)U_n(\omega)\,d\omega \tag{53}$$

Because of the orthonormal properties of $u_n(\tau)$ as given by (14), we see immediately that

$$2\pi\int_{-\infty}^\infty \overline{U}_m(\omega)U_n(\omega)\,d\omega = \begin{cases}1 & \text{if } m = n\\ 0 & \text{if } m \neq n\end{cases} \tag{54}$$

These properties of $U_n(\omega)$ indicate that the set $\{U_n(\omega)\}$ is orthonormal when each term is multiplied by $\sqrt{2\pi}$. Now, if both sides of (52) are multiplied by $\overline{U}_m(\omega)$ and integrated over $(-\infty, \infty)$, we have

$$\int_{-\infty}^\infty F(\omega)\overline{U}_m(\omega)\,d\omega = \int_{-\infty}^\infty \sum_{n=1}^\infty c_nU_n(\omega)\overline{U}_m(\omega)\,d\omega \tag{55}$$

Application of the properties (54) to this equation yields the result

$$c_n = 2\pi\int_{-\infty}^\infty F(\omega)\overline{U}_n(\omega)\,d\omega \tag{56}$$

As noted earlier, these coefficients expressed in the frequency domain are numerically the same as those given by (33) expressed in the time domain.

If we write (32) for $f(-\tau)$ and then transform both sides of the equation, we would get, instead of (52), the equation

$$\overline{F}(\omega) = \sum_{n=1}^\infty c_n\overline{U}_n(\omega) \tag{57}$$

and c_n given by (56) becomes

$$c_n = 2\pi\int_{-\infty}^\infty \overline{F}(\omega)U_n(\omega)\,d\omega \tag{58}$$

which is numerically the same as (56).

Therefore, if we put

$$\varphi_{11}(\tau) = [\varphi_{11}(\tau)]^+ + [\varphi_{11}(\tau)]^- \tag{59}$$

where

$$[\varphi_{11}(\tau)]^+ = \begin{cases} \varphi_{11}(\tau) & \text{for } 0 \leq \tau < \infty \\ 0 & \text{for } -\infty < \tau < 0 \end{cases} \tag{60}$$

and

$$[\varphi_{11}(\tau)]^- = \begin{cases} 0 & \text{for } 0 \leq \tau < \infty \\ \varphi_{11}(\tau) & \text{for } -\infty < \tau < 0 \end{cases} \tag{61}$$

and if $[\Phi_{11}(\omega)]^+$ and $[\Phi_{11}(\omega)]^-$ are the transforms of (60) and (61) respectively, then in accordance with (52) and (57)

$$\begin{aligned} \Phi_{11}(\omega) &= [\Phi_{11}(\omega)]^+ + [\Phi_{11}(\omega)]^- \\ &= \sum_{n=1}^{\infty} c_n U_n(\omega) + \sum_{n=1}^{\infty} c_n \overline{U}_n(\omega) \\ &= \sum_{n=1}^{\infty} c_n [U_n(\omega) + \overline{U}_n(\omega)] \end{aligned} \tag{62}$$

in which

$$\begin{aligned} c_n &= 2\pi \int_{-\infty}^{\infty} [\Phi_{11}(\omega)]^+ \overline{U}_n(\omega)\, d\omega \\ &= 2\pi \int_{-\infty}^{\infty} [\Phi_{11}(\omega)]^- U_n(\omega)\, d\omega \end{aligned} \tag{63}$$

in accordance with (56) and (58). It is shown in Chapter 19, Sec. (6), equations (78) through (85), that (63) can be expressed in the form

$$c_n = \pi \int_{-\infty}^{\infty} \Phi_{11}(\omega)[U_n(\omega) + \overline{U}_n(\omega)]\, d\omega \tag{64}$$

Therefore, with the coefficients evaluated from (64), the power density spectrum $\Phi_{11}(\omega)$ is expressed by (62) in terms of the orthogonal set $\{U_n(\omega) + \overline{U}_n(\omega)\}$ whose first few terms are

$$\begin{aligned} U_1(\omega) + \overline{U}_1(\omega) &= \frac{p\sqrt{2p}}{\pi}\left[\frac{1}{p^2 + \omega^2}\right] \\ U_2(\omega) + \overline{U}_2(\omega) &= \frac{2p\sqrt{p}}{\pi}\left[\frac{2}{p^2 + \omega^2} - \frac{6}{(2p)^2 + \omega^2}\right] \\ U_3(\omega) + \overline{U}_3(\omega) &= \frac{p\sqrt{6p}}{\pi}\left[\frac{3}{p^2 + \omega^2} - \frac{24}{(2p)^2 + \omega^2} + \frac{30}{(3p)^2 + \omega^2}\right] \\ U_4(\omega) + \overline{U}_4(\omega) &= \frac{2p\sqrt{2p}}{\pi}\left[\frac{4}{p^2 + \omega^2} - \frac{60}{(2p)^2 + \omega^2} + \frac{180}{(3p)^2 + \omega^2}\right] \end{aligned} \tag{65}$$

5. Method of Orthonormalization in the Frequency Domain *

The method of orthonormalization given in Sec. 2 of this chapter is a well-known method. It shows clearly the properties of an orthonormal set. Although the method is straightforward, the solution of simultaneous equations becomes a serious difficulty as the number of terms of the orthonormal set increases. We will show that the same set can be obtained with considerable ease by orthonormalization in the frequency domain instead of in the time domain.

When we orthonormalize the sequence $e^{-p\tau}$, $e^{-2p\tau}$, ... obtaining the set $\{u_n(\tau)\}$ given in (19), (25), (30), and (31), whose transforms are the set $\{U_n(\omega)\}$ in (51), we assign one pole to $U_1(\lambda)$ at ($\sigma = p$, $\omega = 0$), two poles to $U_2(\lambda)$ at ($\sigma = p$, $\omega = 0$) and at ($\sigma = 2p$, $\omega = 0$), and three poles to $U_3(\lambda)$ at ($\sigma = p$, $\omega = 0$), ($\sigma = 2p$, $\omega = 0$), and ($\sigma = 3p$, $\omega = 0$), and so on. The poles of these functions are shown in Fig. 1. On the

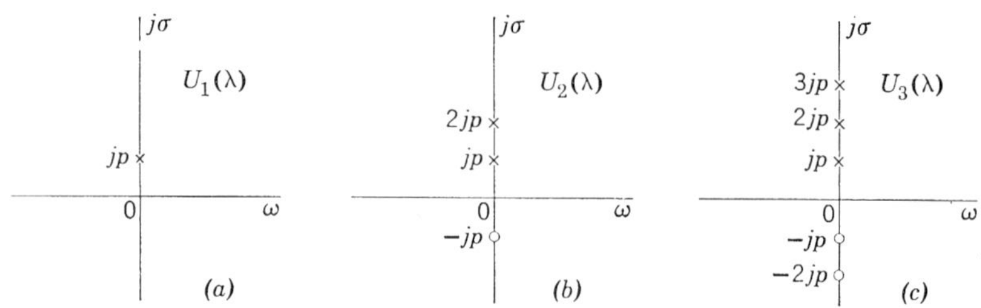

Fig. 1. Poles and zeros of the orthogonal functions $\{U_n(\lambda)\}$.

basis of the manner in which the poles are assigned to the terms of the set and on the basis of the orthonormal properties (54), we can determine the whole set $\{U_n(\lambda)\}$ in an interesting and simple manner.

Let us first write (54) in terms of the complex variable λ. Thus

$$2\pi \int_{-\infty+j0}^{\infty+j0} \overline{U}_m(\lambda) U_n(\lambda)\, d\lambda = \begin{cases} 1 & \text{if } m = n \\ 0 & \text{if } m \neq n \end{cases} \tag{66}$$

We shall perform the integrations by the method of residues. Since the poles are in the finite plane and the path of integration is along the ω-axis, the path can be closed either through a semi-circle in the upper half-plane enclosing all the poles in that half-plane or through a semi-circle in the lower half-plane enclosing all the poles in that half-plane. Integrations along these two paths should yield the same values.

* This method was developed as a result of the author's discussions with graduate students in his class at M.I.T. in 1949.

We start with the first term of the set $U_1(\lambda)$ which has a pole at $(\sigma = p,\ \omega = 0)$, so that it has the form

$$U_1(\lambda) = A_1 \frac{1}{p + j\lambda} \tag{67}$$

The normal property in (66), when applied to (67), is

$$2\pi \int_{-\infty+j0}^{\infty+j0} A_1^2 \frac{1}{p - j\lambda} \frac{1}{p + j\lambda} \, d\lambda = 1 \tag{68}$$

Performing the integration, we have

$$(2\pi)^2 A_1^2 \frac{1}{2p} = 1 \tag{69}$$

so that

$$A_1 = \frac{\sqrt{2p}}{2\pi} \tag{70}$$

and (67) becomes

$$U_1(\lambda) = \frac{\sqrt{2p}}{2\pi} \frac{1}{p + j\lambda} \tag{71}$$

Two factors determine the general form of the next term $U_2(\lambda)$. First, two poles are assigned to it at $(\sigma = p,\ \omega = 0)$ and $(\sigma = 2p,\ \omega = 0)$ so as to correspond to the term $u_2(\tau) = b_1 e^{-pt} + b_2 e^{-2pt}$. Second, it must be orthogonal to $\overline{U}_1(\lambda)$. Since $U_1(\lambda)$ has a pole at $(\sigma = p,\ \omega = 0)$, its conjugate $\overline{U}_1(\lambda)$ must have a pole at $(\sigma = -p,\ \omega = 0)$. In order that the orthogonal condition be satisfied, the product $U_2(\lambda)\overline{U}_1(\lambda)$ is made to have no poles in the lower half-plane so that when we close the path of integration over the lower half-plane we enclose no poles and the integral must be zero. Therefore we must provide $U_2(\lambda)$ with a zero in the lower half-plane at $(\sigma = -p,\ \omega = 0)$ so that it cancels the pole of $\overline{U}_1(\lambda)$. Our conclusion is that $U_2(\lambda)$ should have the form

$$U_2(\lambda) = A_2 \frac{(p - j\lambda)}{(p + j\lambda)(2p + j\lambda)} \tag{72}$$

The zero and poles of this function are shown in Fig. 1*b*. To find A_2, we need the normal property of (66) which in the present case is

$$2\pi \int_{-\infty+j0}^{\infty+j0} A_2^2 \frac{(p + j\lambda)}{(p - j\lambda)(2p - j\lambda)} \frac{(p - j\lambda)}{(p + j\lambda)(2p + j\lambda)} \, d\lambda = 1 \tag{73}$$

This equation reduces to

$$2\pi \int_{-\infty+j0}^{\infty+j0} A_2^2 \frac{d\lambda}{(2p - j\lambda)(2p + j\lambda)} = 1 \tag{74}$$

which yields upon integration

$$(2\pi)^2 A_2^2 \frac{1}{4p} = 1 \tag{75}$$

so that

$$A_2 = \frac{\sqrt{p}}{\pi} \tag{76}$$

and (72) becomes

$$U_2(\lambda) = \frac{\sqrt{p}}{\pi} \frac{(p - j\lambda)}{(p + j\lambda)(2p + j\lambda)} \tag{77}$$

Reasoning in this manner, we find that, since three poles are assigned to $U_3(\lambda)$ at $(\sigma = p, \omega = 0)$, $(\sigma = 2p, \omega = 0)$, and $(\sigma = 3p, \omega = 0)$ and since it must be orthogonal to both $\overline{U}_1(\lambda)$, which has a pole at $(\sigma = -p, \omega = 0)$, and $\overline{U}_2(\lambda)$, which has poles at $(\sigma = -p, \omega = 0)$ and $(\sigma = -2p, \omega = 0)$, the form of $U_3(\lambda)$ must be

$$U_3(\lambda) = A_3 \frac{(p - j\lambda)(2p - j\lambda)}{(p + j\lambda)(2p + j\lambda)(3p + j\lambda)} \tag{78}$$

The zeros and poles of (78) are shown in Fig. 1*c*.

Now we can readily see that the form of $U_n(\lambda)$ must be

$$U_n(\lambda) = A_n \frac{(p - j\lambda)(2p - j\lambda) \cdots ([n-1]p - j\lambda)}{(p + j\lambda)(2p + j\lambda) \cdots (np + j\lambda)} \tag{79}$$

for which A_n is found from the normal property

$$2\pi \int_{-\infty+j0}^{\infty+j0} A_n^2 \frac{(p + j\lambda) \cdots ([n-1]p + j\lambda)}{(p - j\lambda) \cdots (np - j\lambda)} \times \frac{(p - j\lambda) \cdots ([n-1]p - j\lambda)}{(p + j\lambda) \cdots (np + j\lambda)} d\lambda = 1 \tag{80}$$

This equation simplifies to

$$2\pi \int_{-\infty+j0}^{\infty+j0} A_n^2 \frac{d\lambda}{(np - j\lambda)(np + j\lambda)} = 1 \tag{81}$$

and yields

$$(2\pi)^2 A_n^2 \frac{1}{2np} = 1 \tag{82}$$

so that

$$A_n = \frac{\sqrt{2np}}{2\pi} \tag{83}$$

Therefore, the set $\{U_n(\omega)\}$ is

$$\{U_n(\omega)\} = \left\{\frac{\sqrt{2np}}{2\pi} \frac{(p - j\omega)(2p - j\omega) \cdots ([n-1]p - j\omega)}{(p + j\omega)(2p + j\omega) \cdots (np + j\omega)}\right\} \quad \text{for } n = 1, 2, 3, \ldots \tag{84}$$

This orthogonal set is normal if every term is multiplied by $\sqrt{2\pi}$.

6. Laguerre Functions

An important and interesting orthonormal set of functions is obtained by orthonormalization of the sequence

$$(p\tau)^n e^{-p\tau} \qquad \text{for } n = 0, 1, 2, \ldots \tag{85}$$

in the range $(0 \leq \tau < \infty)$. The set is known as Laguerre functions $\{l_n(\tau)\}$ whose first few terms and general expressions are

$$\left.\begin{aligned}
l_0(\tau) &= \sqrt{2p}\, e^{-p\tau} \\
l_1(\tau) &= \sqrt{2p}\, (2p\tau - 1)e^{-p\tau} \\
l_2(\tau) &= \sqrt{2p}\, (2p^2\tau^2 - 4p\tau + 1)e^{-p\tau} \\
l_3(\tau) &= \sqrt{2p}\, (\tfrac{4}{3}p^3\tau^3 - 6p^2\tau^2 + 6p\tau - 1)e^{-p\tau} \\
l_4(\tau) &= \sqrt{2p}\, (\tfrac{2}{3}p^4\tau^4 - \tfrac{16}{3}p^3\tau^3 + 12p^2\tau^2 - 8p\tau + 1)e^{-p\tau} \\
&\vdots \\
l_n(\tau) &= \sqrt{2p}\left[\frac{(2p)^n}{n!}\tau^n - \frac{n(2p)^{n-1}}{(n-1)!}\tau^{n-1}\right. \\
&\qquad + \frac{n(n-1)(2p)^{n-2}}{2!(n-2)!}\tau^{n-2} \\
&\qquad - \frac{n(n-1)(n-2)(2p)^{n-3}}{3!(n-3)!}\tau^{n-3} \\
&\qquad \left. + \cdots + (-1)^n\right] e^{-p\tau}
\end{aligned}\right\} \tag{86}$$

These functions have the properties

$$\int_0^\infty l_m(\tau)l_n(\tau)\,d\tau = \begin{cases} 1 & \text{if } m = n \\ 0 & \text{if } m \neq n \end{cases} \tag{87}$$

If

$$f(\tau) = \begin{cases} \sum_{n=0}^{\infty} c_n l_n(\tau) & \text{for } 0 \le \tau < \infty \\ 0 & \text{for } -\infty < \tau < 0 \end{cases} \tag{88}$$

then

$$c_n = \int_0^\infty f(\tau) l_n(\tau)\,d\tau \tag{89}$$

To find the transforms of the set $\{l_n(\tau)\}$, we apply the integral

$$\frac{1}{2\pi}\int_0^\infty \tau^n e^{-p\tau} e^{-j\omega\tau}\,d\tau = \frac{n!}{2\pi(p + j\omega)^{n+1}} \tag{90}$$

to the last expression in (86). Letting $\{L_n(\omega)\}$ be the transforms of $\{l_n(\tau)\}$, we have

$$\begin{aligned} L_n(\omega) &= \frac{\sqrt{2p}}{2\pi}\left[\frac{(2p)^n}{n!}\frac{n!}{(p+j\omega)^{n+1}} - \frac{n(2p)^{n-1}}{(n-1)!}\frac{(n-1)!}{(p+j\omega)^n}\right. \\ &\quad + \frac{n(n-1)(2p)^{n-2}}{2!(n-2)!}\frac{(n-2)!}{(p+j\omega)^{n-1}} - \frac{n(n-1)(n-2)(2p)^{n-3}}{3!(n-3)!} \\ &\quad \left.\times \frac{(n-3)!}{(p+j\omega)^{n-2}} + \cdots + (-1)^n \frac{1}{p+j\omega}\right] \\ &= \frac{\sqrt{2p}}{2\pi}\frac{1}{(p+j\omega)^{n+1}}\left[(2p)^n - n(2p)^{n-1}(p+j\omega)\right. \\ &\quad + \frac{n(n-1)2p^{n-2}}{2!}(p+j\omega)^2 - \frac{n(n-1)(n-2)(2p)^{n-3}}{3!} \\ &\quad \left.\times (p+j\omega)^3 + \cdots + (-1)^n(p+j\omega)^n\right] \\ &= \frac{\sqrt{2p}}{2\pi}\frac{1}{(p+j\omega)^{n+1}}[2p - (p+j\omega)]^n \\ &= \frac{\sqrt{2p}}{2\pi}\frac{(p-j\omega)^n}{(p+j\omega)^{n+1}} \end{aligned} \tag{91}$$

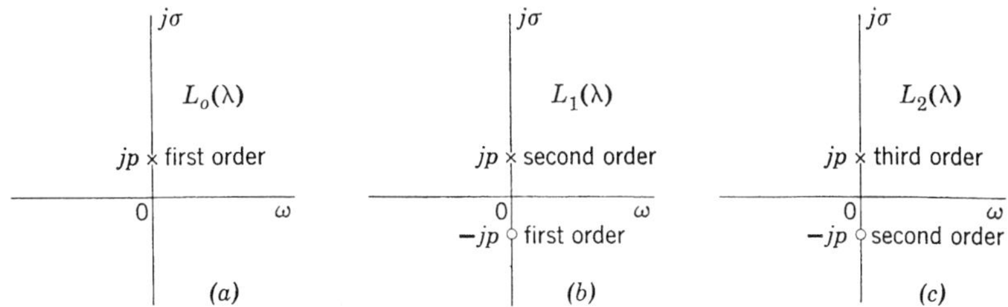

Fig. 2. Poles and zeros of the Laguerre functions $\{L_n(\lambda)\}$.

As shown in Fig. 2 the function $L_n(\lambda)$ has an nth-order zero at $(\sigma = -p, \omega = 0)$ and an $(n+1)$th-order pole at $(\sigma = p, \omega = 0)$.

By the method of orthonormalization in the frequency domain we see that, since $e^{-p\tau}$ corresponds to a pole of the first order at $(\sigma = p, \omega = 0)$, $p\tau e^{-p\tau}$ corresponds to a second-order pole at $(\sigma = p, \omega = 0)$, $(p\tau)^2 e^{-p\tau}$ corresponds to a third-order pole at $(\sigma = p, \omega = 0)$, and so on, the first term of the set $\{L_n(\lambda)\}$ should be of the form

$$L_0(\lambda) = A_0 \frac{1}{p + j\lambda} \tag{92}$$

and the second term should be of the form

$$L_1(\lambda) = A_1 \frac{p - j\lambda}{(p + j\lambda)^2} \tag{93}$$

with the first-order zero at $(\sigma = -p, \omega = 0)$ to cancel the first-order pole of $\bar{L}_0(\lambda)$ at $(\sigma = -p, \omega = 0)$. Similarly the third term should take the form

$$L_2(\lambda) = A_2 \frac{(p - j\lambda)^2}{(p + j\lambda)^3} \tag{94}$$

with a second-order zero at $(\sigma = -p, \omega = 0)$ to cancel the first-order pole of $\bar{L}_0(\lambda)$ and the second-order pole of $\bar{L}_1(\lambda)$ both at $(\sigma = -p, \omega = 0)$. Clearly the general form of the set should be

$$L_n(\lambda) = A_n \frac{(p - j\lambda)^n}{(p + j\lambda)^{n+1}} \tag{95}$$

Since the properties (87) in the frequency domain are

$$2\pi \int_{-\infty+j0}^{\infty+j0} \bar{L}_m(\lambda) L_n(\lambda)\, d\lambda = \begin{cases} 1 & \text{if } m = n \\ 0 & \text{if } m \neq n \end{cases} \tag{96}$$

which are similar to (66) for the set $\{U_n(\lambda)\}$, we apply the normal property in (96) to (95) to obtain

$$2\pi \int_{-\infty+j0}^{\infty+j0} A_n{}^2 \frac{(p+j\lambda)^n}{(p-j\lambda)^{n+1}} \frac{(p-j\lambda)^n}{(p+j\lambda)^{n+1}} \, d\lambda = 1 \tag{97}$$

which reduces to

$$2\pi \int_{-\infty+j0}^{\infty+j0} A_n{}^2 \frac{1}{(p-j\lambda)} \frac{1}{(p+j\lambda)} \, d\lambda = 1 \tag{98}$$

giving

$$(2\pi)^2 A_n{}^2 \frac{1}{2p} = 1 \tag{99}$$

so that we have the Laguerre set

$$\{L_n(\omega)\} = \left\{\frac{\sqrt{2p}}{2\pi} \frac{(p-j\omega)^n}{(p+j\omega)^{n+1}}\right\} \qquad \text{for } n = 0, 1, 2, \ldots \tag{100}$$

which agrees with (91).

If $F(\omega)$ is the Fourier transform of $f(\tau)$ in (88), then

$$F(\omega) = \sum_{n=0}^{\infty} c_n L_n(\omega) \tag{101}$$

where

$$c_n = 2\pi \int_{-\infty}^{\infty} F(\omega) \overline{L}_n(\omega) \, d\omega \tag{102}$$

These expressions for the set $\{L_n(\omega)\}$ are similar to (52) and (56) for the set $\{U_n(\omega)\}$. Since the application of (101) and (102) and of (88) and (89) to the representation of power density spectrums and correlation functions can be carried out in the manner discussed in Secs. 3 and 4, we will not repeat the details here.

7. Legendre Functions

Still another important orthogonal set is obtained from the sequence x^n for $n = 0, 1, 2, \ldots$ and for the interval $(-1 < x < 1)$. The set $\{P_n(x)\}$ from this orthogonalization is known as Legendre polynomials. The first few terms of the set are

$$\begin{aligned} P_0(x) &= 1 \\ P_1(x) &= x \\ P_2(x) &= \tfrac{3}{2}x^2 - \tfrac{1}{2} \\ P_3(x) &= \tfrac{5}{2}x^3 - \tfrac{3}{2}x \\ P_4(x) &= \tfrac{1}{8}(35x^4 - 30x^2 + 3) \\ P_5(x) &= \tfrac{1}{8}(63x^5 - 70x^3 + 15x) \end{aligned} \tag{103}$$

The set has the properties

$$\int_{-1}^{1} P_m(x)P_n(x)\,dx = \begin{cases} \dfrac{2}{2n+1} & \text{for } m = n \\ 0 & \text{for } m \neq n \end{cases} \tag{104}$$

Although this set cannot be directly applied to our problem of expressing correlation functions in terms of orthonormal functions because the range of the Legendre polynomials does not correspond to the range that we need, this difficulty can be overcome by a change of variables.

We will show that by a change of variable the range of the Legendre set goes from $(-1, 1)$ to $(0, 1)$ and that by a second change of variable the range goes from $(0, 1)$ to $(0, \infty)$, which is the desired range. The first change of variable is

$$x = 2y - 1 \tag{105}$$

With this change (104) becomes

$$2\int_{0}^{1} P_m(2y-1)P_n(2y-1)\,dy = \begin{cases} \dfrac{2}{2n+1} & \text{if } m = n \\ 0 & \text{if } m \neq n \end{cases} \tag{106}$$

which shows that $P_n(2y-1)$ are orthogonal functions in the range $(0 < y < 1)$. Furthermore, if we let the second change of variable be

$$y = e^{-p\tau} \tag{107}$$

then (106) becomes

$$\int_{0}^{\infty} pe^{-p\tau}P_m(2e^{-2p\tau}-1)P_n(2e^{-2p\tau}-1)\,d\tau = \begin{cases} \dfrac{1}{2n+1} & \text{if } m = n \\ 0 & \text{if } m \neq n \end{cases} \tag{108}$$

This result states that $\sqrt{p}\, e^{-(p/2)\tau}P_n(2e^{-2p\tau} - 1)$ are orthogonal in the range $(0 < \tau < \infty)$. In order to normalize the Legendre functions that have been produced by changes of variables, we write (108) in the following manner

$$\int_{0}^{\infty} [\sqrt{p}\,\sqrt{2m+1}\, e^{-(p/2)\tau}P_m(2e^{-p\tau}-1)] \times [\sqrt{p}\,\sqrt{2n+1}\, e^{-(p/2)\tau}P_n(2e^{-p\tau}-1)]\,d\tau = \begin{cases} 1 & \text{if } m = n \\ 0 & \text{if } m \neq n \end{cases} \tag{109}$$

If we let

$$v_n(\tau) = \sqrt{p(2n+1)}\, e^{-(p/2)\tau}P_n(2e^{-p\tau}-1) \tag{110}$$

then (109) reads

$$\int_0^\infty v_m(\tau)v_n(\tau)\,d\tau = \begin{cases} 1 & \text{if } m = n \\ 0 & \text{if } m \neq n \end{cases} \tag{111}$$

We now have the orthonormal set $\{v_n(\tau)\}$. As in the discussion of the orthonormal set $\{u_n(\tau)\}$, if $f(\tau)$ is a real function of time satisfying the conditions

$$\left.\begin{aligned} f(\tau) &= 0 \qquad \text{for } \tau < 0 \\ \text{and} \qquad \int_0^\infty f^2(\tau)\,d\tau &< \infty \end{aligned}\right\} \tag{112}$$

then in terms of the set $\{v_n(\tau)\}$ we have

$$f(\tau) = \begin{cases} \sum_{n=0}^{\infty} c_n v_n(\tau) & \text{for } 0 \leq \tau < \infty \\ 0 & \text{for } -\infty < \tau < 0 \end{cases} \tag{113}$$

in which

$$c_n = \int_0^\infty f(\tau)v_n(\tau)\,d\tau \tag{114}$$

The first few terms of $\{v_n(\tau)\}$ are

$$\begin{aligned}
v_0(\tau) &= \sqrt{p}\,e^{-(p/2)\tau} \\
v_1(\tau) &= \sqrt{3p}\,(-1 + 2e^{-p\tau})e^{-(p/2)\tau} \\
v_2(\tau) &= \sqrt{5p}\,(1 - 6e^{-p\tau} + 6e^{-2p\tau})e^{-(p/2)\tau} \\
v_3(\tau) &= \sqrt{7p}\,(-1 + 12e^{-p\tau} - 30e^{-2p\tau} + 20e^{-3p\tau})e^{-(p/2)\tau} \\
v_4(\tau) &= \sqrt{9p}\,(1 - 20e^{-p\tau} + 90e^{-2p\tau} - 140e^{-3p\tau} + 70e^{-4p\tau})e^{-(p/2)\tau} \\
v_5(\tau) &= \sqrt{11p}\,(-1 + 30e^{-p\tau} - 210e^{-2p\tau} + 560e^{-3p\tau} - 630e^{-4p\tau} \\
&\quad + 252e^{-5p\tau})e^{-(p/2)\tau} \\
v_6(\tau) &= \sqrt{13p}\,(1 - 42e^{-p\tau} + 420e^{-2p\tau} - 1680e^{-3p\tau} + 3150e^{-4p\tau} \\
&\quad - 2772e^{-5p\tau} + 924e^{-6p\tau})e^{-(p/2)\tau} \\
v_7(\tau) &= \sqrt{15p}\,(-1 + 56e^{-p\tau} - 756e^{-2p\tau} + 4200e^{-3p\tau} - 11{,}550e^{-4p\tau} \\
&\quad + 16{,}632e^{-5p\tau} - 12{,}012e^{-6p\tau} + 3432e^{-7p\tau})e^{-(p/2)\tau} \\
v_8(\tau) &= \sqrt{17p}\,(1 - 72e^{-p\tau} + 1260e^{-2p\tau} - 9240e^{-3p\tau} + 34{,}650e^{-4p\tau} \\
&\quad - 72{,}072e^{-5p\tau} + 84{,}084e^{-6p\tau} - 51{,}480e^{-7p\tau} \\
&\quad + 12{,}870e^{-8p\tau})e^{-(p/2)\tau}
\end{aligned} \tag{115}$$

Other terms of the set are obtained from (110) with the aid of a table of Legendre polynomials $P_n(x)$.

We observe that the set $\{v_n(\tau)\}$ is clearly the result of orthonormalization in the range $(0 \leq \tau < \infty)$ of the sequence

$$e^{-(np/2)\tau} \qquad \text{for } n = 1, 3, 5, \ldots \tag{116}$$

whose transforms are the sequence

$$\frac{1}{\dfrac{np}{2} + j\omega} \qquad \text{for } n = 1, 3, 5, \ldots \tag{117}$$

In accordance with the method of orthonormalization in the frequency domain, the general expression of the set $\{V_n(\lambda)\}$ should be

$$V_n(\lambda) = A_n \frac{\left[\dfrac{p}{2} - j\lambda\right]\left[\dfrac{3p}{2} - j\lambda\right] \cdots \left[\dfrac{(2n-1)p}{2} - j\lambda\right]}{\left[\dfrac{p}{2} + j\lambda\right]\left[\dfrac{3p}{2} + j\lambda\right] \cdots \left[\dfrac{(2n+1)p}{2} + j\lambda\right]} \qquad \text{for } n = 0, 1, 2, \ldots \tag{118}$$

Here $V_n(\lambda)$ is the Fourier transform of $v_n(\tau)$. Since this expression is similar to (79) for the set $\{U_n(\lambda)\}$ which has been explained in detail, we will not repeat the arguments here. By application of the normal property we can show that

$$A_n = \frac{1}{2\pi} \sqrt{(2n+1)p} \tag{119}$$

Inasmuch as the procedure for obtaining (119) is now familiar to us, no further details are necessary. We now have the orthogonal set

$$\{V_n(\omega)\} = \left\{ \frac{\sqrt{(2n+1)p}\left[\dfrac{p}{2} - j\omega\right]\left[\dfrac{3p}{2} - j\omega\right] \cdots \left[\dfrac{(2n-1)p}{2} - j\omega\right]}{2\pi\left[\dfrac{p}{2} + j\omega\right]\left[\dfrac{3p}{2} + j\omega\right] \cdots \left[\dfrac{(2n+1)p}{2} + j\omega\right]} \right\} \qquad \text{for } n = 0, 1, 2, \ldots \tag{120}$$

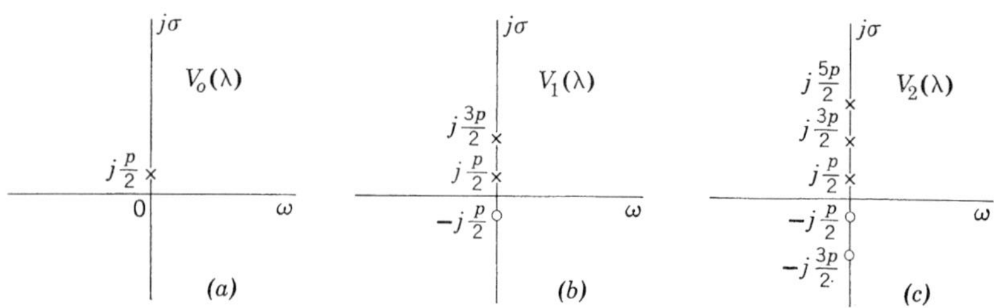

Fig. 3. Poles and zeros of the Legendre functions $\{V_n(\lambda)\}$.

The poles and zeros of the first three terms of this set are shown in Fig. 3. If $F(\omega)$ is the Fourier transform of $f(\tau)$ in (113), we have

$$F(\omega) = \sum_{n=0}^{\infty} c_n V_n(\omega) \tag{121}$$

and

$$c_n = 2\pi \int_{-\infty}^{\infty} F(\omega) \overline{V}_n(\omega) \, d\omega \tag{122}$$

The application of (113), (114), (121), and (122) to correlation functions and power density spectrums is clear in view of the discussions given in Secs. 3 and 4.

chapter 19

Synthesis of Optimum Linear Systems by Means of Orthonormal Functions

Although optimum linear systems can be synthesized in the same manner as conventional filters are, with the aid of standard techniques, we shall consider in this chapter the synthesis of linear systems by means of orthonormal functions because this method is particularly suitable for the synthesis of optimum systems and because it is an interesting method. In spite of the fact that it was developed in the early 1930's, the method has not received as much attention as others in the field of network synthesis. The comparative obscurity is probably due to the fact that the development has been primarily recorded in patent literature. Of historical interest is the fact that a part of the development of this method of network synthesis was the discovery of the relation between the amplitude function and the phase function of a system function and the relation between the real part and imaginary part of this function. These relations, expressed as Hilbert transforms, are an essential part of the method of synthesis by means of orthonormal functions.

The material in this chapter is based upon the joint work, referred to at the beginning of Chapter 18, of N. Wiener and the author.

1. Introduction

The orthonormal functions that we have studied in the last chapter are applicable to the synthesis of optimum systems in addition to the representation of correlation functions and power density spectrums. It is particularly important to note that the synthesis can be carried out in either the time domain or the frequency domain with the same ease.

When the orthonormal set $\{u_n(\tau)\}$ given by (19), (25), (30), and (31) is applied to network synthesis, the unit-impulse response of the system to be synthesized is expressed in terms of the set $\{u_n(t)\}$. We make the simple change of variable from τ to t so that the variable t corresponds to the customary t in the unit-impulse response $h(t)$. Note that the range of $\{u_n(t)\}$, which is $(0 \le t < \infty)$, is the same as that of $h(t)$. Since the method of synthesis considered in this chapter is applicable not only to optimum systems but also to linear systems in general, we shall regard the notations $h(t)$ and $H(\omega)$ as representing all these systems. The first few terms of $\{u_n(t)\}$ are repeated here for convenience.

$$\begin{aligned} u_1(t) &= \sqrt{2p}\, e^{-pt} \\ u_2(t) &= 2\sqrt{p}\,(2e^{-pt} - 3e^{-2pt}) \\ u_3(t) &= \sqrt{6p}\,(3e^{-pt} - 12e^{-2pt} + 10e^{-3pt}) \\ u_4(t) &= 2\sqrt{2p}\,(4e^{-pt} - 30e^{-2pt} + 60e^{-3pt} - 35e^{-4pt}) \end{aligned} \tag{1}$$

If $h(t)$ is represented by the set $\{u_n(t)\}$, we have

$$h(t) = \begin{cases} \sum_{n=1}^{\infty} c_n u_n(t) & \text{for } 0 \le t < \infty \\ 0 & \text{for } -\infty < t < 0 \end{cases} \tag{2}$$

where

$$c_n = \int_0^\infty h(t) u_n(t)\, dt \tag{3}$$

For the synthesis of the system function $H(\omega)$ in the frequency domain, we have the orthogonal set $\{U_n(\omega)\}$ given by (51) or (84), Chapter 18. Since

$$\begin{aligned} H(\omega) &= \int_{-\infty}^{\infty} h(t) e^{-j\omega t}\, dt \\ &= 2\pi \left[\frac{1}{2\pi} \int_0^\infty \sum_{n=1}^{\infty} c_n u_n(t) e^{-j\omega t}\, dt \right] \end{aligned} \tag{4}$$

we have

$$H(\omega) = 2\pi \sum_{n=1}^{\infty} c_n U_n(\omega) \tag{5}$$

where

$$c_n = \int_{-\infty}^{\infty} H(\omega) \overline{U}_n(\omega)\, d\omega \tag{6}$$

By finding a set of networks which corresponds to the set of orthogonal functions $\{U_n(\omega)\}$ and connecting the components in accordance with the expansion (5), we have a method of synthesis. Since the coefficients c_n in both (2) and (5) are the same and their evaluation in the time domain is given by (3) and in the frequency domain by (6), we see that synthesis in either domain results in the same network.

2. Networks Corresponding to the Orthonormal Set of Exponential Functions

Let us consider the networks to which the set $\{U_n(\omega)\}$, given by (84), Chapter 18, correspond. The first term of the set is

$$U_1(\omega) = \frac{\sqrt{2p}}{2\pi} \frac{1}{p + j\omega} \tag{7}$$

which can be the input to output voltage ratio of a simple RC or RL network. Figure 1a represents $2\pi U_1(\omega)$. The factor 2π is accounted for

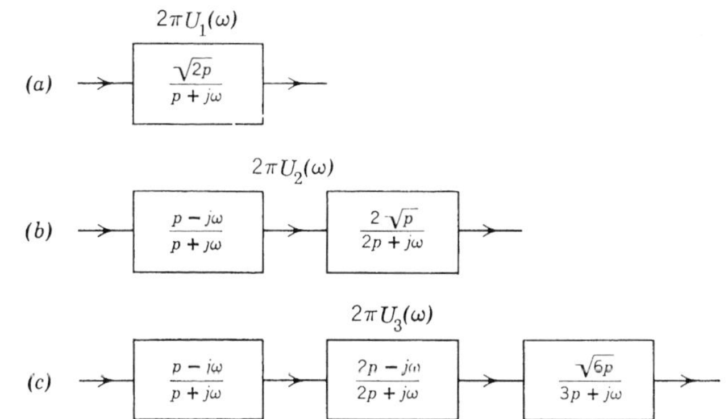

Fig. 1. Networks representing the first three terms of the set $\{U_n(\omega)\}$.

by the fact that the expansion (5) is in terms of $2\pi U_n(\omega)$. The second term of the set is

$$U_2(\omega) = \frac{\sqrt{p}}{\pi} \frac{(p - j\omega)}{(p + j\omega)(2p + j\omega)} \tag{8}$$

which we can write as

$$U_2(\omega) = \left(\frac{p - j\omega}{p + j\omega}\right)\left(\frac{\sqrt{p}}{\pi} \frac{1}{2p + j\omega}\right) \tag{9}$$

so that we can consider $2\pi U_2(\omega)$ as the system function of a pure phase-shift network and an RC or RL network in cascade as shown in Fig. 1*b*. Similarly, the third term of the set is

$$U_3(\omega) = \frac{\sqrt{6p}}{2\pi} \frac{(p - j\omega)(2p - j\omega)}{(p + j\omega)(2p + j\omega)(3p + j\omega)} \tag{10}$$

which we can write as

$$U_3(\omega) = \left(\frac{p - j\omega}{p + j\omega}\right)\left(\frac{2p - j\omega}{2p + j\omega}\right)\left(\frac{\sqrt{6p}}{2\pi}\frac{1}{3p + j\omega}\right) \tag{11}$$

indicating that the system corresponding to this expression consists of two pure phase-shift networks and an RC or RL network in cascade, as shown in Fig. 1*c*.

A form of the phase-shift network is the lattice network shown in Fig. 2. Each of the two parallel branches has an impedance Z_1, and

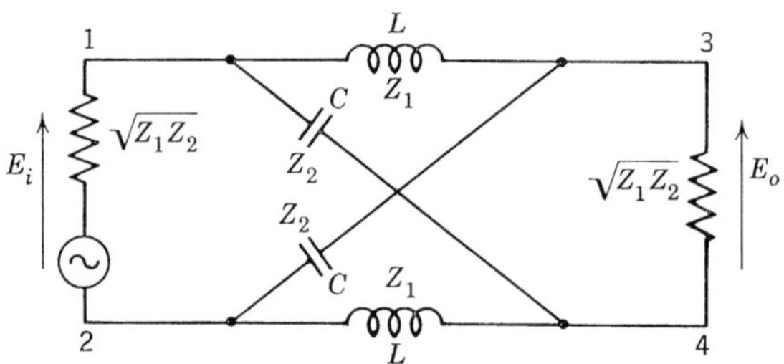

Fig. 2. A form of phase-shift network.

each of the two cross branches has an impedance Z_2. The network terminates in the impedance $\sqrt{Z_1Z_2}$, and has the impedance $\sqrt{Z_1Z_2}$ in series with the source. It can be shown that the ratio of the output voltage across terminals 3 and 4 to the input voltage across terminals 1 and 2 is

$$H_1(\omega) = \frac{E_o}{E_i}$$

$$= \frac{\sqrt{Z_2} - \sqrt{Z_1}}{\sqrt{Z_2} + \sqrt{Z_1}} \tag{12}$$

If Z_1 is purely inductive and Z_2 purely capacitative so that $Z_1 = j\omega L$ and $Z_2 = 1/j\omega C$ where L is the inductance and C the capacitance, then the terminal impedance is a pure resistance; that is, $\sqrt{Z_1Z_2} = \sqrt{L/C}$.

For the values of Z_1 and Z_2 stated, (12) becomes

$$H_1(\omega) = \frac{\dfrac{1}{\sqrt{LC}} - j\omega}{\dfrac{1}{\sqrt{LC}} + j\omega} \tag{13}$$

If a system unit-impulse response is expanded as in (2) or the system function of the system is expanded as in (5), the network corresponding

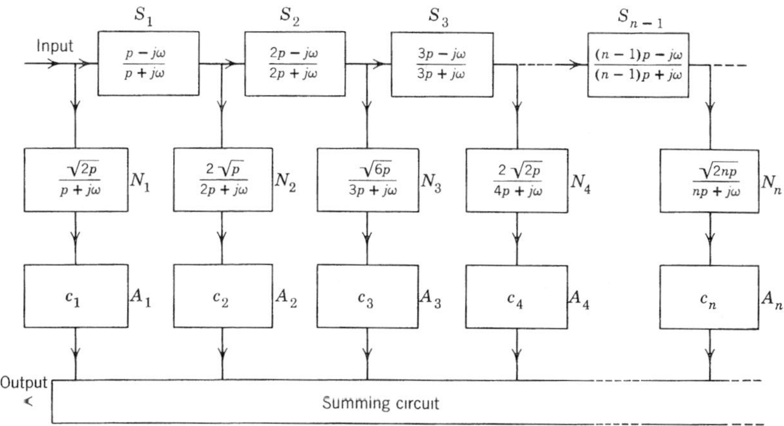

Fig. 3. Synthesis system for expansions in terms of the set $\{U_n(\omega)\}$.

to the expansions is given in Fig. 3. A central portion of the entire system is the phase-shift chain

$$\left(\frac{p - \omega}{p + j\omega}\right)\left(\frac{2p - j\omega}{2p + j\omega}\right)\left(\frac{3p - j\omega}{3p + j\omega}\right) \cdots \left(\frac{np - j\omega}{np + j\omega}\right) \cdots \tag{14}$$

The input voltage of the system is E_i. It is clear that each section of this lattice structure can be made to have a terminal impedance of pure resistance equal to $\sqrt{L/C}$. Of course, the last section of this chain terminates into a pure resistance of this value. From junctions between sections S_1, S_2, ... of this phase-shift chain, voltages are led to RC or RL networks N_1, N_2, ... The input impedances of these networks are assumed to be so much higher than the resistance $\sqrt{L/C}$ that voltage tapping by these networks causes essentially no disturbance to the phase-shift chain. In series with the networks N_1, N_2, ... are amplifiers (or attenuators) $A_1, A_2, \ldots$ whose amplifications (or attenuations) are adjusted to correspond to the values $c_1, c_2, \ldots$ of the expansion. The output voltages of the amplifiers $A_1, A_2, \ldots$ are algebraically summed to form the output voltage E_o of the system.

It is evident from the description of the synthesis system that the branch consisting of N_1 and A_1 in series has the output-to-input-voltage ratio $2\pi c_1 U_1(\omega)$; the branch consisting of S_1, N_2, and A_2 in series has the output-to-input-voltage ratio $2\pi c_2 U_2(\omega)$; the branch consisting of S_1, S_2, N_3, and A_3 in series has the output-to-input-voltage ratio $2\pi c_3 U_3(\omega)$, and so on. Therefore, by providing a sufficiently large number of branches to correspond to exactly the same number of terms in its expansion, a system function or its unit-impulse response may be approximated as closely as desired, at least in theory.

3. Legendre Networks

In the synthesis of linear systems based upon the Legendre functions $\{v_n(t)\}$ as given by (115) in Chapter 18, with τ replaced by t, or upon the Fourier transforms of these functions $\{V_n(\omega)\}$ as given by (120) in Chapter 18, we have that, if the unit-impulse response of the system $h(t)$ is represented by the series

$$h(t) = \begin{cases} \sum_{n=0}^{\infty} c_n v_n(t) & \text{for } 0 \le t < \infty \\ 0 & \text{for } -\infty < t < 0 \end{cases} \tag{15}$$

then

$$c_n = \int_0^{\infty} h(t) v_n(t)\, dt \tag{16}$$

and, if the system function $H(\omega)$ is represented by the series

$$H(\omega) = 2\pi \sum_{n=0}^{\infty} c_n V_n(\omega) \tag{17}$$

then

$$c_n = \int_{-\infty}^{\infty} H(\omega) \overline{V}_n(\omega)\, d\omega \tag{18}$$

We have noted before the important fact that the coefficients c_n in (15) and (17) are the same.

The standard synthesis system for the Legendre functions is substantially the same as that for the set $\{U_n(\omega)\}$. In a manner similar to (7), (9), and (11), we write the first three terms of $\{V_n(\omega)\}$ as given by

(120), Chapter 18, as follows:

$$V_0(\omega) = \frac{\sqrt{p}}{2\pi} \frac{1}{\dfrac{p}{2} + j\omega} \tag{19}$$

$$V_1(\omega) = \left(\frac{\dfrac{p}{2} - j\omega}{\dfrac{p}{2} + j\omega}\right)\left(\frac{\sqrt{3p}}{2\pi} \frac{1}{\dfrac{3p}{2} + j\omega}\right) \tag{20}$$

$$V_2(\omega) = \left(\frac{\dfrac{p}{2} - j\omega}{\dfrac{p}{2} + j\omega}\right)\left(\frac{\dfrac{3p}{2} - j\omega}{\dfrac{3p}{2} + j\omega}\right)\left(\frac{\sqrt{5p}}{2\pi} \frac{1}{\dfrac{5p}{2} + j\omega}\right) \tag{21}$$

These and succeeding terms lead to the system shown in Fig. 4. Because

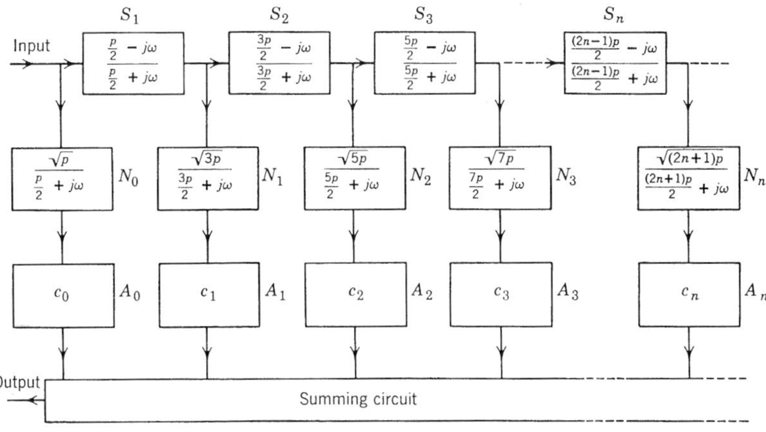

Fig. 4. Synthesis system for expansions in terms of the Legendre functions.

of the great similarity between this system and the system for $\{U_n(\omega)\}$ shown in Fig. 3, further explanation seems superfluous.

4. Laguerre Networks

When applied to network synthesis, the Laguerre functions are $\{l_n(t)\}$ which are given by (86), Chapter 18, with τ replaced by t. The transforms of these functions are $\{L_n(\omega)\}$. [See (100), Chapter 18.] Again, writing as in (2) through (6), we have that, if the unit-impulse response of the system to be synthesized is

$$h(t) = \begin{cases} \sum_{n=0}^{\infty} c_n l_n(t) & \text{for } 0 \le t < \infty \\ 0 & \text{for } -\infty < t < 0 \end{cases} \tag{22}$$

then

$$c_n = \int_0^\infty h(t) l_n(t)\, dt \tag{23}$$

and, if the system function of the system is expressed as

$$H(\omega) = 2\pi \sum_{n=0}^{\infty} c_n L_n(\omega) \tag{24}$$

then

$$c_n = \int_{-\infty}^{\infty} H(\omega) \bar{L}_n(\omega)\, d\omega \tag{25}$$

As in other orthogonal expansions, the coefficients c_n in (22) and (24) are the same coefficients.

The Laguerre set corresponds to a much simpler standard system for synthesis. Arranging the factors in each term of the set $\{L_n(\omega)\}$ with a pure phase-shift term as one of the factors, we have

$$L_0(\omega) = \left(\frac{\sqrt{2p}}{2\pi} \frac{1}{p + j\omega} \right) \tag{26}$$

$$L_1(\omega) = \left(\frac{\sqrt{2p}}{2\pi} \frac{1}{p + j\omega} \right) \left(\frac{p - j\omega}{p + j\omega} \right) \tag{27}$$

$$L_2(\omega) = \left(\frac{\sqrt{2p}}{2\pi} \frac{1}{p + j\omega} \right) \left(\frac{p - j\omega}{p + j\omega} \right) \left(\frac{p - j\omega}{p + j\omega} \right) \tag{28}$$

and so on. Since the first term of the set is common to all terms of the set, we make the network representing it the first section N_o of the synthesis system as shown in Fig. 5. In cascade with this section we have a phase-shift chain of identical sections S_1, S_2, ... We recall that, for the other sets of orthogonal functions considered thus far, the sections in the phase-shift chain differ from one another. The simplicity of the phase-shift chain in the Laguerre network is an advantage. As in previous cases, the input impedances of the networks connected to the

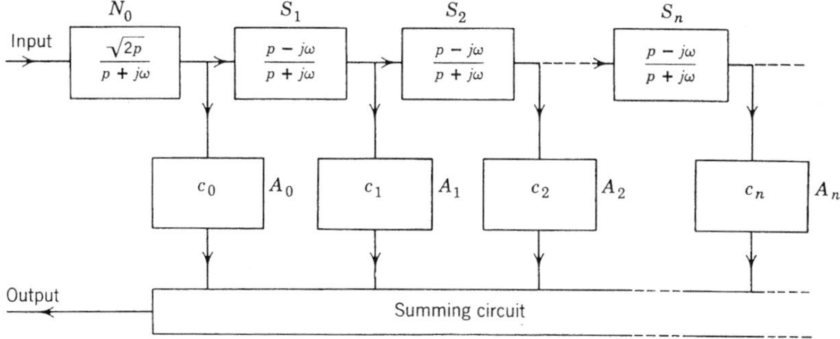

Fig. 5. Synthesis system for expansions in terms of the Laguerre functions.

junctions of the sections of the phase-shift chain are considerably higher than the value of the resistance measured at any junction of the chain so that tapping for the voltages at the junctions does not disturb the voltages throughout the whole chain. The networks A_o, A_1, ... are amplifiers (or attenuators) whose amplifications (or attenuations) are adjusted in accordance with the coefficients of expansion c_o, c_1, ... By taking the algebraic sum of the output voltages of A_o, A_1, ..., we see that the output of the entire synthesis system is formed as indicated in the figure. Clearly, the ratio of the output voltage to the input voltage of the branch consisting of N_o and A_o in series is $2\pi c_o L_o(\omega)$, the ratio of the output voltage to the input voltage of the branch consisting of N_o, S_1, and A_1 in series is $2\pi c_1 L_1(\omega)$, and so on. Thus the number of branches corresponds exactly to the number of terms in the orthogonal expansion.

5. Fourier Networks

We now wish to develop a method of synthesis on the basis of the sequence of functions

$$F_n(\omega) = \frac{1}{2\pi}\left(\frac{p - j\omega}{p + j\omega}\right)^n \qquad \text{for } n = 0, 1, 2, \ldots \tag{29}$$

This method is interesting because it closely resembles the method of Laguerre functions, and it turns out to be a method of Fourier series.

Let us suppose that the system function to be synthesized is represented by the series

$$H(\omega) = 2\pi \sum_{n=0}^{\infty} c_n F_n(\omega) \tag{30}$$

By a change of variables, we can show that (30) is expressible as a

Fourier series. First let us note that

$$\left(\frac{p - j\omega}{p + j\omega}\right)^n = e^{-jn\theta} \tag{31}$$

where

$$\theta = 2 \tan^{-1} \frac{\omega}{p} \tag{32}$$

from which we obtain

$$\omega = p \tan \frac{\theta}{2} \tag{33}$$

Expressing (30) in terms of θ, we have the Fourier series

$$\begin{aligned} H\left(p \tan \frac{\theta}{2}\right) &= 2\pi \sum_{n=0}^{\infty} c_n e^{-jn\theta} \\ &= 2\pi \sum_{n=0}^{\infty} c_n(\cos n\theta - j \sin n\theta) \end{aligned} \tag{34}$$

where the coefficients c_n are

$$c_n = \int_{-\pi}^{\pi} H\left(p \tan \frac{\theta}{2}\right)(\cos n\theta - j \sin n\theta)\, d\theta \tag{35}$$

Letting

$$\begin{aligned} H_e(\omega) &= \text{the even (or real) component of } H(\omega) \\ H_o(\omega) &= \text{the odd (or imaginary) component of } H(\omega) \end{aligned} \tag{36}$$

so that

$$H(\omega) = H_e(\omega) + jH_o(\omega) \tag{37}$$

we can write (35) in the form

$$\begin{aligned} c_n &= \int_{-\pi}^{\pi} \left[H_e\left(p \tan \frac{\theta}{2}\right) + jH_o\left(p \tan \frac{\theta}{2}\right)\right][\cos n\theta - j \sin n\theta]\, d\theta \\ &= 4\int_0^{\pi} H_e\left(p \tan \frac{\theta}{2}\right) \cos n\theta\, d\theta \\ &= 4\int_0^{\pi} H_o\left(p \tan \frac{\theta}{2}\right) \sin n\theta\, d\theta \end{aligned} \tag{38}$$

The last two expressions of (38) are obtained by the application of Hilbert transforms discussed in Sec. 6.

It is interesting that the coefficients for the expansion (30) can be obtained by either the cosine transformation of $H_e[p \tan (\theta/2)]$ or the sine transformation of $H_o[p \tan (\theta/2)]$. Thus the evaluation of the coefficients of expansion has been reduced to harmonic analysis after the change of variables (33) is made. Graphically, the functions $H_e[p \tan (\theta/2)]$ or $H_o[p \tan (\theta/2)]$ can be plotted on tangent-scale paper as shown in Fig. 6 and analyzed with the aid of mechanical or electrical devices.

The standard network system for synthesis by the present method is shown in Fig. 7. It is noted that, if the component N_o in the Laguerre network of Fig. 5 is removed, the network becomes a Fourier network. As in previous systems considered in this chapter the input impedances of the amplifiers or attenuators A_o, A_1, . . . are assumed to be essentially infinite in comparison with the characteristic impedance of the lattice phase-shift chain S_1, S_2, ... The ratio of the output voltage to the input voltage of the branch A_1 is $2\pi c_o$; the ratio of the output voltage to the input voltage of the branch consisting of networks S_1 and A_1 in series is $2\pi c_1 F_1(\omega)$; the ratio of the output voltage to the input voltage of the branch consisting of networks S_1, S_2, and A_2 in series is $2\pi c_2 F_2(\omega)$; and so on. With as many branches as the number of terms in the expansion (30), the summing circuit takes the algebraic sum of the output voltages of the branches to form the output of the whole synthesis system.

6. Hilbert Transforms

To facilitate the evaluation of the coefficients of orthogonal expansions in the synthesis of optimum systems, certain relations between the real and imaginary parts of system functions are necessary. These relations are expressed as Hilbert transforms. To derive these relations, let us consider the unit-impulse response of a linear system $h(t)$ as the sum of an even component $h_e(t)$ and an odd component $h_o(t)$; that is,

$$h(t) = h_e(t) + h_o(t) \tag{39}$$

We assume that neither component is zero for all values of t. Since the singularity functions $u(t)$ and its derivatives $u^{(n)}(t)$ are either even functions or odd functions so that the resolution into two nonzero components as indicated in (39) is impossible, these functions are excluded. Because $h_e(t)$ is an even function, we have

$$h_e(t) = h_e(-t) \tag{40}$$

and, because $h_o(t)$ is an odd function, we have

$$h_o(t) = -h_o(-t) \tag{41}$$

Therefore

$$h(-t) = h_e(t) - h_o(t) \tag{42}$$

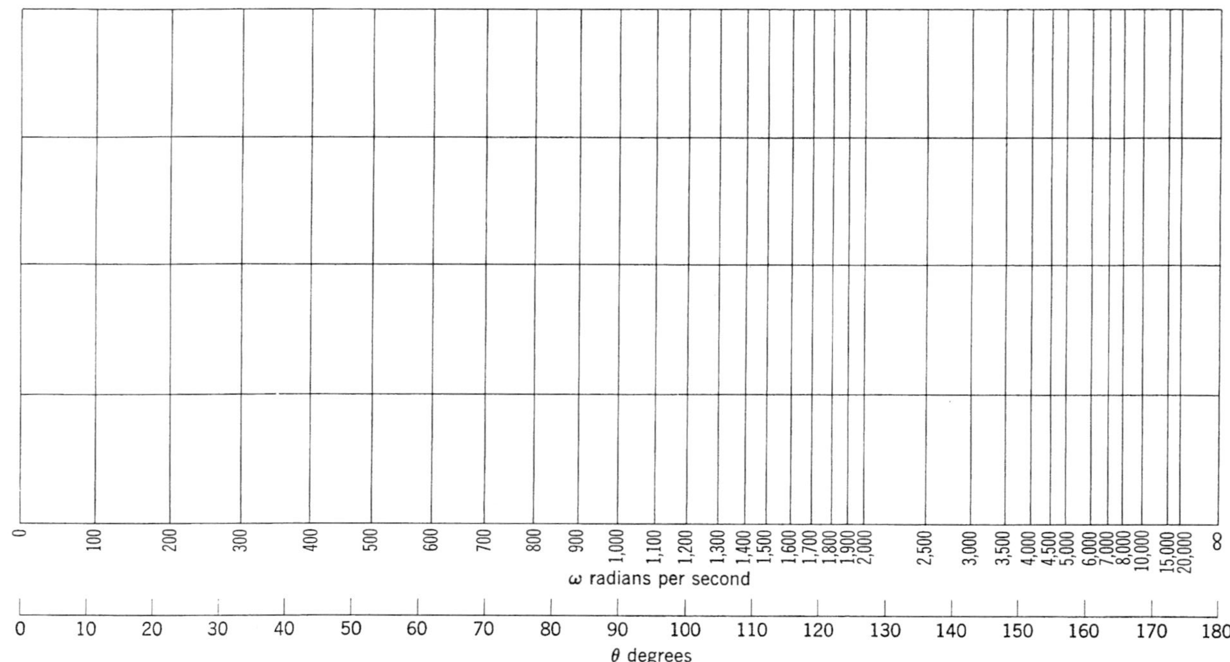

Fig. 6. Change of variable for synthesis in accordance with a Fourier expansion. $\omega = p \tan(\theta/2)$. Drawn for $p = 1000$.

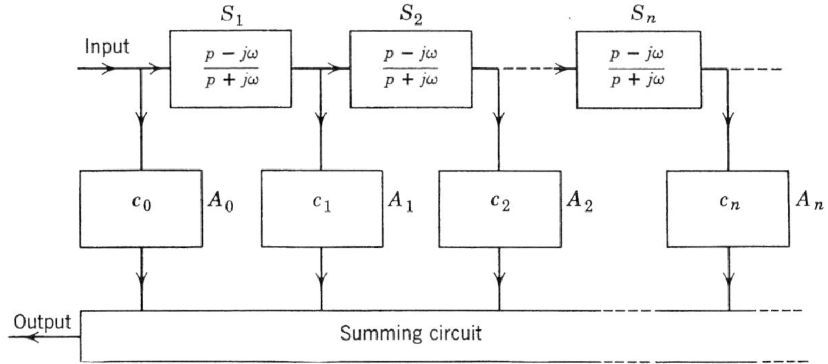

Fig. 7. Synthesis system for Fourier expansions.

Solving (39) and (42) for $h_e(t)$ and $h_o(t)$ in terms of $h(t)$ and $h(-t)$, we find

$$h_e(t) = \tfrac{1}{2}[h(t) + h(-t)] \tag{43}$$

and

$$h_o(t) = \tfrac{1}{2}[h(t) - h(-t)] \tag{44}$$

The resolution of $h(t)$ into its even and odd components is illustrated in Fig. 8.

Since the system function $H(\omega)$ in terms of $h(t)$ is

$$H(\omega) = \int_{-\infty}^{\infty} h(t)e^{-j\omega t}\,dt \tag{45}$$

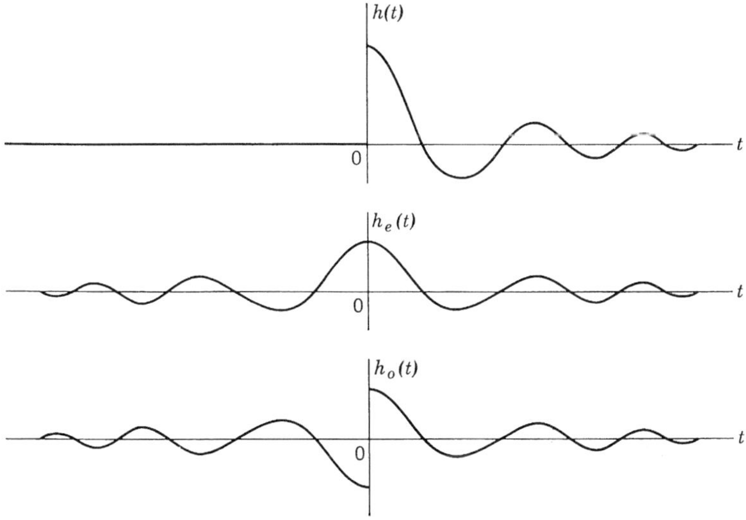

Fig. 8. Resolution of a unit-impulse response into even and odd components.

we have, with (39) in (45),

$$\begin{aligned} H(\omega) &= \int_{-\infty}^{\infty} [h_e(t) + h_o(t)][\cos \omega t - j \sin \omega t]\, dt \\ &= \int_{-\infty}^{\infty} h_e(t) \cos \omega t\, dt - j \int_{-\infty}^{\infty} h_o(t) \sin \omega t\, dt \\ &= H_e(\omega) + jH_o(\omega) \qquad (46) \end{aligned}$$

Here the real component of $H(\omega)$

$$H_e(\omega) = \int_{-\infty}^{\infty} h_e(t) \cos \omega t\, dt \qquad (47)$$

is also the even component of $H(\omega)$, and the imaginary component of $H(\omega)$

$$H_o(\omega) = -\int_{-\infty}^{\infty} h_o(t) \sin \omega t\, dt \qquad (48)$$

is also the odd component of $H(\omega)$. That (47) is an even function of ω is evident from the fact that $\cos(-\omega t) = \cos \omega t$, and (48) is an odd function of ω, from the fact that $\sin(-\omega t) = -\sin \omega t$. The inverse of (47) and (48) are respectively

$$h_e(t) = \frac{1}{2\pi} \int_{-\infty}^{\infty} H_e(\omega) \cos \omega t\, d\omega \qquad (49)$$

and

$$h_o(t) = -\frac{1}{2\pi} \int_{-\infty}^{\infty} H_o(\omega) \sin \omega t\, d\omega \qquad (50)$$

We assume that the transforms (47) through (50) exist.

Inasmuch as

$$h(t) = 0 \qquad \text{for } t < 0 \qquad (51)$$

we have

$$h_e(t) = h_o(t) \qquad \text{for } t \geq 0 \qquad (52)$$

Expressing (52) in terms of (49) and (50), we have

$$\int_{-\infty}^{\infty} H_e(\omega) \cos \omega t\, d\omega = -\int_{-\infty}^{\infty} H_o(\omega) \sin \omega t\, d\omega \qquad \text{for } t \geq 0 \qquad (53)$$

Furthermore, since

$$h_e(t) = -h_o(t) \qquad \text{for } t < 0 \qquad (54)$$

equation (53) can be made to hold for all values of t if $\sin \omega t$ in the right-hand integral of (53) is replaced by $\sin \omega|t|$. Hence

$$\int_{-\infty}^{\infty} H_e(\omega) \cos \omega t \, d\omega = -\int_{-\infty}^{\infty} H_o(\omega) \sin \omega|t| \, d\omega \qquad \text{for } -\infty < t < \infty \tag{55}$$

Taking the cosine transform of (49), we have

$$\int_{-\infty}^{\infty} h_e(t) \cos \omega t \, dt = \frac{1}{2\pi} \int_{-\infty}^{\infty} \cos \omega t \, dt \int_{-\infty}^{\infty} H_e(u) \cos ut \, du \tag{56}$$

and, applying (47), we see that (56) becomes

$$H_e(\omega) = \frac{1}{2\pi} \int_{-\infty}^{\infty} \cos \omega t \, dt \int_{-\infty}^{\infty} H_e(u) \cos ut \, du \tag{57}$$

Since the extreme right-hand integral is related to $H_o(\omega)$ by (55), we can establish a relationship between $H_e(\omega)$ and $H_o(\omega)$ by applying (55) to (57). In so doing, we obtain

$$H_e(\omega) = -\frac{1}{2\pi} \int_{-\infty}^{\infty} \cos \omega t \, dt \int_{-\infty}^{\infty} H_o(u) \sin u|t| \, du \tag{58}$$

which can also be written as

$$H_e(\omega) = -\frac{2}{\pi} \int_{0}^{\infty} \cos \omega t \, dt \int_{0}^{\infty} H_o(u) \sin ut \, du \tag{59}$$

This is an important equation. It shows that, given the imaginary part of a system function, the real part of the function can be determined from it. In other words, given the imaginary part of a system function, the real part of the function cannot be arbitrarily assigned. This equation has an inverse.

To derive the inverse of (59), we take the sine transform of (50) to get

$$\int_{-\infty}^{\infty} h_o(t) \sin \omega t \, dt = -\frac{1}{2\pi} \int_{-\infty}^{\infty} \sin \omega t \, dt \int_{-\infty}^{\infty} H_o(u) \sin ut \, du \tag{60}$$

and then apply (48) to simplify (60) to

$$H_o(\omega) = \frac{1}{2\pi} \int_{-\infty}^{\infty} \sin \omega t \, dt \int_{-\infty}^{\infty} H_o(u) \sin ut \, du \tag{61}$$

In order that we may relate $H_o(\omega)$ in (61) to $H_e(\omega)$ by means of (55), we first replace t by $|t|$ in (61). Thus

$$H_o(\omega) = \frac{1}{2\pi} \int_{-\infty}^{\infty} \sin \omega|t| \, dt \int_{-\infty}^{\infty} H_o(u) \sin u|t| \, du \tag{62}$$

This change does not alter the value of the double integral because

$$\frac{1}{2\pi}\int_{-\infty}^{\infty} \sin \omega(-t)\, dt \int_{-\infty}^{\infty} H_o(u) \sin u(-t)\, du$$
$$= \frac{1}{2\pi}\int_{-\infty}^{\infty} \sin \omega t\, dt \int_{-\infty}^{\infty} H_o(u) \sin ut\, du \qquad (63)$$

Now, replacing the extreme right-hand integral of (62) by an integral involving $H_e(\omega)$ in accordance with (55), we obtain

$$H_o(\omega) = -\frac{1}{2\pi}\int_{-\infty}^{\infty} \sin \omega|t|\, dt \int_{-\infty}^{\infty} H_e(u) \cos ut\, du \qquad (64)$$

A more convenient form of (64) is

$$H_o(\omega) = -\frac{2}{\pi}\int_{0}^{\infty} \sin \omega t\, dt \int_{0}^{\infty} H_e(u) \cos ut\, du \qquad (65)$$

This is the inverse of (59), which states that the odd (or imaginary) component $H_o(\omega)$ of a system function $H(\omega)$ is the sine transform of the cosine transform of the even (or real) component $H_e(\omega)$ of the system function. In an analogous manner the expression (59) states that the even (or real) component of the system function is the cosine transform of the sine transform of the odd (or imaginary) component of the system function. Transforms of this type are known as Hilbert transforms. The relations between the real and the imaginary components of a system function are extremely important in the theory of linear systems.

As an example, consider

$$H(\omega) = \frac{1}{1 + j\omega}$$
$$= \frac{1}{1+\omega^2} - j\frac{\omega}{1+\omega^2} \qquad (66)$$

so that

$$H_e(\omega) = \frac{1}{1+\omega^2} \qquad (67)$$

and

$$H_o(\omega) = -\frac{\omega}{1+\omega^2} \qquad (68)$$

Let us first determine the imaginary component of $H(\omega)$ from its real component. By (65) we have

$$
\begin{aligned}
H_o(\omega) &= -\frac{2}{\pi}\int_0^\infty \sin \omega t\, dt \int_0^\infty \frac{1}{1+u^2}\cos ut\, du \\
&= -\frac{2}{\pi}\int_0^\infty \frac{\pi}{2} e^{-t} \sin \omega t\, dt \\
&= -\frac{\omega}{1+\omega^2} \qquad (69)
\end{aligned}
$$

which agrees with (68). To illustrate the determination of the real component by the imaginary component, we apply (59). We have

$$
\begin{aligned}
H_e(\omega) &= -\frac{2}{\pi}\int_0^\infty \cos \omega t\, dt \int_0^\infty \frac{-u}{1+u^2}\sin ut\, du \\
&= \frac{2}{\pi}\frac{\pi}{2}\int_0^\infty e^{-t} \cos \omega t\, dt \\
&= \frac{1}{1+\omega^2} \qquad (70)
\end{aligned}
$$

This result is in agreement with (67).

To establish an important relation in the synthesis of linear systems by means of orthonormal functions, let us consider two system functions $H(\omega)$ and $G(\omega)$ whose real parts are $H_e(\omega)$ and $G_e(\omega)$, and imaginary parts are $H_o(\omega)$ and $G_o(\omega)$ respectively, so that

$$H(\omega) = H_e(\omega) + jH_o(\omega) \qquad (71)$$

and

$$G(\omega) = G_e(\omega) + jG_o(\omega) \qquad (72)$$

In particular, let us consider the integral

$$\int_0^\infty H_e(\omega)G_e(\omega)\, d\omega \qquad (73)$$

Substitution of the Hilbert transform (59) for $H_e(\omega)$ into (73) leads to the equation

$$\int_0^\infty H_e(\omega)G_e(\omega)\, d\omega = -\frac{2}{\pi}\int_0^\infty G_e(\omega)\, d\omega \int_0^\infty \cos \omega t\, dt \int_0^\infty H_o(u) \sin ut\, du \qquad (74)$$

Through inversion of the order of integration, (74) becomes

$$\int_0^\infty H_e(\omega)G_e(\omega)\,d\omega = -\frac{2}{\pi}\int_0^\infty H_o(u)\,du\int_0^\infty \sin ut\,dt\int_0^\infty G_e(\omega)\cos \omega t\,d\omega \tag{75}$$

Since, in accordance with the Hilbert transform (65),

$$G_o(\omega) = -\frac{2}{\pi}\int_0^\infty \sin \omega t\,dt\int_0^\infty G_e(u)\cos ut\,du \tag{76}$$

equation (75) is simplified to

$$\int_0^\infty H_e(\omega)G_e(\omega)\,d\omega = \int_0^\infty H_o(u)G_o(u)\,du \tag{77}$$

This is an interesting result. It states that the integral over $(0, \infty)$ of the product of the real parts of two system functions is equal to the integral over $(0, \infty)$ of the product of their imaginary parts.

The relation (77) is extremely helpful in the evaluation of coefficients of expansions in the representation of power density spectrums by orthonormal functions and in the synthesis of networks by means of these functions. As a first application we refer back to (63), Chapter 18. The coefficients

$$c_n = 2\pi\int_{-\infty}^\infty [\Phi_{11}(\omega)]^+\overline{U}_n(\omega)\,d\omega \tag{78}$$

can be written as

$$c_n = 2\pi\int_{-\infty}^\infty \{[\Phi_{11}(\omega)]_e{}^+ + j[\Phi_{11}(\omega)]_o{}^+\}\{U_{ne}(\omega) - jU_{no}(\omega)\}\,d\omega \tag{79}$$

where the subscript e denotes the even (or real) part and the subscript o denotes the odd (or imaginary) part of the functions concerned. Equation (79) is clearly

$$c_n = 2\pi\int_{-\infty}^\infty \{[\Phi_{11}(\omega)]_e{}^+U_{ne}(\omega) + [\Phi_{11}(\omega)]_o{}^+U_{no}(\omega)\}\,d\omega \tag{80}$$

By the theorem (77) we have

$$\int_{-\infty}^\infty [\Phi_{11}(\omega)]_e{}^+U_{ne}(\omega)\,d\omega = \int_{-\infty}^\infty [\Phi_{11}(\omega)]_o{}^+U_{no}(\omega)\,d\omega \tag{81}$$

so that (80) simplifies to

$$c_n = 4\pi\int_{-\infty}^\infty [\Phi_{11}(\omega)]_e{}^+U_{ne}(\omega)\,d\omega \tag{82}$$

Inasmuch as

$$[\Phi_{11}(\omega)]_e^{+} = \tfrac{1}{2}\Phi_{11}(\omega) \tag{83}$$

and

$$U_{ne}(\omega) = \tfrac{1}{2}[U_n(\omega) + \overline{U}_n(\omega)] \tag{84}$$

the coefficients (82) finally become

$$c_n = \pi \int_{-\infty}^{\infty} \Phi_{11}(\omega)[U_n(\omega) + \overline{U}_n(\omega)]\, d\omega \tag{85}$$

This is equation (64) in Chapter 18.

In the synthesis of networks by means of the set of functions $\{U_n(\omega)\}$, the coefficients are given by (6)

$$c_n = \int_{-\infty}^{\infty} H(\omega)\overline{U}_n(\omega)\, d\omega \tag{86}$$

which can be written as

$$\begin{aligned} c_n &= \int_{-\infty}^{\infty} [H_e(\omega) + jH_o(\omega)][U_{ne}(\omega) - jU_{no}(\omega)]\, d\omega \\ &= \int_{-\infty}^{\infty} [H_e(\omega)U_{ne}(\omega) + H_o(\omega)U_{no}(\omega)]\, d\omega \end{aligned} \tag{87}$$

By the theorem (77) the coefficients are

$$\begin{aligned} c_n &= 2\int_{-\infty}^{\infty} H_e(\omega)U_{ne}(\omega)\, d\omega \\ &= 2\int_{-\infty}^{\infty} H_o(\omega)U_{no}(\omega)\, d\omega \end{aligned} \tag{88}$$

Here we repeat that $U_{ne}(\omega)$ is the even (or real) part of $U_n(\omega)$ and $U_{no}(\omega)$ is the odd (or imaginary) part of $U_n(\omega)$.

Clearly in the synthesis of systems by means of the Legendre networks, the coefficients (18) are

$$\begin{aligned} c_n &= \int_{-\infty}^{\infty} H(\omega)\overline{V}_n(\omega)\, d\omega \\ &= 2\int_{-\infty}^{\infty} H_e(\omega)V_{ne}(\omega)\, d\omega \\ &= 2\int_{-\infty}^{\infty} H_o(\omega)V_{no}(\omega)\, d\omega \end{aligned} \tag{89}$$

where $V_{ne}(\omega)$ is the even (or real) part of $V_n(\omega)$ and $V_{no}(\omega)$ is the odd (or imaginary) part of $V_n(\omega)$. Similarly, in the synthesis of systems by means of the Laguerre networks, the coefficients (25) are

$$c_n = \int_{-\infty}^{\infty} H(\omega)\overline{L}_n(\omega)\,d\omega$$

$$= 2\int_{-\infty}^{\infty} H_e(\omega)L_{ne}(\omega)\,d\omega$$

$$= 2\int_{-\infty}^{\infty} H_o(\omega)L_{no}(\omega)\,d\omega \tag{90}$$

Here $L_{ne}(\omega)$ is the even (or real) part of $L_n(\omega)$, and $L_{no}(\omega)$ is the odd (or imaginary) part of $L_n(\omega)$.

7. Approximation Mean-Square Error

To determine the mean-square error caused by the inaccuracy in the representation of an optimum system by a finite expansion in terms of orthonormal functions, or by other approximations, we begin with the expression for the variation in the minimum mean-square error. The expression is (49), Sec. 4, Chapter 14. Since it is shown, in the section referred to, that the variation of mean-square error given by the expression is the variation from minimum mean-square error, we write the expression as

$$\overline{\delta\varepsilon^2_{h(\text{opt})}(t)} = \epsilon^2\int_{-\infty}^{\infty} \eta(\tau)\,d\tau \int_{-\infty}^{\infty} \eta(\sigma)\,d\sigma\,\varphi_{ii}(\tau - \sigma) \tag{91}$$

Inasmuch as the unit-impulse response of the approximate system is the unit-impulse response of the optimum system plus an error and the error is actually a variation of the optimum system unit-impulse response, we have

$$h_{\text{app}}(t) = h_{\text{opt}}(t) + \epsilon\eta(t) \tag{92}$$

where $h_{\text{app}}(t)$ is the approximate expression for $h_{\text{opt}}(t)$. Writing the variation as

$$\Delta h(t) = \epsilon\eta(t) \tag{93}$$

and calling the variation in minimum mean-square error as the *approximation mean-square error* $\overline{\varepsilon^2_{\text{app}}(t)}$, we have

$$\overline{\delta \varepsilon^2_{h(\text{opt})}(t)} = \overline{\varepsilon^2_{\text{app}}(t)} \tag{94}$$

and

$$\overline{\varepsilon^2_{\text{app}}(t)} = \int_{-\infty}^{\infty} \Delta h(\tau)\, d\tau \int_{-\infty}^{\infty} \Delta h(\sigma)\, d\sigma\, \varphi_{ii}(\tau - \sigma) \tag{95}$$

in accordance with (91).

With the change of variable $t = \sigma - \tau$, (95) becomes

$$\overline{\varepsilon^2_{\text{app}}(t)} = \int_{-\infty}^{\infty} \Delta h(\tau)\, d\tau \int_{-\infty}^{\infty} \Delta h(\tau + t)\, dt\, \varphi_{ii}(t) \tag{96}$$

When the order of integration is inverted, (96) is

$$\overline{\varepsilon^2_{\text{app}}(t)} = \int_{-\infty}^{\infty} \varphi_{ii}(t)\, dt \int_{-\infty}^{\infty} \Delta h(\tau)\, \Delta h(\tau + t)\, d\tau \tag{97}$$

Letting $\varphi_{\Delta h \Delta h}(t)$ be the autocorrelation function of $\Delta h(t)$, that is, letting

$$\varphi_{\Delta h \Delta h}(t) = \int_{-\infty}^{\infty} \Delta h(\tau)\, \Delta h(\tau + t)\, d\tau \tag{98}$$

where

$$\Delta h(t) = h_{\text{app}}(t) - h_{\text{opt}}(t) \tag{99}$$

we express the approximation mean-square error (97) in the form

$$\overline{\varepsilon^2_{\text{app}}(t)} = \int_{-\infty}^{\infty} \varphi_{ii}(t) \varphi_{\Delta h \Delta h}(t)\, dt \tag{100}$$

Index

DOVER PHOENIX EDITIONS

An Introduction to the Approximation of Functions, Theodore J. Rivlin. Unabridged republication of the 1969 edition. 160pp. 49554-X

An Essay on the Foundations of Geometry, Bertrand Russell. Unabridged republication of the 1897 edition. 224pp. 49555-8

Elements of Number Theory, I. M. Vinogradov. Unabridged republication of the first edition, 1954. 240pp. 49530-2

Asymptotic Expansions for Ordinary Differential Equations, Wolfgang Wasow. Unabridged republication of the 1976 corrected, slightly enlarged reprint of the original 1965 edition. 384pp. 49518-3

Physics

Semiconductor Statistics, J. S. Blakemore. Unabridged, corrected, and slightly enlarged republication of the 1962 edition. 141 illustrations. xviii+318pp. 49502-7

Wave Propagation in Periodic Structures, L. Brillouin. Unabridged republication of the 1946 edition. 131 illustrations. 272pp. 49556-6

The Conceptual Foundations of the Statistical Approach in Mechanics, Paul and Tatiana Ehrenfest. Unabridged republication of the 1959 edition. 128pp. 49504-3

The Analytical Theory of Heat, Joseph Fourier. Unabridged republication of the 1878 edition. 20 figures. 496pp. 49531-0

States of Matter, David L. Goodstein. Unabridged republication of the 1975 edition. 154 figures. 4 tables. 512pp. 49506-X

The Principles of Mechanics, Heinrich Hertz. Unabridged republication of the 1900 edition. 320pp. 49557-4

Thermodynamics of Small Systems, Terrell L. Hill. Unabridged and corrected republication in one volume of the two-volume edition published in 1963–1964. 32 illustrations. 408pp. 6½ x 9¼. 49509-4

Theoretical Physics, A. S. Kompaneyets. Unabridged republication of the 1961 edition. 56 figures. 592pp. 49532-9

Quantum Mechanics, H. A. Kramers. Unabridged republication of the 1957 edition. 14 figures. 512pp. 49533-7

The Theory of Electrons, H. A. Lorentz. Unabridged reproduction of the 1915 edition. 9 figures. 352pp. 49558-2

The Principles of Physical Optics, Ernst Mach. Unabridged republication of the 1926 edition. 279 figures. 10 portraits. 336pp. 49559-0

The Scientific Papers of James Clerk Maxwell, James Clerk Maxwell. Unabridged republication of the 1890 edition. 197 figures. 39 tables. Total of 1,456pp. *Volume I* (640pp.) 49560-4; *Volume II* (816pp.) 49561-2

Vectors and Tensors in Crystallography, Donald E. Sands. Unabridged and corrected republication of the 1982 edition. xviii+228pp. 49516-7

Principles of Mechanics and Dynamics, Sir William (Lord Kelvin) Thompson and Peter Guthrie Tait. Unabridged republication of the 1912 edition. 168 diagrams. Total of 1,088pp. *Volume I* (528pp.) 49562-0; *Volume II* (560pp.) 49563-9

Treatise on Irreversible and Statistical Thermophysics: An Introduction to Nonclassical Thermodynamics, Wolfgang Yourgrau, Alwyn van der Merwe, and Gough Raw. Unabridged, corrected republication of the 1966 edition. xx+268pp. 49519-1

Engineering

Principles of Aeroelasticity, Raymond L. Bisplinghoff and Holt Ashley. Unabridged, corrected republication of the original 1962 edition. xi+527pp. 49500-0

Statics of Deformable Solids, Raymond L. Bisplinghoff, James W. Mar, and Theodore H. H. Pian. Unabridged and corrected Dover republication of the edition published in 1965. 376 illustrations. xii+322pp. 6½ x 9¼. 49501-9